INTERMEDIATE ALGEBRA

FOURTH EDITION

INTERMEDIATE ALGEBRA

FOURTH EDITION

MERVIN L. KEEDY
PURDUE UNIVERSITY

MARVIN L. BITTINGER
INDIANA UNIVERSITY
PURDUE UNIVERSITY AT INDIANAPOLIS

ADDISON-WESLEY PUBLISHING COMPANY

Reading, Massachusetts ▪ Menlo Park, California
London ▪ Amsterdam ▪ Don Mills, Ontario ▪ Sydney

SPONSORING EDITOR:	*Patricia Mallion*
PRODUCTION EDITOR:	*Robert Hartwell Fiske*
TEXT DESIGNERS:	*Marshall Henrichs and Vanessa Piñeiro*
ILLUSTRATOR:	*VAP Group Limited*
COVER DESIGNER:	*Richard Hannus*
ART COORDINATOR:	*Susanah H. Michener*
PRODUCTION MANAGER:	*Herbert Nolan*

The text of this book was composed in Century Schoolbook by Syntax International.

Library of Congress Cataloging in Publication Data

Keedy, Mervin Laverne.
　　Intermediate algebra.

　　Includes index.
　　1. Algebra.　I. Bittinger, Marvin L.　II. Title.
QA154.2.K434 1982　　　512.9　　　82–13816
ISBN 0–201–14790–4

The success of the first three editions of this text has been most gratifying. We wish to thank the many instructors and students who have both praised and criticized our product. The criticism has helped us tremendously in making significant improvements in the book.

WHAT'S NEW IN THE FOURTH EDITION?

In addition to the computerized testbank and videotapes mentioned in the list of supplements that follows this preface, there are the following new features.

- *Extension Exercises.* At the end of each chapter is a set of extension exercises, keyed to the sections of the chapter. These exercises require the student to go beyond the individual objectives of the sections. They may, for example, ask the student to synthesize the objectives of several sections.

- *Aside Comments.* A new feature of this edition is the use of aside comments to students. Many of them are cautions or suggestions for avoiding common errors. These comments are enclosed in colored octagons.

- *Polishing of Presentation.* We have further improved the development in spots that have been troublesome for some learners. The changes are numerous, and many of them are not minor. We expect that the improvements, based on users' suggestions, will make this edition a good deal more teachable than the previous editions.

FEATURES OF THE BOOK

One of the important features of this book is its format. Most pages have an outer margin, used for material of several types, including developmental exercises and statements of objectives.

- *Objectives.* The objectives for each section are stated in the margin. They have been made easy to identify with a domino symbol ▪. That same symbol is placed in the text and also in the exercises and tests beside the corresponding material, making the text easy to read and usable for self-study.

- *Margin Exercises.* The exercises in the margins are designed to get the student *involved actively* in the development, an important feature. Those exercises are always very much like the examples in the text and also like the homework exercises.

- *Exercise Sets.* The exercises for each section are placed at the end of the section on tear-out sheets. The spacing of the answer lines matches that in the *Answer Booklet* to make for quick and easy grading. The exercises are carefully indexed to the objectives, and identified by the domino symbols ▪▪. The exercises are also paired, so that each even-numbered exercise is very much like the one that precedes it. They are also carefully graded, the easier exercises coming first. Answers to the odd-numbered exercises are in the back of the book.

Special, optional calculator exercises and optional challenge exercises have also been included at the ends of certain exercise sets.

- *Readiness Checks.* A readiness check is a short set of exercises that can be used to determine whether a student has acquired the skills that are prerequisite to success with a chapter. Readiness checks have been placed at the beginnings of all chapters except the first one. They are indexed to places in the text where the topic concerned is covered. The readiness checks are optional.

- *Tests and Reviews.* At the end of each chapter there is a set of questions that can be used as a test or a review, and there is a final exam at the end of the book. The items are again indexed to the places in the text where the material concerned is covered. Additional test materials are provided in the *Instructor Manual with Tests* and the *computerized testbank*.

TEACHING MODES

This book can be used in many ways. We list a few.

- *As an ordinary textbook.* To use the book this way, the instructor and students simply ignore the margin exercises.

- *For a modified lecture.* To bring student-centered activity into the class, the instructor stops lecturing and has the students do margin exercises at appropriate times.

- *For a no-lecture class.* The instructor makes assignments that students do on their own, including working exercise sets. During the following class period the instructor answers questions, and students have an extra day or two to polish their work before handing it in. In the meantime, they are working on the next assignment. This method provides individualization while keeping a class together. It minimizes the number of instructor hours required and has been found to work well with large classes.

- *In a learning laboratory.* This book has a low reading level and high understandability, so it is quite suitable for use in a learning laboratory or any kind of learning situation that is essentially self-study.

INTENDED AUDIENCE AND SCOPE

This book is one of a series of basic mathematics texts, which include the following:

> *Arithmetic*
> *Introductory Algebra*
> *Intermediate Algebra*
> *College Algebra* (or *Algebra and Trigonometry*)

The first book in the series, *Arithmetic*, covers arithmetical skills as a prelude to an introductory algebra course. By appropriate choices of topics it can also be used for a terminal course in arithmetic. *Introductory Algebra*, the second book of the series, is for students who have not studied algebra. It is a beginner's introduction, although it may also be appropriate for those needing a review. The present book, *Intermediate Algebra*, is intended for use by students who have finished a first course in algebra in high school or have taken a comparable course, such as that provided by the second book of this series. *Intermediate Algebra* starts "at the beginning," but proceeds faster than an introductory book does, and covers the usual topics of intermediate algebra, including quadratics, exponential functions, and logarithms. It provides adequate preparation for a course in finite mathematics or brief calculus. The last book, *College Algebra*, provides the level of algebraic maturity usually desired as a prerequisite to the mainstream calculus.

A Placement Test is available to help place students in the appropriate book of the series.

ACKNOWLEDGMENTS

We wish to express great thanks to Don Campbell, of Cochise College, for his suggestions regarding the alternate factoring method in Chapter 4. Our thanks also go to the following people who reviewed the manuscript: Jack A. Golden, of the University of Montana; Donald R. Johnson, of Scottsdale Community College; and Jacky Peterson, formerly of Arizona State University and now in private industry. To the following participants in the Focus Group, we are also grateful: Laura Cameron, of the University of New Mexico; Marianna McClymonds, of Phoenix College; Jacky Peterson; Dennis H. Sorge, of Purdue University; James W. Stark Jr., of Lansing Community College; and John D. Whitesitt, of Southern Oregon State College.

M.L.K.
M.L.B.
January, 1983

SUPPLEMENTS

This book is one of a series. The series is accompanied by the following kinds of supplements.

■ *Computerized Testbank.*   Test items can be randomly selected and a test constructed by a computer, using the testbank. The bank can be used with almost all computer hardware currently available. The bank contains approximately ten test items for each objective considered to be a terminal objective. The test question format is multiple choice, but the questions are worded in such a way that an ordinary constructed-response test can also be constructed, merely by leaving off the answer choices. The computerized testbank is available to users in three ways:

1. *On computer tape.*   In this mode the bank can be used with most large computing systems. Users may have to specify the kind of tape their system accepts.

2. *On mini-floppy discs for TRS-80 microcomputers.*

3. *On mini-floppy discs for Apple II microcomputers.*

Final copies of tests can be produced from our programs where a letter quality printer is available (such as NEC spinwriter, DIABLO, or CUME) by using the printout and writing in only some special characters and certain graphs. If a printer is not available, the computer will still generate tests. Notation such as fractions and exponents, requiring spacing other than the usual one-line-at-a-time, will be printed in coded form. That notation is then to be added and the tests can then be typed and duplicated.

The authors owe very special thanks to John Whitesitt, of Southern Oregon State College, for his work in designing and producing the computerized testbank.

■ *Printed Testbank.*   The same test items appear in the printed form of the testbank as in the computerized testbank. The printed bank is usable by those who do not wish to use a computer or who do not have access to a computer.

■ *Videotape Cassettes.*   Color TV is used to help students over the rough spots. A lecturer (John Jobe, of Oklahoma State University) speaks to students and works out examples, with lucid explanations. These video cassettes can be used to supplement lectures or in individualized learning situations, such as learning laboratories.

■ *Audiotape Cassettes.*   In the tape recorder, the student has the voice of a friendly tutor, who helps him or her by talking through examples and problems, using the book itself as a visual device, to make a truly audiovisual presentation. These audiotapes are especially useful in individualized learning situations, such as learning laboratories, and are an excellent help for students who have missed a number of classes.

■ *Instructor's Manual with Tests.*   In this booklet there are five alternative forms of each chapter test and five alternative forms of the final examination. They are classroom ready, and all answers are provided.

■ *Answer Booklet.*   This booklet contains answers to *all* exercises. Answers to the odd-numbered exercises are in the text. The booklet is a great time-saver for the instructor, and it can also be made available to students, if desired, at low cost.

■ *Student's Guide to Exercises.*   Complete worked-out solutions are provided in this booklet. There are solutions to about 70 percent of the odd-numbered exercises in the exercise sets. That represents a change from the third edition, where most of the worked-out exercises were margin exercises, as found in the paperback books of the series. The change was made because it was found that help with exercises from the exercise sets themselves was preferred by students.

■ *Placement Test.*   To help place a student properly in one of the algebra books of the series, a test has been constructed. The test is handy because it is short, although its brevity may make it appear less useful than it actually is. It was thoroughly evaluated statistically and is known to give a good screening with respect to placement into one of four categories into which the books of the series fall.

I ARITHMETIC, Fourth Edition

II INTRODUCTORY ALGEBRA, Fourth Edition
ALGEBRA: AN INTRODUCTORY COURSE*

III INTERMEDIATE ALGEBRA, Fourth Edition
ALGEBRA: AN INTERMEDIATE COURSE*

IV COLLEGE ALGEBRA, Third Edition
FUNDAMENTAL COLLEGE ALGEBRA, Second Edition*
ALGEBRA AND TRIGONOMETRY, Third Edition
FUNDAMENTAL ALGEBRA AND TRIGONOMETRY, Second Edition*
TRIGONOMETRY: TRIANGLES AND FUNCTIONS, Third Edition

*These texts are hardcover versions of the paperbound texts.

More precise placement within a specific book can be made by using an alternative form of the final examination as a pretest.

CONTENTS

REAL NUMBERS AND THEIR PROPERTIES

1

LINEAR EQUATIONS

2

SYSTEMS OF EQUATIONS

3

xii   Contents

EQUATIONS OF SECOND DEGREE AND THEIR GRAPHS

8

EXPONENTIAL AND LOGARITHMIC FUNCTIONS

9

INEQUALITIES AND SETS

10

REAL NUMBERS AND THEIR PROPERTIES

1

As a building is built, unit by unit, on a firm foundation, so algebra builds, starting with the real numbers.

OBJECTIVES

After finishing Section 1.1, you should be able to:

:•: Identify various kinds of real numbers.

:••: Convert from decimal to fractional notation, and vice versa.

:•••: Find percent notation for rational numbers, and convert from percent notation to decimal or fractional notation.

::••:: Determine which of two numbers is greater and indicate which, using < or >.
Determine the absolute value of a number.

1. Which of the following are natural numbers?

$$1, \quad \frac{2}{3}, \quad -5, \quad 0, \quad 17$$

2. Which of the following are whole numbers?

$$1, \quad \frac{2}{3}, \quad -5, \quad 0, \quad 17$$

3. Which of the following are integers?

$$1, \quad \frac{2}{3}, \quad -5, \quad 0, \quad 17$$

1.1 NUMBERS AND ALGEBRA

Numbers are important in algebra as well as arithmetic. The most important set of numbers in ordinary algebra is the set of real numbers.

:•: SOME SUBSETS OF THE REAL NUMBERS

There is just one real number for each point of a line.

The *positive* numbers are shown to the right of 0. The *negative* numbers are shown to the left. Zero is neither positive nor negative. There are several kinds of real numbers.

> *Natural numbers:* **Those used for counting: 1, 2, 3, and so on.**

When we include 0 with this set we have the system of whole numbers.

> *Whole numbers:* **The set of natural numbers, with 0 included: 0, 1, 2, 3, and so on.**

We extend the set of whole numbers to the integers by including some negative numbers.

> *Integers:* **0, 1, −1, 2, −2, 3, −3, and so on.**

DO EXERCISES 1–3 (IN THE MARGIN AT THE LEFT).

> *Rational numbers:* **This set consists of the integers and all quotients of integers (excluding division by zero).**

Thus the rational numbers can be named by fractional symbols with integers for numerator and denominator. The following are rational numbers:

$$\frac{4}{5}, \quad \frac{-4}{7}, \quad \frac{9}{1}, \quad 6, \quad -4, \quad 0, \quad \frac{78}{-5}, \quad -\frac{2}{3}, \quad 1.5, \quad -0.12$$

$$\left(-\frac{2}{3} \text{ can be named } \frac{-2}{3} \text{ or } \frac{2}{-3}\right), \quad \left(1.5 \text{ can be named } \frac{3}{2}\right).$$

Real numbers that are not rational are called *irrational*. They will be considered in a later chapter.

> In arithmetic you usually write $2\frac{1}{2}$, rather than $\frac{5}{2}$. In algebra you will find that the so-called *improper* symbols such as $\frac{5}{2}$ are more useful.

●● DECIMAL AND FRACTIONAL NOTATION

Rational numbers can also be named using *decimal* notation.

Example 1 Find decimal notation for $\frac{5}{8}$.

$\frac{5}{8}$ means $5 \div 8$, so we divide.

```
      0.625
   8)5.000
     4 8
     ───
       20
       16
       ──
        40
        40
        ──
```

Decimal notation for $\frac{5}{8}$ is thus 0.625. We also call 0.625 a *decimal*.

Decimal notation for some numbers repeats.

Example 2 Find decimal notation for $\frac{6}{11}$.

We divide:
```
        0.5454 . . .
    11)6.0000
        5 5
        ──
         50
         44
         ──
          60
          55
          ──
           50
           44
           ──
            6
```

Repeating decimal notation can be abbreviated by writing a bar over the repeating part, in this case, $0.\overline{54}$.

DO EXERCISES 4–6 IN THE MARGIN AT THE RIGHT.

Example 3 Find fractional notation for 3.6.

$3.6 = \dfrac{36}{10}$ The last decimal place is *tenths* so the denominator is 10.

Example 4 Find fractional notation for 0.75.

$0.75 = \dfrac{75}{100}$ The denominator is 100 because the last decimal place is *hundredths*.

DO EXERCISES 7 AND 8.

Find decimal notation.

4. $\dfrac{7}{8}$

5. $\dfrac{7}{11}$

6. $\dfrac{9}{7}$ (This requires a lengthy division.)

Find fractional notation.

7. 5.9

8. 0.125

Find decimal notation and fractional notation.

9. 3.5%

10. 38.4%

Find decimal notation and percent notation.

11. $\dfrac{3}{8}$

12. $\dfrac{17}{20}$

13. Find percent notation for 0.0035.

14. Find decimal notation for 235%.

●●● PERCENT NOTATION

We can also name rational numbers using percent notation. Let us recall what percent means.

> **The symbol % means " × 0.01" or " × $\dfrac{1}{100}$."**

We use this definition to change to or from percent notation.

Example 5   Find decimal notation and fractional notation for 67.9%.

$$67.9\% = 67.9 \times 0.01 \qquad \text{Replacing \% by } \times 0.01$$
$$= 0.679 \qquad \text{Multiplying. This is decimal notation.}$$
$$= \frac{679}{1000} \qquad \text{Changing to fractional notation}$$

Example 6   Find decimal notation and percent notation for $\dfrac{5}{8}$.

We divide 5 by 8.

$$8\overline{)5.000} \quad .625$$

Decimal notation is 0.625.

$$\frac{5}{8} = 0.625 \times 100 \times 0.01 \qquad \text{Multiplying by } 100 \times 0.01, \text{ or } 1$$
$$= (0.625 \times 100) \times 0.01$$
$$= 62.5 \times 0.01$$
$$= 62.5\% \qquad \text{Replacing } \times 0.01 \text{ by \%}$$

DO EXERCISES 9–12.

Percents can be less than 1% or greater than 100%.

Examples   Find percent notation.

7. $0.005 = 0.005 \times 100 \times 0.01$
$$= (0.005 \times 100) \times 0.01$$
$$= 0.5 \times 0.01$$
$$= 0.5\%$$

8. $1.35 = 1.35 \times 100 \times 0.01$
$$= 135 \times 0.01$$
$$= 135\%$$

Examples   Find decimal notation.

9. $120\% = 120 \times 0.01$
$$= 1.2$$

10. $0.25\% = 0.25 \times 0.01$
$$= 0.0025$$

DO EXERCISES 13 AND 14.

There is a table of decimal, fractional, and percent equivalents inside the back cover. If you do not already know those facts, you should memorize them.

⠒⠒ ORDER

The number line shows the real numbers lined up, in order. We say that one number is *greater* than another if it is to the right of the other on the number line. A number is *less* than another if it occurs to the left of the other. The symbol < means "is less than." The symbol > means "is greater than."

Examples

11. $-5 < 2$

12. $-7 < 0$

13. $2.314 < 2.315$

14. $1.2 > -1.2$

15. $-6 > -8$

16. $-\dfrac{4}{5} < -\dfrac{1}{3}$

DO EXERCISES 15–18.

⠒⠒ ABSOLUTE VALUE

The absolute value of a number is its distance from 0 on a number line. The absolute value of a number a can be named $|a|$. Note on the number line above that 3 and -3 are both a distance of three units from 0.

Examples   Simplify.

17. $|-7|$ The distance of -7 from 0 is 7, so $|-7| = 7$.

18. $|5|$ The distance of 5 from 0 is 5, so $|5| = 5$.

19. $|0|$ The distance of 0 from 0 is 0, so $|0| = 0$.

Absolute value is simple, so try not to make it difficult. If a number is positive or zero, then when you take its absolute value, just leave it alone. If a number is negative, to find its absolute value, just make it positive.

DO EXERCISES 19–22.

Insert < or > to make true sentences.

15. -5 -4

16. $-\dfrac{1}{4}$ $\dfrac{1}{2}$

17. 87 67

18. -9.8 -4.5

Simplify.

19. $|2|$

20. $\left|\dfrac{3}{2}\right|$

21. $|0|$

22. $\left|-\dfrac{1}{4}\right|$

From time to time throughout this book, you will find some *extras*, such as the one below. These are optional. You may find them of interest, or you may wish to skip them. Some require the use of a calculator. The very simplest calculator will usually be good enough.

Any calculations or exercises marked ▦ are to be done with a calculator.

SOMETHING EXTRA

REPEATING DECIMALS

Given a repeating decimal, how can we find fractional notation?

Example 1   Find fractional notation for $0.666\overline{6}$.

We first let n represent the number. Then we multiply by 10, moving the decimal point one place.

$$10n = 6.666\overline{6}$$

Now $$n = 0.666\overline{6}$$ This is the original number.

$$9n = 6$$ We subtract n from $10n$ on the left, to get $9n$. On the right, when we subtract, we get 6.

Next, we divide by 9:

$$n = \frac{6}{9}, \quad \text{or} \quad \frac{2}{3}.$$

▦ We check the answer by dividing: $2 \div 3 = 0.666\overline{6}$.

Example 2   Find fractional notation for $0.83333\overline{3}$.

First we multiply by 10. Letting n represent the number, we get

$$10n = 8.3333\overline{3}.$$

Now the repeating part starts at the decimal point, as in Example 1.

We multiply by 10 again, to get

$$100n = 83.3333\overline{3}.$$

Next, we subtract $10n$ from $100n$:

$$100n = 83.333333\overline{3}$$
$$10n = 8.333333\overline{3}$$
$$90n = 75$$

Now we divide by 90:

$$n = \frac{75}{90}, \quad \text{or} \quad \frac{5}{6}.$$

Check: ▦ Divide 5 by 6: $5 \div 6 = 0.833\overline{3}$.

EXERCISES   Find fractional notation.

1. $0.555\overline{5}$ 2. $0.444\overline{4}$ 3. $0.166\overline{6}$ 4. $0.83333\overline{3}$

5. $0.1212\overline{12}$ 6. $0.123123\overline{123}$ 7. $0.12555\overline{5}$ 8. $0.42333\overline{3}$

NAME   CLASS   ANSWERS

EXERCISE SET 1.1

You should review the objectives for this section before doing these exercises.

■● Given the numbers $-5, 0, 2, -\frac{1}{2}, -3, \frac{7}{8}, 14, -\frac{8}{3}, 2.43, 7\frac{1}{2}$:

1. Name the natural numbers.   **2.** Name the whole numbers.

3. Name the rational numbers.   **4.** Name the integers.

■● ● Find decimal notation.

5. $\frac{3}{8}$ **6.** $\frac{1}{8}$ **7.** $\frac{5}{3}$ **8.** $\frac{7}{6}$

Find fractional notation.

9. 2.7   **10.** 13.91   **11.** 0.145   **12.** 0.0213

■● ● ● Find percent notation.

13. $\frac{4}{5}$ **14.** $\frac{7}{25}$ **15.** $\frac{3}{8}$ **16.** $\frac{1}{8}$

17. $\frac{12}{5}$ **18.** $\frac{14}{5}$ **19.** $\frac{3}{80}$ **20.** $\frac{3}{160}$

21. 0.412   **22.** 0.325   **23.** 0.001   **24.** 0.0025

25. 1.25   **26.** 2.34

ANSWERS

1. _____
2. _____
3. _____
4. _____
5. _____
6. _____
7. _____
8. _____
9. _____
10. _____
11. _____
12. _____
13. _____
14. _____
15. _____
16. _____
17. _____
18. _____
19. _____
20. _____
21. _____
22. _____
23. _____
24. _____
25. _____
26. _____

ANSWERS

27. _____

28. _____

29. _____

30. _____

31. _____

32. _____

33. _____

34. _____

35. _____

36. _____

37. _____

38. _____

39. _____

40. _____

41. _____

42. _____

43. _____

44. _____

45. _____

46. _____

47. _____

48. _____

Find decimal and fractional notation.

27. 56.7%   **28.** 47.5%   **29.** 6.85%   **30.** 8.23%

31. 120%   **32.** 180%   **33.** 0.5%   **34.** 0.1%

:: Insert > or < to make true sentences.

35. 55   40   **36.** 31   94   **37.** 2.3   3.2   **38.** 4.7   47

39. −10   −4   **40.** −6   −20   **41.** −2.3   −3.2   **42.** −16   −1.6

:: Simplify.

43. |16| **44.** $\left|\dfrac{7}{8}\right|$ **45.** |−32|

46. $\left|-\dfrac{2}{3}\right|$ **47.** |0| **48.** |3 − 3|

1.2 ADDITION AND SUBTRACTION

•　ADDITION

To explain addition of real numbers we can use a number line. To perform the addition $a + b$, we start at a and then move according to b. If b is positive, we move to the right. If b is negative, we move to the left.

Examples

1. $6 + (-4) = 2$

Start at 6. Move 4 units to the left. The answer is 2.

2. $-3 + 2 = -1$

Start at -3. Move 2 units to the right. We get -1.

3. A halfback made a gain of 7 yards, followed by a gain of -5 yards. What was the net gain?

   $7 + (-5) = 2$, so the net gain was 2 yards.

DO EXERCISES 1–4.

What happens when we add two negative numbers?

Examples Add.

4. $-5 + (-9) = -14$

5. $-10 + (-21) = -31$

6. $-9.4 + (-3.2) = -12.6$

7. $-\dfrac{9}{6} + \left(-\dfrac{11}{6}\right) = -\dfrac{20}{6},\quad \text{or} \quad -\dfrac{10}{3}$

8. $-\dfrac{1}{2} + \left(-\dfrac{1}{4}\right) = -\dfrac{1}{2} \cdot \dfrac{2}{2} + \left(-\dfrac{1}{4}\right) = -\dfrac{2}{4} + \left(-\dfrac{1}{4}\right) = -\dfrac{3}{4}$

> Here we have multiplied by 1 to get a common denominator.

9. $-\dfrac{1}{5} + \left(-\dfrac{1}{15}\right) = -\dfrac{1}{5} \cdot \dfrac{3}{3} + \left(-\dfrac{1}{15}\right) = -\dfrac{3}{15} + \left(-\dfrac{1}{15}\right) = -\dfrac{4}{15}$

> **The sum of two negative numbers is negative. Add their absolute values and then make the answer negative.**

DO EXERCISES 5–11.

OBJECTIVES

After finishing Section 1.2, you should be able to:

•　Add real numbers.

••　Find additive inverses of real numbers.

•••　Subtract real numbers.

::　Identify uses of the commutative and associative laws of addition.

Add, using a number line.

1. $-3 + 5$

2. $9 + (-5)$

3. $6 + (-10)$

4. $-5 + 5$

Add.

5. $-7 + (-11)$

6. $-8 + (-9)$

7. $-20 + (-9)$

8. $-8.9 + (-9.7)$

9. $-\dfrac{6}{5} + \left(-\dfrac{23}{5}\right)$

10. $-\dfrac{2}{5} + \left(-\dfrac{3}{10}\right)$

11. $-\dfrac{2}{3} + \left(-\dfrac{5}{6}\right)$

Add.

12. $-7 + 7$

13. $16 + (-16)$

14. $0 + 0$

Add.

15. $4 + (-7)$

16. $23 + (-11)$

17. $-7.8 + 4.5$

18. $-\dfrac{7}{3} + \dfrac{14}{3}$

19. $\dfrac{3}{8} + \left(-\dfrac{5}{8}\right)$

20. $\dfrac{1}{4} + \left(-\dfrac{3}{8}\right)$

21. $-\dfrac{3}{5} + \dfrac{7}{10}$

What happens when we add a positive number and a negative number? There are several different cases.

Examples   Add.

10. $-6 + 6 = 0$

11. $-10 + 10 = 0$

12. $7 + (-7) = 0$

13. $4.2 + (-4.2) = 0$

> If a positive number and a negative number have the same absolute value, their sum is 0.

DO EXERCISES 12–14.

Now suppose we add a positive number and a negative number with different absolute values. The answer may be positive or negative.

Examples   Add.

14. $-5 + 7 = 2$

15. $-5 + 1 = -4$

16. $6.2 + (-9.8) = -3.6$

17. $-5.89 + 2.11 = -3.78$

18. $\dfrac{1}{4} + \left(-\dfrac{1}{2}\right) = \dfrac{1}{4} + \left(-\dfrac{2}{4}\right) = -\dfrac{1}{4}$

19. $-\dfrac{3}{8} + \dfrac{1}{2} = -\dfrac{3}{8} + \dfrac{4}{8} = \dfrac{1}{8}$

> To add a positive and a negative number with different absolute values, find the difference of the absolute values. If the negative number has the greater absolute value, the answer is negative. If the positive number has the greater absolute value, the answer is positive.

DO EXERCISES 15–21.

●● ADDITIVE INVERSES

The additive inverse of a number is the number opposite to it on the number line. The additive inverse or simply *inverse* (sometimes called the "opposite") of a number x is named $-x$.

Examples   Find the additive inverse.

20. 5

We reflect to the opposite side of 0. The additive inverse of 5 is -5. (-5 is read as "the inverse of 5." It is also read *negative 5*.)

21. -2

We reflect to the opposite side of -2. The additive inverse of -2 is 2. In symbols, $-(-2) = 2$.

> This symbol should be read as "the inverse of negative 2," or "the inverse of the inverse of 2." You should not say "negative negative 2" or "minus minus 2."

Examples

22. Find $-a$ when a is 8.

$-8 = -8$

23. Find $-a$ when a is -2.

$-(-2) = 2$

24. Find $-a$ when a is 0.

$-0 = 0$

25. Find $-(-a)$ when a is -3.

$-(-(-3)) = -3$

Every real number has an additive inverse.
1. **The additive inverse of a number** x **can be named** $-x$.
2. **The additive inverse of a positive number is negative. The inverse of a negative number is positive. The inverse of 0 is 0.**
3. **A number and its additive inverse have the same absolute value. A number and its inverse add to 0.**
4. **Replacing a number by its additive inverse is sometimes called** *changing the sign*.

DO EXERCISES 22–26.

●●● SUBTRACTION

Subtraction is defined as follows.

The difference $a - b$ **is the number which when added to** b **gives** a.

To see that the definition is right, think of $5 - 3$. When you subtract, you get 2. To check, you add 2 to 3, to see if 2 is the number which when added to 3 gives 5.

Using the definition we can explain subtraction on a number line.

Example 26 Subtract: $-3 - 5$.

To subtract 5 from -3 we find the number which when added to 5 gives -3. We start at 5 and see that a move of 8 units to the left gives us -3. The answer is -8.

DO EXERCISES 27–29.

Compare the following.

a) $3 - 8 \quad\; = -5$
$3 + (-8) = -5$

b) $-2 - (-9) = 7$
$-2 + 9 \quad\; = 7$

The pattern above shows us how to subtract without a number line. We first change the subtraction to an addition.

22. Find $-a$ when a is 9.

23. Find $-a$ when a is -23.

24. Find $-(-a)$ when a is -5.

25. Add $12 + (-12)$.

26. Change the sign.
 a) 11
 b) -17
 c) 0
 d) x
 e) $-x$

Draw a number line. Then do these subtractions. Check by adding.

27. $-2 - 3$

28. $5 - 8$

29. $-2 - (-5)$

Subtract

30. $8 - (-9)$

31. $-10 - 6$

32. $-23.7 - 5.9$

33. $-\dfrac{11}{12} - \left(-\dfrac{23}{12}\right)$

34. $-\dfrac{3}{5} - \dfrac{3}{10}$

35. $\dfrac{2}{3} - \left(-\dfrac{5}{6}\right)$

36. a) $17 - 23$
    b) $-17 - 23$
    c) $-17 - (-23)$

Do these calculations.

37. $(4 \times 10) + 13$

38. $4 \times (10 + 13)$

Tell which property of real numbers is illustrated by each sentence.

39. $4 + \left(9 + \dfrac{1}{4}\right) = (4 + 9) + \dfrac{1}{4}$

40. $4.5 + 9.67 = 9.67 + 4.5$

41. $\left(9 + \dfrac{1}{2}\right) + \dfrac{1}{2} = 9 + \left(\dfrac{1}{2} + \dfrac{1}{2}\right)$

> **For any real numbers a and b, $a - b = a + (-b)$. (We can subtract by adding the additive inverse of the subtrahend.)**

Examples   Subtract.

27. $5 - (-3) = 5 + 3 = 8$ Changing the sign of -3 and changing the subtraction to addition

28. $-7 - 3 = -7 + (-3) = -10$

29. $-19.4 - 5.6 = -19.4 + (-5.6) = -25$

30. $-\dfrac{4}{3} - \left(-\dfrac{2}{5}\right) = -\dfrac{4}{3} + \dfrac{2}{5} = -\dfrac{20}{15} + \dfrac{6}{15} = -\dfrac{14}{15}$

DO EXERCISES 30–36.

■■ THE COMMUTATIVE AND ASSOCIATIVE LAWS

An important property of real numbers is that the order in which they are added makes no difference. For example, $4 + 7 = 7 + 4$.

> **For any numbers a and b, $a + b = b + a$. This is the *Commutative Law of Addition*. (It says that numbers can be added in any order.)**

Numbers can also be grouped in various ways for adding. Before we state the law let's consider the use of parentheses to show grouping. Calculations inside parentheses are to be done first.

Example 31

$(3 \times 7) + 10$ means $21 + 10$, or 31,   but

$3 \times (7 + 10)$ means 3×17, or 51.

DO EXERCISES 37 AND 38.

The way we group, or place parentheses, can affect an answer, unless there are only additions to do. In that case we can group, or place parentheses, as we please.

> **For any real numbers a, b, and c, $a + (b + c) = (a + b) + c$. This is the *Associative Law of Addition*.**

Examples   Tell which property of real numbers is illustrated by each sentence (look for regrouping or reordering).

32. $4.5 + \left(9 + \dfrac{1}{2}\right) = (4.5 + 9) + \dfrac{1}{2}$ Associative law of addition (grouping has been changed)

33. $7.8 + \dfrac{1}{4} = \dfrac{1}{4} + 7.8$ Commutative law of addition (order has been changed)

DO EXERCISES 39–41.

NAME

CLASS

EXERCISE SET 1.2

▪ • Add.

1. $-12 + (-16)$

2. $-11 + (-18)$

3. $-8 + (-8)$

4. $-6 + (-6)$

5. $8 + (-3)$

6. $9 + (-4)$

7. $11 + (-7)$

8. $12 + (-8)$

9. $-16 + 9$

10. $-23 + 8$

11. $-24 + 0$

12. $-34 + 0$

13. $-8.4 + 9.6$

14. $-6.3 + 8.2$

15. $-2.62 + (-6.24)$

16. $-5.83 + (-7.43)$

17. $-\frac{2}{7} + \frac{3}{7}$

18. $-\frac{5}{6} + \frac{1}{6}$

19. $-\frac{11}{12} + \left(-\frac{5}{12}\right)$

20. $-\frac{3}{8} + \left(-\frac{7}{8}\right)$

21. $\frac{2}{5} + \left(-\frac{3}{10}\right)$

22. $-\frac{3}{4} + \frac{1}{8}$

▪ • • Find $-a$ when a is

23. 7

24. -8

25. -2.8

26. 0

Change the sign.

27. 10

28. -8

29. 0

30. $-2x$

ANSWERS	
1.	
2.	
3.	
4.	
5.	
6.	
7.	
8.	
9.	
10.	
11.	
12.	
13.	
14.	
15.	
16.	
17.	
18.	
19.	
20.	
21.	
22.	
23.	
24.	
25.	
26.	
27.	
28.	
29.	
30.	

ANSWERS

●●● Subtract.

31. _____

31. $5 - 7$ **32.** $9 - 12$ **33.** $-5 - 7$ **34.** $-9 - 12$

32. _____

33. _____

35. $-6 - (-11)$ **36.** $-7 - (-12)$ **37.** $10 - (-5)$ **38.** $28 - (-16)$

34. _____

35. _____

36. _____

39. $15.8 - 27.4$ **40.** $17.2 - 34.9$ **41.** $-18.01 - 11.24$ **42.** $-19.04 - 15.76$

37. _____

38. _____

43. $-\dfrac{21}{4} - \left(-\dfrac{7}{4}\right)$ **44.** $-\dfrac{16}{5} - \left(-\dfrac{3}{5}\right)$ **45.** $-\dfrac{1}{2} - \left(-\dfrac{1}{12}\right)$ **46.** $-\dfrac{3}{4} - \left(-\dfrac{3}{2}\right)$

39. _____

40. _____

41. _____

⚅ Tell which property of real numbers is illustrated by each sentence.

42. _____

47. $1.23 + 3.21 = 3.21 + 1.23$ **48.** $8 + (-2.3) = -2.3 + 8$

43. _____

44. _____

49. $9 + (2 + 5) = (9 + 2) + 5$ **50.** $(2.6 + 5.2) + 3.4 = 2.6 + (5.2 + 3.4)$

45. _____

46. _____

All exercises below a line like this are extra and optional.

47. _____

Any exercises marked **▦** are to be worked with a calculator.

48. _____

51. What laws or properties of real numbers are illustrated by this sentence?

49. _____

   $(3 + 2) + (-5) = (-5 + 3) + 2$

50. _____

52. ▦ Add.

51. _____

   475.362
   $-$123.814
   30.105
   $-$649.914

52. _____

1.3 MULTIPLICATION AND DIVISION

● MULTIPLICATION OF REAL NUMBERS

We know how to multiply positive numbers. What happens when we multiply a positive number by a negative number?

DO EXERCISE 1.

> **When we multiply a positive number and a negative number, we multiply their absolute values. Then we make the answer negative.**

Examples Multiply.

1. $-3 \cdot 5 = -15$

2. $6 \cdot (-7) = -42$

3. $-1.2 \times 4.5 = -5.4$

4. $3 \cdot \left(-\dfrac{1}{2}\right) = \dfrac{3}{1} \cdot \left(-\dfrac{1}{2}\right) = -\dfrac{3}{2}$

DO EXERCISES 2–4.

What happens when we multiply two negative numbers?

DO EXERCISE 5.

> **To multiply two negative numbers, we multiply their absolute values. The answer is positive.**

Examples Multiply.

5. $-3 \cdot (-5) = 15$

6. $-5.2(-10) = 52$

7. $-8.8 \times (-3.5) = 30.8$

8. $\left(-\dfrac{3}{4}\right) \cdot \left(-\dfrac{5}{2}\right) = \dfrac{15}{8}$

DO EXERCISES 6–8.

The commutative and associative laws hold for all real numbers.

> **For any real numbers a, b, and c,**
>
> $a \cdot b = b \cdot a$ **(The Commutative Law of Multiplication)**
>
> $a \cdot (b \cdot c) = (a \cdot b) \cdot c$ **(The Associative Law of Multiplication)**

The number 1 is special. It is called the *multiplicative identity*.

> **For any real number a, $a \cdot 1 = a$. (The number 1 is the *multiplicative identity*.)**

The number 1 can be named in many ways with fractional notation. Here are some examples.

$$\frac{3}{3}, \quad \frac{-5}{-5}, \quad \frac{1}{1}, \quad \frac{x}{x}, \quad \frac{3/4}{3/4}$$

(Any number, except zero, divided by itself is 1.)

OBJECTIVES

After finishing Section 1.3, you should be able to:

● Multiply positive and negative numbers.

●● Divide positive and negative numbers.

●●● Recognize impossible divisions (divisions by 0).

❚❚ Find the reciprocal of a number, and divide by multiplying by a reciprocal.

1. Look for a pattern and complete.

$4 \cdot 5 = 20$	$-2 \cdot 5 =$
$3 \cdot 5 = 15$	$-3 \cdot 5 =$
$2 \cdot 5 =$	$-4 \cdot 5 =$
$1 \cdot 5 =$	$-5 \cdot 5 =$
$0 \cdot 5 =$	$-6 \cdot 5 =$
$-1 \cdot 5 =$	

Multiply.

2. $-4 \cdot 6$

3. -8.1×3.5

4. $-\dfrac{4}{5} \cdot 10$

5. Look for a pattern and complete.

$4(-5) = -20$	$-1(-5) =$
$3(-5) = -15$	$-2(-5) =$
$2(-5) =$	$-3(-5) =$
$1(-5) =$	$-4(-5) =$
$0(-5) =$	$-5(-5) =$

Multiply.

6. $-8(-9)$

7. $\left(-\dfrac{4}{5}\right) \cdot \left(-\dfrac{2}{3}\right)$

8. $-4.7(-9.1)$

ANSWERS ON PAGE A–1

Multiply.

9. $\dfrac{3}{4} \cdot \dfrac{7}{7}$

10. $\dfrac{4}{3} \cdot \dfrac{-2}{-2}$

11. $\dfrac{-9}{2} \cdot \dfrac{-1}{-1}$

12. $\dfrac{14}{-5} \cdot \dfrac{-1}{-1}$

13. $\dfrac{-17}{-25} \cdot \dfrac{-1}{-1}$

Divide.

14. $\dfrac{-28}{-14}$

15. $\dfrac{125}{-5}$

16. $\dfrac{-75}{25}$

17. $\dfrac{-4.2}{-21}$

Examples   Multiply.

9. $\dfrac{2}{3} \cdot \dfrac{5}{5} = \dfrac{10}{15}$ Since we multiplied by 1, we know that $\dfrac{2}{3} = \dfrac{10}{15}$.

10. $\dfrac{-4}{5} \cdot \dfrac{-1}{-1} = \dfrac{4}{-5}$ Since we multiplied by 1, we know that

$$\dfrac{-4}{5} = \dfrac{4}{-5}.$$

DO EXERCISES 9–13.

⬛⬛ DIVISION

The definition of division parallels the one for subtraction.

> **The quotient p/q is defined to be the number (if it exists) which when multiplied by q gives p.**

Using this definition, we can see how to handle signs when dividing.

Examples

11. $\dfrac{10}{-2} = -5$, because $-5 \cdot (\; -2 \;) = 10$

12. $\dfrac{-32}{4} = -8$, because $-8 \cdot \; 4 \; = -32$

13. $\dfrac{-25}{-5} = 5$, because $5 \cdot (\; -5 \;) = -25$

> **In division, if both numbers are negative, the answer is positive. If one is positive and one is negative, the answer is negative.**

Examples   Divide.

14. $\dfrac{40}{-4} = -10$ **15.** $\dfrac{-10}{5} = -2$

16. $\dfrac{-10}{-40} = \dfrac{1}{4}$, or 0.25 **17.** $\dfrac{-10}{-3} = \dfrac{10}{3}$, or $3.33\overline{3}$

DO EXERCISES 14–17.

⬛⬛⬛ DIVISION BY ZERO

We do not divide by zero. Let's see why. By the definition of division, $n/0$ would be some number which when multiplied by 0 gives n. But when any number is multiplied by 0 the result is 0. The only possibility for n would be 0.

Consider 0/0. We might say that it is 5 because $5 \cdot 0 = 0$. We might also say that it is -8 because $-8 \cdot 0 = 0$. In fact, 0/0 could be any number at all. So we agree not to divide by 0.

Examples   Which of the following divisions are possible?

18. $\dfrac{7}{0}$ Not possible; division by 0.

19. $\dfrac{0}{7}$ Possible; the quotient is 0 because $0 \cdot 7 = 0$.

20. $\dfrac{4}{x - x}$ Not possible; $x - x = 0$ for any x.

DO EXERCISES 18–21.

⠿ DIVISION AND RECIPROCALS

Two numbers whose product is 1 are called *reciprocals* (or *multiplicative inverses*) of each other.

> **Every nonzero real number a has a reciprocal $1/a$. The reciprocal of a positive number is positive. The reciprocal of a negative number is negative.**

Examples   Find the reciprocal of each number.

21. $\dfrac{4}{5}$ The reciprocal is $\boxed{\dfrac{5}{4}}$, because $\dfrac{4}{5} \cdot \boxed{\dfrac{5}{4}} = 1$.

22. 8   The reciprocal is $\boxed{\dfrac{1}{8}}$, because $8 \cdot \boxed{\dfrac{1}{8}} = 1$

23. $-\dfrac{2}{3}$ The reciprocal is $\boxed{-\dfrac{3}{2}}$, because $-\dfrac{2}{3} \cdot \left(\boxed{-\dfrac{3}{2}} \right) = 1$.

24. 0.25   The reciprocal is $\boxed{\dfrac{1}{0.25}}$ or $\boxed{4}$, because $0.25 \times \boxed{4} = 1$.

(To find $\dfrac{1}{0.25}$, divide 1 by 0.25)

> Remember that the reciprocal of a negative number is negative. Do *not* change sign when taking the reciprocal of a number.

DO EXERCISES 22–27.

The number 1 and reciprocals can be used to explain division with fractional notation. We write fractional notation to express the division, and then multiply by 1.

Which of these divisions are possible?

18. $\dfrac{8}{0}$

19. $\dfrac{0}{9}$

20. $\dfrac{17}{2x - 2x}$

21. $\dfrac{3x - 3x}{x - x}$

Find the reciprocal of each number.

22. $\dfrac{3}{8}$

23. $-\dfrac{4}{5}$

24. 18

25. -4.3

26. 0.5

27. Complete this table.

	Additive inverse	Reciprocal
$\frac{2}{3}$	$-\frac{2}{3}$	$\frac{3}{2}$
$\frac{4}{5}$		
$-\frac{3}{4}$		
0.25		
8		
-5		
0		

ANSWERS ON PAGES A–1 AND A–2

Divide by multiplying by 1.

28. $\dfrac{3}{2} \div \left(-\dfrac{5}{8}\right)$

29. $\dfrac{1}{2} \div \dfrac{4}{5}$

30. $-\dfrac{1}{4} \div \dfrac{5}{6}$

Divide by multiplying by a reciprocal.

31. $-\dfrac{3}{4} \div \dfrac{7}{8}$

32. $-\dfrac{12}{5} \div \left(-\dfrac{7}{15}\right)$

Example 25   Divide: $\dfrac{2}{5} \div \dfrac{7}{3}$.

We first write fractional notation for the division: $\dfrac{\frac{2}{5}}{\frac{7}{3}}$.

We then find the reciprocal of the denominator, $\frac{3}{7}$. Next, we multiply by 1, as follows:

$$\frac{\frac{2}{5}}{\frac{7}{3}} = \frac{\frac{2}{5}}{\frac{7}{3}} \times \frac{\frac{3}{7}}{\frac{3}{7}} \qquad \text{Multiplying by 1}$$

$$= \frac{\frac{2}{5} \times \frac{3}{7}}{\frac{7}{3} \times \frac{3}{7}} = \frac{\frac{6}{35}}{1} = \frac{6}{35}. \qquad \begin{array}{l}\text{Multiplying numerators and}\\\text{denominators and simplifying}\end{array}$$

DO EXERCISES 28–30.

In the example above we wound up multiplying the dividend by the reciprocal of the divisor. This always happens.

> **To do the division $\dfrac{a}{b} \div \dfrac{c}{d}$ we can do the multiplication $\dfrac{a}{b} \times \dfrac{d}{c}$ (that is, we can multiply by the reciprocal of the divisor).**

Examples   Divide by multiplying by the reciprocal of the divisor.

26. $\dfrac{1}{4} \div \dfrac{3}{5} = \dfrac{1}{4} \times \dfrac{5}{3} = \dfrac{5}{12}$

27. $\dfrac{2}{3} \div \left(-\dfrac{4}{9}\right) = \dfrac{2}{3} \times \left(-\dfrac{9}{4}\right) = -\dfrac{18}{12}, \quad \text{or} \quad -\dfrac{3}{2}$

To explain the simplification in Example 27, we can use the idea of multiplying by 1.

Example 28   Simplify: $-\dfrac{18}{12}$.

We first factor numerator and denominator:

$$-\frac{18}{12} = -\frac{3 \cdot 6}{2 \cdot 6}.$$

Next, we factor the fraction, with one factor equal to 1:

$$-\frac{3 \cdot 6}{2 \cdot 6} = -\frac{3}{2} \cdot \frac{6}{6}.$$

Then we leave off the factor of 1, to get $-\frac{3}{2}$.

DO EXERCISES 31 AND 32.

NAME   CLASS

EXERCISE SET 1.3

● Multiply.

1. $3(-7)$ **2.** $5(-8)$ **3.** $-2 \cdot 4$ **4.** $-5 \cdot 9$

5. $-8(-2)$ **6.** $-7(-3)$ **7.** $-9 \cdot 14$ **8.** $-8 \cdot 17$

9. $-6(-5.7)$ **10.** $-7(-6.1)$ **11.** $-4.2(-6.3)$ **12.** $-7.4(-9.6)$

13. $-3\left(-\dfrac{2}{3}\right)$ **14.** $-5\left(-\dfrac{3}{5}\right)$ **15.** $-3(-4)(5)$ **16.** $-6(-8)(9)$

17. $-\dfrac{3}{5} \cdot \dfrac{4}{7}$ **18.** $-\dfrac{5}{4} \cdot \dfrac{11}{3}$ **19.** $-\dfrac{9}{11} \cdot \left(-\dfrac{11}{9}\right)$ **20.** $-\dfrac{13}{7} \cdot \left(-\dfrac{5}{2}\right)$

21. $-\dfrac{2}{3} \cdot \left(-\dfrac{2}{3}\right) \cdot \left(-\dfrac{2}{3}\right)$ **22.** $-\dfrac{4}{5} \cdot \left(-\dfrac{4}{5}\right) \cdot \left(-\dfrac{4}{5}\right)$

●● Divide.

23. $\dfrac{-8}{4}$ **24.** $\dfrac{-16}{2}$ **25.** $\dfrac{56}{-8}$ **26.** $\dfrac{63}{-7}$

27. $\dfrac{-77}{-11}$ **28.** $\dfrac{-48}{-6}$ **29.** $\dfrac{-5.4}{-18}$ **30.** $\dfrac{-8.4}{-12}$

ANSWERS

1.
2.
3.
4.
5.
6.
7.
8.
9.
10.
11.
12.
13.
14.
15.
16.
17.
18.
19.
20.
21.
22.
23.
24.
25.
26.
27.
28.
29.
30.

ANSWERS

31. _____

32. _____

33. _____

34. _____

35. _____

36. _____

37. _____

38. _____

39. _____

40. _____

41. _____

42. _____

43. _____

44. _____

45. _____

46. _____

47. _____

48. _____

49. _____

50. _____

51. _____

52. _____

53. _____

54. _____

55. _____

● ● ●   Which of these divisions are possible?

31. $\dfrac{9}{0}$

32. $\dfrac{8}{0}$

33. $\dfrac{0}{16}$

34. $\dfrac{0}{28}$

35. $\dfrac{9}{y - y}$

36. $\dfrac{2x - 2x}{2x - 2x}$

● ●
● ●   Find the reciprocal of each.

37. $\dfrac{3}{4}$

38. $\dfrac{9}{10}$

39. $-\dfrac{7}{8}$

40. $-\dfrac{5}{6}$

41. 15

42. -75

43. 4.4

44. -5.5

Divide.

45. $\dfrac{2}{7} \div \left(-\dfrac{11}{3}\right)$

46. $\dfrac{3}{5} \div \left(-\dfrac{6}{7}\right)$

47. $-\dfrac{10}{3} \div \left(-\dfrac{2}{15}\right)$

48. $-\dfrac{12}{5} \div \left(-\dfrac{3}{10}\right)$

49. $18.6 \div (-3.1)$

50. $39.9 \div (-13.3)$

51. $(-75.5) \div (-15.1)$

52. $(-12.1) \div (-0.11)$

Compute.

53. ▦ $-80{,}397 \times (-583)$

54. ▦ -0.56004×0.003

55. The reciprocal of an electric resistance is called *conductance*. When two resistors are connected in parallel, the conductance is the sum of the conductances, $\dfrac{1}{r_1} + \dfrac{1}{r_2}$. Find the conductance of two resistors of 10 ohms and 5 ohms when connected in parallel.

1.4 THE DISTRIBUTIVE LAWS AND THEIR USE

▪• SOME AGREEMENTS

When a letter can represent various numbers, we call it a *variable*. When we write two or more letters together we agree that this means multiplication. We make the same agreement when variables and numerals occur together.

Examples

1. Prt means $P \cdot r \cdot t$.

2. $5xy$ means $5 \cdot x \cdot y$.

We also agree that in an expression like $(4 \cdot 10) + (9 \cdot 2)$, we can omit the parentheses. Thus $4 \cdot 10 + 9 \cdot 2$ means $(4 \cdot 10) + (9 \cdot 2)$. In other words, we do the multiplications first. We make a similar agreement for subtraction. That is, $4 \cdot 10 - 9 \cdot 2$ means $(4 \cdot 10) - (9 \cdot 2)$. The multiplications are to be done first.

Example 3 Evaluate the following expression when $x = 4$ and $y = 5$. Then simplify.

$$8x + 3y = 8 \cdot 4 + 3 \cdot 5 \qquad \text{Substituting}$$
$$= 32 + 15 \qquad \text{Multiplying}$$
$$= 47 \qquad \text{Adding}$$

DO EXERCISES 1 AND 2.

THE DISTRIBUTIVE LAWS

Compare.

a) $\boxed{5}\,(4 + 8) = \boxed{5} \cdot 12 = 60$

   $\boxed{5} \cdot 4 + \boxed{5} \cdot 8 = 20 + 40 = 60$

b) $\boxed{-4}\,(9 + 5) = \boxed{-4} \cdot 14 = -56$

   $\boxed{-4} \cdot 9 + (\boxed{-4}) \cdot 5 = -36 + (-20) = -56$

We can either add and then multiply, or multiply and then add.

> *The Distributive Law of Multiplication over Addition:*
> **For any numbers** a, b, **and** c, $a(b + c) = ab + ac$.

> Note that we cannot omit the parentheses in the expression $a(b + c)$.
> If we did, we would have $ab + c$, which by our agreement means $(ab) + c$.

ANSWERS ON PAGE A-2

OBJECTIVES

After finishing Section 1.4, you should be able to:

▪• Evaluate an expression like $ab + cd$ when numbers are specified for the letters.

▪•• Use the distributive laws to multiply expressions like 5 and $(x + 9)$.

▪▪▪ Use the distributive laws to factor expressions like $4x - 12$.

▪▪ Collect like terms.

Evaluate each expression when $a = 5$, $b = -2$, $c = 4$, and $d = 10$.

1. $ab + cd$

2. $ab - cd$

Simplify and compare.

3. a) $5(6 + 4)$
   b) $5 \cdot 6 + 5 \cdot 4$

4. a) $59(100 - 1)$
   b) $59 \cdot 100 - 59 \cdot 1$

Multiply.

5. $5(x + 9)$

6. $8(y - 10)$

7. $a(x + y - z)$

8. $10\left(4x - 6y + \dfrac{1}{2}z\right)$

There is another distributive law.

> *The Distributive Law of Multiplication over Subtraction:*
> **For any numbers a, b, and c, $a(b - c) = ab - ac$.**

We can subtract and then multiply, or we can multiply and then subtract.

DO EXERCISES 3 AND 4.

●● MULTIPLYING

The distributive laws are the basis of multiplying in algebra, as well as in arithmetic. In the following examples, note that we multiply each number or letter inside the parentheses by the factor outside.

Examples   Multiply.

4. $4(x - 2) = 4 \cdot x - 4 \cdot 2 = 4x - 8$

5. $b(s - t + f) = b s - b t + b f$

6. $-3(y + 4) = -3 \cdot y + (-3) \cdot 4 = -3y - 12$

7. $-2x(y - 1) = -2x \cdot y - (-2x) \cdot 1 = -2xy + 2x$

DO EXERCISES 5–8.

●●● FACTORING

The reverse of multiplying is called *factoring*. To *factor* an expression means to rewrite it as a product.

Examples   Factor.

8. $8 x + 8 y = 8(x + y)$

9. $c x - c y = c(x - y)$

The parts of an expression such as $3x + 4y - 2z$ are called the *terms* of the expression. In this case the terms are $3x$, $4y$, and $-2z$.

Whenever the terms of an expression have a factor in common, we can "remove" that factor using the distributive laws. We proceed as in Examples 8 and 9, but we may have to factor some of the terms first in order to display the common factor.

Example 10   Factor: $4x + 8$.

$4x + 8 = 4x + 4 \cdot 2$ Factoring 8 to display the common factor

$\quad\quad = 4(x + 2)$ Using the distributive law

Generally, we try to factor out the largest factor common to all the terms. In the following example, we might factor out 3, but there is a larger factor common to the terms, 9. So we factor out the 9.

Example 11   Factor: $9x + 27y$.

$$9x + 27y = 9x + 9 \cdot (3y) = 9(x + 3y)$$

We often have to supply a factor of 1 when factoring out a common factor, as in the next example, which is from a formula about simple interest.

Example 12   Factor: $P + Prt$.

$$P + Prt = P \cdot \boxed{1} + Prt \qquad \text{Writing } P \text{ as a product of } P \text{ and } 1$$

$$= P(\boxed{1} + rt) \qquad \text{Using the distributive law}$$

↑

> It is a common error to omit this 1. If you do that, when you factor out P, you might leave out an entire term.

DO EXERCISES 9–13.

▪▪ COLLECTING LIKE TERMS

If two terms have the same letter, or letters, we say that they are *like terms*, or *similar terms*. If two terms have no letter at all but are just numbers, they are similar terms. We can simplify by *collecting** like terms, using the distributive laws.

Examples   Collect like terms.

13. $3\,\boxed{x} + 5\,\boxed{x} = (3 + 5)\,\boxed{x} = 8\,\boxed{x}$

14. $\boxed{x} - 3\,\boxed{x} = 1 \cdot \boxed{x} - 3 \cdot \boxed{x} = (1 - 3)\,\boxed{x} = -2\,\boxed{x}$

15. $2x + 3y - 5x - 2y$

$\qquad = 2x + 3y + (-5x) + (-2y) \qquad$ Subtraction is the same as adding an inverse.

$\qquad = 2x + (-5x) + 3y + (-2y) \qquad$ Using the commutative and associative laws

$\qquad = (2 - 5)x + (3 - 2)y \qquad$ Using the distributive law

$\qquad = -3x + y \qquad$ Adding

16. $3x + 2x + 5 + 7 = (3 + 2)x + (5 + 7) = 5x + 12$

You need not write the intermediate steps.

DO EXERCISES 14–18.

---

* Also called "combining" like terms.

Factor.

9. $9x + 9y$

10. $ac - ay$

11. $6x - 12$

12. $35x - 25y + 15w + 5$

13. $bs + bt - bw$

Collect like terms.

14. $9x + 11x$

15. $5x - 12x$

16. $5x + x$

17. $x - 7x$

18. $22x - 2.5y + 1.4x + 6.4y$

19. The population of a city is P. After an 8% decrease, what is the population?

The following example shows one way in which factoring is useful for solving practical problems.

Example 17 A certain amount of money is invested. This is called *principal* and we will represent this amount by P. The investment pays 7% interest per year. *Interest* is the money you are paid for your investment. Find an expression for the value of the investment one year later.

$$\text{Principal} + \text{Interest} = P + 7\% \cdot P$$
$$= 1 \cdot P + 0.07P$$
$$= (1 + 0.07)P$$
$$= 1.07P$$

Example 17 is concerned with a percent of increase. Margin exercise 19 is concerned with a percent of decrease, but is similar to Example 17.

DO EXERCISE 19.

SOMETHING EXTRA

CALCULATOR CORNER: COMPUTING INTEREST

You open a savings account, depositing $115. The interest rate is $4\frac{1}{2}\%$. How much money will be in your account at the end of a year? To calculate the interest, you take $4\frac{1}{2}\%$ of $115.

$$4.5\% \times 115 = 4.5 \times 0.01 \times 115 = \$5.18$$

Add this to your $115 to get $120.18.

After 60 days, how much will be in your account? To figure this, assume that a year has 360 days. Thus the interest will be $\frac{60}{360}$ of a year's interest.

$$\text{60-day interest} = \frac{60}{360} \times 4.5\% \times 115$$

$$= \frac{1}{6} \times 4.5 \times 0.01 \times 115 = \$0.86$$

Add this to your $115, to get $115.86.

EXERCISES

Suppose you deposit $312.60 at $5\frac{1}{4}\%$.

1. Calculate your interest for 120 days.

2. How much will be in your account at the end of 240 days?

NAME   CLASS   ANSWERS

EXERCISE SET 1.4

⬛• Evaluate each expression when $x = -2$, $y = 3$, and $z = -4$.

1. $xy + x$ **2.** $xy - y$ **3.** $xz + yz$

4. $xy - xz$ **5.** $yz + ry$ **6.** $xz + xy$

The expression $P + Prt$ gives the value of an account of P dollars principal, invested at a rate r (in percent) for a time t (in years). Find the value of an account under the following conditions.

7. $P = \$120$ **8.** $P = \$500$
   $r = 6\%$ $r = 8\%$
   $t = 1$ yr $t = \frac{1}{2}$ yr

⬛⬛ Multiply.

9. $3(a + 1)$ **10.** $8(x + 1)$ **11.** $4(x - y)$ **12.** $9(a - b)$

13. $-5(2a + 3b)$ **14.** $-2(3c + 5d)$ **15.** $2a(b - c + d)$ **16.** $5x(y - z + w)$

17. $2\pi r(h + 1)$ **18.** $P(1 + rt)$ **19.** $\frac{1}{2}h(a + b)$ **20.** $\pi r(1 + s)$

ANSWERS

1. _____
2. _____
3. _____
4. _____
5. _____
6. _____
7. _____
8. _____
9. _____
10. _____
11. _____
12. _____
13. _____
14. _____
15. _____
16. _____
17. _____
18. _____
19. _____
20. _____

ANSWERS

21. _____

22. _____

23. _____

24. _____

25. _____

26. _____

27. _____

28. _____

29. _____

30. _____

31. _____

32. _____

33. _____

34. _____

35. _____

36. _____

37. _____

38. _____

39. _____

40. _____

41. _____

42. _____

43. _____

44. _____

45. _____

46. _____

47. _____

48. _____

49. _____

50. _____

51. _____

52. _____

53. _____

54. _____

55. _____

56. _____

●●● Factor.

21. $8x + 8y$

22. $7a + 7b$

23. $9p - 9$

24. $12x - 12$

25. $7x - 21$

26. $6y - 36$

27. $xy + x$

28. $ab + a$

29. $2x - 2y + 2z$

30. $3x + 3y - 3z$

31. $3x + 6y - 3$

32. $4a + 8b - 4$

33. $ab + ac - ad$

34. $xy - xz + xw$

35. $\pi rr + \pi rs$

36. $\frac{1}{2}ah + \frac{1}{2}bh$

■■ Collect like terms.

37. $4a + 5a$

38. $9x + 3x$

39. $8b - 11b$

40. $9c - 12c$

41. $14y + y$

42. $13x + x$

43. $12a - a$

44. $15x - x$

45. $t - 9t$

46. $x - 6x$

47. $5x - 3x + 8x$

48. $3x - 11x + 2x$

49. $5x - 8y + 3x$

50. $9a - 10b + 4a$

51. $7c + 8d - 5c + 2d$

52. $12a + 3b - 5a + 6b$

53. $4x - 7 + 18x + 25$

54. $13p + 5 - 4p + 7$

55. $13x + 14y - 11x - 47y$

56. $17a + 17b - 12a - 38b$

1.5 MULTIPLYING BY −1 AND SIMPLIFYING

● MULTIPLYING BY −1

What happens when we multiply a number by −1?

Examples

1. $-1 \cdot 9 = -9$ 2. $-1 \cdot (-6) = 6$ 3. $-1 \cdot 0 = 0$

DO EXERCISES 1–3.

For any number a, $-1 \cdot a = -a$. (Negative 1 times a is the additive inverse of a.)*

From this fact, we know that we can replace − by −1 or the reverse, in any expression. In that way, we can rename additive inverses.

Examples

4. $-(3x) = -1(3x)$ Replacing − by −1

$= (-1 \cdot 3)x$ Using associativity

$= -3x$ Multiplying

5. $-(-9y) = -1(-9y)$ Replacing − by −1

$= [-1(-9)]y$ Using associativity

$= 9y$ Multiplying

DO EXERCISES 4 AND 5.

Examples   Find the additive inverse and name it without parentheses.

6. $4 + x$. The inverse is $-(4 + x)$.

$-(4 + x) = -1(4 + x)$ Replacing − by −1

$= \boxed{-1} \cdot 4 + (\boxed{-1}) \cdot x$ Multiplying

$= -4 + (-x)$ Replacing $-1 \cdot x$ by $-x$

$= -4 - x$ Adding an inverse is the same as subtracting.

7. $3x - 2y + 4$. The inverse is $-(3x - 2y + 4)$.

$-(3x - 2y + 4) = -1(3x - 2y + 4)$

$= \boxed{-1} \cdot 3x - (\boxed{-1})2y + (\boxed{-1})4$

$= -3x + [-(-2y)] + (-4)$

$= -3x + 2y - 4$

8. $a - b$. The inverse is $-(a - b)$.

$-(a - b) = -1(a - b) = -1 \cdot a - (-1) \cdot b$

$= -a + [-(-1)b] = -a + b = b - a$

DO EXERCISES 6–10.

---

* Or, *changing the sign* is also the same as multiplying by −1.

OBJECTIVES

After finishing Section 1.5, you should be able to:

● Rename an additive inverse without parentheses.

●● Identify the terms of an expression.

●●● Simplify expressions by removing parentheses and collecting like terms.

●●●● Simplify expressions containing parentheses within parentheses.

Multiply.

1. $-1 \cdot 24$

2. $-1 \cdot 0$

3. $-1 \cdot (-10)$

Rename each additive inverse without parentheses.

4. $-(9x)$

5. $-(-24t)$

Rename each additive inverse without parentheses.

6. $-(7 - y)$

7. $-(x - y)$

8. $-(9x + 6y + 11)$

9. $-(23x - 7y - 2)$

10. $-(-3x - 2y - 1)$

ANSWERS ON PAGE A–2

Rename each additive inverse without parentheses.

11. $-(-2x - 5z + 24)$

12. $-(3x - 2y)$

13. $-\left(\dfrac{1}{4}t + 41w - 5d - 23\right)$

14. What are the terms of
$-5x - 7y + 67t - \frac{4}{5}$?

Remove parentheses and simplify.

15. $6x - (3x + 8)$

16. $6y - 4 - (2y - 5)$

17. $6x - (9y - 4) - (8x + 10)$

> Example 8 illustrates something that you should remember, because it is a shortcut. The inverse of an expression $a - b$ is $b - a$; that is, $-(a - b) = b - a$.

The above examples show we can rename an additive inverse by multiplying every term by -1. We could also say that we change the sign of every term inside the parentheses. Thus we can skip some steps.

Example 9 Rename without parentheses: $-(-9t + 7z - \dfrac{1}{4}w)$.

$$-\left(-9t + 7z - \frac{1}{4}w\right) = 9t - 7z + \frac{1}{4}w \qquad \begin{array}{l}\text{Changing the sign}\\\text{of every term}\end{array}$$

DO EXERCISES 11–13.

■ ■ ●   TERMS

What do we mean by the *terms* of an expression? When there are only addition signs, it is easy to tell, but if there are some subtraction signs, we might have reason to wonder. If there are subtraction signs, we can always rename using addition signs.

Example 10   What are the terms of $3x - 4y + 2z$?

$3x - 4y + 2z = 3x + (-4y) + 2z$ Renaming using additions

Thus the terms are $3x$, $-4y$, and $2z$.

DO EXERCISE 14.

● ● ●   REMOVING CERTAIN PARENTHESES

In some expressions commonly encountered in algebra there are parentheses preceded by subtraction signs. These parentheses can be removed by changing the sign of *every* term inside.

Examples   Remove parentheses and simplify.*

11. $6x - (4x + 2) = 6x + [-(4x + 2)]$ Subtracting is adding the additive inverse.

$\qquad\qquad\qquad = 6x + (-4x) + (-2)$ Renaming the inverse

$\qquad\qquad\qquad = 6x - 4x - 2$ Subtracting is adding the inverse.

$\qquad\qquad\qquad = 2x - 2$ Collecting like terms

12. $3y - 4 - (9y - 7) = 3y - 4 - 9y + 7$

$\qquad\qquad\qquad\qquad = 6y + 3, \quad\text{or}\quad 3 - 6y$

> If parentheses are preceded by an addition sign, *no* signs are changed when they are removed. For example, $2x + (5x - 7) = 2x + 5x - 7$.

DO EXERCISES 15–17.

---

* *Simplify* means to rewrite using a minimal amount of symbolism.

We now consider subtracting an expression consisting of several terms preceded by a number.

Examples   Remove parentheses and simplify.

13. $x - 3(x + y) = x + [-3(x + y)]$ Subtracting is adding the inverse.

$$= x + [-3x - 3y]$$ Multiplying -3 by $x + y$
$$= x - 3x - 3y$$ Removing parentheses
$$= -2x - 3y$$ Collecting like terms

14. $3y - 2(4y - 5) = 3y + [-2(4y - 5)]$ Subtraction is adding the inverse.

$$= 3y + [-8y + 10]$$ Multiplying -2 by $4y - 5$
$$= 3y - 8y + 10$$ Removing parentheses
$$= -5y + 10$$ Collecting like terms.

A common error is to forget to change this sign. *Remember*: When multiplying like this, change the sign of *every* term inside the parentheses.

DO EXERCISES 18 AND 19.

▚ PARENTHESES WITHIN PARENTHESES

When parentheses occur within parentheses, we may make them of different shapes, such as [ ] (also called "brackets") and { } (usually called "braces"). All of these have the same meaning. When parentheses occur within parentheses, computations in the *innermost* ones are to be done first.

Example 15   Simplify: $5 - \{6 - [3 - (7 + 3)]\}$.

$$5 - \{6 - [3 - (7 + 3)]\} = 5 - \{6 - [3 - 10]\}$$ Adding $7 + 3$
$$= 5 - \{6 - (-7)\}$$ Subtracting $3 - 10$
$$= 5 - 13$$ Subtracting $6 - (-7)$
$$= -8$$

Example 16   Simplify: $7 - [3(2 - 5) - 4(2 + 3)]$.

$$7 - [3(2 - 5) - 4(2 + 3)] = 7 - [3(-3) - 4(5)]$$ Doing the calculations in the innermost parentheses
$$= 7 - [-9 - 20]$$
$$= 7 - [-29]$$
$$= 36$$

DO EXERCISES 20 AND 21.

When expressions with parentheses contain variables, we still work from the inside out when simplifying.

Remove parentheses and simplify.

18. $x - 2(y + x)$

19. $3x - 5(2y - 4x)$

Simplify.

20. $6 - \{5 - [2 - (8 + 20)]\}$

21. $5 + \{6 - [2 + (5 - 2)]\}$

ANSWERS ON PAGE A-2

Simplify.

22. $15x - \{2[2(x - 5) - 6(x + 3)] + 4\}$

Example 17   Simplify: $6y - \{4[3(y - 2) - 4(y + 2)] - 3\}$.

$6y - \{4[3y - 6 - 4y - 8] - 3\}$ Multiplying to remove the innermost parentheses

$6y - \{4[-y - 14] - 3\}$ Collecting like terms in the inner parentheses

$6y - \{-4y - 56 - 3\}$ Multiplying to remove the inner parentheses

$6y - \{-4y - 59\}$ Collecting like terms

$6y + 4y + 59$ Removing parentheses

$10y + 59$ Collecting like terms

DO EXERCISES 22 AND 23.

SOMETHING EXTRA

AN APPLICATION: SUCCESSIVE DISCOUNTS

Wholesalers often give retailers *successive* discounts. For example, there may be a 20% discount on an item, followed by a 10% additional discount for buying a large quantity, followed by a 5% discount for paying within 30 days.

Example 1   Find the cost of an item listed at $100 when successive discounts of 20%, 10%, and 5% are given.

After the discount of 20%, the cost is 80% of $100, or $80.
After the discount of 10%, the cost is 90% of $80, or $72.
After the discount of 5%, the cost is 95% of $72, or $68.40.

What happens if the discounts are given in a different order?

Example 2   Find the cost of an item listed at $100 when successive discounts of 10%, 5%, and 20% are given.

After the discount of 10%, the cost is 90% of $100, or $90.
After the discount of 5%, the cost is 95% of $90, or $85.50.
After the discount of 20%, the cost is 80% of $85.50, or $68.40.

23. $9a + \{3a - 2[(a - 4) - (a + 2)]\}$

In Examples 1 and 2 we got the same answer. Will this always be the case? Let's see.

We can always multiply the list price by several percentages, as above.

$[(\$100 \times 80\%) \times 90\%] \times 95\%$.

Does it matter how we use parentheses to group? Does it matter in which order we multiply? The commutative and associative laws tell us that it does not. Therefore, successive discounts can be figured in any order.

EXERCISES

Find the cost, given the list price and the successive discounts.

1. $500, 20%, 15%, 5%   **2.** $450, 10%, 10%, 20%

3. $58, 5%, 5%, 25%   **4.** $210, 10%, 10%, 10%

NAME	CLASS	ANSWERS

EXERCISE SET 1.5

⬤ Rename each additive inverse without parentheses.

1. $-(-4b)$ **2.** $-(-5x)$ **3.** $-(a + 2)$ **4.** $-(b + 9)$

5. $-(b - 3)$ **6.** $-(x - 8)$ **7.** $-(t - y)$ **8.** $-(r - s)$

Find the additive inverse.

9. $a + b + c$ **10.** $x + y + z$

11. $8x - 6y + 13$ **12.** $9a - 7b + 24$

13. $-2c + 5d - 3e + 4f$ **14.** $-4x + 8y - 5w + 9z$

⬤ ⬤ What are the terms of the following?

15. $4a - 5b + 6$ **16.** $5x - 9y + 12$

17. $2x - 3y \quad 2z$ **18.** $5a - 7b - 9c$

⬤⬤⬤ Simplify.

19. $a + (2a + 5)$ **20.** $x + (5x + 9)$

21. $4m - (3m - 1)$ **22.** $5a - (4a - 3)$

ANSWERS

1. _____
2. _____
3. _____
4. _____
5. _____
6. _____
7. _____
8. _____
9. _____
10. _____
11. _____
12. _____
13. _____
14. _____
15. _____
16. _____
17. _____
18. _____
19. _____
20. _____
21. _____
22. _____

ANSWERS

23. _____

24. _____

25. _____

26. _____

27. _____

28. _____

29. _____

30. _____

31. _____

32. _____

33. _____

34. _____

35. _____

36. _____

23. $3d - 7 - (5 - 2d)$

24. $8x - 9 - (7 - 5x)$

25. $-2(x + 3) - 5(x - 4)$

26. $-9(y + 7) - 6(y - 3)$

27. $5x - 7(2x - 3)$

28. $8y - 4(5y - 6)$

⁝⁝ Simplify.

29. $9a - [7 - 5(7a - 3)]$

30. $12b - [9 - 7(5b - 6)]$

31. $5\{-2 + 3[4 - 2(3 + 5)]\}$

32. $7\{-7 + 8[5 - 3(4 + 6)]\}$

33. $2y + \{7[3(2y - 5) - (8y + 7)] + 9\}$

34. $7b - \{6[4(3b - 7) - (9b + 10)] + 11\}$

35. ▦ Remove parentheses and simplify.

$(0.0079x - 0.000843y) - (-0.00793x - 0.000853y)$

36. Simplify.

$-\{-[-(-9)]\}$

1.6 EXPONENTIAL NOTATION

◼◦ WHOLE-NUMBER EXPONENTS

Exponential notation is a shorthand device. For $3 \cdot 3 \cdot 3 \cdot 3$ we write 3^4. The latter is called *exponential notation*. In 3^4 the number 3 is called the *base* and the 4 is called the *exponent*.

> **Exponential notation a^n, where n is an integer greater than 1, means**
>
> $$\underbrace{a \cdot a \cdot a \cdots a \cdot a}_{n \textbf{ factors.}}.$$

> *Caution!* a^n does *not* mean to multiply n times. For example, 3^2 means $3 \cdot 3$, or 9. We multiply only once.

Examples   Write exponential notation.

1. $7 \cdot 7 \cdot 7$ *Answer:* 7^3

2. xxx *Answer:* x^3

3. $2x \cdot 2x \cdot 2x \cdot 2x$ *Answer:* $(2x)^4$

DO EXERCISES 1–3.

Examples   Rewrite without exponents.

4. 5^2 *Answer:* $5 \cdot 5$, or 25

5. $(4y)^2$ *Answer:* $4y \cdot 4y$, or $16yy$

DO EXERCISES 4–6.

Examples   Simplify.

6. $(4x)^2 = 4x \cdot 4x = 16x^2$

7. $(-3y)^3 = (-3y)(-3y)(-3y) = -27y^3$

DO EXERCISES 7 AND 8.

In general, an exponent tells how many times the base occurs as a factor. What happens when the exponent is 1 or 0? We cannot have the base occurring as a factor 1 time or 0 times. Look for a pattern.

$$10^3 = 10 \cdot 10 \cdot 10 = 1000$$
$$10^2 = 10 \cdot 10 = 100$$
$$10^1 = \;?$$
$$10^0 = \;?$$

In order for the pattern to continue, 10^1 would have to be 10 and 10^0 would have to be 1. We shall *agree* that exponents of 1 and 0 have that meaning.

> **For any number a, we agree that a^1 means a.**
> **For any nonzero number a, we agree that a^0 means 1.**

Write exponential notation.

1. $8 \cdot 8 \cdot 8 \cdot 8$

2. mmm

3. $4y \cdot 4y \cdot 4y \cdot 4y \cdot 4y$

Rewrite without exponents.

4. 3^4

5. y^2

6. $(5x)^4$

Simplify.

7. $(5y)^2$

8. $(-2x)^3$

Rewrite without exponents.

9. 8^1

10. $(-31)^1$

11. 3^0

12. $(-7)^0$

13. y^0, where $y \neq 0$

Rewrite without negative exponents.

14. 10^{-4}

15. $(-4)^{-3}$

Rewrite with negative exponents.

16. $\dfrac{1}{4^3}$

17. $\dfrac{1}{(-5)^4}$

> *Caution!* a^0 does *not* mean a times 0! It means 1. We will see later why we do not allow the base a to be zero.

Examples   Rewrite each of the following without an exponent.

8. 4^1 *Answer:*   4

9. $(-9y)^1$ *Answer:* $-9y$

10. 6^0 *Answer:*   1

11. $(-37.4)^0$ *Answer:*   1

DO EXERCISES 9–13.

◖◗ NEGATIVE INTEGERS AS EXPONENTS

How shall we define negative integers as exponents? Look for a pattern.

$$10^2 = 100$$
$$10^1 = 10$$
$$10^0 = 1$$
$$10^{-1} =$$
$$10^{-2} =$$

In order for the pattern to continue, 10^{-1} would have to be $\frac{1}{10}$ and 10^{-2} would have to be $\frac{1}{100}$. This leads to the following agreement.

> If n is any integer, a^{-n} is given the meaning $1/a^n$. In other words, a^n and a^{-n} are reciprocals.

Examples   Rewrite without using a negative exponent.

12. 7^{-3} *Answer:* $\dfrac{1}{7^3}$

13. $(-2)^{-3}$ *Answer:* $\dfrac{1}{(-2)^3}$, or $\dfrac{1}{-8}$

> *Caution!* A negative exponent does *not* necessarily indicate that an answer is negative! For example, 3^{-2} means $\dfrac{1}{3^2}$, which is $\dfrac{1}{9}$.

DO EXERCISES 14 AND 15.

Examples   Rewrite using a negative exponent.

14. $\dfrac{1}{5^2}$ *Answer:* 5^{-2}

15. $\dfrac{1}{(-7)^4}$ *Answer:* $(-7)^{-4}$

DO EXERCISES 16 AND 17.

NAME

CLASS

EXERCISE SET 1.6

⬛⬤ Write exponential notation.

1. $4 \cdot 4 \cdot 4 \cdot 4 \cdot 4$ **2.** $6 \cdot 6 \cdot 6$ **3.** $5 \cdot 5 \cdot 5 \cdot 5 \cdot 5 \cdot 5$

4. $x \cdot x \cdot x \cdot x$ **5.** $mmmm$ **6.** $ttttt$

7. $3a \cdot 3a \cdot 3a \cdot 3a$ **8.** $5x \cdot 5x \cdot 5x \cdot 5x \cdot 5x$

9. $5 \cdot 5 \cdot c \cdot c \cdot c \cdot d \cdot d \cdot d \cdot d$ **10.** $2 \cdot 2 \cdot 2 \cdot r \cdot r \cdot r \cdot r \cdot t \cdot t$

Rewrite without exponents.

11. 2^5 **12.** 7^3 **13.** $(-3)^4$ **14.** $(-8)^2$ **15.** x^4

16. y^6 **17.** $(4b)^3$ **18.** $(3x)^4$ **19.** $(ab)^4$ **20.** $(xyz)^3$

Rewrite without exponents.

21. 5^1 **22.** $(\sqrt{6})^1$ **23.** $\left(\dfrac{7}{8}\right)^1$ **24.** $\left(\dfrac{9}{7}\right)^1$

25. $(\sqrt{8})^0$ **26.** $(-4)^0$ **27.** $(3z)^0$ **28.** $(5xy)^0$

●● Rewrite without using negative exponents.

29. 6^{-3} **30.** 8^{-4} **31.** 9^{-5}

32. 16^{-2} **33.** $(-11)^{-1}$ **34.** $(-4)^{-3}$

Rewrite using negative exponents.

35. $\dfrac{1}{3^4}$ **36.** $\dfrac{1}{9^2}$ **37.** $\dfrac{1}{10^3}$

38. $\dfrac{1}{12^5}$ **39.** $\dfrac{1}{(-16)^2}$ **40.** $\dfrac{1}{(-8)^6}$

---

41. How can you place $+$ and $\times$ in the blanks to make a true sentence?

$1 \underline{\quad} 2 \underline{\quad} 3 \underline{\quad} 4 \underline{\quad} 5 \underline{\quad} 6 \underline{\quad} 7 \underline{\quad} 8 \underline{\quad} 9 = 100$

1.7 PROPERTIES OF EXPONENTS

◼• MULTIPLICATION AND DIVISION

To see how to multiply, or simplify, in an expression such as $a^3 \cdot a^2$, recall the definition of exponential notation.

$a^3 \cdot a^2$ means $(a \cdot a \cdot a) \cdot (a \cdot a)$ or a^5.

The exponent in a^5 is the *sum* of those in $a^3 \cdot a^2$. In general, the exponents are added when we multiply, but note that the base must be the same in all parts.

Let us consider a case in which one exponent is positive and one is negative.

$$b^5 \cdot b^{-2} = b \cdot b \cdot b \cdot b \cdot b \cdot \frac{1}{b \cdot b} \quad \text{By the definition of exponents}$$

$$= \frac{b \cdot b}{b \cdot b} \cdot b \cdot b \cdot b$$

$$= b^3$$

Adding exponents again gives the correct result. Let us now consider a case in which both exponents are negative.

$$c^{-3} \cdot c^{-2} = \frac{1}{c \cdot c \cdot c} \cdot \frac{1}{c \cdot c} \quad \text{By the definition of exponents}$$

$$= \frac{1}{c \cdot c \cdot c \cdot c \cdot c} \quad \text{Multiplying}$$

$$= c^{-5} \quad \text{Simplifying}$$

Again, adding exponents gives the correct result. This is always the case, even if exponents are negative or zero.

> In multiplication with exponential notation, we can add exponents if the bases are the same.
>
> $$a^m a^n = a^{m+n}$$

Examples Multiply and simplify.

1. $x^4 \cdot x^3 = x^{4+3} = x^7$

2. $4^5 \cdot 4^{-3} = 4^{5+(-3)} = 4^2$

3. $(-2)^{-3}(-2)^7 = (-2)^{-3+7} = (-2)^4 = 16$

4. $(8x^4y^{-2})(-3x^{-3}y) = 8 \cdot (-3) \cdot x^4 \cdot x^{-3} \cdot y^{-2} \cdot y^1$

$$= -24x^{4-3}y^{-2+1}$$

$$= -24xy^{-1} \quad \text{or} \quad -\frac{24x}{y}$$

> The answer $\frac{-24x}{y}$ would also be correct here.

There is no standard way of simplifying in cases like Example 4. In some problems it is better to have all exponents positive, but in some it is not. We give answers in two ways. You should also be able to give answers in two ways.

DO EXERCISES 1–6.

Multiply and simplify.

1. $8^{-3}8^7$

2. $y^7 y^{-2}$

3. $(9x^4)(-2x^7)$

4. $(-3x^{-4})(25x^{-10})$

5. $(5x^{-3}y^4)(-2x^{-9}y^{-2})$

6. $(4x^{-2}y^4)(15x^2y^{-3})$

ANSWERS ON PAGE A–3

Divide and simplify.

7. $\dfrac{4^8}{4^5}$

8. $\dfrac{5^4}{5^{-2}}$

9. $\dfrac{10^{-8}}{10^{-2}}$

10. $\dfrac{42y^7x^6}{-21y^{-3}x^{10}}$

11. $\dfrac{33a^5b^{-2}}{22a^2b^{-4}}$

We consider a division.

$$\frac{8^5}{8^3} \quad \text{means} \quad \frac{8 \cdot 8 \cdot 8 \cdot 8 \cdot 8}{8 \cdot 8 \cdot 8}. \quad \text{This simplifies to } 8 \cdot 8, \text{ or } 8^2.$$

We can obtain the result by subtracting exponents. This is always the case, even if exponents are negative or zero.

> **In division with exponential notation, we can subtract exponents if the bases are the same.**
>
> $$\frac{a^m}{a^n} = a^{m-n}$$

Examples   Divide and simplify.

5. $\dfrac{5^7}{5^3} = 5^{7-3} = 5^4 \qquad$ Subtracting exponents

6. $\dfrac{5^7}{5^{-3}} = 5^{7-(-3)} = 5^{7+3} \qquad$ Subtracting exponents (adding an inverse)

$\qquad\qquad = 5^{10}$

7. $\dfrac{9^{-2}}{9^5} = 9^{-2-5} = 9^{-7}, \quad \text{or} \quad \dfrac{1}{9^7}$

8. $\dfrac{7^{-4}}{7^{-5}} = 7^{-4-(-5)} = 7^{-4+5} = 7^1 = 7$

9. $\dfrac{16x^4y^7}{-8x^3y^9} = \dfrac{16}{-8} \cdot \dfrac{x^4}{x^3} \cdot \dfrac{y^7}{y^9} = -2xy^{-2}, \quad \text{or} \quad -\dfrac{2x}{y^2}$

> The answer $\dfrac{-2x}{y^2}$ would also be correct here.

10. $\dfrac{14x^7y^{-3}}{4x^5y^{-5}} = \dfrac{14}{4} \cdot \dfrac{x^7}{x^5} \cdot \dfrac{y^{-3}}{y^{-5}} = \dfrac{7}{2}x^2y^2$

> In exercises such as Examples 1–10 above, it may help to think as follows: After writing the base, write the top exponent. Then write a subtraction sign. Then write the bottom exponent. Then do the subtraction. For example,
>
> $$\frac{x^{-3}}{x^{-5}} = x^{-3-(-5)}$$
>
> Writing the base and the top exponent | Writing a subtraction sign | Writing the bottom exponent

DO EXERCISES 7–11.

We do not define 0^0. Now we can see why. Note the following. We know that 0^0 is equal to 0^{1-1}. But 0^{1-1} is also equal to $0/0$. We have already seen that we must leave $0/0$ undefined, so we also leave 0^0 undefined.

◨◧ RAISING POWERS TO POWERS

Consider the expression $(\boxed{5^2})^4$. This means

$$\boxed{5^2} \cdot \boxed{5^2} \cdot \boxed{5^2} \cdot \boxed{5^2}, \quad \text{or} \quad 5^8.$$

We can obtain the result by multiplying the exponents.

Consider $(\boxed{8^{-2}})^3$. This means

$$\boxed{\frac{1}{8^2}} \cdot \boxed{\frac{1}{8^2}} \cdot \boxed{\frac{1}{8^2}}, \quad \text{or} \quad \frac{1}{8^6}, \quad \text{which is} \quad 8^{-6}.$$

Again, we could obtain the result by multiplying the exponents. This works, in general, whether the exponents are positive, negative, or zero.

> **To raise a power to a power we can multiply exponents.**
>
> $$(a^m)^n = a^{m \cdot n}$$

Examples   Simplify.

11. $(3^5)^7 = 3^{5 \cdot 7} = 3^{35}$

12. $(x^{-5})^4 = x^{-5 \cdot 4} = x^{-20}$

DO EXERCISES 12 AND 13.

We consider cases in which there are several factors inside the parentheses. Consider $(5^2 \cdot 7^3)^4$. This means

$$(5^2 \cdot 7^3) \cdot (5^2 \cdot 7^3) \cdot (5^2 \cdot 7^3) \cdot (5^2 \cdot 7^3), \quad \text{or} \quad 5^8 \cdot 7^{12}.$$

Again, we can obtain the result by multiplying the exponents. This is true, in general, for positive, negative, or zero exponents.

> **To raise an expression with several factors to a power, raise each factor to the power, multiplying exponents.**
>
> $$(a^m b^n)^p = a^{m \cdot p} \cdot b^{n \cdot p}$$

Examples   Simplify.

13. $(3x^2y^{-2})^3 = 3^3(x^2)^3(y^{-2})^3 = 3^3x^6y^{-6}, \quad \text{or} \quad 27x^6y^{-6}, \quad \text{or} \quad \dfrac{27x^6}{y^6}$

14. $(5x^3y^{-5}z^2)^4 = 5^4(x^3)^4(y^{-5})^4(z^2)^4,$

   or $\quad 625x^{12}y^{-20}z^8, \quad \text{or} \quad \dfrac{625x^{12}z^8}{y^{20}}$

DO EXERCISES 14–17.

Simplify.

12. $(3^7)^6$

13. $(8^2)^{-7}$

Simplify.

14. $(2xy)^3$

15. $(4x^{-2}y^7)^2$

16. $(-2x^4y^2)^5$

17. $(10x^{-4}y^7z^{-2})^3$

ANSWERS ON PAGE A–3

Simplify.

18. $\left(\dfrac{x^{-3}}{y^4}\right)^{-3}$

We now consider raising a quotient to a power. Let's take as an example $\left(\dfrac{3^2}{5^3}\right)^4$. This means

$$\left(\frac{3^2}{5^3}\right) \cdot \left(\frac{3^2}{5^3}\right) \cdot \left(\frac{3^2}{5^3}\right) \cdot \left(\frac{3^2}{5^3}\right), \quad \text{or} \quad \frac{3^8}{5^{12}}.$$

Once more, we see that we can obtain the result by multiplying the exponents. This is true, in general, for positive, negative, or zero exponents.

> **To raise a quotient to a power, raise both numerator and denominator to the power, multiplying exponents.**
>
> $$\left(\frac{a^m}{b^n}\right)^p = \frac{a^{m \cdot p}}{b^{n \cdot p}}$$

Examples   Simplify.

15. $\left(\dfrac{x^2}{y^{-3}}\right)^{-5} = \dfrac{x^{2 \cdot (-5)}}{y^{-3 \cdot (-5)}} = \dfrac{x^{-10}}{y^{15}}, \quad \text{or} \quad \dfrac{1}{x^{10}y^{15}}$

16. $\left(\dfrac{2x^3y^{-2}}{3y^4}\right)^5 = \dfrac{(2x^3y^{-2})^5}{(3y^4)^5} = \dfrac{2^5 x^{3 \cdot 5} y^{-2 \cdot 5}}{3^5 y^{4 \cdot 5}}$

$\qquad = \dfrac{32x^{15}y^{-10}}{243y^{20}} = \dfrac{32x^{15}}{243y^{30}}, \quad \text{or} \quad \dfrac{32}{243}x^{15}y^{-30}$

19. $\left(\dfrac{3x^2y^{-3}}{2y^5}\right)^2$

DO EXERCISES 18 AND 19.

SOMETHING EXTRA

CALCULATOR CORNER: DIVIDING AND RECIPROCALS

Note what happens when you multiply by a reciprocal.

Divide.	Find a reciprocal.	Multiply by the number you found in the middle column.
$4.32 \div 4.5 =$ _____	$\dfrac{1}{4.5} =$ _____	$4.32 \times \dfrac{1}{4.5} =$ _____
$845 \div 325 =$ _____	$\dfrac{1}{325} =$ _____	$845 \times \dfrac{1}{325} =$ _____
$897 \div 11.78 =$ _____	$\dfrac{1}{11.78} =$ _____	$897 \times \dfrac{1}{11.78} =$ _____
$0.035 \div 11.3 =$ _____	$\dfrac{1}{11.3} =$ _____	$0.035 \times \dfrac{1}{11.3} =$ _____

NAME   CLASS   ANSWERS

EXERCISE SET 1.7

●   Multiply and simplify.

1. $5^6 \cdot 5^3$ **2.** $6^2 \cdot 6^6$ **3.** $8^{-6} \cdot 8^2$ **4.** $9^{-5} \cdot 9^3$

5. $8^{-2} \cdot 8^{-4}$ **6.** $9^{-1} \cdot 9^{-6}$ **7.** $b^2 \cdot b^{-5}$ **8.** $a^4 \cdot a^{-3}$

9. $a^{-3} \cdot a^4 \cdot a^2$ **10.** $x^{-8} \cdot x^5 \cdot x^3$ **11.** $(2x)^3(3x)^2$ **12.** $(9y)^2(2y)^3$

13. $(14m^2n^3)(-2m^3n^2)$ **14.** $(6x^5y^{-2})(-3x^2y^3)$

15. $(-2x^{-3})(7x^{-8})$ **16.** $(6x^{-4}y^3)(-4x^{-8}y^{-2})$

Divide and simplify.

17. $\dfrac{6^8}{6^3}$ **18.** $\dfrac{7^9}{7^4}$ **19.** $\dfrac{4^3}{4^{-2}}$ **20.** $\dfrac{5^8}{5^{-3}}$

21. $\dfrac{10^{-3}}{10^6}$ **22.** $\dfrac{12^{-4}}{12^8}$ **23.** $\dfrac{9^{-4}}{9^{-6}}$ **24.** $\dfrac{2^{-7}}{2^{-5}}$

1. _____
2. _____
3. _____
4. _____
5. _____
6. _____
7. _____
8. _____
9. _____
10. _____
11. _____
12. _____
13. _____
14. _____
15. _____
16. _____
17. _____
18. _____
19. _____
20. _____
21. _____
22. _____
23. _____
24. _____

ANSWERS

25. _____

26. _____

27. _____

28. _____

29. _____

30. _____

31. _____

32. _____

33. _____

34. _____

35. _____

36. _____

37. _____

38. _____

39. _____

40. _____

41. _____

42. _____

43. _____

44. _____

45. _____

46. _____

47. _____

48. _____

25. $\dfrac{a^3}{a^{-2}}$

26. $\dfrac{y^4}{y^{-5}}$

27. $\dfrac{9a^2}{(-3a)^2}$

28. $\dfrac{24a^5b^3}{-8a^4b}$

29. $\dfrac{-24x^6y^7}{18x^{-3}y^9}$

30. $\dfrac{14a^4b^{-3}}{-8a^8b^{-5}}$

31. $\dfrac{-18x^{-2}y^3}{-12x^{-5}y^5}$

32. $\dfrac{-14a^{14}b^{-5}}{-18a^{-2}b^{-10}}$

● ● Simplify.

33. $(4^3)^2$

34. $(5^4)^5$

35. $(8^4)^{-3}$

36. $(9^3)^{-4}$

37. $(6^{-4})^{-3}$

38. $(7^{-8})^{-5}$

39. $(3x^2y^2)^3$

40. $(2a^3b^4)^5$

41. $(-2x^3y^{-4})^{-2}$

42. $(-3a^2b^{-5})^{-3}$

43. $(-6a^{-2}b^3c)^{-2}$

44. $(-8x^{-4}y^5z^2)^{-4}$

45. $\left(\dfrac{4^{-3}}{3^4}\right)^3$

46. $\left(\dfrac{5^2}{4^{-3}}\right)^{-3}$

47. $\left(\dfrac{2x^3y^{-2}}{3y^{-3}}\right)^3$

48. $\left(\dfrac{-4x^4y^{-2}}{5x^{-1}y^4}\right)^{-4}$

EXTENSION EXERCISES

SECTION 1.1

Find decimal notation.

1. $\dfrac{3}{14}$

2. $\dfrac{17}{128}$

3. $\dfrac{1403}{3300}$

4. $\dfrac{49}{200}$

5. $\dfrac{2,635,269}{12,345,679}$

6. $\dfrac{41}{333,000}$

7. All rational numbers can be named using decimal notation. Without dividing, is it possible to know if the decimal notation for the rational number $\frac{p}{q}$ will be repeating? If yes, explain.

8. Given that the decimal notation for a rational number $\frac{p}{q}$ is repeating, what is the greatest number of digits that can be in the repeating part?

9. Place the following rational numbers in increasing order.

$$1.1\%, \ \frac{1}{11}, \ \frac{2}{7}, \ 0.3\%, \ 0.11, \ \frac{1}{8}\%, \ 0.009, \ \frac{99}{1000}, \ 0.286, \ \frac{1}{8}, \ \frac{9}{100}, \ 1\%$$

10. a) What is the largest integer that is less than $-\frac{29}{5}$?
    b) What is the smallest integer that is greater than $-\frac{29}{5}$?
    c) What integer is nearest to $-\frac{29}{5}$?

SECTION 1.2

Compute.

11. $(8 - 10) - (-6 + 2)$

12. $|8 - 10| - |-6 + 2|$

13. $-10 + 18 - 6 - 14 + 31$

14. $-10 + 18 - (6 - 14) + 31$

15. $-10 + 18 - 6 - (14 + 31)$

16. $|(8 - |-9|)| + |-6 + 12|$

17. $|-6 - (-3)| - |3 - 6|$

18. Find a number that is 27 less than -6.

19. What number can be added to 11.7 to obtain $-7\frac{3}{4}$?

SECTION 1.3

Compute.

20. $\dfrac{(-7 - 3) \div (-5)}{(4 - 8)(-2 + 5)}$

21. $|-0.09 \div 0.3| \cdot |0.3 - 0.9|$

22. $\left[\dfrac{1}{2} \div \left(-\dfrac{2}{3}\right)\right] \cdot \left(\dfrac{1}{4} - 4\right)$

Place parentheses in each expression so that it has the given value.

23. $3 \cdot 3 + 10 - 8; 15$

24. $7 \cdot 6 - 1 + 4 \cdot 3 - 12; -1$

25. $20 \div 2 \cdot 5 - 2 + 14; 44$

SECTION 1.4

Evaluate each expression when $a = -1$, $b = 7$, $c = -6$, and $d = \frac{1}{2}$.

26. $\dfrac{d(ab - c)}{2a}$

27. $\dfrac{ad + bd}{b - a} - \dfrac{ac}{d}$

SECTION 1.5

Simplify.

28. $\dfrac{2}{3}[2(x + y) + 4(x + 4y)]$

29. $-4[3(x - y - z) - 3(2x + y - 5z)]$

30. $[-(7a - b) - (a + 5b)] - \left[2\left(a + \dfrac{1}{2}b\right) + 3\left(7a - \dfrac{5}{3}b\right)\right]$

31. $0.01\{0.1(x - 2y) - [0.001(3x + y) - (0.2x - 0.1y)]\} - (x - y)$

SECTION 1.6

Simplify.

32. $(-2)^0 - (-2)^3 - (-2)^{-1} + (-2)^4 - (-2)^{-2}$

33. $2(6^1 \cdot 6^{-1} - 6^{-1} \cdot 6^0)$

34. $\dfrac{(-8)^{-2} \cdot (8 - 8^0)}{2^{-6}}$

35. $\left[\dfrac{1}{(-3)^{-2}} - (-3)^1\right] \cdot \left[(-3)^2 + (-3)^{-2}\right]$

36. Evaluate when $x = 1$ and $y = -2$.

$$(x - y)(x^{y-x} - y^{x-y})$$

37. True or false.

$$-5^2 = (-5)^2$$

SECTION 1.7

Simplify.

38. $\dfrac{(2^{-2})^{-4} \cdot (2^3)^{-2}}{(2^{-2})^2 \cdot (2^5)^{-3}}$

39. $\{[(8^{-2})^3]^{-4}\}^5 \cdot [(8^0)^{-2}]^6$

40. $\left[\dfrac{(-3x^{-2}y^5)^{-3}}{(2x^4y^{-8})^{-2}}\right]^2$

41. $\left[\left(\dfrac{a^{-2}}{b^7}\right)^{-3} \cdot \left(\dfrac{a^4}{b^{-3}}\right)^2\right]^{-1}$

NAME   SCORE   ANSWERS

TEST OR REVIEW—CHAPTER 1

Review the material in the chapter. Then test yourself by doing these exercises. If you miss an item, restudy the section indicated to the left of the item.

[1.1, ● ●]   **1.** Find decimal notation for $-\dfrac{13}{5}$.

2. Find fractional notation for 3.26.

[1.1, ● ● ●]   Find percent notation.

   3. 0.036   **4.** $\dfrac{3}{16}$

   Find decimal notation

   5. 5.71%   **6.** $\dfrac{3}{5}\%$

[1.1, ■ ■]   **7.** Insert < or > to make a true sentence.

     -4.5 -8.7

[1.1, ■ ● ■]   Simplify.

   8. $\left|\dfrac{7}{8}\right|$ **9.** $|-13.4|$ **10.** $|0|$

[1.2, ●]   Add.

   11. $7 + (-9)$ **12.** $-5.3 + (-7.8)$ **13.** $-\dfrac{5}{2} + \left(-\dfrac{7}{4}\right)$

[1.2, ● ●]   Find $-a$ when a is as given. (In other words, find the additive inverse of each number.)

   14. 8   **15.** -13 **16.** 0

[1.2, ● ● ●]   Subtract.

   17. $-6 - (-5)$ **18.** $-18.2 - 11.5$ **19.** $\dfrac{19}{4} - \left(-\dfrac{3}{2}\right)$

[1.3, ●]   Multiply.

   20. $(-4.1)(8.2)$ **21.** $-\dfrac{4}{5}\left(-\dfrac{15}{16}\right)$ **22.** $-6(-4)(-11)$

ANSWERS

1. _____
2. _____
3. _____
4. _____
5. _____
6. _____
7. _____
8. _____
9. _____
10. _____
11. _____
12. _____
13. _____
14. _____
15. _____
16. _____
17. _____
18. _____
19. _____
20. _____
21. _____
22. _____

ANSWERS

23. _____

24. _____

25. _____

26. _____

27. _____

28. _____

29. _____

30. _____

31. _____

32. _____

33. _____

34. _____

35. _____

36. _____

37. _____

38. _____

39. _____

40. _____

41. _____

42. _____

43. _____

[1.3, ● ▪]   Divide.

23. $\dfrac{-10}{2}$ **24.** $\dfrac{-75}{-5}$ **25.** $\dfrac{7}{-5}$

[1.3, ▪● ●▪]   Divide.

26. $-\dfrac{4}{3} \div \dfrac{5}{7}$ **27.** $-\dfrac{11}{3} \div \left(-\dfrac{7}{6}\right)$ **28.** $39.9 \div (-26.6)$

[1.4, ● ▪]   Multiply.

29. $-2(3a - 4b)$ **30.** $3\pi r(s + 1)$

[1.4, ●●●]   Factor.

31. $ab - ac + 2ad$ **32.** $2ah + h$

[1.4, ▪● ●▪]   Collect like terms.

33. $6y - 8x + 4y + 3x$ **34.** $4a - 7 + 17a + 21$

[1.5, ●]   **35.** Find the additive inverse of $-9x + 7y - 22$.

[1.5, ●●●]   Simplify.

36. $-3(x + 2) - 4(x - 5)$ **37.** $9y - 5(4y - 6)$

Simplify.

[1.5, ▪● ●▪]   **38.** $4x - [6 - 3(2x - 5)]$ **39.** $3a - 2[5(2a - 5) -$
$(8a + 4)] + 10$

[1.7, ●]   Multiply or divide and simplify.

40. $(3a^4b^{-2})(-2a^5b^{-3})$ **41.** $\dfrac{-12x^3y^{-4}}{8x^7y^{-6}}$

[1.7, ● ▪]   Simplify.

42. $(-3a^{-3}b^2c)^{-4}$ **43.** $\left(\dfrac{-3x^3y^{-3}}{5x^{-2}y^5}\right)^{-3}$

LINEAR EQUATIONS

2

Mathematics can be found to apply in very unexpected situations. A linear equation gives a relation between temperature and the speed at which crickets chirp.

READINESS CHECK: SKILLS FOR CHAPTER 2

Add.

1. $-8 + (-4)$ **2.** $-8.2 + 3.6$

Find $-a$ when a is:

3. -7 **4.** 0

5. 8

Subtract.

6. $8 - (-5)$ **7.** $-11 - 9$

Multiply.

8. $-5 \cdot 2$ **9.** $-2.3 \times (-20)$

Find the reciprocal of:

10. 5 **11.** $-\dfrac{4}{5}$

12. Factor: $5x - 10y + 15$.

13. Multiply: $3(2x - 3y + 4)$.

14. Collect like terms: $x + 3x - 5x$.

15. Evaluate $y - xz$ when $x = -2$, $y = 3$, and $z = -4$.

OBJECTIVES

After finishing Section 2.1, you should be able to:

- ◧ Tell what an equation means and what is meant by a solution of an equation.

- ◨◨ Solve equations using the addition principle.

- ◨◨◨ Solve equations using the multiplication principle.

- ⬛⬛ Solve equations using the principles together.

2.1 SOLVING EQUATIONS

◧ EQUATIONS

By an *equation* we mean a number sentence with $=$ for its verb.

Examples

1. $1 + 10 = 11$ is an equation. It is true.

2. $7 - 8 = 9 - 13$ is an equation. It is false.

3. $x - 9 = 3$ is an equation. It is neither true nor false until we determine what x represents.

DO EXERCISES 1–3 ON PAGE 49.

What is the meaning of an equation?

An equation *says* that the symbols on either side of the equals sign represent the same thing.

Example 4 $7 - 8 = 9 - 13$ says that $7 - 8$ and $9 - 13$ represent the same number. In this case, the equation is false.

DO EXERCISES 4 AND 5.

SOLUTIONS OF EQUATIONS

If an equation contains a variable it may be neither true nor false. Some replacements may make it true and some may make it false.

> The replacements making an equation true are called its *solutions.* When we find them, no matter how, we say that we have *solved* the equation. There are various procedures for solving equations.

Example 5 The number 5 makes the equation $2x + 3 = 13$ true, because if we substitute 5 into the left side we get $2 \cdot 5 + 3$, or 13.

DO EXERCISES 6 AND 7.

●● THE ADDITION PRINCIPLE

One of the principles we use in solving equations concerns adding. Consider the equation $a = b$. It says that a and b represent the same number. Suppose this is true and then add a number c to a. We will get the same answer if we add c to b, because a and b are the same number.

> *The Addition Principle.* **If $a = b$ is true, then $a + c = b + c$ is true for any number c.**

When we use the addition principle, we sometimes say that we "add the same number on both sides of an equation."

Example 6 Solve: $x + 6 = -15$.

$$x + 6 + (\boxed{-6}) = -15 + (\boxed{-6})$$

Using the addition principle, we added -6.

$$\left. \begin{array}{r} x + 0 = -15 + (-6) \\ x = -21 \end{array} \right\}$$ Simplifying

Check:
$$\begin{array}{c|c} x + 6 = -15 \\ \hline -21 + 6 & -15 \\ -15 & \end{array}$$

The solution is -21.

In Example 6, why did we add -6? Because we wanted the variable x alone on one side of the equation. Then the other side shows what calculations to do to find the answer. When we added the inverse of 6, that got rid of the 6 on the left.

Consider the following equations.

a) $3 + 4 = 7$
b) $5 - 1 = 2$
c) $21 + 2 = 24$
d) $x - 5 = 12$
e) $9 - x = x$
f) $13 + 2 = 15$

1. Which equations are true?

2. Which equations are false?

3. Which equations are neither true nor false?

Tell what each equation means.

4. $89 + 2 = 3 + 4$

5. $6 - 7 = 4 - 5$

6. Consider the replacements 5, 8, 4, 0, and 2. Which of them make the equation $2x + 3 = 11$ false?

7. Find the replacement that makes $3x + 2 = 11$ true.

ANSWERS ON PAGE A–4

Solve, using the addition principle.

8. $x + 9 = 2$

9. $x + \dfrac{1}{4} = -\dfrac{3}{4}$

10. $13 = -25 + y$

11. $y - 61.4 = 78.9$

Example 7   Solve: $y - 4.7 = 13.9$.

$$y - 4.7 + \boxed{4.7} = 13.9 + \boxed{4.7} \qquad \text{Using the addition}$$
$$y + 0 = 13.9 + 4.7 \qquad\qquad \text{principle}$$
$$y = 13.9 + 4.7$$
$$y = 18.6 \qquad \text{Doing the calculation}$$
$$13.9 + 4.7$$

The number 18.6 checks, so it is the solution.

DO EXERCISES 8–11.

●●●   THE MULTIPLICATION PRINCIPLE

Suppose an equation $a = b$ is true and we multiply a by a number c. We will get the same answer if we multiply b by c. This is because a and b are the same number.

> *The Multiplication Principle.*   **If an equation $a = b$ is true, then an equation $a \cdot c = b \cdot c$ is true for any number c.**

Example 8   Solve: $4x = 9$.

$$\boxed{\dfrac{1}{4}} \cdot 4x = \boxed{\dfrac{1}{4}} \cdot 9 \qquad \begin{array}{l}\text{Using the multiplication}\\ \text{principle, we multiply by } \frac{1}{4}, \text{ the}\\ \text{reciprocal of 4.}\end{array}$$

$$1 \cdot x = \dfrac{9}{4} \Bigg\}$$
$$x = \dfrac{9}{4} \Bigg\} \quad \text{Simplifying}$$

Check: $\dfrac{4x = 9}{\dfrac{4 \cdot \frac{9}{4}}{9} \Bigg| 9}$

The solution is $\frac{9}{4}$.

In Example 8, why did we choose to multiply by $\frac{1}{4}$? Because we wanted x alone on one side of the equation. We multiplied by the reciprocal of 4. Then we got $1 \cdot x$, which simplified to x. This got rid of the 4 on the left.

Example 9   Solve: $22 = -\dfrac{4}{5}x$.

$$\boxed{-\dfrac{5}{4}} \cdot 22 = \boxed{-\dfrac{5}{4}} \cdot \left(-\dfrac{4}{5}\right)x \qquad \begin{array}{l}\text{Multiplying by } -\frac{5}{4}, \text{ the}\\ \text{reciprocal of } -\frac{4}{5}\end{array}$$

$$-\dfrac{55}{2} = 1 \cdot x$$

$$-\dfrac{55}{2} = x$$

Check:

$$22 = -\tfrac{4}{5}x$$

22	$-\tfrac{4}{5} \cdot \left(-\tfrac{55}{2}\right)$
	22

The solution is $\tfrac{55}{2}$.

DO EXERCISES 12–15.

▍▍ USING THE PRINCIPLES TOGETHER

Let's see how we can use the addition and multiplication principles together.

Example 10   Solve: $3x - 4 = 13$.

$$3x - 4 + \boxed{4} = 13 + \boxed{4} \qquad \text{Using the addition principle, we add 4.}$$

$$3x + (-4) + 4 = 13 + 4 \qquad \text{Subtraction is done by adding an inverse.}$$

$$3x = 17 \qquad \text{Simplifying}$$

$$\boxed{\tfrac{1}{3}} \cdot 3x = \boxed{\tfrac{1}{3}} \cdot 17 \qquad \begin{array}{l}\text{Using the multi-}\\ \text{plication principle,}\\ \text{we multiply by } \tfrac{1}{3}.\end{array}$$

$$x = \frac{17}{3}$$

Check:

$$3x - 4 = 13$$

$3 \cdot \tfrac{17}{3} - 4$	13
$17 - 4$	
13	

The solution is $\tfrac{17}{3}$, or $5\tfrac{2}{3}$.

In Example 10 we used the addition principle first. That usually makes the work simpler.

> There are various ways to solve equations. With the method we are developing, the idea is to find an equation so simple that we can *see* what the solution is. Then we check to see if that solution is also a solution of the original equation.

DO EXERCISE 16.

When there are fractions or decimals in an equation, it helps to eliminate them at the outset. We do that by using the multiplication principle. The process is called *clearing* of decimals or fractions.

Example 11   Clear the equation $\tfrac{3}{4}x + \tfrac{1}{2} = \tfrac{3}{2}$ of fractions.

We multiply on both sides by the least common multiple of the denominators, in this case, 4.

$$\boxed{4}\left(\frac{3}{4}x + \frac{1}{2}\right) = \boxed{4} \cdot \frac{3}{2} \qquad \text{Multiplying by 4}$$

$$4 \cdot \frac{3}{4}x + 4 \cdot \frac{1}{2} = 4 \cdot \frac{3}{2} \qquad \text{Using the distributive law}$$

$$3x + 2 = 6 \qquad \text{Simplifying}$$

Solve, using the multiplication principle.

12. $8x = 10$

13. $-\dfrac{3}{7}y = 21$

14. $-4x = -\dfrac{6}{7}$

15. $-12.6 = 4.2y$

Solve.

16. $-4 + 9x = 8$

17. Clear of fractions.

$$\frac{2}{3} - \frac{5}{6}y = \frac{1}{3}$$

Example 12   Clear the equation $12.4 - 5.12x = 3.14$ of decimals.

We multiply on both sides by a power of ten to get rid of the decimal points.

$$\boxed{100} \, (12.4 - 5.12x) = \boxed{100} \, (3.14) \qquad \text{Multiplying by 100}$$

$$100 \times 12.4 - 100 \times 5.12x = 100 \times 3.14 \qquad \text{Using the distributive law}$$

$$1240 - 512x = 314 \qquad \text{Simplifying}$$

DO EXERCISES 17 AND 18.

18. Clear of decimals.

$$6.3x - 9.1 = 3$$

When there are like terms in an equation we combine them on each side. Then if there are still like terms on opposite sides, we get them on the same side using the addition principle.

Example 13   Solve: $8x + 6 - 2x = -4x - 14$.

$$6x + 6 = -4x - 14 \qquad \text{Combining like terms on the left}$$

$$6x + 4x + 6 = -14 \qquad \text{Adding } 4x \text{ on both sides}$$

$$10x + 6 = -14 \qquad \text{Combining like terms on the left}$$

$$10x = -20 \qquad \text{Adding } -6 \text{ on both sides}$$

$$\frac{1}{10} \cdot 10x = \frac{1}{10} \cdot (-20) \qquad \text{Multiplying on both sides by } \frac{1}{10}$$

$$x = -2 \qquad \text{Simplifying}$$

Solve.

19. $\dfrac{5}{2}x + \dfrac{9}{2}x = 21$

Check:

$$\begin{array}{c|c} \multicolumn{2}{c}{8x + 6 - 2x = -4x - 14} \\ \hline 8(-2) + 6 - 2(-2) & -4(-2) - 14 \\ -16 + 6 + 4 & 8 - 14 \\ -6 & -6 \end{array}$$

The solution is -2.

Here is a summary of the method of solving equations, as used in the preceding examples:

20. $1.4x - 0.9x + 0.7 = -2.2$

1. **Clear of fractions or decimals if that is needed.**
2. **Collect like terms on both sides of the equation, if necessary.**
3. **Get all like terms with letters on one side and all other terms on the other side.**
4. **Collect like terms again, if necessary.**
5. **Use the multiplication principle to get the letter alone on one side.**

21. $-4x + 2 + 5x = 3x - 15$

DO EXERCISES 19–21.

NAME

CLASS

ANSWERS

EXERCISE SET 2.1

Remember to review the objectives before doing the exercises.

●● Solve, using the addition principle. Don't forget to check.

1. $x + 5 = 14$

2. $y + 7 = 19$

3. $x - 18 = -22$

4. $y - 19 = -26$

5. $-8 + y = 15$

6. $-9 + t = 17$

7. $-12 + z = -51$

8. $-37 + x = -89$

9. $p - 296 = 839$

10. $z - 149 = 573$

●●● Solve, using the multiplication principle. Don't forget to check.

11. $5x = 20$

12. $3x = 21$

13. $-4x = 88$

14. $-11y = 121$

15. $4 = 24t$

16. $8 = 40y$

17. $-3x = -96$

18. $-8y = -120$

19. $48y = -288$

20. $39t = -273$

ANSWERS

1. _____

2. _____

3. _____

4. _____

5. _____

6. _____

7. _____

8. _____

9. _____

10. _____

11. _____

12. _____

13. _____

14. _____

15. _____

16. _____

17. _____

18. _____

19. _____

20. _____

⋮⋮ Solve, using the principles together. Don't forget to check.

21. $4x - 12 = 60$

22. $4x - 6 = 70$

23. $5x - 10 = 45$

24. $6z - 7 = 11$

25. $9t + 4 = -104$

26. $5x + 7 = -108$

27. $-\dfrac{7}{3}x + \dfrac{2}{3} = -18$

28. $-\dfrac{9}{2}y + 4 = -\dfrac{91}{2}$

29. $\dfrac{6}{5}x + \dfrac{4}{10}x = \dfrac{32}{10}$

30. $\dfrac{9}{5}y + \dfrac{4}{10}y = \dfrac{66}{10}$

31. $0.9y - 0.7y = 4.2$

32. $0.8t - 0.3t = 6.5$

33. $8x + 48 = 3x - 12$

34. $15x + 20 = 8x - 22$

35. $7y - 1 = 23 - 5y$

36. $3x - 15 = 15 - 3x$

37. $4x - 3 = 5 + 12x$

38. $9t - 4 = 14 + 15t$

39. $5 - 4a = a - 13$

40. $8 - 5x = x - 16$

41. $3m - 7 = -7 - 4m - m$

42. $5x - 8 = -8 + 3x - x$

43. $5x + 3 = 11 - 4x + x$

44. $6y + 20 = 10 + 3y + y$

Solve. Remember to check. (With a calculator, you need not clear of decimals.)

45. ▦ $0.0008x = 0.0000056$

46. ▦ $43.008z = 1.201135$

2.2 MORE ON SOLVING EQUATIONS

◼️• EQUATIONS CONTAINING PARENTHESES

Equations containing parentheses can often be solved by first multiplying to remove parentheses and then proceeding as before.

Example 1   Solve: $3(7 - 2x) = 14 - 8(x - 1)$.

$21 - 6x = 14 - 8x + 8$ Multiplying to remove parentheses

$21 - 6x = 22 - 8x$ Combining like terms

$8x - 6x = 22 + (-21)$ Adding -21 and also $8x$

$2x = 1$ Combining like terms

$x = \dfrac{1}{2}$ Multiplying by $\dfrac{1}{2}$

Check:

$$
\begin{array}{c|c}
3(7 - 2x) = 14 - 8(x - 1) \\
\hline
3(7 - 2 \cdot \frac{1}{2}) & 14 - 8(\frac{1}{2} - 1) \\
3(7 - 1) & 14 - 8(-\frac{1}{2}) \\
3 \cdot 6 & 14 + 4 \\
18 & 18
\end{array}
$$

The solution is $\frac{1}{2}$.

> On the right side of the equation, note that $14 - 8(x - 1) = 14 - 8x + 8$.

DO EXERCISES 1 AND 2.

◼️•• THE PRINCIPLE OF ZERO PRODUCTS

Equations can have more than one solution. We now look at an equation-solving principle that will help to solve some such equations.

When we multiply two numbers, the product will be zero if one of the factors is zero. Furthermore, if a product is zero, then one of the factors must be zero. This gives us another principle for solving.

> *The Principle of Zero Products.*   **For any numbers a and b, if $a \cdot b = 0$, then $a = 0$ or $b = 0$; and if $a = 0$ or $b = 0$, then $a \cdot b = 0$.**

Example 2   Solve: $(x + 4)(x - 2) = 0$.

Here we have a product that is zero. The equation will become true when either factor is zero. Hence it is true when

$x + 4 = 0$ or $x - 2 = 0$.

Now, solving the equations separately, we get

$x = -4$ or $x = 2$.

There are two solutions, -4 and 2. This is so because when x is replaced by either number, the sentence becomes true.

The check is on the following page.

OBJECTIVES

After finishing Section 2.2, you should be able to:

◼️• Solve certain equations containing parentheses.

◼️•• Solve equations using the principle of zero products.

Solve.

1. $30 + 7(x - 1) = 3(2x + 7)$

2. $3(y - 1) - 1 = 2 - 5(y + 5)$

Solve.

3. $(x - 19)(x + 5) = 0$

4. $x(3x - 17) = 0$

5. $(9x + 2)(-6x + 3) = 0$

Try to solve.

6. $4 + 7x = 7x + 9$

7. $3 + 9x = 9x + 3$

Check: We show that -4 is a solution.

$$
\frac{(x + 4)(x - 2) = 0}{\begin{array}{c|c} (-4 + 4)(-4 - 2) & 0 \\ 0 \cdot (-6) & \\ 0 & \end{array}}
$$

We show that 2 is also a solution.

$$
\frac{(x + 4)(x - 2) = 0}{\begin{array}{c|c} (2 + 4)(2 - 2) & 0 \\ 6 \cdot 0 & \\ 0 & \end{array}}
$$

> **To solve using the principle of zero products:**
>
> 1. You *must* have a product on one side of the equation and 0 on the other.
> 2. Set each factor of the product equal to zero. This gives several simple equations.
> 3. Solve the simple equations.

Example 3   Solve: $7x(4x + 2) = 0$.

$$7x(4x + 2) = 0$$

$7x = 0$ or $4x + 2 = 0$ Using the principle of zero products.

$\left. \begin{array}{ccc} x = 0 & \text{or} & 4x = -2 \\ x = 0 & \text{or} & x = -\frac{1}{2} \end{array} \right\}$ We solve the equations separately. For $7x = 0$ we multiply by $\frac{1}{7}$. For $4x = -2$ we multiply by $\frac{1}{4}$.

The solutions are 0 and $-\frac{1}{2}$.

DO EXERCISES 3–5.

SPECIAL CASES

There are equations with no solution.

Example 4   Solve: $-8x + 5 = 14 - 8x$.

$-8x + 5 = 14 - 8x$

$\qquad 5 = 14$ Adding $8x$, we get a false equation.

No matter what number we try for x we get a false sentence. Thus the equation has no solution.

There are equations for which any real number is a solution.

Example 5   Solve: $-8x + 5 = 5 - 8x$.

$-8x + 5 = 5 - 8x$

$\qquad 5 = 5$ Adding $8x$, we get a true equation.

Replacing x by any real number gives a true sentence. Thus any real number is a solution.

DO EXERCISES 6 AND 7.

NAME   CLASS   ANSWERS

EXERCISE SET 2.2

■ Solve.

1. $2(x + 6) = 8x$ **2.** $3(y + 5) = 8y$ **3.** $80 = 10(3t + 2)$

4. $27 = 9(5y - 2)$ **5.** $180(n - 2) = 900$ **6.** $210(x - 3) = 840$

7. $5y - (2y - 10) = 25$ **8.** $8x - (3x - 5) = 40$

9. $7(3x + 6) = 11 - (x + 2)$ **10.** $9(2x + 8) = 20 - (x + 5)$

11. $\dfrac{1}{8}(16y + 8) - 17 = -\dfrac{1}{4}(8y - 16)$ **12.** $\dfrac{1}{6}(12t + 48) - 20 = -\dfrac{1}{8}(24t - 144)$

1. _____

2. _____

3. _____

4. _____

5. _____

6. _____

7. _____

8. _____

9. _____

10. _____

11. _____

12. _____

◐ ◑ Solve.

13. $(x + 2)(x - 5) = 0$ **14.** $(x + 4)(x - 8) = 0$

13. _____

14. _____

15. $(y - 8)(y - 9) = 0$ **16.** $(t - 3)(t - 7) = 0$

15. _____

16. _____

17. _____

17. $(2x - 3)(3x - 2) = 0$ **18.** $(3y - 4)(4y - 1) = 0$

18. _____

19. _____

19. $m(m - 8) = 0$ **20.** $p(p - 5) = 0$

20. _____

21. _____

21. $0 = (2x + 8)(3x - 9)$ **22.** $0 = (4x + 16)(5x - 10)$

22. _____

23. _____

24. _____

Solve.

25. _____

23. $x(x - 1)(x + 2) = 0$ **24.** $y(y - 4)(y + 2) = 0$

25. ▦ $4.62(x + 6.01) = 7.92x$ **26.** ▦ $4.83y - (2.61y - 9.88) = 23.65$

26. _____

2.3 SOLVING PROBLEMS

⬛• TRANSLATING AND SOLVING

The first step in solving a problem is to translate to mathematical language. Very often this means translating to an equation. Drawing a picture usually helps a great deal. We solve the equation and then check to see if we have a solution to the problem.

Example 1 A 28-foot rope is cut into two pieces. One piece is 3 feet longer than the other. How long are the pieces?

Drawing a picture

28 ft

One piece   3 ft longer

The picture can help in translating. Here is one way to do it.

$$\underbrace{\text{Length of one piece}}_{x} \text{ plus } \underbrace{\text{length of other}}_{(x+3)} \text{ is } 28$$

$$x \quad + \quad (x+3) \quad = 28 \qquad \text{Translation}$$

We used x for the length of one piece and $x + 3$ for the length of the other.

Now we solve:
$$x + (x + 3) = 28$$
$$2x + 3 = 28$$
$$2x = 25$$
$$x = \tfrac{25}{2}, \quad \text{or} \quad 12\tfrac{1}{2}.$$

Do we have an answer to the *problem*? If one piece is $12\tfrac{1}{2}$ ft long, the one that is 3 ft longer must be $15\tfrac{1}{2}$ ft long. The lengths of the pieces add up to 28 ft, so this checks.

Steps to Use in Solving a Problem
1. **Translate to an equation. (Always draw a picture if it makes sense to do so.) Carefully keep track of what each variable you use stands for.**
2. **Solve the equation.**
3. **Check the answer in the original problem.**

DO EXERCISE 1.

Example 2 Five plus twice a number is seven times the number. What is the number?

This time it does not make sense to draw a picture.

① 5 plus $\underbrace{\text{twice a number}}$ is $\underbrace{\text{seven times the number}}$.

$$5 \quad + \quad 2x \quad = \quad 7 \quad \cdot \quad x \qquad \text{Translation}$$

We have used x to represent the unknown number. Notice that "is" translates to $=$.

2. If 7 times a certain number is subtracted from 6, the result is 5 times the number. What is the number? (*Note:* 7 times a certain number subtracted from 6 translates to $6 - 7x$.)

② Now solve:

$$5 + 2x = 7x$$
$$5 = 7x - 2x$$
$$5 = 5x$$
$$1 = x.$$

③ Now check: Twice 1 is 2. If we add 5, we get 7. This is $7 \cdot 1$. This checks so the answer to the *problem* is 1.

DO EXERCISE 2.

Example 3 What percent of 76 is 13.68?

$$y \quad \% \quad \times 76 = 13.68 \qquad \text{Translating* (a picture is not needed here)}$$

$$\left. \begin{array}{l} y \times (0.01) \times 76 = 13.68 \\ 0.76y = 13.68 \\ y = \dfrac{13.68}{0.76}, \quad \text{or} \quad 18 \end{array} \right\} \text{Solving}$$

Check: $18\% \cdot 76 = 0.18 \times 76$
$$= 13.68$$

The answer checks. The solution of the problem is 18%.

DO EXERCISE 3.

3. What percent of 84 is 11.76?

> When solving a problem, remember these two things.
>
> 1. Read the problem carefully, more than once.
> 2. Make a list of the known information.

Example 4 One angle of a triangle is five times as large as the first angle. The measure of the third angle is 2° less than that of the first angle. How large are the angles?

We draw a picture. We use x for the measure of the first angle. The second is five times as large so its measure will be $5x$. The third angle has a measure 2° less than that of the first angle, so its measure will be $x - 2$.

Now, to translate, we need to recall a geometric fact. The measures of the angles of any triangle add up to 180°.

$$\underbrace{\text{Measure of first angle}}_{x} \; \overset{\text{plus}}{+} \; \underbrace{\text{Measure of second angle}}_{5x} \; \overset{\text{plus}}{+} \; \underbrace{\text{Measure of third angle}}_{(x - 2)} \; \overset{\text{is } 180°}{= 180}$$

---

* Notice that *what* translates to a variable, *of* translates to a multiplication sign, and *is* translates to an equals sign.

Now we solve: $x + 5x + (x - 2) = 180$

$$7x - 2 = 180$$

$$7x = 182$$

$$x = 26$$

The solution of the equation is 26.

The angles will then have measures as follows:

First angle	$x = 26°$,
Second angle	$5x = 130°$,
Third angle	$x - 2 = 24°$.

These add up to 180° so they give an answer to the *problem*.

> *Important! Write down* what your letters represent. In Example 4, you should write that x = the measure of the first angle. Then write that $5x$ = the measure of the second angle and $x - 2$ = the measure of the third angle.

DO EXERCISE 4.

Example 5   It has been found that the world record for the 10,000-meter run has been decreasing steadily since 1940. The record is 30.18 minutes minus 0.12 times the number of years since 1940. If the record continues to decrease in this way, what will it be in 1985?

Record is 30.18 minutes minus 0.12 times number of years since 1940

$$R \quad = 30.18 \quad - \quad 0.12 \quad \cdot \quad t$$

$$R = 30.18 - 0.12t$$

In 1985, t will be 45, so we have

$$R = 30.18 - 0.12 \cdot 45. \qquad \text{This is the translation.}$$

Solving, we get 24.78. This checks in the problem, so 24.78 minutes is the answer. (In this case it is only a *prediction*.)

DO EXERCISE 5.

Example 6   On December 17, 1974, the pilots of Pan American Airlines shocked the business world by taking a pay cut of 11% to a new salary of $48,950 per year. What was their former salary?

Former salary minus 11% of former salary is new salary

$$x \quad - \quad 11\% \quad \cdot \quad x \quad = \quad 48{,}950 \quad \text{Translation}$$

We have used x to represent the former salary. We solve.

$$x - 11\% \cdot x = 48{,}950$$

$$1x - 0.11x = 48{,}950 \qquad \text{Replacing 11\% by 0.11}$$

$$\left.\begin{array}{r} (1 - 0.11)x = 48{,}950 \\ 0.89x = 48{,}950 \end{array}\right\} \quad \text{Collecting like terms}$$

$$x = 55{,}000 \qquad \text{Multiplying by } \frac{1}{0.89}$$

4. The first angle of a triangle is twice as large as the second. The measure of the third angle is 20° greater than that of the second angle. How large are the angles?

5. The County Cab Company charges sixty cents plus eleven cents per kilometer as the fare. What will be the total cost of a 12-km ride?

ANSWERS ON PAGE A–4

6. A clothing store drops the price of suits 25% to a sale price of $93. What was the former price?

Check: 11% of 55,000 is 6050. Subtracting this from 55,000 we get 48,950. This checks, so the former salary was $55,000. This is the answer to the *problem*.

Example 6 and Exercise 6 are both concerned with a percent of decrease.

DO EXERCISE 6.

Example 7   The sum of two consecutive integers is 35. What are the integers? (Consecutive integers are next to each other, such as 4 and 5.)

$$\underbrace{\text{First integer}}_{x} + \underbrace{\text{second integer}}_{(x+1)} = 35$$
$$\qquad\qquad\qquad\qquad\text{Translation}$$

Since the integers are consecutive, we know that one of them is 1 greater than the other; thus we call one of them x and the other $x + 1$.

7. The sum of two consecutive odd integers is 36. What are the integers? (*Note:* Consecutive odd integers are next to each other, such as 5 and 7. If x is an odd integer, the next consecutive *odd* integer is $x + 2$.)

Solve: $x + (x + 1) = 35$

$$2x + 1 = 35$$
$$2x = 34$$
$$x = 17$$

Check: The answers are 17 and 18. These are both integers and consecutive. Their sum is 35, so the answers check in the problem.

DO EXERCISE 7.

Example 8   The perimeter of a rectangle (distance around it) is 322 meters (m). The length is 25 m greater than the width. Find the dimensions.

We first draw a picture.

8. The length of a rectangle is twice the width. The perimeter is 39 meters. Find the dimensions. (*Hint:* The perimeter is the distance around the rectangle, not just halfway around.)

We have used w to represent the width. Then $w + 25$ represents the length.

$$\underbrace{\text{Twice the width}}_{2w} + \underbrace{\text{twice the length}}_{2(w+25)} = 322$$
$$\qquad\qquad\qquad\qquad\text{Translation}$$

Next we solve and get $w = 68$ (width) and $w + 25 = 93$ (length). Finally we check: The length is 25 more than the width. The perimeter is $2 \cdot 68 + 2 \cdot 93$, which is 322. The answer checks. The dimensions of the rectangle are 68 m and 93 m.

DO EXERCISE 8.

NAME

CLASS

ANSWERS

EXERCISE SET 2.3

For these exercises you may find the list of formulas on the inside front cover helpful.

▪ ● ▪ Translate to equations and solve.

1. A 12-in. piece of tubing is cut into two pieces. One piece is 4 in. longer than the other. How long are the pieces?

2. A 30-foot piece of wire is cut into two pieces. One piece is 2 feet longer than the other. How long are the pieces?

3. A piece of wire four meters long is cut into two pieces so that one piece is two-thirds as long as the other. Find the length of each piece.

4. A piece of rope five meters long is cut into two pieces so that one piece is three-fifths as long as the other. Find the length of each piece.

5. A babysitting service charges $2.50 per day plus $1.75 per hour. What is the cost of a seven-hour babysitting job?

6. The cost of renting a rug shampooer is $3.25 per hour plus $2.75 for the shampoo. Find the cost of shampooing if the time involved is 3.5 hours.

1. _____

2. _____

3. _____

4. _____

5. _____

6. _____

7. Klunker Car Rental charges $16 per day plus 15¢ per mile. Find the cost of renting a car for a one-day trip of 290 miles.

8. A phone company charges 30¢ per long-distance call plus 20¢ per minute. Find the cost of an 18-minute long-distance call.

7. _____

8. _____

9. Eight plus five times a number is seven times the number. What is the number?

10. Six plus three times a number is four times the number. What is the number?

9. _____

11. Five more than three times a number is the same as ten less than six times the number. What is the number?

12. Six more than nine times a number is the same as two less than ten times the number. What is the number?

10. _____

11. _____

13. A pro shop in a bowling alley drops the price of bowling balls 24% to a sale price of $34.20. What was the former price?

14. An appliance store drops the price of a certain type of television 18% to a sale price of $410. What was the former price?

12. _____

13. _____

14. _____

15. Money is borrowed at 9% simple interest. After one year $708.50 pays off the loan. How much was originally borrowed?

16. Money is borrowed at 7% simple interest. After one year $856 pays off the loan. How much was originally borrowed?

15. _____

17. The second angle of a triangle is three times the first, and the third is 12° less than twice the first. Find the measures of the angles.

18. The second angle of a triangle is four times the first, and the third is 5° more than twice the first. Find the measures of the angles.

16. _____

17. _____

19. The perimeter of a college basketball court is 96 m and the length is 14 m more than the width. What are the dimensions?

20. The perimeter of a certain soccer field is 310 m. The length is 65 m more than the width. What are the dimensions?

18. _____

19. _____

20. _____

ANSWERS

21. Find three consecutive integers such that the sum of the first, twice the second, and three times the third is 80.

22. Find two consecutive even integers such that two times the first plus three times the second is 76.

21. _____

23. After a person gets a 20% raise in salary the new salary is $9600. What was the old salary? (*Hint:* What number plus 20% of that number is 9600?)

24. A person gets a 17% raise, bringing the salary to $21,645. What was the salary before the raise?

22. _____

23. _____

25. The total cost for tuition plus room and board at State University is $2584. Tuition costs $704 more than room and board. What is the tuition fee?

26. The cost of a private pilot course is $1275. The flight portion costs $625 more than the ground school portion. What is the cost of each?

24. _____

25. _____

26. _____

2.4 FORMULAS

■ SOLVING FORMULAS

A formula is a kind of recipe, or rule, for doing a certain kind of calculation. Formulas are often given by equations. The formula

$$C = \pi D$$

tells us that to find the circumference of a circle, we should multiply the diameter by π. Suppose that we know the circumference and want to find the diameter. To do this, we can get D alone on one side of the equation. We "solve" for D.

Example 1   Solve the circle formula for D.

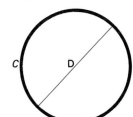

$$C = \pi D \longleftarrow \text{We want this letter alone.}$$

$$\frac{1}{\pi} \cdot C = \frac{1}{\pi} \cdot \pi D \qquad \text{Multiplying by } \frac{1}{\pi}$$

$$\frac{C}{\pi} = D$$

When we solve a formula for a given letter, we do the same things we would do to solve any equation. To see this, compare Example 1 with the following.

Example 1A   Solve the equation $3 = 2x$ for x.

$$3 = 2x$$

$$\frac{1}{2} \cdot 3 = \frac{1}{2} \cdot 2x \qquad \text{Multiplying by } \frac{1}{2}$$

$$\frac{3}{2} = x$$

> The idea when solving a formula for a certain letter is to get that letter alone on one side of the equation. To do that, proceed as in solving any equation.

DO EXERCISE 1.

Example 2   Solve for b: $A = \frac{5}{2}(b - 20)$.

$$A = \frac{5}{2}(b - 20)$$

$$\frac{2}{5}A = b - 20 \qquad \text{Multiplying by } \frac{2}{5}$$

$$\frac{2}{5}A + 20 = b \qquad \text{Adding 20}$$

DO EXERCISE 2.

After finishing Section 2.4, you should be able to:

■ Solve formulas for a specified letter.

1. A formula for the area of a triangle is $A = \frac{1}{2}bh$. Solve for b.

2. Solve for c: $P = \frac{3}{5}(c + 10)$.

ANSWERS ON PAGE A–4

3. Solve for m: $H = 2r + 3m$.

Example 3   Solve for r: $H = 2r + 3m$.

$$H = 2r + 3m \qquad \text{We want this letter alone.}$$

$$H - 3m = 2r \qquad \text{Adding } -3m$$

$$\frac{H - 3m}{2} = r \qquad \text{Multiplying by } \frac{1}{2}$$

Compare Example 3 with an ordinary equation with numbers.

Example 3A   Solve for x: $11 = 2x + 3 \cdot 5$.

$$11 = 2x + 3 \cdot 5$$

$$11 - 3 \cdot 5 = 2x \qquad \text{Adding } -3 \cdot 5$$

$$\frac{11 - 3 \cdot 5}{2} = x \qquad \text{Multiplying by } \frac{1}{2}$$

In this case we can simplify further, whereas in Example 3 we could not.

DO EXERCISE 3.

Example 4   The formula $A = P + Prt$ tells amount to which a principal P, in dollars, will grow at simple interest at a rate r in t years. Solve the formula for P.

$$A = P + Prt \qquad \text{We want this letter alone.}$$

$$A = P(1 + rt) \qquad \text{Factoring (or collecting like terms)}$$

$$A \cdot \frac{1}{1 + rt} = P(1 + rt) \cdot \frac{1}{1 + rt} \qquad \text{Multiplying by } \frac{1}{1 + rt}$$

$$\frac{A}{1 + rt} = P \qquad \text{Simplifying}$$

4. Solve for Q: $T = Q + Qiy$.

Compare Example 4 with an equation in which there are numbers.

Example 4A   Solve $7 = x + 4 \cdot 5 \cdot x$ for x.

$$7 = x(1 + 4 \cdot 5) \qquad \text{Factoring, or collecting like terms}$$

$$7 \cdot \frac{1}{1 + 4 \cdot 5} = x(1 + 4 \cdot 5) \cdot \frac{1}{1 + 4 \cdot 5} \qquad \text{Multiplying by } \frac{1}{1 + 4 \cdot 5}$$

$$\frac{7}{1 + 4 \cdot 5} = x \qquad \text{Simplifying}$$

In this case we can simplify further, to $\frac{7}{21} = x$, whereas in Example 4 we could not simplify further.

DO EXERCISE 4.

NAME	CLASS

EXERCISE SET 2.4

■ ● Solve.

1. $A = lw$, for l
    (an area formula)

2. $A = lw$, for w

3. $W = EI$, for I
    (an electricity formula)

4. $W = EI$, for E

5. $F = ma$, for m
    (a physics formula)

6. $F = ma$, for a

7. $I = Prt$, for t
    (an interest formula)

8. $I = Prt$, for P

9. $E - mc^2$, for m
    (a relativity formula)

10. $E = mc^2$, for c^2

11. $P = 2l + 2w$, for l
    (a perimeter formula)

12. $P = 2l + 2w$, for w

ANSWERS

1. _____

2. _____

3. _____

4. _____

5. _____

6. _____

7. _____

8. _____

9. _____

10. _____

11. _____

12. _____

13. $c^2 = a^2 + b^2$, for a^2
(a geometry formula)

14. $c^2 = a^2 + b^2$, for b^2

13. _____

15. $A = \pi r^2$, for r^2
(a circle formula)

16. $A = \pi r^2$, for π

14. _____

15. _____

17. $W = \dfrac{11}{2}(h - 40)$, for h

18. $C = \dfrac{5}{9}(F - 32)$, for F

(a temperature formula)

16. _____

17. _____

19. $V = \dfrac{4}{3}\pi r^3$, for r^3

20. $V = \dfrac{4}{3}\pi r^3$, for π

(a volume formula)

18. _____

19. _____

21. $A = \dfrac{1}{2}h(a + b)$, for h

22. $A = \dfrac{1}{2}h(a + b)$, for b

(an area formula)

20. _____

21. _____

23. $F = \dfrac{mv^2}{r}$, for m

24. $F = \dfrac{mv^2}{r}$, for v^2

(a physics formula)

22. _____

23. _____

24. _____

25. _____

25. In Exercise 7, you solved the formula $I = Prt$ for t. Now use it to find how long it will take a deposit of $75 to earn $3 interest when invested at 5% simple interest.

26. In Exercise 8, you solved the formula $I = Prt$ for P. Now use it to find how much principal would be needed to earn $6 in two-thirds of a year at 6% simple interest.

26. _____

2.5 GRAPHS

POINTS AND ORDERED PAIRS

On a number line each point is the graph of a number. On a plane each point is the graph of a number pair. We use two perpendicular number lines, called *axes*. The point where they cross is called the *origin* and is labeled 0. The arrows show the positive directions.

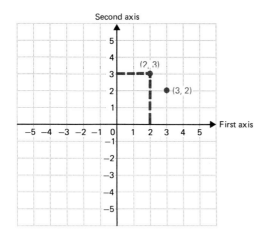

• PLOTTING POINTS

Note that (2, 3) and (3, 2) give different points. These are called *ordered pairs* of numbers since it makes a difference which number comes first.

Example 1 Plot the point $(-4, 3)$.

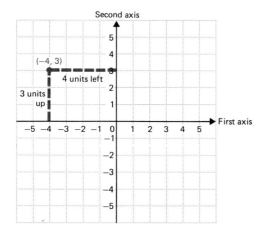

The first number, -4, tells us the distance in the first, or horizontal, direction. We go 4 units *left*. The second number tells us the distance in the second, or vertical, direction. We go 3 units *up*.

DO EXERCISES 1–9.

The numbers in an ordered pair are called *coordinates*. In $(-4, 3)$, the *first coordinate* is -4 and the *second coordinate** is 3.

---

* The first coordinate is sometimes called the *abscissa* and the second coordinate is called the *ordinate*.

After finishing Section 2.5, you should be able to:

• Plot points on a plane, given their coordinates.

•• Determine whether an ordered pair of numbers is a solution of an equation with two variables.

••• Graph equations of the type $y = mx$.

Plot these points on the graph below.

1. (6, 4)

2. (4, 6)

3. $(-3, 5)$

4. $(-4, -3)$

5. $(4, -2)$

6. (0, 3)

7. (3, 0)

8. $(0, -4)$

9. $(-4, 0)$

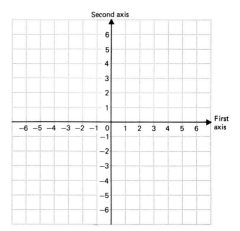

10. What can you say about the coordinates of a point in the third quadrant?

11. What can you say about the coordinates of a point in the fourth quadrant?

12. Determine whether $(9, -2)$ is a solution of the equation $y = 4x + 17$.

13. Determine whether $(0, 23)$ is a solution of $y = 4x + 17$.

14. Determine whether $(2, -2)$ is a solution of $2y + 6x = 8$.

QUADRANTS

The axes divide the plane into four regions, as shown, called *quadrants*. In region I (the *first* quadrant) both coordinates of a point are positive. In region II (the *second* quadrant) the first coordinate is negative and the second coordinate is positive.

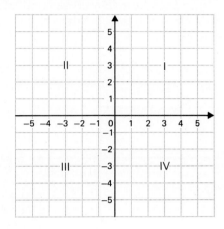

DO EXERCISES 10 AND 11.

■ ■   SOLUTIONS OF EQUATIONS

If an equation has two variables, its solutions are pairs of numbers. We usually take the variables in alphabetical order, hence we get ordered pairs of numbers for solutions.

Example 2   Determine whether the following ordered pairs are solutions of the equation $y = 3x - 1$: $(-1, -4)$, $(7, 5)$.

$$
\begin{array}{c|c}
\multicolumn{2}{l}{y = 3x - 1} \\
\hline
-4 & 3(-1) - 1 \\
\hline
-4 & -3 - 1 \\
 & -4
\end{array}
$$

We substitute -1 for x and -4 for y (alphabetical order of variables).

The equation becomes true: $(-1, -4)$ is a solution.

$$
\begin{array}{c|c}
\multicolumn{2}{l}{y = 3x - 1} \\
\hline
5 & 3 \cdot 7 - 1 \\
5 & 21 - 1 \\
 & 20
\end{array}
$$

We substitute.

The equation becomes false: $(7, 5)$ is not a solution.

DO EXERCISES 12–14.

■ ■ ■   GRAPHING EQUATIONS OF THE TYPE $y = mx$

To *graph* an equation means to make a drawing of its solutions. If an equation has a graph that is a line, we can graph it by plotting a few points and then drawing a line through them.

Example 3   Graph the equation $y = x$.

We will use alphabetical order. Thus the first axis is the x-axis and the second axis is the y-axis.

Next, we find some ordered pairs that are solutions of the equation. In this case, it is easy. Here are a few:

$$(0, 0), \quad (1, 1), \quad (5, 5), \quad (-1, -1), \quad (-6, -6).$$

Now we plot these points. We can see that if we were to plot a million solutions, the dots that we draw would merge into a solid line. We see the pattern, so we can draw the line with a ruler. The line is the graph of the equation $y = x$. We label the line $y = x$ on the graph paper.

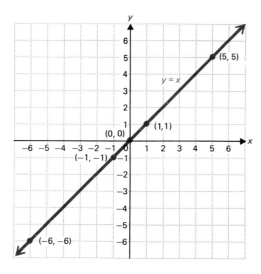

Example 4   Graph the equation $y = 2x$.

We find some ordered pairs that are solutions. This time we shall keep them in a table. To find an ordered pair, we can choose *any* number for x and then determine y. For example, if we choose 3 for x, then $y = 2(3)$ (substituting into the equation of the line), or 6. We make some negative choices for x, as well as positive ones. If a number takes us off of the graph paper, we usually do not use it.

x	y
0	0
1	2
3	6
-2	-4
-3	-6

Next, we plot these points. If we had enough of them they would make a solid line. We can draw the line with a ruler, and we label it $y = 2x$.

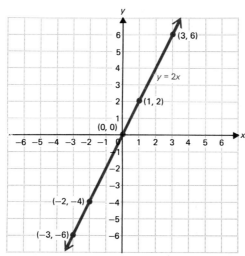

15. Graph the equation $y = \frac{1}{2}x$.

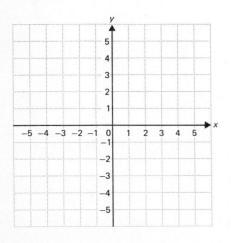

16. Graph the equation $y = 3x$.

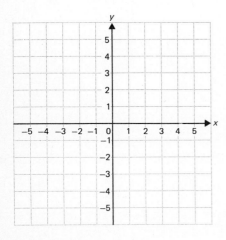

17. Graph the equation $y = -2x$.

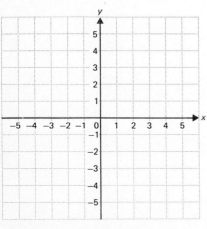

Example 5 Graph the equation $y = -5x$.

To find an ordered pair, we choose any number for x and then determine y. For example, if we choose 3 for x, we get $y = -5(3)$, or -15.

When we choose -2 for x, we get $y = -5(-2)$, or 10. We find several ordered pairs, plot them, and draw the line.

x	y
3	-15
-2	10
0	0
1	-5

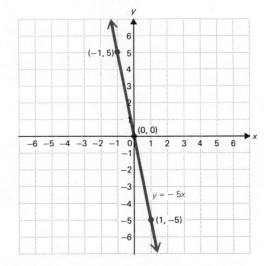

DO EXERCISES 15–17.

SOMETHING EXTRA

AN APPLICATION: FIELD GOAL COMPLETIONS

The linear equation

$$P = 0.792t + 23$$

gives an estimate of the average percentage P of field goal completions by college basketball players t years after 1940.

Example Use the equation to estimate the average percentage of field goal completions in 1990.

$$P = 0.792t + 23$$
$$= 0.792 \cdot (50) + 23 \qquad \text{Substituting (note that 1990}$$
$$\text{is 50 years after 1940)}$$
$$= 62.6$$

The answer is that 62.6 percent of field goals attempted in 1990 should be completed.

EXERCISES

Use the equation to estimate the average percentage of field goal completions in the year

1. 1985   **2.** 1995   **3.** 2000   **4.** 2005

NAME

CLASS

EXERCISE SET 2.5

● Plot the following points.

1. $A(5, 3)$, $B(2, 4)$, $C(0, 2)$, $D(0, -6)$, $E(3, 0)$, $F(-2, 0)$, $G(1, -3)$, $H(-5, 3)$, $J(-4, 4)$

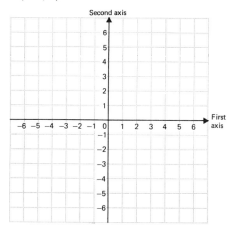

2. $A(3, 5)$, $B(1, 5)$, $C(0, 4)$, $D(0, -4)$, $E(5, 0)$, $F(-5, 0)$, $G(1, -5)$, $H(-7, 4)$, $J(-5, 5)$

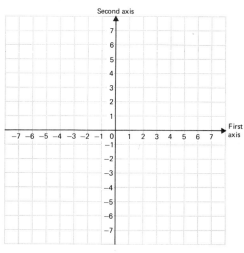

1. See graph.

2. See graph.

3. Plot the points $M(2, 3)$, $N(5, -3)$, and $P(-2, -3)$. Draw $\overline{MN}$, $\overline{NP}$, and $\overline{MP}$. ($\overline{MN}$ means the line segment from M to N). What kind of geometric figure is formed? What is its area?

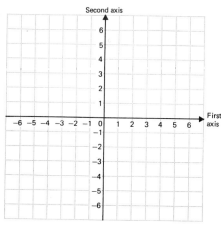

4. Plot the points $Q(-4, 3)$, $R(5, 3)$, $S(2, -1)$, and $T(-7, -1)$. Draw $\overline{QR}$, $\overline{RS}$, $\overline{ST}$, and $\overline{TQ}$. What kind of figure is formed? What is its area?

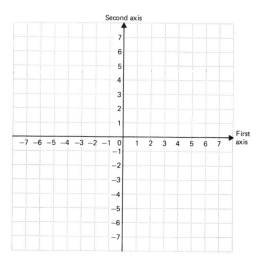

3. _____

4. _____

ANSWERS

5. Determine whether $(10, 3)$ is a solution of $3x - 7y = 9$.

6. Determine whether $(-1, 1)$ is a solution of $5y + 2x = -3$.

5. _____

6. _____

7. Determine whether $(4, -2)$ is a solution of $4y - 5x = 7 + 3x$.

8. Determine whether $(8, -5)$ is a solution of $3x + 7y = 5 - 2x$.

7. _____

8. _____

9. See graph. _____

Use graph paper. For each equation, make a table of values for x and y. Then graph the equation.

9. $y = 5x$

10. $y = \dfrac{1}{3}x$

11. $y = -4x$

10. See graph. _____

11. See graph. _____

12. See graph. _____

12. $y = -3x$

13. $y = 0.25x$

14. $y = -1.5x$

13. See graph. _____

14. See graph. _____

2.6 GRAPHS OF LINEAR EQUATIONS

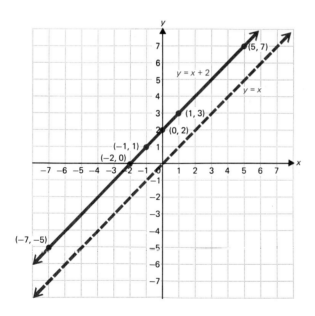 **GRAPHS OF EQUATIONS OF THE TYPE** $y = mx + b$

We know that the graph of any equation $y = mx$ is a straight line through the origin. What will happen if we add a number b on the right-hand side to get an equation $y = mx + b$?

Example 1   Graph $y = x + 2$ and compare it with the graph of $y = x$.

We first make a table of values.

x	y (or $x + 2$)
0	2
1	3
5	7
-1	1
-2	0
-7	-5

Then graph and compare. The graph of $y = x + 2$ is a line. It looks just like the graph of $y = x$, but it is moved up 2 units.

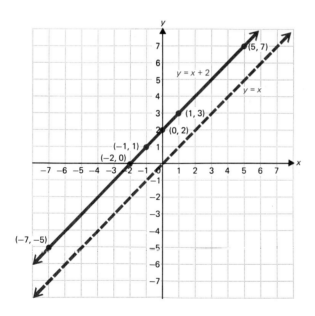

DO EXERCISES 1 AND 2.

OBJECTIVES

After finishing Section 2.6, you should be able to:

 • Given an equation $y = mx + b$, determine the slope and y-intercept and graph the equation.

 •• Graph linear equations with a missing variable.

 ••• Determine which equations are linear.

1. Graph $y = x + 3$ and compare with the graph of $y = x$.

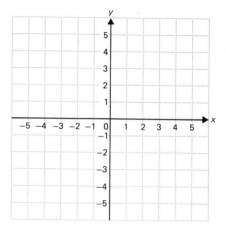

2. Graph $y = x - 1$ and compare with the graph of $y = x$.

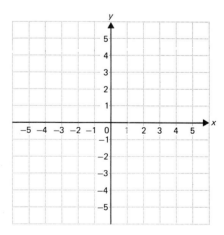

3. Graph $y = 2x + 1$ and compare with the graph of $y = 2x$.

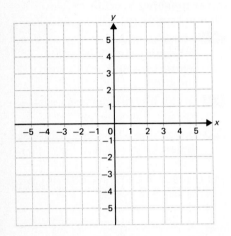

4. Graph $y = 2x - 4$ and compare with the graph of $y = 2x$.

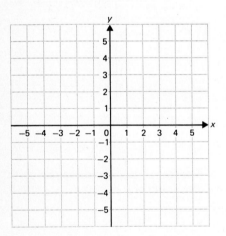

Find the y-intercept.

5. $y = -5.4x + 7$

6. $y = 12.5x - 17$

Example 2   Graph $y = 2x - 3$ and compare it with the graph of $y = 2x$.

We first make a table of values. Then graph and compare.

x	y (or $2x - 3$)
0	-3
1	-1
3	3
-2	-7

The graph of $y = 2x - 3$ is a line. It looks just like the graph of $y = 2x$, but it is moved down 3 units.

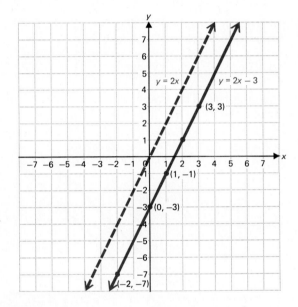

DO EXERCISES 3 AND 4.

The graph of an equation $y = mx$ is a line through the origin. Any equation $y = mx + b$ also has a graph that is a line. It is parallel to $y = mx$ but moved up or down. The graph of $y = mx + b$ goes through the point $(0, b)$. This point is called the *y-intercept*. We also refer to the number b as the y-intercept.

Examples   Find the y-intercept.

3. $y = -5x + 4$

   The y-intercept is $(0, 4)$, or simply 4.

4. $y = 5.3x - 12$

   The y-intercept is $(0, -12)$, or simply -12.

DO EXERCISES 5 AND 6.

We now look at the number m in the equation $y = mx + b$. The number m is called the *slope*, and it shows how the line slants. The greater the slope, the steeper the line slants from left to right. If the slope is negative, the line slants downward from left to right.

Example 5   Determine the slope and y-intercept for $y = \frac{2}{3}x + 4$ and draw its graph.

The *y-intercept* is the point $(0, 4)$, so we know one point on the graph. The *slope* is the number $\frac{2}{3}$. It tells us that from *any* point of the graph, the line slants *up* 2 units for every 3 *over* to the right. We start at the y-intercept. We go *up* 2 units and *over* 3. This gives us the point $(3, 6)$ and we can now draw the graph.

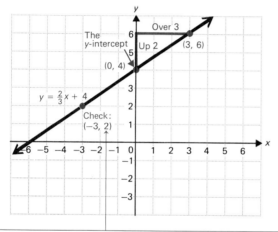

Important: As a check we choose some other value for x, say -3, and determine y (in this case, 2). We plot that point and see if it is on the line. If it is not, there has been some error.

DO EXERCISES 7 AND 8.

In Example 5 the slope is $\frac{2}{3}$. We can find some other names for $\frac{2}{3}$ by multiplying by 1 and then find other points of the graph.

$$\frac{2}{3} = \frac{2}{3} \cdot \frac{2}{2} = \frac{4}{6} \quad \leftarrow \begin{array}{l} \text{Thus from any point of the graph we can find} \\ \leftarrow \text{another point by going } up \text{ 4 units and } over \text{ 6.} \end{array}$$

$$\frac{2}{3} = \frac{2}{3} \cdot \frac{3}{3} = \frac{6}{9} \quad \begin{array}{l} \text{Thus we can find a point by going } up \text{ 6 and} \\ over \text{ 9, from any point of the graph.} \end{array}$$

When numbers are negative, we reverse directions.

$$\frac{2}{3} = \frac{2}{3} \cdot \frac{-1}{-1} = \frac{-2}{-3} \quad \begin{array}{l} \leftarrow \text{Thus from any point of the graph we can} \\ \leftarrow \text{find another point by going } down \text{ 2 units} \\ \text{and then } to \text{ the left 3 units.} \end{array}$$

If the slope of a line is negative, it slants downward from left to right.

Example 6   Graph $y = -\frac{1}{2}x + 5$.

The y-intercept is $(0, 5)$. The slope is $-\frac{1}{2}$, or $\frac{-1}{2}$. From the y-intercept we go *down* 1 unit and to the right 2 units. That gives us the point $(2, 4)$. We can now draw the graph.

Determine the slope and y-intercept and then graph. Be sure to find a third point as a check.

7. $y = \frac{3}{4}x - 2$

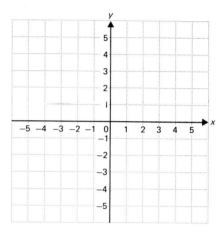

8. $y = \frac{3}{2}x + 1$

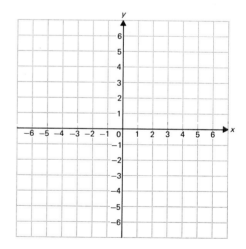

Determine the slope and *y*-intercept and then graph. Be sure to find a third point as a check.

9. $y = -\dfrac{3}{5}x + 5$

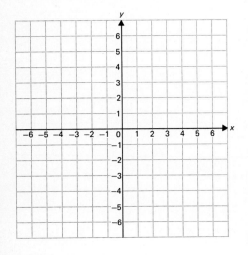

10. $y = -\dfrac{5}{3}x - 4$

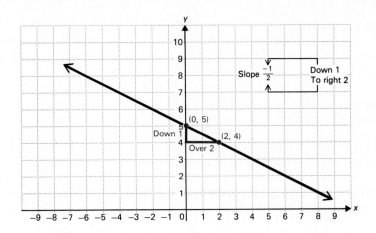

As a check, we rename the slope and find another point:

$$\frac{-1}{2} = \frac{-1}{2} \cdot \frac{-3}{-3} = \frac{3}{-6}.$$

Thus we can go *up* 3 units and then *to the left* 6 units. This gives the point $(-6, 8)$.

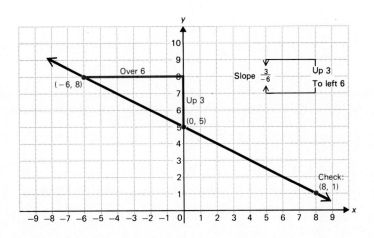

We can also check by choosing a number for *x* and finding *y*. If we choose 8 for *x*, we find that *y* is 1. Thus the point $(8, 1)$ should be on the graph.

DO EXERCISES 9 AND 10.

All the equations we have been graphing so far have graphs that are straight lines. Therefore, they are called *linear* equations. We know that any equation $y = mx + b$ is linear, but there are other equations that have straight lines for graphs

The *slope-intercept* **equation of a line is** $y = mx + b$.
Any equation $y = mx + b$ **has a graph that is a straight line.**
It goes through the point (0, *b*) and has slope *m*.

●● GRAPHS THAT ARE PARALLEL TO AN AXIS

Some equations have graphs that are parallel to one of the axes. These equations have a missing variable. In the following example, x is missing.

Example 7 Graph $y = 3$.

Since x is missing, any number for x will do. Thus all ordered pairs $(x, 3)$ are solutions. The graph is a line parallel to the x-axis.

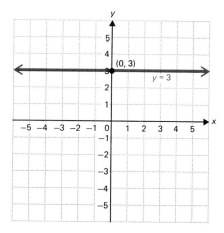

DO EXERCISES 11 AND 12.

In the following example, y is missing and the graph turns out to be parallel to the y-axis.

Example 8 Graph $x = -2$.

Since y is missing, any number for y will do. Thus all ordered pairs $(-2, y)$ are solutions. The graph is a line parallel to the y-axis.

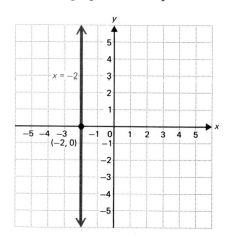

The graph of an equation $x = a$, where a is a constant, is a vertical line through the point $(a, 0)$.

The graph of an equation $y = b$, where b is a constant, is a horizontal line through the point $(0, b)$.

DO EXERCISE 13.

Graph.

11. $y = 4$

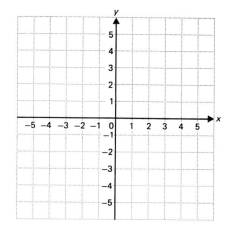

12. $3y + 6 = 0$
  (*Hint:* Solve for y.)

13. $x = -5$

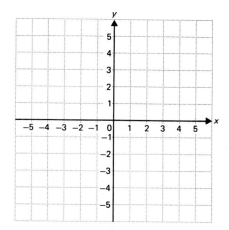

ANSWERS ON PAGE A–7

Which equations are linear?

14. $2y - 4x = 6$

15. $2y - 5 = 3y$

16. $4x - 5 = y^2 + 2$

■■■ WHICH EQUATIONS ARE LINEAR?

A *linear* equation is one having a straight line for its graph. From the examples in this section, we know which equations are linear.

> A *linear equation* is one that is equivalent to an equation $y = mx + b$ or an equation $x = c$, where m, b, and c are constants.

Keep in mind that the constants m, b, and c can be 0. Therefore an equation $y = mx$ is linear. Also, the equation $y = 0$ is linear. (It has a slope of 0 and its y-intercept is also 0.)

Example 9   Show that the equation $2x - 3y = 6$ is linear. Determine the slope and the y-intercept.

To show that the equation is equivalent to an equation $y = mx + b$, we solve for y:

$$2x = 6 + 3y \qquad \text{Adding } 3y$$

$$2x - 6 = 3y \qquad \text{Adding } -6$$

$$\frac{1}{3}(2x - 6) = \frac{1}{3} \cdot 3y \qquad \text{Multiplying by } \frac{1}{3}$$

$$\frac{2}{3}x - 2 = y \quad \text{or} \quad y = \frac{2}{3}x - 2$$

Since we have an equation $y = mx + b$, we know that the graph is a straight line. The slope is $\frac{2}{3}$ and the y-intercept is -2.

Examples   Which of these equations are linear?

10. $5x + 10 = 0$

Since there is no y in the equation, we solve for x to see if we get an equation $x = c$:

$$5x = -10 \qquad \text{Adding } -10$$

$$\frac{1}{5} \cdot 5x = \frac{1}{5}(-10) \qquad \text{Multiplying by } \frac{1}{5}$$

$$x = -2.$$

Since we have an equation $x = c$, we know that the equation is linear. The graph is parallel to the y-axis, 2 units to the left of that axis.

11. $2x^2 = 3y - 4$

We solve for y:

$$2x^2 + 4 = 3y \qquad \text{Adding 4}$$

$$\frac{1}{3}(2x^2 + 4) = \frac{1}{3} \cdot 3y \qquad \text{Multiplying by } \frac{1}{3}$$

$$\frac{2}{3}x^2 + \frac{4}{3} = y, \quad \text{or} \quad y = \frac{2}{3}x^2 + \frac{4}{3}.$$

The equation is not linear because it has an x^2-term.

DO EXERCISES 14–16.

NAME	CLASS

EXERCISE SET 2.6

Use graph paper for all graphs required on this page.

⚫ Determine the slope and y-intercept.

1. $y = 4x + 5$

2. $y = 5x + 3$

3. $y = -2x - 6$

4. $y = -5x - 7$

5. $y = -\dfrac{6}{16}x - 0.2$

6. $y = \dfrac{15}{7}x + 2.2$

7. $y = 15.3x + 4.04$

8. $y = -0.314x + 18.9$

Determine the slope and y-intercept. Then draw a graph. Be sure to use a third point as a check.

9. $y = \dfrac{5}{2}x + 1$

10. $y = \dfrac{2}{5}x - 4$

11. $y = -\dfrac{5}{2}x - 4$

12. $y = -\dfrac{2}{5}x + 3$

13. $y = 2x - 5$

$\left(\textit{Hint:} \text{ Think of 2 as } \dfrac{2}{1}.\right)$

14. $y = -2x + 4$

15. $y = \dfrac{1}{3}x + 6$

16. $y = -3x + 6$

17. $y = -0.25x + 2$

18. $y = 1.5x - 3$

ANSWERS

1. _____

2. _____

3. _____

4. _____

5. _____

6. _____

7. _____

8. _____

9. _____

10. _____

11. _____

12. _____

13. _____

14. _____

15. _____

16. _____

17. _____

18. _____

◉◉ Graph.

19. $3y - 9 = 0$

20. $6y + 24 = 0$

19. See graph.

20. See graph.

21. See graph.

22. See graph.

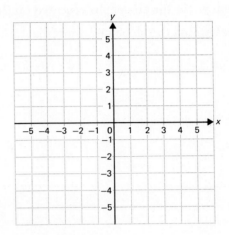

23. _____

21. $3x + 15 = 0$

22. $2x - 10 = 0$

24. _____

25. _____

26. _____

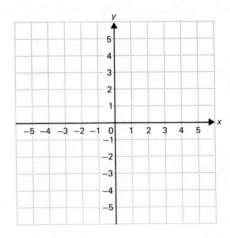

27. _____

◉◉◉ Determine which equations are linear.

28. _____

23. $3x + 5y + 15 = 0$

24. $5x - 3y = 15$

25. $3x - 12 = 0$

26. $16 = -4y$

29. _____

27. $2x + 4y^2 = 19$

28. $3y - 5x^2 - 4 = 0$

30. _____

29. $5x - 4xy = 12$

30. $3y = 7xy - 5$

2.7 ANOTHER LOOK AT LINEAR EQUATIONS AND GRAPHS

▪ GRAPHING USING INTERCEPTS

In many cases we can graph a linear equation very quickly by finding the x-intercept and the y-intercept and then connecting those points. To do this, it is not necessary to solve the equation for either y or x.

Example 1   Graph $3x + 2y = 12$.

This equation is linear. (We could show that it is by solving for y.) To find the y-intercept, we let $x = 0$ and see what y must be. If we just cover up the $3x$ or ignore it, that amounts to letting x be 0. So we cover up $3x$ and see $2y = 12$. Then y must be 6. The y-intercept is $(0, 6)$.

To find the x-intercept, we can cover up the y-term and look at the rest of the equation. We have

$$3x = 12, \quad \text{or} \quad x = 4.$$

The x-intercept is $(4, 0)$.

We plot these points and draw the line, using a third point as a check.

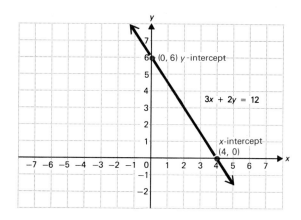

> An x-intercept is a point $(a, 0)$. To find a, let $y = 0$ in the equation.
>
> A y-intercept is a point $(0, b)$. To find b, let $x = 0$ in the equation.

DO EXERCISE 1.

We summarize the quickest procedure for graphing linear equations.

> *To Graph Linear Equations*
> 1. **Is there a variable missing? If so, solve for the other one. The graph will be a line parallel to an axis.**
> 2. **If no variable is missing, find the intercepts. Graph using the intercepts if that is feasible.**
> 3. **If the intercepts are too close together, choose another point farther from the origin.**
> 4. **In any case, use a third point as a check.**

DO EXERCISE 2 AND EXERCISE 3 ON THE NEXT PAGE

ANSWERS ON PAGE A–9

After finishing Section 2.7, you should be able to:

▪ Graph linear equations using intercepts.

▪▪ Given two points of a line, find its slope.

▪▪▪ Identify lines with zero slope and lines that do not have a slope.

Graph, using intercepts.

1. $4y - 12 = -6x$

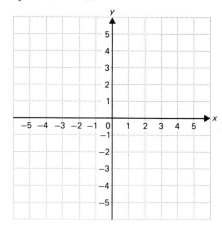

Graph.

2. $2y + x = 0$

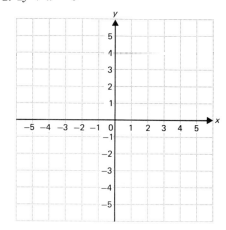

3. $8x + 4y = 1$

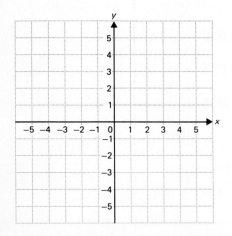

4. Two points of a line are $(1, 2)$ and $(6, 9)$. Find the slope of the line.

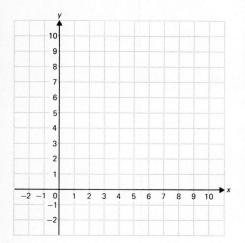

5. For the line of Examples 2 and 3, find the slope using the points $(-3, 0)$ and $(1, 3)$.

●● FINDING SLOPE

If we know the coordinates of *any* two points of a line, we can find its slope. We divide the vertical distance (the rise) by the horizontal distance (the run). That is, slope = rise/run.

Example 2   Two points of a line are $(1, 3)$ and $(5, 6)$. Find the slope of the line.

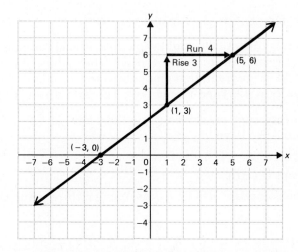

We graph the points and draw the line. Then we mark the rise (the distance *up*) and the run (the distance *over*) by drawing a right triangle. The rise is 3 (from 3 to 6) and the run is 4 (from 1 to 5).

$$\text{Slope} = \frac{\text{rise}}{\text{run}} = \frac{3}{4}$$

DO EXERCISE 4.

We can choose *any* two points of a line to find its slope.

Example 3   Find the slope of the line in Example 2 by using a different pair of points.

We can use *any* two points. Let's choose $(-3, 0)$ and $(5, 6)$.

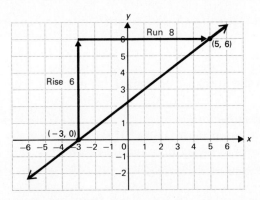

The rise is 6 (from 0 to 6) and the run is 8 (from -3 to 5).

$$\text{Slope} = \frac{\text{rise}}{\text{run}} = \frac{6}{8} = \frac{3}{4}$$

DO EXERCISE 5.

Example 4   Two points of a line are $(-5, -3)$ and $(6, 2)$. Find the slope of the line.

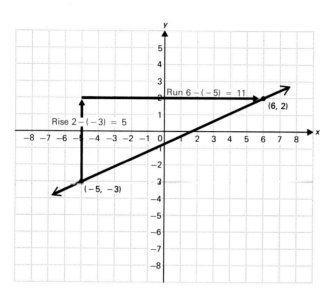

6. Two points of a line are $(-5, 3)$ and $(3, -10)$. Find the slope of the line.

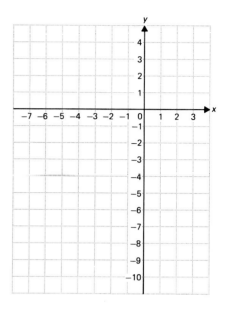

We graph the points and draw the line. We mark the rise and the run. We find them by subtracting.

$$\text{Rise} = 2 - (-3) = 5 \text{ (from } -3 \text{ to } 2),$$
$$\text{Run} = 6 - (-5) = 11 \text{ (from } -5 \text{ to } 6),$$

$$\text{Slope} = \frac{\text{rise}}{\text{run}} = \frac{5}{11}.$$

DO EXERCISE 6.

In Example 4, what will happen if we subtract coordinates in the opposite order? For the rise we get $-3 - 2$, or -5. For the run we get $-5 - 6 = -11$. For the slope we get

$$\text{Slope} = \frac{\text{rise}}{\text{run}} = \frac{-5}{-11} = \frac{5}{11}.$$

This is the same answer we got before. We can subtract either way, as long as we subtract in the same order both times.

> **If (x_1, y_1) and (x_2, y_2) are two points of a line, the slope m of the line can be calculated as follows:**
>
> $$m = \frac{y_2 - y_1}{x_2 - x_1} \quad \text{or} \quad \frac{y_1 - y_2}{x_1 - x_2}.$$

Examples   Find the slope of the line through each pair of points.

5. $(5, -2)$ and $(-7, -5)$

$$\text{Slope} = \frac{y_2 - y_1}{x_2 - x_1} = \frac{-5 - (-2)}{-7 - 5} = \frac{-5 + 2}{-12}$$

$$= \frac{-3}{-12} = \frac{1}{4} \quad \text{or} \quad 0.25$$

Find the slope of the line through each pair of points.

7. $(4, -3)$ and $(-8, -7)$

8. $(4.2, -6.8)$ and $(-8.3, 2.1)$

Find the slope, if it exists, of the line through each pair of points.

9. $(6, -4)$ and $(-10, -4)$

10. $(8, -5)$ and $(8, 13)$

11. Which equations have zero slope and which have no slope?

a) $3y = 6$

b) $5x = 13$

c) $5y + x = 5y + 4$

d) $y + 2 = 3x - 12$

e) $3x - 7 = 3y + 3x$

We could also have subtracted in the opposite order, as follows:

$$\frac{y_1 - y_2}{x_1 - x_2} = \frac{-2 - (-5)}{5 - (-7)} = \frac{3}{12} = \frac{1}{4}$$

6. $(-4.2, 5.9)$ and $(8.4, -2.2)$

$$\text{Slope} = \frac{y_2 - y_1}{x_2 - x_1} = \frac{-2.2 - 5.9}{8.4 - (-4.2)}$$

$$= \frac{-8.1}{12.6}$$

$$\approx -0.643 \qquad \approx \text{ means "is approximately equal to."}$$

DO EXERCISES 7 AND 8.

●●● ZERO SLOPE AND LINES WITHOUT SLOPE

If two points have the same second coordinate, what about the slope of the line joining them? Since $y_2 = y_1$, we have

$$\frac{y_2 - y_1}{x_2 - x_1} = \frac{0}{x_2 - x_1} = 0.$$

Every horizontal line has a slope of 0.

Suppose two different points are on a vertical line. Then they have the same first coordinate. Since $x_2 = x_1$, we have

$$\frac{y_2 - y_1}{x_2 - x_1} = \frac{y_2 - y_1}{0}.$$

Since we cannot divide by 0 this is undefined.

Vertical lines do not have a slope.

DO EXERCISES 9–11.

NAME	CLASS	ANSWERS

EXERCISE SET 2.7

■ ● Graph these equations, using intercepts and a third point as a check. Use graph paper.

1. $5y = -15 + 3x$

2. $7x = 3y - 21$

3. $6x - 7 + 3y = 9x - 2y + 8$
(*Hint:* Collect like terms before
graphing.)

4. $7x - 8 + 4y = 8y + 4x + 4$

5. $1.4y - 3.5x = -9.8$

6. $3.6x - 2.1y = 22.68$

■ ● ● For each pair of points, find the slope, if it exists, of the line containing them.

7. $(3, 8)$ and $(9, -4)$

8. $(17, -12)$ and $(-9, -15)$

9. $(-8, -7)$ and $(-9, -12)$

10. $(14, 3)$ and $(2, 12)$

11. $(-16.3, 12.4)$ and $(-5.2, 8.7)$

12. $(14.4, -7.8)$ and $(-12.5, -17.6)$

13. $(3.2, -12.8)$ and $(3.2, 2.4)$

14. $(-1.5, 7.6)$ and $(-1.5, 8.8)$

●●● Which equations have zero slope and which have no slope?

15. $3x = 12 + y$ **16.** $5y - 12 = 3x$ **17.** $5x - 6 = 15$

18. $-12 = 4x - 7$ **19.** $5y = 6$ **20.** $19 = -6y$

21. $y - 6 = 14$ **22.** $3y - 5 = 8$ **23.** $12 - 4x = 9 + x$

24. $15 + 7x = 3x - 5$ **25.** $2y - 4 = 35 + x$ **26.** $2x - 17 + y = 0$

27. $3y + x = 3y + 2$ **28.** $x - 4y = 12 - 4y$

29. $3y - 2x = 5 + 9y - 2x$ **30.** $17y + 4x + 3 = 7 + 4x$

2.8 OTHER EQUATIONS OF LINES (Optional)

● POINT-SLOPE EQUATIONS

Suppose we know one point on a line and also its slope. We can find an equation of the line. Suppose the known point is (x_1, y_1) and the slope is m. Then consider any other point (x, y) of the line.

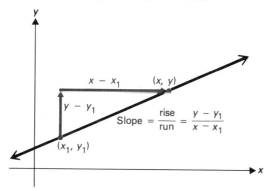

From this drawing we see that the slope of the line is $\dfrac{y - y_1}{x - x_1}$.

The slope is also m. This gives us the equation

$$\frac{y - y_1}{x - x_1} = m. \tag{1}$$

Multiplying by $x - x_1$ gives us the equation we are looking for.

> **The** *point-slope equation* **of a line is**
> $$y - y_1 = m(x - x_1).$$

Note that Equation (1) is not true when $x = x_1$ and $y = y_1$, but the point-slope equation is. Thus Equation (1) is not an equation of the entire line.

> Equation (1) is easy to remember. You can easily get the point-slope equation from it by multiplying by $x - x_1$.

Example 1 Find an equation of the line containing the point $(\frac{1}{2}, -1)$ with slope 5.

We substitute in the equation $y - y_1 = m(x - x_1)$:

$$y - (-1) = 5\left(x - \frac{1}{2}\right) \quad \text{Substituting}$$

$$y + 1 = 5x - \frac{5}{2} \quad \text{Simplifying}$$

> *Caution!* A common error here is forgetting to multiply *both* terms by 5.

$$y = 5x - \frac{7}{2}$$

OBJECTIVES

After finishing Section 2.8, you should be able to:

● Given the slope of a line and a point on the line, find an equation of the line.

●● Given two points of a line, find an equation of the line.

●●● Given a line and a point, find an equation of the line parallel to the given line and containing the point, and an equation of the line perpendicular to the given line and containing the point.

1. Find an equation of the line containing the point $(1, -2)$, with slope 3.

Example 2   Find an equation of the line with x-intercept $(4, 0)$ and slope 2.

$$y - 0 = 2(x - 4) \qquad \text{Substituting}$$
$$y = 2x - 8 \qquad \text{Simplifying}$$

DO EXERCISES 1 AND 2.

Vertical lines do not have point-slope equations, because they do not have slopes. All other lines do.

⊙⊙ TWO-POINT EQUATIONS

Suppose we know two points of a nonvertical line. Call them (x_1, y_1) and (x_2, y_2). We can find the slope m:

$$m = \frac{y_2 - y_1}{x_2 - x_1}.$$

We substitute in the point-slope equation, and get the two-point equation. From this we can write an equation of a nonvertical line whenever we know two points on the line.

> The *two-point equation* of a nonvertical line is
> $$y - y_1 = \frac{y_2 - y_1}{x_2 - x_1}(x - x_1).$$

2. Find an equation of the line containing the point $(-5, \frac{3}{4})$, with slope $-\frac{4}{3}$.

Example 3   Find an equation of the line containing the points $(2, 3)$ and $(1, -4)$.

Let us take $(2, 3)$ to be (x_1, y_1) and $(1, -4)$ to be (x_2, y_2). We substitute:

$$y - 3 = \frac{-4 - 3}{1 - 2}(x - 2)$$
$$y - 3 = 7(x - 2). \qquad \text{Simplifying}$$

If we wish, we can simplify further to get the slope-intercept equation:

$$y - 3 = 7x - 14$$
$$y = 7x - 11.$$

We can also take the points in opposite order. If we take $(1, -4)$ to be (x_1, y_1) and $(2, 3)$ to be (x_2, y_2), we get

$$y - (-4) = \frac{3 - (-4)}{2 - 1}(x - 1)$$
$$y + 4 = 7(x - 1).$$

Simplifying further to get the slope-intercept equation, we obtain

$$y + 4 = 7x - 7$$
$$y = 7x - 11.$$

Note that in Example 3, what we did amounted to finding the slope and then substituting it into the *point-slope* equation. If you wish to think of it that way you need memorize only that formula, and not the *two-point* formula above.

DO EXERCISES 3 AND 4.

●●● PARALLEL AND PERPENDICULAR LINES

If two lines are vertical they are parallel. How can we tell whether nonvertical lines are parallel? The answer is simple: We look at their slopes.

Two nonvertical lines are parallel if they have the same slope.

If one line is vertical and another is horizontal, they are perpendicular. Otherwise, how can we tell whether two lines are perpendicular?

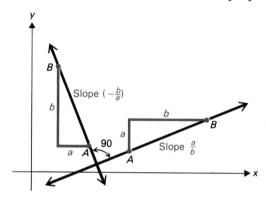

Consider a line $\overleftrightarrow{AB}$ as shown, with slope a/b. Then think of rotating the figure 90° to get a line perpendicular to $\overleftrightarrow{AB}$. For the new line the rise and run are interchanged, but the run is now negative. Thus the slope of the new line is $-b/a$. Let us multiply the slopes:

$$\frac{a}{b}\left(-\frac{b}{a}\right) = -1.$$

This is the condition under which lines will be perpendicular.

Two lines are perpendicular if the product of their slopes is −1. (If one line has slope m, the slope of a line perpendicular to it is $-1/m$. That is, we take the reciprocal and change the sign.)

Example 4   Given the line $8y = 7x - 24$ and the point $(-1, 2)$.

a) Find an equation of a line containing the point and parallel to the line.

We find the slope:

$$y = \frac{7}{8}x - 3. \qquad \text{Solving for } y$$

3. Find an equation of the line containing the points $(4, -3)$ and $(1, 2)$.

4. Find an equation of the line containing the points $(-3, -5)$ and $(-4, 12)$.

5. Find an equation of the line parallel to the line $8x = 7y - 24$ and containing the point $(2, -1)$.

We use the point-slope equation:

$$y - 2 = \frac{7}{8}[x - (-1)] \quad \text{or} \quad y = \frac{7}{8}x + \frac{23}{8}. \qquad \text{Substituting } \frac{7}{8} \text{ for the slope and } (-1, 2) \text{ for the point}$$

b) Find an equation of the line containing the point and perpendicular to the line.

To find the slope we take the reciprocal of $\frac{7}{8}$ and then change the sign. The slope is $-\frac{8}{7}$.

$$y - 2 = -\frac{8}{7}[x - (-1)] \quad \text{or} \quad y = -\frac{8}{7}x + \frac{6}{7}. \qquad \text{Substituting } -\frac{8}{7} \text{ for the slope and } (-1, 2) \text{ for the point}$$

DO EXERCISES 5 AND 6.

SOMETHING EXTRA

AN APPLICATION: DEPRECIATION—STRAIGHT LINE METHOD (PART I)*

A company buys a copying machine for $5200 and will *depreciate* it over a period of 8 years. This means they will calculate how its value decreases during that time. At the end of 8 years, the trade-in or salvage value will be $1300. Thus its value decreases by $3900. This is the *depreciation*. If they allow the same depreciation each year, $\frac{1}{8}$, or 12.5%, of the $3900 each year, the calculation is called *straight-line* depreciation. From the graph you can see why this name is used. Each year the depreciation is 12.5% of $3900, or $487.50. We subtract this from $5200 to get the book value after one year. We subtract again to get the book value after two years, and so on.

6. Find an equation of the line perpendicular to the line $2x - 4y = 12$ and containing the point $(5, 4)$.

EXERCISES

Find the book value of the copying machine for each of the remaining years.

---

* For more on depreciation, see pp. 252, 390, and 404.

EXERCISE SET 2.8

◼• Find an equation of the line having the given slope and containing the given point.

1. $m = 4$, $(3, 2)$

2. $m = 5$, $(5, 4)$

3. $m = -2$, $(4, 7)$

4. $m = -3$, $(7, 3)$

5. $m = 3$, $(-2, -4)$

6. $m = 1$, $(-5, -7)$

7. $m = -2$, $(8, 0)$

8. $m = -3$, $(-2, 0)$

9. $m = 0$, $(0, -7)$

10. $m = 0$, $(0, 4)$

◼◼ Find an equation of the line containing each pair of points.

11. $(1, 4)$ and $(5, 6)$

12. $(2, 5)$ and $(4, 7)$

13. $(-1, -1)$ and $(2, 2)$

14. $(-5, -5)$ and $(7, 7)$

15. $(-2, 0)$ and $(0, 5)$

16. $(0, -3)$ and $(7, 0)$

17. $(-2, -3)$ and $(-4, -6)$

18. $(-4, -7)$ and $(-2, -1)$

19. $(0, 0)$ and $(5, 2)$

20. $(0, 0)$ and $(-3, 4)$

1. _____

2. _____

3. _____

4. _____

5. _____

6. _____

7. _____

8. _____

9. _____

10. _____

11. _____

12. _____

13. _____

14. _____

15. _____

16. _____

17. _____

18. _____

19. _____

20. _____

●●● Write an equation of the line containing the given point and parallel to the given line.

21. $(3, 7)$, $x + 2y = 6$

22. $(0, 3)$, $2x - y = 7$

21. _____

22. _____

23. $(2, -1)$, $5x - 7y = 8$

24. $(-4, -5)$, $2x + y = -3$

23. _____

24. _____

Write an equation of the line containing the given point and perpendicular to the given line.

25. $(2, 5)$, $2x + y = 3$

26. $(4, 1)$, $x - 3y = 9$

25. _____

26. _____

27. _____

27. $(3, -2)$, $3x + 4y = 5$

28. $(-3, -5)$, $5x - 2y = 4$

28. _____

29. _____

29. Find an equation of the line containing $(4, -2)$ and parallel to the line containing $(-1, 4)$ and $(2, -3)$.

30. Find an equation of the line containing $(-1, 3)$ and perpendicular to the line containing $(3, -5)$ and $(-2, 7)$.

30. _____

31. Use slopes to show that the triangle with vertices $(-2, 7)$, $(6, 9)$, and $(3, 4)$ is a right triangle.

31. _____

2.9 APPLICATIONS OF LINEAR EQUATIONS (Optional)

◼️• FITTING EQUATIONS TO DATA

There are many situations in the world to which linear equations can be applied. How do we know when we have such a situation?

Example 1 Plot the following data and determine whether a linear equation gives an approximate fit.

Crickets are known to chirp faster when the temperature is higher. Here are some measurements made at different temperatures.

Temperature °C	6	8	10	15	20
Number of chirps per min	11	29	47	75	107

We make a graph with a T (temperature) axis and an N (number per minute) axis and plot these data. We see that they do lie approximately on a straight line, so we can use a linear equation in this situation.

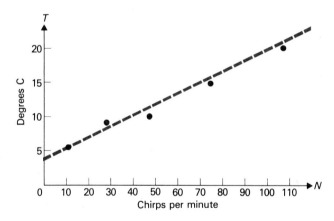

Example 2 Wind friction, or *resistance*, increases with speed. Here are some measurements made in a wind tunnel. Plot the data and determine whether a linear equation will give an approximate fit.

Velocity, km/h	10	21	34	40	45	52
Force of resistance, kg	3	4.2	6.2	7.1	15.1	29.0

We make a graph with a V (velocity) axis and an F (force) axis and plot the data. They do not lie on a straight line, even approximately. Therefore we cannot use a linear equation in this situation.*

---

* The situation in Example 2 might fit an equation that is not linear. We shall consider such problems later.

OBJECTIVE

After finishing Section 2.9, you should be able to:

◼️• Find a linear equation that fits two data points and use the equation to make predictions.

1. (*Records in the 100-meter dash*). It has been found that running records change with time according to linear equations. In 1920 the record for the 100-meter dash was 10.43 seconds. In 1970 it was 9.93 seconds.

a) Fit a linear equation to the data points. (Use R for the record and t for the number of years since 1920.)

When we have a situation that a linear equation fits, we can find an equation and use it to solve problems and make predictions.

Example 3   When crickets chirp 40 times per minute, the temperature is 10°C. When they chirp 112 times per minute, the temperature is 20°C. We know that a linear equation fits this situation. **(a)** Find a linear equation that fits the data. **(b)** From your equation, find the temperature when crickets chirp 76 times per minute; 100 times per minute.

a) To find the equation, we use the two known ordered pairs (40, 10°) and (112, 20°). We call these *data points*. We find the two-point equation, using these data points.

$$T - T_1 = \frac{T_2 - T_1}{N_2 - N_1}(N - N_1)$$

$$T - 10 = \frac{20 - 10}{112 - 40}(N - 40) \qquad \text{Substituting}$$

$$T - 10 = \frac{10}{72}(N - 40)$$

$$T - 10 = \frac{5}{36}N - \frac{5}{36} \cdot 40$$

$$T = \frac{5}{36}N + \frac{40}{9}, \quad \text{or} \quad \frac{5N + 160}{36} \qquad \text{Solving for } T$$

b) Use your equation to predict the record in 1982; in 1990.

b) Using this as a formula, we find T when $N = 76$.

$$T = \frac{5 \cdot 76 + 160}{36} = 15°$$

When $N = 100$, we get

$$T = \frac{5 \cdot 100 + 160}{36} \approx 18.3°. \quad \approx \text{ means "is approximately equal to."}$$

c) In what year will the record be 9.0 seconds?

DO EXERCISE 1.

Equations give decent results only within certain limits. A negative number of chirps per minute would be meaningless. Also, imagine what would happen to a cricket at $-40°C$ or at $100°C$!

ANSWERS ON PAGE A–10

NAME   CLASS

EXERCISE SET 2.9

•

1. (*Life expectancy of females in the United States*). In 1950, the life expectancy of females was 72 years. In 1970, it was 75 years. Let E represent the life expectancy and t the number of years since 1950 ($t = 0$ gives 1950 and $t = 10$ gives 1960).

 a) Fit a linear equation to the data points. [They are (0, 72) and (20, 75).]

 b) Use the equation of (a) to predict the life expectancy of females in 1980; in 1985.

2. (*Life expectancy of males in the United States*). In 1950, the life expectancy of males was 65 years. In 1970, it was 68 years. Let E represent life expectancy and t the number of years since 1950.

 a) Fit a linear equation to the data points.

 b) Use the equation of (a) to predict the life expectancy of males in 1980; in 1985.

3. (*Weight vs. height*). It has been shown experimentally that a person's height H is related to that person's weight W by a linear equation. An average-size person who is 70 in. tall weighs 165 lb, and a person 67 in. tall weighs 145 lb.

 a) Fit a linear equation to the data points.

 b) Use the equation of (a) to estimate the height of an average-size person weighing 130 lb. (This equation is valid only for heights above 40 in.)

4. (*Natural gas demand*). In 1950, natural gas demand in the United States was 20 quadrillion joules. In 1960, the demand was 22 quadrillion joules. Let D represent the demand for natural gas t years after 1950.

 a) Fit a linear equation to the data points.

 b) Use the equation of (a) to predict the natural gas demand in 1980; in 2000.

1. a) _____

b) _____

2. a) _____

b) _____

3. a) _____

b) _____

4. a) _____

b) _____

5. (*Records in the 1500-meter run*). In 1930, the record for the 1500-meter run was 3.85 minutes. In 1950, it was 3.70 minutes. Let R represent the record in the 1500-meter run and t the number of years since 1930.

 a) Fit a linear equation to the data points.

 b) Use the equation of (a) to predict the record in 1980; in 1984.

 c) When will the record be 3.3 minutes?

6. (*Records in the 400-meter run*). In 1930, the record for the 400-meter run was 46.8 seconds. In 1970, it was 43.8 seconds. Let R represent the record in the 400-meter run and t the number of years since 1930.

 a) Fit a linear equation to the data points.

 b) Use the equation of (a) to predict the record in 1980; in 1990.

 c) When will the record be 40 seconds?

7. (*The cost of a taxi ride*). The cost of a taxi ride for 2 miles in Browntown is $1.75. For 3 miles the cost is $2.00.

 a) Fit a linear equation to the data.

 b) Use the equation to find the cost of a 7-mile ride.

8. (*The cost of renting a car*). If you rent a car for one day and drive it 100 miles, the cost is $30. If you drive it 150 miles, the cost is $37.50.

 a) Fit a linear equation to the data points.

 b) Use the equation to find how much it will cost to rent the car for one day if you drive it 200 miles.

EXTENSION EXERCISES

SECTION 2.1

Solve.

1. $-\dfrac{3}{4}x + \dfrac{1}{8} = -2$

2. $y - \dfrac{1}{3}y - 15 = 0$

3. $\dfrac{a}{5} - \dfrac{a}{25} = 3.1$

4. $\dfrac{3x}{2} + \dfrac{5x}{3} - \dfrac{13x}{6} - \dfrac{2}{3} = \dfrac{5}{6}$

5. $8b - 11 - 3b + 1 - b = 15 - b$

6. $10 - x + 2 = 2x - 6 - 9x + 7$

7. $\dfrac{11}{2}x + \dfrac{1}{2} - \dfrac{3}{4}x - x - \dfrac{5}{8} = 2x + \dfrac{7}{8} - 4x - x - \dfrac{1}{4}$

SECTION 2.2

Solve.

8. $2x - 4 - (x + 1) - 3(x - 2) = 6(2x - 3) - 3(6x - 1) - 8$

9. $5(x - 3) + 4(x + 6) = 9\left(\dfrac{1}{3}x + 1\right)$

10. $2x - 10 - \left(6x + \dfrac{1}{2}\right) = 13 - 4x$

11. $\dfrac{1}{7}(a - 3)(7a + 4) = 0$

12. $24\left(\dfrac{x}{6} - \dfrac{1}{3}\right) = x - 24$

13. $0.5(x - 2) - 2(x - 5) = 0.4(x - 5) - 5(x - 2)$

SECTION 2.3

14. A student's scores on five tests are 93%, 89%, 72%, 80%, and 96%. What must the student score on the sixth test so that the average will be 88%?

15. The yearly changes in the population census of a city for three consecutive years are, respectively, 20% increase, 30% increase, and 20% decrease. What is the total percent change from the beginning to the end of the third year, to the nearest percent?

16. Three numbers are such that the second is 6 less than 3 times the first, and the third is 2 more than $\frac{2}{3}$ the second. The sum of the three numbers is 172. Find the largest number.

17. An appliance store is having a sale on 13 TV models. They are displayed left to right in order of increasing prices. The price of each TV differs by $20 from that of each adjacent TV. For the price of the TV at the extreme right a customer can buy both the second and the seventh models. What is the price of the least expensive TV?

18. A tank at a marine exhibit contains 2000 gallons of seawater. The seawater is 7.5% salt. How many gallons, to the nearest gallon, of freshwater must be added to the tank so that the mixture contains only 7% salt?

SECTION 2.4

Solve.

19. $s = v_i t + \dfrac{1}{2}at^2$, for a　　**20.** $A = \pi rs + \pi r^2$, for s　　**21.** $\dfrac{P_1 V_1}{T_1} = \dfrac{P_2 V_2}{T_2}$, for V_1　　**22.** $\dfrac{P_1 V_1}{T_1} = \dfrac{P_2 V_2}{T_2}$, for T_2

SECTION 2.5

23. If $(-10, -2)$, $(-3, 4)$, and $(6, 4)$ are the coordinates of three of the four vertices of a parallelogram, what are the coordinates of the fourth vertex?

24. Graph $y = 6x$, $y = 3x$, $y = \frac{1}{2}x$, $y = -6x$, $y = -3x$, $y = -\frac{1}{2}x$ on the same set of axes and compare the slants of the lines. What does the number in front of the x tell you about the slant of the line?

25. Which of the following equations has $\left(-\frac{1}{3}, \frac{1}{4}\right)$ as a solution?

a) $-\frac{3}{2}x - 3y = -\frac{1}{4}$ b) $8y - 15x = \frac{7}{2}$ c) $0.16y = -0.09x + 0.1$ d) $2(-y + 2) - \frac{1}{4}(3x - 1) = 4$

SECTION 2.6

Determine the slope and y-intercept of each line.

26. $2x + 5y + 2 = 5x - 10y - 8$ **27.** $\frac{1}{8}y = -x - \frac{7}{16}$ **28.** $0.4y - 0.004x = -0.04$ **29.** $x = -\frac{7}{3}y - \frac{2}{11}$

SECTION 2.7

30. Determine a so that the slope of the line through this pair of points has the given value.

$$(-2, 3a), (4, -a); \qquad m = -\frac{5}{12}$$

31. Find the slope of the line that contains the given pair of points.

a) $(5b, -6c) (b, -c)$ b) $(b, d) (b, d + e)$ c) $(c + f, a + d) (c - f, -a - d)$

32. A line contains the points $(-100, 4)$ and $(0, 0)$. List four more points of the line.

33. Find the slopes of the sides of the triangle formed by joining the midpoints of the sides of the triangle whose vertices are $A(-8, -2)$, $B(-2, 6)$, and $C(4, -10)$. $\left[\textit{The midpoint formula:} \text{ If the endpoints of a segment are } (x_1, y_1) \text{ and } (x_2, y_2), \text{ then the coordinates of the midpoint are } \left(\frac{x_1 + x_2}{2}, \frac{y_1 + y_2}{2}\right).\right]$

SECTION 2.8

34. Find an equation of the line containing each pair of points.

a) $(-0.2, 0.7) (-0.7, -0.3)$ b) $\left(\frac{1}{11}, \frac{1}{2}\right) \left(-\frac{10}{11}, -2\right)$

35. Which pairs of the following four equations represent perpendicular lines?

(1) $7y - 3x = 21$ (2) $-3x - 7y = 12$ (3) $7y + 3x = 21$ (4) $3y + 7x = 12$

36. Find the value of a so that the graphs of $5y = ax + 5$ and $\frac{1}{4}y = \frac{1}{10}x - 1$ are parallel.

37. Find the value of k so that the graphs of $x + 7y = 70$ and $y + 3 = kx$ are perpendicular.

38. Write an equation of the line that has y-intercept $(0, \frac{5}{7})$ and is parallel to the graph of $6x - 3y = 1$.

39. Write an equation of the line that has x-intercept $(-1.2, 0)$ and is perpendicular to the graph of $6x - 3y = 1$.

40. Write an equation of the line that has x-intercept $(-3, 0)$ and y-intercept $(0, \frac{2}{5})$.

NAME   SCORE

TEST OR REVIEW—CHAPTER 2

Review the material in the chapter. Then test yourself by doing these exercises. If you miss an item, restudy the section indicated to the left of the item.

1. _____

Solve.

[2.1,]   **1.** $-9 + y = 13$

[2.1,]   **2.** $-12x = -8$

[2.1,]   **3.** $0.7x - 0.1 = 2.1 - 0.3x$

2. _____

[2.2,]   **4.** $5(3x + 6) = 6 - (x + 8)$

[2.2,]   **5.** $(t + 1)(t - 3) = 0$

[2.4,]   **6.** Solve $h = \dfrac{2}{11}W + 40$, for W.

[2.6,]   **7.** Solve for y. Then determine the slope and y-intercept, and graph.

3. _____

$$2x + 3y = 6$$

4. _____

[2.7,]   **8.** Graph using intercepts.

$$3y = 15 - 5x$$

5. _____

6. _____

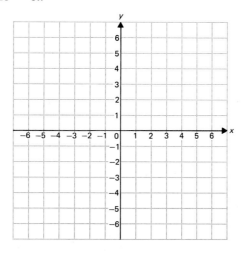

7. _____

8. See graph.

[2.7, ● ●]   **9.** Find the slope of the line containing the following points, if it exists.

(3, 5) and (10, − 4)

9. _____

[2.8, ●]   **10.** Find an equation of the line having the given slope and containing the given point.

$m = 2$, $(−2, −3)$

10. _____

[2.8, ● ●]   **11.** Find an equation of the line containing the following pair of points.

$(−2, 3)$ and $(−4, 6)$

11. _____

[2.8, ●●●]   **12.** Find an equation of the line containing the given point and parallel to the given line.

$(3, −1)$, $5x + 7y = 8$

12. _____

[2.8, ●●●]   **13.** Find an equation of the line containing the given point and perpendicular to the given line.

$(5, 2)$, $3x + y = 5$

13. _____

[2.3, ●]   Translate to equations and solve.

14. Five more than three times a number is five less than four times the number. Find the number.

15. The price of a saw went up 12% to $14.00. What was the price before the increase?

14. _____

15. _____

SYSTEMS OF EQUATIONS

3

Some problems involving mixing chemicals can be solved using systems of equations.

OBJECTIVES

After finishing Section 3.1, you should be able to:

◼ Determine whether an ordered pair is a solution of a system of two equations.

◼◼ Solve systems of two linear equations graphically.

1. Determine whether (20, 40) is a solution of the system

$$2x = y$$

$$4y + x = 180.$$

2. Determine whether $(-2, 3)$ is a solution of the system

$$2a - 5b = 7$$

$$5b + 3a = -4.$$

3.1 SYSTEMS OF EQUATIONS IN TWO VARIABLES

◼ **IDENTIFYING SOLUTIONS**

A pair of linear equations is called a *system* of linear equations. A solution of such a system is an ordered pair that makes *both* equations true.

Example 1 Determine whether (4, 7) is a solution of the system

$$x + y = 11$$

$$3x - y = 5$$

We use alphabetical order of the variables. Thus we replace x by 4 and y by 7.

$x + y = 11$	$3x - y = 5$
$4 + 7 \mid 11$	$3 \cdot 4 - 7 \mid 5$
11	$12 - 7$
	5

The pair (4, 7) makes both equations true, so it is a solution of the *system*. We sometimes describe such a solution by saying that $x = 4$ and $y = 7$.

DO EXERCISES 1 AND 2.

GRAPHING SYSTEMS OF EQUATIONS

Recall that the graph of an equation is a picture of its solution set. If the graph of an equation is a line, then every point on that line corresponds to an ordered pair that is a solution of the equation. If we graph a *system* of two linear equations, the point at which the lines intersect will be a solution of *both* equations.

Example 2   Solve this system graphically:

$$y - x = 1$$
$$y + x = 3.$$

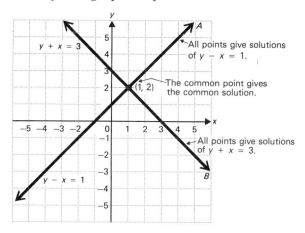

All ordered pairs from line A are solutions of the first equation. All ordered pairs from line B are solutions of the second equation. The point of intersection has coordinates that make *both* equations true. The solution seems to be the point $(1, 2)$. Graphing is not perfectly accurate, so solving by graphing may give only approximate answers. The pair $(1, 2)$ does check, so it is a solution.

DO EXERCISES 3 AND 4.

Some systems of equations have no solution. That happens when the graphs are parallel lines. Such equations are called *inconsistent*, because they are never *both* true at once. Some equations have only one solution, as in Example 2. Some have an infinite (unending) set of solutions. That happens when the equations have exactly the same graph. Such equations in two variables are called *dependent*. The three possibilities are shown in the following graphs.

No solution.
Graphs are parallel.
Equations are *inconsistent* because there is no solution.

Many solutions.
Equations have the same graph.
Equations are *dependent* (but they are consistent because there is at least one solution).

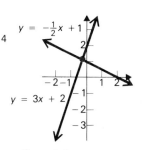

One solution.
Graphs intersect.
Equations are *consistent* (and not dependent).

Solve graphically

3. $-2x + y = 1$
         $3x + y = 1$

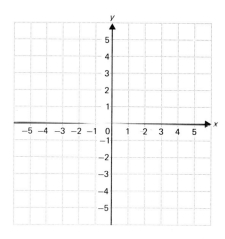

4. $x = 2y$
         $4y + x = 6$

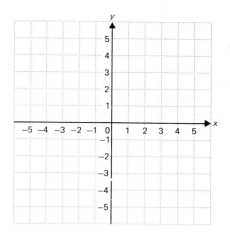

ANSWERS ON PAGE A–11

■■■■■■■■■■■■■■■■■■■■■■■■■■■■■■■■

SOMETHING EXTRA

AN APPLICATION: ESTIMATING ANNUAL PERCENTAGE RATE (APR)

Suppose you get a $1000 car loan for one year at 12%. The lender may do the following: Take 12% of $1000, which is $120, then add this on, to get $1120. If you are to make 4 payments, you will repay $\frac{1}{4}$ of this amount ($280) each time. This is called *add-on* interest, but 12% is not the true rate of interest. The true rate is called the *annual percentage rate*, or APR.

As you make payments your principal decreases: $1120, $840, $560, $280. (*Note:* You do not keep the borrowed amount for the entire time.) If r is the true rate of interest (APR), the total interest you pay is estimated by

$$1120 \cdot \frac{1}{4}r + 840 \cdot \frac{1}{4}r + 560 \cdot \frac{1}{4}r + 280 \cdot \frac{1}{4}r.$$

We get this by using the formula $I = Prt$, and $t = \frac{1}{4}$ yr in each case, since there are 4 payments in a year. Let's do some algebra, factoring out $\frac{1}{4}r$. The interest is then given by

$$\$120 = (1120 + 840 + 560 + 280) \cdot \frac{1}{4}r.$$

Note that $(1120 + 840 + 560 + 280) \cdot \frac{1}{4}$ is the *average* principal. If we calculate it, we get $700. Now we have

$$\$120 = 700r.$$

We solve for r and get

$$r = \frac{120}{700}, \quad \text{or} \quad 17.14\%.$$

This is an estimate of the APR.

The APR can always be calculated as follows.

$$\text{APR} = \frac{\text{Total interest}}{\text{Average principal}}.$$

It is illegal in the United States for a lender not to inform you of the APR.

EXERCISES

Find the APR. Assume that these are car loans at the given add-on rate for one year with 12 equal payments.

1. Loan $1000 at 12%   **2.** Loan $4000 at 13%

NAME   CLASS

EXERCISE SET 3.1

■ ● Determine whether the given ordered pair is a solution of the system of equations. Remember to use alphabetical order of variables.

1. $(1, 2)$; $4x - y = 2$
 $10x - 3y = 4$

2. $(-1, -2)$; $2x + y = -4$
 $x - y = 1$

3. $(2, 5)$; $y = 3x - 1$
 $2x + y = 4$

4. $(-1, -2)$; $x + 3y = -7$
 $3x - 2y = 12$

5. $(1, 5)$; $x + y = 6$
 $y = 2x + 3$

6. $(5, 2)$; $a + b = 7$
 $2a - 8 = b$

7. $(2, -7)$; $3a + b = -1$
 $2a - 3b = -8$

8. $(2, 1)$; $3p + 2q = 5$
 $4p + 5q = 2$

1. _____

2. _____

3. _____

4. _____

5. _____

6. _____

7. _____

8. _____

██　Use graph paper. Solve each system graphically. Be sure to check.

9. $x + y = 4$
　　$x - y = 2$

10. $x - y = 3$
　　$x + y = 5$

11. $2x - y = 4$
　　$5x - y = 13$

12. $3x + y = 5$
　　$x - 2y = 4$

9. _____

10. _____

11. _____

13. $4x - y = 9$
　　$x - 3y = 16$

14. $2y = 6 - x$
　　$3x - 2y = 6$

15. $a = 1 + b$
　　$b = -2a + 5$

16. $x = y - 1$
　　$2x = 3y$

12. _____

13. _____

14. _____

15. _____

17. $2u + v = 3$
　　$2u = v + 7$

18. $2b + a = 11$
　　$a - b = 5$

19. $y = -\dfrac{1}{3}x - 1$

　　$4x - 3y = 18$

20. $y = -\dfrac{1}{4}x + 1$

　　$2y = x - 4$

16. _____

17. _____

18. _____

21. $6x - 2y = 2$
　　$9x - 3y = 1$

22. $y - x = 5$
　　$2x - 2y = 10$

19. _____

20. _____

21. _____

22. _____

3.2 SOLVING SYSTEMS OF EQUATIONS

Graphing is not an efficient or accurate method for solving systems of equations. We now consider more efficient methods.

▌•▐ THE SUBSTITUTION METHOD

Example 1   Solve this system:

$$x + y = 5$$
$$x = y + 1.$$

The second equation says that x and $y + 1$ name the same thing. In the first equation we can substitute $y + 1$ for x.

$$x + y = 5$$
$$(y + 1) + y = 5 \qquad \text{Substituting } y + 1 \text{ for } x$$

> It can be very important to use parentheses here.

We solve this last equation, using methods learned earlier.

> It is easy to be careless in removing these parentheses, especially if subtraction is involved.

$$(y + 1) + y = 5$$
$$2y + 1 = 5 \qquad \text{Removing parentheses and collecting like terms}$$
$$2y = 4 \qquad \text{Adding } -1 \text{ on both sides}$$
$$y = 2 \qquad \text{Multiplying on both sides by } \tfrac{1}{2}$$

We return to the original pair of equations, and substitute 2 for y in either of them. Calculation will be easier if we choose the second one.

$$x = y + 1$$
$$x = 2 + 1$$
$$x = 3$$

We obtain the ordered pair (3, 2). We check to be sure it is a solution.

Check:

$x + y = 5$		$x = y + 1$	
$3 + 2$	5	3	$2 + 1$
	5		3

Since (3, 2) checks, it is the solution.

DO EXERCISES 1 AND 2.

Suppose neither equation of a pair has a variable alone on one side. We then solve one equation for one of the variables.

Example 2   Solve this system:

$$2x + y = 6$$
$$3x + 4y = 4.$$

First solve one equation for one variable. We solve the first equation for y because that will make the work simpler:

$$y = 6 - 2x.$$

OBJECTIVES

After finishing Section 3.2, you should be able to:

▌•▐ Solve systems of equations in two variables by the substitution method.

▌••▐ Solve systems of equations in two variables by the addition method.

Solve by the substitution method.

1. $x + y = 6$
   $\quad y = x + 2$

2. $\quad\quad y = 7 - x$
   $2x - y = 8$

   (*Caution:* Use parentheses when you substitute. Then be very careful about removing them.)

Solve. Use the substitution method.

3. $2y + x = 1$
$y - 2x = 8$

Then we substitute $6 - 2x$ for y in the second equation and solve for x:

$3x + 4(6 - 2x) = 4$ Substituting $6 - 2x$ for y

$3x + 24 - 8x = 4$ Multiplying to remove parentheses

$3x - 8x = 4 - 24$

$-5x = -20$

$x = 4.$

Now we can substitute 4 for x in either equation and solve for y. We use the first equation.

$2x + y = 6$

$2 \cdot 4 + y = 6$

$y = -2$

We obtain $(4, -2)$. This checks so it is the solution.

DO EXERCISES 3 AND 4.

● ●   THE ADDITION METHOD

The *addition method* for solving systems of equations makes use of the *addition principle*. Some systems are much easier to solve using the addition principle.

Consider this system:

$2x - 3y = 0$

$-4x + 3y = -1.$

The first equation says that $2x - 3y$ and 0 represent the same number. We use the addition principle with the second equation. We add the "same thing" on both sides. On the left we call it $2x - 3y$. On the right we call it 0:

$-4x + 3y + (2x - 3y) = -1 + 0$

or

$-2x = -1.$

Using this method we try to get an equation with only one variable. We often say that we have *eliminated* a variable.

Example 3   Solve this system:

$2x - 3y = 0$

$-4x + 3y = -1.$

We have

$2x - 3y = 0$

$\underline{-4x + 3y = -1}$

$-2x + 0 = -1$ Adding

We have made one variable "disappear." Now we solve this equation for x:

$-2x = -1$

$x = \dfrac{1}{2}.$

4. $8x + 5y = 184$
$x - y = -3$

We substitute $\frac{1}{2}$ for x in the first equation and solve for y:

$$2 \cdot \frac{1}{2} - 3y = 0 \qquad \text{Substituting}$$

$$\left. \begin{array}{r} 1 - 3y = 0 \\ y = \frac{1}{3} \end{array} \right\} \qquad \text{Solving for } y$$

We obtain $(\frac{1}{2}, \frac{1}{3})$. This checks, so it is the solution.

DO EXERCISES 5 AND 6.

Note in Example 3 that a term in one equation and a term in the other were additive inverses of each other. Thus their sum was 0. That enabled us to eliminate a variable.

In order to eliminate a variable we sometimes must multiply one or both of the equations by something before adding.

Example 4 Solve this system:

$$3x + 3y = 15$$
$$2x + 6y = 22.$$

If we add, we do not eliminate a variable. However, if the $3y$ in the first equation were $-6y$, we could. So we multiply by -2 on both sides of the first equation.

$$\begin{array}{r} -6x - 6y = -30 \qquad \text{Multiplying by } -2 \text{ on both sides} \\ 2x + 6y = 22 \\ \hline -4x + 0 = -8 \qquad \text{Adding} \\ x = 2 \qquad \text{Solving for } x \end{array}$$

$$2 \cdot 2 + 6y = 22 \qquad \text{Substituting 2 for } x \text{ in second equation}$$
$$4 + 6y = 22$$
$$y = 3 \qquad \text{Solving for } y$$

We obtain $(2, 3)$, or $x = 2$, $y = 3$. This checks, so it is the solution.

DO EXERCISE 7.

Sometimes we have to multiply twice to make two terms become additive inverses.

Example 5 Solve this system:

$$2x + 3y = 12$$
$$5x + 7y = 29.$$

From first equation: $\qquad 10x + 15y = 60 \qquad$ Multiplying by 5
From second equation: $\quad -10x - 14y = -58 \qquad$ Multiplying by -2
$$\begin{array}{r} \hline 0 + y = 2 \qquad \text{Adding} \\ y = 2 \qquad \text{Solving for } y \end{array}$$

Solve. Use the addition method.

5. $\begin{aligned} 5x + 3y &= 17 \\ -5x + 2y &= 3 \end{aligned}$

6. $\begin{aligned} -3a + 2b &= 0 \\ 3a - 4b &= -1 \end{aligned}$

Solve. Use the addition method.

7. $\begin{aligned} 2y + 3x &= 12 \\ -4y + 5x &= -2 \end{aligned}$

ANSWERS ON PAGE A–12

8. Solve. Use the addition method.

$$3x + 5y = 30$$
$$5x + 3y = 34$$

We then have

$$2x + 3 \cdot 2 = 12 \qquad \text{Substituting 2 for } y \text{ in first equation}$$
$$2x = 6$$
$$x = 3 \qquad \text{Solving for } x$$

We obtain $(3, 2)$, or $x = 3$, $y = 2$. This checks, so it is the solution.

DO EXERCISE 8.

When decimals or fractions occur, we *clear* before solving.

Example 6 Clear of decimals:

$$0.5x + 0.3y = 1.9$$
$$1.2x - 0.3y = 12.4.$$

$$5x + 3y = 19 \qquad \text{Multiplying on both sides by 10}$$
$$12x - 3y = 124 \qquad \text{Multiplying on both sides by 10}$$

We can now finish solving as in Example 4.

DO EXERCISE 9.

9. Clear of decimals.

$$1.32x + 0.12y = 0.1$$
$$4.2x - 0.6y = 1.2$$

(*Hint:* Multiply the first equation by 100 and the second one by 10.)

Example 7 Clear of fractions:

$$\frac{1}{2}x + \frac{2}{3}y = 1$$

$$\frac{3}{4}x - \frac{1}{3}y = 2.$$

$$6\left(\frac{1}{2}x + \frac{2}{3}y\right) = 6 \cdot 1 \qquad \text{Multiplying on both sides by 6}$$

$$12\left(\frac{3}{4}x - \frac{1}{3}y\right) = 12 \cdot 2 \qquad \text{Multiplying by 12}$$

We have multiplied each equation by the LCM of the denominators. Now we simplify.

First equation: $\qquad 6 \cdot \frac{1}{2}x + 6 \cdot \frac{2}{3}y = 6$

$$3x + 4y = 6$$

Second equation: $\qquad 12 \cdot \frac{3}{4}x - 12 \cdot \frac{1}{3}y = 24$

$$9x - 4y = 24$$

We can now finish solving as in Example 3.

10. Clear of fractions.

$$\frac{3}{5}x + \frac{2}{3}y = \frac{1}{3}$$

$$\frac{3}{4}x - \frac{1}{3}y = \frac{1}{4}$$

DO EXERCISE 10.

WHICH METHOD IS BETTER?

The substitution method is usually better when a variable has a coefficient of 1, as in the following examples:

$$8x + 4y = 180 \qquad\qquad x + y = 561$$
$$-x + y = 3; \qquad\qquad x = 2y.$$

Both methods work. When in doubt, use the addition method.

NAME

CLASS

ANSWERS

EXERCISE SET 3.2

• Solve using the substitution method.

1. $3x + 5y = 3$
   $x = 8 - 4y$

2. $2x - 3y = 13$
   $y = 5 - 4x$

3. $9x - 2y = 3$
   $3x - 6 = y$

4. $x = 3y - 3$
   $x + 2y = 9$

5. $5m + n = 8$
   $3m - 4n = 14$

6. $4x + y = 1$
   $x - 2y = 16$

7. $4x + 12y = 4$
   $-5x + y = 11$

8. $-3b + a = 7$
   $5a + 6b = 14$

• • Solve using the addition method.

9. $x + 3y = 7$
   $-x + 4y = 7$

10. $x + y = 9$
    $2x - y = -3$

11. $2x + y = 6$
    $x - y = 3$

12. $x - 2y = 6$
    $-x + 3y = -4$

1. _____

2. _____

3. _____

4. _____

5. _____

6. _____

7. _____

8. _____

9. _____

10. _____

11. _____

12. _____

ANSWERS

13. $9x + 3y = -3$
$2x - 3y = -8$

14. $6x - 3y = 18$
$6x + 3y = -12$

15. $5x + 3y = 19$
$2x - 5y = 11$

16. $3x + 2y = 3$
$9x - 8y = -2$

13. _____

14. _____

15. _____

17. $5r - 3s = 24$
$3r + 5s = 28$

18. $5x - 7y = -16$
$2x + 8y = 26$

19. $0.3x - 0.2y = 4$
$0.2x + 0.3y = 1$

16. _____

17. _____

18. _____

20. $0.7x - 0.3y = 0.5$
$-0.4x + 0.7y = 1.3$

21. $\frac{1}{2}x + \frac{1}{3}y = 4$

$\frac{1}{4}x + \frac{1}{3}y = 3$

22. $\frac{2}{3}x + \frac{1}{7}y = -11$

$\frac{1}{7}x - \frac{1}{3}y = -10$

19. _____

20. _____

21. _____

22. _____

23. ▦ Solve: $3.5x - 2.1y = 106.2$
$4.1x + 16.7y = -106.28.$

23. _____

3.3   SOLVING PROBLEMS

● To solve problems, we often translate to a system of equations in two variables. That can make the translation much easier than if we use just one equation.

Example 1   Eight times a certain number added to five times a second number is 184. The first number minus the second number is -3. Find the numbers.

There are two statements in the problem. We translate the first one.

8 times a certain number added to 5 times a second number is 184.

$$8x \qquad\qquad + \qquad\qquad 5y \qquad\qquad = 184$$

Here x represents the first number and y represents the second. Now we translate the second statement, remembering to use x and y.

The first number minus the second number is -3.

$$x \qquad\qquad - \qquad\qquad y \qquad\qquad = -3$$

We now have a system of equations:

$$8x + 5y = 184$$
$$x - y = -3.$$

We solve this system by either substitution or addition. We'll use the substitution method this time.

$x = y - 3$ Adding y on both sides of the second equation

$8(y - 3) + 5y = 184$ Substituting into the first equation

$8y - 24 + 5y = 184$

$13y = 208$

> Remember to use parentheses when you substitute. Then remove them carefully.

$$y = \frac{208}{13}, \quad \text{or} \quad 16$$

We go back to the second of the original equations.

$x - 16 = -3$ Substituting the value of y we just obtained

$x = 13.$

Now we must check in the *original problem*. Since $8 \cdot 13 = 104$ and $5 \cdot 16 = 80$, the sum is 184. The difference between 13 and 16 is -3, so we have the answer.

DO EXERCISE 1.

Example 2   One day Glovers, Inc., sold 20 pairs of gloves. The cloth gloves brought $4.95 per pair and the pigskin gloves sold for $7.50 per pair. The company took in $137.25. How many of each kind were sold?

We let x = the number of pairs of cloth gloves sold and y = the number of pairs of pigskin gloves sold. The total was 20 pairs, so we have

$x + y = 20.$

After finishing Section 3.3, you should be able to:

● Solve problems by translating them to systems of two equations and solving the system.

1. One number is 4 times another number. Their sum is 175. Find the numbers.

2. A store sold 30 sweatshirts. They sold white ones for $8.95 and red ones for $9.50. They took in $272.90. How many of each color did they sell?

The amount taken in for cloth gloves was $4.95x$. The amount taken in for pigskin gloves was $7.50y$. These amounts are in dollars. The total taken in was $137.25, so we have

$$4.95x + 7.50y = 137.25.$$

We can multiply on both sides by 100 to clear of decimals.

$$495x + 750y = 13{,}725.$$

We now have the following system of equations. We solve it, using the substitution method or the addition method. In this case, the substitution method is simpler.

$$x + \quad y = 20$$
$$495x + 750y = 13{,}725$$

After we solve, we find that $x = 5$ and $y = 15$. This must be checked in the original problem. Remember, x is the number of cloth gloves sold and y is the number of pigskin gloves sold.

Money from cloth gloves: $\quad \$4.95x = 4.95 \times 5 = \$24.75$

Money from pigskin gloves: $\quad \$7.50y = 7.50 \times 15 = \$112.50$

$$\text{Total} \quad \overline{\$137.25}$$

Number of gloves $x + y = 5 + 15 = 20$.

The numbers check in the original problem, so the business sold 5 pairs of cloth gloves and 15 pairs of pigskin gloves.

DO EXERCISE 2.

Example 3 (*A mixture problem*)   Solution A is 2% alcohol. Solution B is 6% alcohol. A service station owner wants to mix the two to get 60 liters of a solution that is 3.2% alcohol. How many liters of each should the owner use?

We can keep information in a table. That will help in translating.

	Amount of solution	Percent of alcohol	Amount of alcohol in solution
A	x liters	2%	$2\%x$
B	y liters	6%	$6\%y$
Mixture	60 liters	3.2%	0.032×60, or 1.92 liters

Note that we have used x for the number of liters of A and y for the number of liters of B. To get the amount of alcohol, we multiply by the percentages, of course. If we add x and y in the first column we get 60, and this gives us one equation:

$$x + y = 60.$$

If we add the amounts of alcohol in the third column we get 1.92, and this gives us another equation:

$$2\%x + 6\%y = 1.92.$$

After changing percents to decimals and clearing, we have this system:

$$x + y = 60$$
$$2x + 6y = 192.$$

Now we solve the system

$$x + y = 60$$
$$2x + 6y = 192.$$

$$
\begin{array}{ll}
-2x - 2y = -120 & \text{Multiplying in the first equation by } -2 \\
\underline{2x + 6y = 192} & \\
0 + 4y = 72 & \text{Adding} \\
y = 18 & \text{Solving for } y \\
x + 18 = 60 & \text{Substituting into the first equation of the system} \\
x = 42 &
\end{array}
$$

Remember, x — the number of liters of 2% solution; y = the number of liters of 6% solution.

Total number of liters of mixture: $x + y = 42 + 18 = 60$

Amount of alcohol: $2\% \times 42 + 6\% \times 18 = 0.02 \times 42 + 0.06 \times 18$
$$= 1.92 \text{ liters}$$

Percentage of alcohol in mixture: $\dfrac{1.92}{60} = 0.032, \quad \text{or} \quad 3.2\%$

The numbers check in the original problem, so the answer is that the owner should use 42L of 2% alcohol and 18L of 6% alcohol.

DO EXERCISE 3.

Example 4 (*An interest problem*) Two investments are made totaling $4800. In the first year they yield $412 in simple interest. Part of the money is invested at 8% and the rest at 9%. Find the amount invested at each rate of interest.

Keeping information in a table will help in translating. The columns in the table come from the formula for simple interest: $I = Prt$.

	Principal	Rate of interest	Time	Interest
First investment	x	8%	1 yr	8%x
Second investment	y	9%	1 yr	9%y
Total	$4800			$412

Note that we have used x and y for the numbers of dollars invested. They total $4800, and this gives us one equation:

$$x + y = 4800.$$

Look at the last column. The interest, or *yield*, totals $412. This gives us a second equation:

$$8\%x + 9\%y = 412, \quad \text{or} \quad 0.08x + 0.09y = 412.$$

After we multiply on both sides to clear of decimals, we have

$$8x + 9y = 41,200.$$

3. A gardener has two kinds of solutions containing weedkiller and water. One is 5% weedkiller and the other is 15% weedkiller. The gardener needs 100 liters of a 12% solution and wants to make it by mixing. How much of each solution should be used?

ANSWER ON PAGE A–12

4. $3700 is invested for a year at simple interest, yielding $297. Part of the money is invested at 7% and the rest at 9%. How much was invested at each rate?

We solve the system

$$x + y = 4800$$
$$8x + 9y = 41,200.$$

We find that $x = 2000$ and $y = 2800$. These numbers check in the problem, so $2000 is invested at 8% and $2800 at 9%.

DO EXERCISE 4.

Making a drawing is often helpful. Whenever it makes sense to do so, you should make a drawing before trying to translate.

Example 5   The perimeter of a rectangular field is 194 yards. The length is 2 more than four times the width. Find the length and the width.

First we make a drawing. We use l for length and w for width. We translate the first statement.

$$\underbrace{\text{The perimeter}} \underset{\downarrow}{\text{is}} \underset{\downarrow}{194.}$$
$$2l + 2w \quad = 194$$

We translate the second statement.

$$\underbrace{\text{The length}} \underset{\downarrow}{\text{is}} \underset{\downarrow}{2} \underbrace{\text{more than}} \underbrace{\text{four}} \underset{\downarrow}{\text{times}} \underbrace{\text{the width.}}$$
$$l \quad = 2 \quad + \quad 4 \quad \cdot \quad w$$

5. The perimeter of a rectangle is 90 ft. The length is 20 ft greater than the width. Find the length and the width.

We now have the system of equations

$$2l + 2w = 194$$
$$l = 2 + 4w.$$

It is probably easier to use the substitution method, substituting $2 + 4w$ for l in the first equation.

$$
\begin{aligned}
2(2 + 4w) + 2w &= 194 & &\text{Substituting} \\
4 + 8w + 2w &= 194 & &\text{Multiplying to remove parentheses on the left} \\
4 + 10w &= 194 & &\text{Collecting like terms} \\
10w &= 190 & &\text{Adding } -4 \\
w &= \frac{190}{10}, \text{ or } 19 & &\text{Multiplying by } \frac{1}{10}
\end{aligned}
$$

Then we substitute 19 for w in the second equation:

$$l = 2 + 4 \cdot 19 = 2 + 76 = 78.$$

The numbers 78 and 19 check in the original problem, so we know that the length is 78 yards and the width is 19 yards.

DO EXERCISE 5.

NAME

CLASS

EXERCISE SET 3.3

■ Solve.

1. The sum of a certain number and a second number is −42. The first number minus the second is 52. Find the numbers.

2. The sum of two numbers is −63. The first number minus the second is −41. Find the numbers.

1. _____

2. _____

3. The difference between two numbers is 16. Three times the larger number is nine times the smaller. What are the numbers?

4. The difference between two numbers is 11. Twice the smaller number plus three times the larger is 123. What are the numbers?

3. _____

4. _____

5. The perimeter of a rectangular field is 628 m. The length of the field exceeds its width by 6 m. Find the dimensions.

6. The perimeter of a lot is 750 ft. The width is one-fourth the length. Find the dimensions.

5. _____

6. _____

ANSWERS

7. One day a store sold 30 sweatshirts. White ones cost $9.95 and yellow ones cost $10.50. In all, $310.60 worth of sweatshirts were sold. How many of each color were sold?

8. One week a business sold 40 scarves. White ones cost $4.95 and printed ones cost $7.95. In all, $282 worth of scarves were sold. How many of each kind were sold?

7. _____

9. Soybean meal is 16% protein; corn meal is 9% protein. How many pounds of each should be mixed together to get a 350-pound mixture that is 12% protein?

10. A chemist has one solution of acid and water that is 25% acid and a second that is 50% acid. How many gallons of each should be mixed together to get 10 gallons of a solution that is 40% acid?

8. _____

9. _____

11. Two investments are made totaling $8800. For a certain year these investments yield $663 in simple interest. Part of the $8800 is invested at 7% and part at 8%. Find the amount invested at each rate.

12. Two investments are made totaling $15,000. For a certain year these investments yield $1432 in simple interest. Part of the $15,000 is invested at 9% and part at 10%. Find the amount invested at each rate.

10. _____

11. _____

A common error is failure to read the examples and work the margin exercises. Many of these problems are very much like the examples in the text. If you get stuck, go back and work through the examples and the margin exercises

12. _____

13. For a certain year $3900 is received in interest from two investments. A certain amount is invested at 5% and $10,000 more than this is invested at 6%. Find the amount invested at each rate.

14. For a certain year $876 is received from two investments. A certain amount is invested at 7%, and $1200 more than this is invested at 8%. Find the amount invested at each rate.

13. _____

15. Ann Teak is twice as old as her son. Ten years ago Ann was three times as old as her son. What are their present ages? (*Hint:* Ten years ago Ann was $x - 10$ years old and her son was $y - 10$ years old.)

16. Gerry Atric is twice as old as his youngest daughter. In eight years Gerry's age will be three times what his daughter's age was six years ago. What are their present ages?

14. _____

15. _____

17. John is 8 years older than his sister Sue. Four years ago Sue was $\frac{2}{3}$ as old as John. How old are they now?

18. Paula is 12 years older than her brother Bob. Four years from now Bob will be $\frac{2}{3}$ as old as Paula. How old are they now?

16. _____

17. _____

18. _____

19. The perimeter of a rectangle is 86 inches. The length is 19 in. greater than the width. Find the length and the width.

20. The perimeter of a rectangle is 388 ft. The length is 82 ft greater than the width. Find the length and the width.

19. _____

20. _____

21. One day a store sold 45 pens, one kind at $8.50 and another kind at $9.75. In all, $398.75 was taken in. How many of each kind were sold?

22. At a club play, 117 tickets were sold. Adult tickets cost $1.25 and children's tickets cost $0.75. In all, $129.75 was taken in. How many of each kind of ticket were sold?

21. _____

22. _____

23. $1150 is invested, part of it at 6% and part of it at 5%. The total yield was $65.75. How much was invested at each rate?

24. $27,000 is invested, part of it at 6% and part of it at 7%. The total yield was $1765. How much was invested at each rate?

23. _____

24. _____

25. An automobile radiator contains 16 liters of antifreeze and water. This mixture is 30% antifreeze. How much of this mixture should be drained and replaced with pure antifreeze so that there will be 50% antifreeze?

25. _____

3.4 SYSTEMS OF EQUATIONS IN THREE VARIABLES

● IDENTIFYING SOLUTIONS

A solution of a system of equations in three variables is an ordered triple that makes *all three* equations true.

Example 1   Determine whether $(\frac{3}{2}, -4, 3)$ is a solution of the system

$$4x - 2y - 3z = 5$$
$$-8x - y + z = -5$$
$$2x + y + 2z = 5.$$

We substitute $(\frac{3}{2}, -4, 3)$ into the equations, using alphabetical order.

$$
\begin{array}{c|c}
4x - 2y - 3z = 5 & 5 \\
\hline
4 \cdot \frac{3}{2} - 2(-4) - 3 \cdot 3 & \\
6 + 8 - 9 & \\
5 &
\end{array}
\qquad
\begin{array}{c|c}
-8x - y + z = -5 & -5 \\
\hline
-8 \cdot \frac{3}{2} - (-4) + 3 & \\
-12 + 4 + 3 & \\
-5 &
\end{array}
$$

$$
\begin{array}{c|c}
2x + y + 2z = 5 & 5 \\
\hline
2 \cdot \frac{3}{2} + (-4) + 2 \cdot 3 & \\
3 - 4 + 6 & \\
5 &
\end{array}
$$

The triple makes *all three* equations true, so it is a solution.

DO EXERCISE 1.

●● SOLVING

The substitution method is not ordinarily convenient for solving systems of equations with three variables. We consider only the addition method.

Example 2   Solve:

$$
\begin{array}{ll}
x + y + z = 4 & (1) \\
x - 2y - z = 1 & (2) \\
2x - y - 2z = -1. & (3)
\end{array}
$$

a) We first use *any* two of the three equations to get an equation in two variables. Let us use equations (1) and (2) and add to eliminate z:

$$
\begin{array}{ll}
x + y + z = 4 & (1) \\
x - 2y - z = 1 & (2) \\
\hline
2x - y = 5 & (4)
\end{array}
$$

b) We use a different pair of equations and eliminate the *same variable* that we did in (a). Let us use equations (1) and (3) and eliminate z.

$$
\begin{array}{ll}
x + y + z = 4 & (1) \\
2x - y - 2z = -1 & (3)
\end{array}
$$

> *Caution!* A common error is to eliminate a different variable the second time.

$$
\begin{array}{ll}
2x + 2y + 2z = 8 & \text{Multiplying equation (1) by 2} \\
2x - y - 2z = -1 & \\
\hline
4x + y = 7 \quad (5) & \text{Adding}
\end{array}
$$

OBJECTIVES

After finishing Section 3.4, you should be able to:

● Determine whether an ordered triple is a solution of a system of three equations in three variables.

●● Solve systems of three equations in three variables.

1. Consider the system

$$
\begin{array}{l}
4x + 2y + 5z = 6 \\
2x - y + z = 5 \\
x + 2y - z = 0.
\end{array}
$$

a) Determine whether $(1, 2, 3)$ is a solution.

b) Determine whether $(2, -1, 0)$ is a solution.

ANSWERS ON PAGE A–12

2. Solve. Don't forget to check.

$$4x - y + z = 6$$
$$-3x + 2y - z = -3$$
$$2x + y + 2z = 3$$

c) Now we solve the resulting system of equations (4) and (5). That will give us two of the numbers.

$$2x - y = 5 \qquad (4)$$
$$\underline{4x + y = 7} \qquad (5)$$
$$6x \quad\quad = 12$$
$$x = 2$$

> We now have two equations in two unknowns. Had we eliminated different letters in parts (a) and (b), this would not be the case.

We use either equation (4) or (5) to find y. We use (5):

$$4x + y = 7$$
$$4 \cdot 2 + y = 7$$
$$8 + y = 7$$
$$y = -1$$

d) Then we use any of the original three equations and substitute to find the third number z. Let us use equation (1) and substitute our two numbers in it.

$$x + y + z = 4 \qquad (1)$$
$$2 + (-1) + z = 4 \qquad \text{Substituting 2 for } x \text{ and } -1 \text{ for } y$$
$$1 + z = 4$$
$$z = 3$$

We obtain $(2, -1, 3)$ and check in *all three* equations.

$$\frac{x + y + z = 4}{2 + (-1) + 3 \;\Big|\; 4} \qquad \frac{x - 2y - z = 1}{2 - 2(-1) - 3 \;\Big|\; 1}$$
$$\qquad\qquad 4 \;\Big| \qquad\qquad\qquad 1 \;\Big|$$

$$\frac{2x - y - 2z = -1}{2 \cdot 2 - (-1) - 2 \cdot 3 \;\Big|\; -1}$$
$$\qquad\qquad -1 \;\Big|$$

> *Caution!* You must check in *all three* of the original equations.

The solution is $(2, -1, 3)$.

In summary,

> **The procedure for using the addition method is as follows.**
>
> **a)** We first use *any* two of the three equations to get an equation in two variables.
>
> **b)** We use a different pair of equations and get another equation in *the same two* variables. That is, we eliminate the same variable that we did in (a).
>
> **c)** Now we solve the resulting system (pair) of equations. That will give us two of the numbers.
>
> **d)** Then we use any of the original three equations to find the third number.

DO EXERCISE 2.

Example 3 Solve:
$$4x - 2y - 3z = 5 \qquad (1)$$
$$-8x - y + z = -5 \qquad (2)$$
$$2x + y + 2z = 5. \qquad (3)$$

a) We first use *any* two of the three equations to get an equation in two variables. Let us use (2) and (3) since the *y*-terms are additive inverses of each other. We add:

$$
\begin{array}{rl}
-8x - y + z = -5 & (2) \\
\underline{2x + y + 2z = \;\;\;5} & (3) \\
-6x \qquad + 3z = \;\;\;0 & (4)
\end{array}
$$

b) We use another pair of equations to get another equation in the same two variables. That is, we eliminate the same variable that we did in (a). We use equations (1) and (3) and eliminate *y*.

$$
\begin{array}{rl}
4x - 2y - 3z = 5 & (1) \\
2x + y + 2z = 5 & (3)
\end{array}
$$

$$
\begin{array}{rll}
4x - 2y - 3z = 5 & & \\
\underline{4x + 2y + 4z = 10} & & \text{Multiplying by 2} \\
8x \qquad + z = 15 & (5) & \text{Adding}
\end{array}
$$

c) Now we solve the resulting system of equations (4) and (5). That will give us two of the numbers.

$$
\begin{array}{rl}
-6x + 3z = \;\;0 & (4) \\
8x + \;\;z = 15 & (5)
\end{array}
$$

$$
\begin{array}{rl}
-6x + 3z = \;\;\;\;0 & \\
\underline{-24x - 3z = -45} & \text{Multiplying by } -3^* \\
-30x \qquad\quad = -45 &
\end{array}
$$

$$
x = \frac{-45}{-30}, \quad \text{or} \quad \frac{3}{2}
$$

We use equation (5) to find *z*:

$$
8x + z = 15
$$

$$
8 \cdot \frac{3}{2} + z = 15
$$

$$
12 + z = 15
$$

$$
z = 3
$$

d) Then we use any of the original equations and substitute to find the third number *y*. We use equation (3) since the coefficient of *y* is 1.

$$
2x + y + 2z = 5
$$

$$
2 \cdot \frac{3}{2} + y + 2 \cdot 3 = 5 \qquad \text{Substituting } \frac{3}{2} \text{ for } x \text{ and 3 for } z
$$

$$
3 + y + 6 = 5
$$

$$
y + 9 = 5
$$

$$
y = -4
$$

We obtain $(\frac{3}{2}, -4, 3)$. This checks, so it is the solution.

DO EXERCISE 3.

---

* We could have multiplied equation (4) by $-\frac{1}{3}$.

3. Solve. Don't forget to check.

$$
\begin{array}{r}
2x + y - 4z = 0 \\
x - y + 2z = 5 \\
3x + 2y + 2z = 3
\end{array}
$$

ANSWER ON PAGE A–12

4. Solve. Don't forget to check.

$$x + y + z = 100$$
$$x - y \quad\ = -10$$
$$x \quad\ - z = -30$$

Example 4 Solve:

$$x + y + z = 180 \quad (1)$$
$$x \quad\ - z = -70 \quad (2)$$
$$2y - z = 0. \quad (3)$$

Note that there is no y in equation (2).

We use equations (1) and (3) and eliminate y.

$$x + y + z = 180 \quad (1)$$
$$2y - z = 0 \quad (3)$$

$$-2x - 2y - 2z = -360 \quad \text{Multiplying by } -2$$
$$2y - z = 0$$
$$\overline{-2x \quad\quad - 3z = -360} \quad (4)$$

Now we solve the resulting system of equations (2) and (4).

$$x - z = -70 \quad (2) \qquad 2x - 2z = -140$$
$$-2x - 3z = -360 \quad (4) \qquad \underline{-2x - 3z = -360}$$
$$-5z = -500$$
$$z = 100$$

Continuing as we did in the previous examples, we get the solution (30, 50, 100).

DO EXERCISE 4.

SOMETHING EXTRA

CALCULATOR CORNER: NUMBER PATTERNS

Do the following calculations. Look for a pattern in the answers.

$$4 \times 6$$
$$44 \times 6$$
$$444 \times 6$$
$$4444 \times 6$$

Now see if you can figure out what 44,444 × 6 is without using your calculator. Then check your answer by doing the actual calculation.

EXERCISES

In each of the following, do the first four calculations and look for a pattern. Do the last calculation without using your calculator, and then check your answer.

1.	2.	3.
3×5	77×78	9×89
33×5	777×78	99×889
333×5	7777×78	999×8889
3333×5	$77,777 \times 78$	$9999 \times 88,889$
$33,333 \times 5$	$777,777 \times 78$	$99,999 \times 888,889$

NAME   CLASS   ANSWERS

EXERCISE SET 3.4

■ ●

1. Determine whether $(1, -2, 3)$ is a
   solution of the system

   $x + y + z = 2$
   $x - 2y - z = 2$
   $3x + 2y + z = 2.$

2. Determine whether $(2, -1, 2)$ is a
   solution of the system

   $-x - 2y + 3z = 6$
   $x + y - 2z = 5$
   $2x - y - z = 7.$

1. _____

2. _____

3. _____

● ● Solve.

3. $x + y + z = 6$
    $2x - y + 3z = 9$
    $-x + 2y + 2z = 9$

4. $2x - y + z = 10$
   $4x + 2y - 3z = 10$
   $x - 3y + 2z = 8$

5. $2x - y - 3z = -1$
   $2x - y + z = -9$
   $x + 2y - 4z = 17$

4. _____

5. _____

6. $x - y + z = 6$
    $2x + 3y + 2z = 2$
    $3x + 5y + 4z = 4$

7. $2x - 3y + z = 5$
   $x + 3y + 8z = 22$
   $3x - y + 2z = 12$

8. $6x - 4y + 5z = 31$
   $5x + 2y + 2z = 13$
   $x + y + z = 2$

6. _____

7. _____

8. _____

9. $3a - 2b + 7c = 13$
$\quad a + 8b - 6c = -47$
$\quad 7a - 9b - 9c = -3$

10. $\quad x + y + z = 0$
$\quad 2x + 3y + 2z = -3$
$\quad -x + 2y - 3z = -1$

11. $2x + 3y + z = 17$
$\quad x - 3y + 2z = -8$
$\quad 5x - 2y + 3z = 5$

9. _____

10. _____

11. _____

12. $2x + y - 3z = -4$
$\quad 4x - 2y + z = 9$
$\quad 3x + 5y - 2z = 5$

13. $2x + y + z = -2$
$\quad 2x - y + 3z = 6$
$\quad 3x - 5y + 4z = 7$

14. $2x + y + 2z = 11$
$\quad 3x + 2y + 2z = 8$
$\quad x + 4y + 3z = 0$

12. _____

13. _____

14. _____

15. Solve: $w - 2x + y - 2z = -7$
$\quad\quad\quad 4w + x + 2y - z = -1$
$\quad\quad\quad 2w + 4x + y - 2z = 5$
$\quad\quad\quad w + x - 2y - z = 3.$

15. _____

3.5 SOLVING PROBLEMS

■ Many problems can be solved by first translating to a system of three equations.

Example 1 The sum of three numbers is 4. The first number minus twice the second minus the third is 1. Twice the first number minus the second minus twice the third is −1. Find the numbers.

Let us call the three numbers x, y, and z. There are three statements in the problem. We translate:

The sum of the three numbers is 4.

$$x + y + z = 4$$

The first number minus twice the second minus the third is 1.

$$x - 2y - z = 1$$

Twice the first number minus the second minus twice the third is −1.

$$2x - y - 2z = -1$$

We now have a system of three equations:

$$x + y + z = 4$$
$$x - 2y - z = 1$$
$$2x - y - 2z = -1.$$

We solved this system in Example 2 on p. 125 and found that $x = 2$, $y = -1$, and $z = 3$. These numbers check in the original problem, so we have a solution.

DO EXERCISE 1.

Example 2 In a triangle the largest angle is 70° greater than the smallest angle. The largest angle is twice as large as the remaining angle. Find the measure of each angle.

We first make a drawing, and label the angles. We let x represent the smallest angle, z the largest angle, and y the remaining angle.

We know that the sum of the measures of the angles is 180° in any triangle. This gives us the equation

$$x + y + z = 180.$$

OBJECTIVE

After finishing Section 3.5, you should be able to:

■ Solve problems by translating to a system of three equations in three variables.

1. The sum of three numbers is 3. The first number minus twice the second minus the third is 4. Twice the first number plus the other two is 5. Find the numbers.

2. One angle of a triangle is twice as large as a second angle. The remaining angle is 20° greater than the first angle. Find the measures of the angles.

There are two statements in the problem. We translate them.

The largest angle is 70° greater than the smallest angle.

$$z \quad = 70 \quad + \quad x$$

The largest angle is twice as large as the remaining angle.

$$z \quad = \quad 2y$$

We now have a system:

$$
\begin{array}{lll}
x + y + z = 180 & & x + y + z = 180 \\
x + 70 = z & \text{or} & x - z = -70 \\
2y = z & & 2y - z = 0.
\end{array}
$$

We solved this system in Example 4 on p. 128 and found that $x = 30$, $y = 50$, and $z = 100$. These numbers check, so we have a solution. The angles measure 30°, 50°, and 100°.

DO EXERCISE 2.

Example 3   In a factory there are three machines A, B, and C. When all three are running, they produce 222 suitcases per day. If A and B work but C does not, they produce 159 suitcases per day. If B and C work but A does not, they produce 147 suitcases per day. What is the daily production of each machine?

3. There are three machines, A, B, and C, in a factory. When all three work, they produce 287 screws per hour. When only A and C work, they produce 197 screws per hour. When only A and B work, they produce 202 screws per hour. How many screws per hour can each produce alone?

Let us use x, y, and z for the number of suitcases produced daily by the machines A, B, and C, respectively. There are three statements. We translate them.

When all three are running, they produce 222 suitcases per day.

$$x + y + z \quad = \quad 222$$

When A and B work, they produce 159 suitcases per day.

$$x + y \quad = \quad 159$$

When B and C work, they produce 147 suitcases per day.

$$y + z \quad = \quad 147$$

We now have a system of three equations:

$$
\begin{array}{l}
x + y + z = 222 \\
x + y = 159 \\
y + z = 147.
\end{array}
$$

We solve and get $x = 75$, $y = 84$, and $z = 63$. These numbers check, so the answer to the problem is that A produces 75 suitcases per day, B produces 84 suitcases per day, and C produces 63 suitcases per day.

DO EXERCISE 3.

Solving systems of three or more equations is important in many applications. Systems of equations arise very often in the use of statistics, for example, in such fields as the social sciences. They also come up in problems of business, science, and engineering.

NAME _____   CLASS _____   ANSWERS

EXERCISE SET 3.5

1. The sum of three numbers is 105. The third is 11 less than 10 times the second. Twice the first is 7 more than 3 times the second. Find the numbers.

2. The sum of three numbers is 57. The second is 3 more than twice the first. The third is 6 more than the first. Find the numbers.

1. _____

2. _____

3. The sum of three numbers is 5. The first number minus the second plus the third is 1. The first minus the third is 3 more than the second. Find the numbers.

4. The sum of three numbers is 26. Twice the first minus the second is 2 less than the third. The third is the second minus three times the first. Find the numbers.

3. _____

4. _____

5. In triangle ABC, the measure of angle B is 2° more than three times the measure of angle A. The measure of angle C is 8° more than the measure of angle A. Find the angle measures.

6. In triangle ABC, the measure of angle B is three times the measure of angle A. The measure of angle C is 30° greater than the measure of angle A. Find the angle measures.

5. _____

6. _____

7. In triangle ABC, the measure of angle B is twice the measure of angle A. The measure of angle C is 80° more than that of angle A. Find the angle measures.

8. In triangle ABC, the measure of angle B is three times that of angle A. The measure of angle C is 20° more than that of angle A. Find the angle measures.

7. _____

8. _____

9. Paula sells magazines part time. On Thursday, Friday, and Saturday she sold $66 worth. On Thursday she sold $3 more than on Friday. On Saturday she sold $6 more than on Thursday. How much did she take in each day?

10. John Gray picked strawberries on three days. He picked a total of 87 quarts. On Tuesday he picked 15 quarts more than on Monday. On Wednesday he picked 3 quarts fewer than on Tuesday. How many quarts did he pick each day?

9. _____

11. Linda has a total of 255 on three tests. The sum of the scores on the first and second tests exceeds her third score by 61. Her first score exceeds her second by 6. Find the three scores.

12. Fred, Jane, and Mary made a total bowling score of 575. Fred's score was 15 more than Jane's. Mary's was 20 more than Jane's. Find the scores.

10. _____

11. _____

12. _____

13. In a factory there are three polishing machines, A, B, and C. When all three of them are working, 5700 lenses can be polished in one week. When only A and B are working, 3400 lenses can be polished in one week. When only B and C are working, 4200 lenses can be polished in one week. How many lenses can be polished in a week by each machine?

14. Sawmills A, B, and C can produce 7400 board feet of lumber per day. A and B together can produce 4700 board feet, while B and C together can produce 5200 board feet. How much can each mill produce by itself?

13. _____

14. _____

15. When three pumps, A, B, and C, are running together, they can pump 3700 gallons per hour. When only A and B are running, 2200 gallons per hour can be pumped. When only A and C are running, 2400 gallons per hour can be pumped. What is the pumping capacity of each pump?

16. Three welders, A, B, and C, can weld 37 linear feet per hour when working together. A and B together can weld 22 linear feet per hour, while A and C together can weld 25 linear feet per hour. How much can each weld alone?

15. _____

16. _____

3.6 DETERMINANTS (Optional)

● EVALUATING DETERMINANTS

The following is called a *determinant*:

$$\begin{vmatrix} a_1 & b_1 \\ a_2 & b_2 \end{vmatrix}.$$

To evaluate a determinant, we do two multiplications and subtract.

Example 1 Evaluate:

$$\begin{vmatrix} 2 & -5 \\ 6 & 7 \end{vmatrix}.$$

We multiply and subtract as follows:

$$\begin{vmatrix} 2 & -5 \\ 6 & 7 \end{vmatrix} = 2 \cdot 7 - 6 \cdot (-5) = 14 + 30 = 44.$$

Determinants are defined according to the pattern shown in Example 1.

The determinant $\begin{vmatrix} a_1 & b_1 \\ a_2 & b_2 \end{vmatrix}$ **is defined to mean** $a_1 b_2 - a_2 b_1$.

> The value of a determinant is a *number*. In Example 1, the value is 44.

DO EXERCISES 1 AND 2.

●● THIRD-ORDER DETERMINANTS

The determinants above are called *second-order*. A *third-order* determinant is defined as follows.

Note the minus sign here.

$$\begin{vmatrix} a_1 & b_1 & c_1 \\ a_2 & b_2 & c_2 \\ a_3 & b_3 & c_3 \end{vmatrix} = a_1 \begin{vmatrix} b_2 & c_2 \\ b_3 & c_3 \end{vmatrix} - a_2 \begin{vmatrix} b_1 & c_1 \\ b_3 & c_3 \end{vmatrix} + a_3 \begin{vmatrix} b_1 & c_1 \\ b_2 & c_2 \end{vmatrix}$$

Note that the a's come from the first column.

Note also that the second-order determinants above can be obtained by crossing out the row and column in which the a occurs.

For a_1: $\begin{vmatrix} a_1 & b_1 & c_1 \\ a_2 & b_2 & c_2 \\ a_3 & b_3 & c_3 \end{vmatrix}$ For a_2: $\begin{vmatrix} a_1 & b_1 & c_1 \\ a_2 & b_2 & c_2 \\ a_3 & b_3 & c_3 \end{vmatrix}$

For a_3: $\begin{vmatrix} a_1 & b_1 & c_1 \\ a_2 & b_2 & c_2 \\ a_3 & b_3 & c_3 \end{vmatrix}$

OBJECTIVES

After finishing Section 3.6, you should be able to:

● Evaluate second-order determinants.

●● Evaluate third-order determinants.

●●● Solve systems of equations using Cramer's rule.

Evaluate these determinants.

1. $\begin{vmatrix} 3 & 2 \\ 4 & 1 \end{vmatrix}$

2. $\begin{vmatrix} 5 & -2 \\ -1 & -1 \end{vmatrix}$

Evaluate these determinants.

3. $\begin{vmatrix} 2 & -1 & 1 \\ 1 & 2 & -1 \\ 3 & 4 & -3 \end{vmatrix}$

Example 2   Evaluate this third-order determinant.

$$\begin{vmatrix} -1 & 0 & 1 \\ -5 & 1 & -1 \\ 4 & 8 & 1 \end{vmatrix} = -1\begin{vmatrix} 1 & -1 \\ 8 & 1 \end{vmatrix} - (-5)\begin{vmatrix} 0 & 1 \\ 8 & 1 \end{vmatrix} + 4\begin{vmatrix} 0 & 1 \\ 1 & -1 \end{vmatrix}$$

We calculate as follows.

$$-1\begin{vmatrix} 1 & 1 \\ 8 & 1 \end{vmatrix} - (-5)\begin{vmatrix} 0 & 1 \\ 8 & 1 \end{vmatrix} + 4\begin{vmatrix} 0 & 1 \\ 1 & 1 \end{vmatrix}$$

$$= -1[1 \cdot 1 - 8(-1)] + 5(0 \cdot 1 - 8 \cdot 1) + 4[0 \cdot (-1) - 1 \cdot 1]$$
$$= -1(9) + 5(-8) + 4(-1)$$
$$= -9 - 40 - 4$$
$$= -53$$

DO EXERCISES 3 AND 4.

⦿⦿⦿ SOLVING SYSTEMS BY DETERMINANTS

Here is a system of two equations in two variables:

$$a_1x + b_1y = c_1$$
$$a_2x + b_2y = c_2.$$

We form three determinants, which we call D, D_x, and D_y.

$$D = \begin{vmatrix} a_1 & b_1 \\ a_2 & b_2 \end{vmatrix}$$ In D we have the coefficients of x and y.

4. $\begin{vmatrix} 3 & 2 & 2 \\ -2 & 1 & 4 \\ 4 & -3 & 3 \end{vmatrix}$

$$D_x = \begin{vmatrix} c_1 & b_1 \\ c_2 & b_2 \end{vmatrix}$$ To form D_x we use D, but we replace the x-coefficients with the constants on the right side of the equations.

$$D_y = \begin{vmatrix} a_1 & c_1 \\ a_2 & c_2 \end{vmatrix}$$ To form D_y we use D, but we replace the y-coefficients with the constants on the right.

It is important that the replacement be done *without changing the order of the columns.* Then the solution of the system can be found as follows. This is known as *Cramer's rule*:

$$x = \frac{D_x}{D}, \qquad y = \frac{D_y}{D}.$$

Example 3 Solve, using Cramer's rule:

$$3x - 2y = 7$$
$$3x + 2y = 9.$$

We compute D, D_x, and D_y.

$$D = \begin{vmatrix} 3 & -2 \\ 3 & 2 \end{vmatrix} = 3 \cdot 2 - 3 \cdot (-2) = 6 + 6 = 12$$

$$D_x = \begin{vmatrix} 7 & -2 \\ 9 & 2 \end{vmatrix} = 7 \cdot 2 - 9(-2) = 14 + 18 = 32$$

$$D_y = \begin{vmatrix} 3 & 7 \\ 3 & 9 \end{vmatrix} = 3 \cdot 9 - 3 \cdot 7 = 27 - 21 = 6$$

Then $x = \dfrac{D_x}{D} = \dfrac{32}{12}$, or $\dfrac{8}{3}$ and $y = \dfrac{D_y}{D} = \dfrac{6}{12} = \dfrac{1}{2}$.

The solution is $(\frac{8}{3}, \frac{1}{2})$.

DO EXERCISE 5.

Cramer's rule for three equations is very similar to that for two.

$$a_1 x + b_1 y + c_1 z = d_1$$
$$a_2 x + b_2 y + c_2 z = d_2$$
$$a_3 x + b_3 y + c_3 z = d_3$$

$$D = \begin{vmatrix} a_1 & b_1 & c_1 \\ a_2 & b_2 & c_2 \\ a_3 & b_3 & c_3 \end{vmatrix}$$

$$D_x = \begin{vmatrix} d_1 & b_1 & c_1 \\ d_2 & b_2 & c_2 \\ d_3 & b_3 & c_3 \end{vmatrix}$$

$$D_y = \begin{vmatrix} a_1 & d_1 & c_1 \\ a_2 & d_2 & c_2 \\ a_3 & d_3 & c_3 \end{vmatrix}$$

D is again the determinant of the coefficients of x, y, and z. This time we have one more determinant, D_z. We get it by using D and replacing the z-coefficients by the constants on the right:

$$D_z = \begin{vmatrix} a_1 & b_1 & d_1 \\ a_2 & b_2 & d_2 \\ a_3 & b_3 & d_3 \end{vmatrix}.$$

The solution of the system is given by

$$x = \frac{D_x}{D}, \qquad y = \frac{D_y}{D}, \qquad z = \frac{D_z}{D}.$$

5. Solve, using Cramer's rule.

$$20x - 15y = 75$$
$$x + 3y = 0$$

ANSWER ON PAGE A–13

6. Solve, using Cramer's rule.

$$x - 3y - 7z = 6$$
$$2x + 3y + z = 9$$
$$4x + y = 7$$

Example 4   Solve, using Cramer's rule:

$$x - 3y + 7z = 13$$
$$x + y + z = 1$$
$$x - 2y + 3z = 4.$$

We compute D, D_x, D_y, and D_z:

$$D = \begin{vmatrix} 1 & -3 & 7 \\ 1 & 1 & 1 \\ 1 & -2 & 3 \end{vmatrix} = -10; \qquad D_x = \begin{vmatrix} 13 & -3 & 7 \\ 1 & 1 & 1 \\ 4 & -2 & 3 \end{vmatrix} = 20;$$

$$D_y = \begin{vmatrix} 1 & 13 & 7 \\ 1 & 1 & 1 \\ 1 & 4 & 3 \end{vmatrix} = -6; \qquad D_z = \begin{vmatrix} 1 & -3 & 13 \\ 1 & 1 & 1 \\ 1 & -2 & 4 \end{vmatrix} = -24.$$

Then $x = \dfrac{D_x}{D} = \dfrac{20}{-10} = -2;$

$$y = \frac{D_y}{D} = \frac{-6}{-10} = \frac{3}{5};$$

$$z = \frac{D_z}{D} = \frac{-24}{-10} = \frac{12}{5}.$$

The solution is $(-2, \frac{3}{5}, \frac{12}{5})$.

In Example 4 we would not have needed to evaluate D_z. Once we found x and y we could have substituted them into one of the equations to find z. In practice, we use determinants to find only two of the numbers; then we find the third by substitution into an equation.

DO EXERCISE 6.

In using Cramer's rule, we divide by D. If D should be 0, we could not do that. If $D = 0$ and at least one of the other determinants is not 0, then the system does not have a solution, and we say that it is *inconsistent*. If $D = 0$ and all of the other determinants are also 0, then there is an infinite set of solutions. In that case we say that the system is *dependent*.

ANSWER ON PAGE A–13

NAME

CLASS

ANSWERS

EXERCISE SET 3.6

◉ Evaluate.

1. $\begin{vmatrix} 2 & 7 \\ 1 & 5 \end{vmatrix}$

2. $\begin{vmatrix} 3 & 2 \\ 2 & -3 \end{vmatrix}$

3. $\begin{vmatrix} 6 & -9 \\ 2 & 3 \end{vmatrix}$

4. $\begin{vmatrix} 3 & 2 \\ -7 & 5 \end{vmatrix}$

◉◉ Evaluate.

5. $\begin{vmatrix} 0 & 2 & 0 \\ 3 & -1 & 1 \\ 1 & -2 & 2 \end{vmatrix}$

6. $\begin{vmatrix} 3 & 0 & -2 \\ 5 & 1 & 2 \\ 2 & 0 & -1 \end{vmatrix}$

7. $\begin{vmatrix} -1 & -2 & -3 \\ 3 & 4 & 2 \\ 0 & 1 & 2 \end{vmatrix}$

8. $\begin{vmatrix} 1 & 2 & 2 \\ 2 & 1 & 0 \\ 3 & 3 & 1 \end{vmatrix}$

9. $\begin{vmatrix} 3 & 2 & -2 \\ -2 & 1 & 4 \\ -4 & -3 & 3 \end{vmatrix}$

10. $\begin{vmatrix} 2 & -1 & 1 \\ 1 & 2 & -1 \\ 3 & 4 & -3 \end{vmatrix}$

1. _____

2. _____

3. _____

4. _____

5. _____

6. _____

7. _____

8. _____

9. _____

10. _____

●●● Solve, using Cramer's rule.

11. $3x - 4y = 6$
$5x + 9y = 10$

12. $5x + 8y = 1$
$3x + 7y = 5$

13. $-2x + 4y = 3$
$3x - 7y = 1$

14. $5x - 4y = -3$
$7x + 2y = 6$

11. _____

12. _____

13. _____

15. $3x + 2y - z = 4$
$3x - 2y + z = 5$
$4x - 5y - z = -1$

16. $3x - y + 2z = 1$
$x - y + 2z = 3$
$-2x + 3y + z = 1$

17. $2x - 3y + 5z = 27$
$x + 2y - z = -4$
$5x - y + 4z = 27$

14. _____

15. _____

16. _____

18. $x - y + 2z = -3$
$x + 2y + 3z = 4$
$2x + y + z = -3$

19. $r - 2s + 3t = 6$
$2r - s - t = -3$
$r + s + t = 6$

20. $a - 3c = 6$
$b + 2c = 2$
$7a - 3b - 5c = 14$

17. _____

18. _____

19. _____

20. _____

21. Show that an equation of the line through points (x_1, y_1) and (x_2, y_2) can be written

$$\begin{vmatrix} x & y & 1 \\ x_1 & y_1 & 1 \\ x_2 & y_2 & 1 \end{vmatrix} = 0.$$

21. _____

3.7 BUSINESS AND ECONOMIC APPLICATIONS (Optional)

● BREAK-EVEN ANALYSIS

When a company manufactures a product it invests money (this is its *cost*). When it sells the product it takes in money (called *revenue*). Hopefully it makes a profit. Profit is computed as follows:

Profit = Revenue − Cost

or

$$P = R - C.$$

If R is greater than C, the company makes a profit. If C is greater than R, the company has a loss. If $R = C$, the company breaks even.

There are two kinds of costs. First, there are costs like rent, insurance, buying machinery, and so on. These costs, which must be paid whether a product is produced or not, are called *fixed costs*. Then, when a product is being produced there are costs for labor, materials, marketing, and so on. These are called *variable costs*. They vary according to the amount of product being produced. Adding these together, we find the *total cost* of producing a product.

Example 1   Snowy Electronics, Inc., is planning to make a new kind of radio. Fixed costs the first year will be $90,000, and it will cost $15 to produce each radio (variable costs). Each radio will sell for $26. Determine the profit and break-even point.

a) We know that $P = R - C$, and that

C = (Fixed costs) plus (Variable costs)   or
$C = 90,000 + 15x,$

where x is the number of radios produced.

$R = 26x$ ($26 times the number of radios sold. We are assuming that all radios produced are sold.)

So we have

$P = R - C$
$P = 26x - [90,000 + 15x],$ or
$P = 11x - 90,000.$

b) The break-even point occurs when revenue equals total cost. So we set $R = C$ and solve:

$26x = 90,000 + 15x$
$11x = 90,000$
$x = 8181.8.$

The firm will need to produce and sell 8182 radios in order to break even. (8181 radios produce a slight loss, and 8182 a slight profit.) It is interesting to graph the equations in a problem like this.

OBJECTIVES

After finishing Section 3.7, you should be able to:

● Given information on revenue and cost, find the profit and the break-even point.

●● Given supply and demand curves, find the equilibrium point.

1. Suppose, in Example 1, that

$$C = 80,000 + 20x$$

and

$$R = 36x.$$

a) Find the profit.

$C = $ (Fixed costs) plus (Variable costs)

$$C = 90,000 + 15x \qquad (1)$$
$$R = 26x \qquad\qquad (2)$$

C and R are both in dollars

Equation (1) has an intercept on the \$-axis of 90,000 and has a slope of 15.

Equation (2) has a graph that goes through the origin, with a slope of 26.

Profit occurs where the revenue is greater than the cost. Loss occurs where revenue is less than cost. The break-even point occurs where the graphs cross.

DO EXERCISE 1.

b) Find the break-even point.

⚆⚆ SUPPLY AND DEMAND

DEMAND VARIES WITH PRICE

As the price of coffee varied over a period of years, the amount sold varied. The table and graph both show how the amount *demanded* by consumers went down as the price went up.

Demand schedule	
Price, P, per kg	Quantity, Q, in millions of kg
\$1	25
2	20
3	15
4	10
5	5

SUPPLY VARIES WITH PRICE

As the price of coffee varied, the amount available varied. The amount of coffee on the market (the *supply*) went up as the price went up.

Supply schedule	
Price, P, per kg	Quantity, Q, in millions of kg
$2.00	5
2.50	10
3.00	15
3.50	20
4.00	25

THE EQUILIBRIUM POINT

Let us look at the graphs together. As supply increases, demand decreases. As supply decreases, demand increases. The point of intersection is called the *equilibrium point*. At that price, the amount that the seller will supply is the same amount that the consumer will buy.

Any ordered pair of coordinates from this graph is (price, quantity), because the *first* axis is the *price* axis and the *second* axis is the *quantity* axis.

Example 2   Find the equilibrium point when demand D and supply S are given by

$$D = 1000 - 60P \quad \text{and} \quad S = 200 + 4P.$$

a) We first set $D = S$ and solve:

$$1000 - 60P = 200 + 4P$$
$$1000 - 64P = 200 \quad \text{Adding} -4P$$
$$-64P = -800 \quad \text{Adding} -1000$$
$$P = \frac{-800}{-64} = 12.5.$$

Thus the equilibrium price, P_E, is $12.50 per unit.

b) To find the equilibrium quantity we substitute P_E into either D or S. We use S:

$$200 + 4P = 200 + 4(12.5) = 200 + 50 = 250.$$

Thus the equilibrium quantity is 250 units, and the equilibrium point is ($12.50, 250).

DO EXERCISE 2.

2. Demand and supply are given by

$$D = 1000 - 50P$$

and

$$S = 300 + 4P.$$

Find the equilibrium point.

SOMETHING EXTRA

AN APPLICATION: STOPPING DISTANCE

The distance required to stop a car, motorcycle, or other vehicle after the brakes are applied can be estimated with a formula like this:

$$s = At^2 + Bt + C,$$

where s is the stopping distance, t is the time from applying the brakes to the stop, and A, B, and C are some numbers. These numbers depend on the vehicle and must be determined experimentally.

Example   An experiment was done with a motorcycle, and the following data were found.

s	t
38 ft	1 sec
66 ft	2 sec
86 ft	3 sec

Find the stopping distance formula.

We substitute these values of s and t and get three equations:

$$38 = A \cdot 1^2 + B \cdot 1 + C,$$
$$66 = A \cdot 2^2 + B \cdot 2 + C,$$
$$86 = A \cdot 3^2 + B \cdot 3 + C.$$

We solve these equations for A, B, and C and find that $A = -4$, $B = 40$, and $C = 2$. Thus the stopping formula is

$$s = -4t^2 + 40t + 2.$$

EXERCISES

1. An experiment is done with a car, and the following data are found: $s = 49$ ft when $t = 1$ sec, $s = 84$ when $t = 2$, and $s = 109$ when $t = 3$. Find the stopping formula for this car.

2. If it takes the car in Exercise 1 four seconds to stop after the brakes are applied, how far will it travel during that time?

NAME

CLASS

ANSWERS

EXERCISE SET 3.7

● For each of the following, (a) find the profit, and (b) find the break-even point.

1. $C = 45x + 600,000$
   $R = 65x$

2. $C = 25x + 360,000$
   $R = 70x$

3. $C = 10x + 120,000$
   $R = 60x$

4. $C = 30x + 49,500$
   $R = 85x$

5. $C = 20x + 10,000$
   $R = 100x$

6. $C = 40x + 22,500$
   $R = 85x$

●● Find the equilibrium point.

7. $D = 2000 - 60P$
   $S = 460 + 94P$

8. $D = 1000 - 10P$
   $S = 250 + 5P$

9. $D = 760 - 13P$
   $S = 430 + 2P$

1. a) _____

b) _____

2. a) _____

b) _____

3. a) _____

b) _____

4. a) _____

b) _____

5. a) _____

b) _____

6. a) _____

b) _____

7. _____

8. _____

9. _____

ANSWERS

10. $D = 800 - 43P$
$S = 210 + 16P$

11. $D = 7500 - 25P$
$S = 6000 + 5P$

12. $D = 8800 - 30P$
$S = 7000 + 15P$

10. _____

11. _____

12. _____

13. a) _____

b) _____

c) _____

d) _____

e) _____

14. a) _____

b) _____

c) _____

d) _____

e) _____

13. A manufacturer is planning a new type of lamp. For the first year, the fixed costs for setting up the production line are $22,500. Variable costs for producing each lamp are estimated to be $40. The sales department projects that 3000 lamps can be sold during the first year. The revenue from each lamp is to be $85.

a) Find the total cost C of producing x lamps.

b) Find the total revenue R from the sale of x lamps.

c) Find total profit P.

d) What profit or loss will the company realize if expected sales of 3000 lamps occur?

e) Find the break-even point.

14. A clothing firm is planning a new line of sport coats. For the first year, the fixed costs for setting up the production line are $10,000. Variable costs for producing each coat are $20. The sales department projects that 2000 coats can be sold during the first year. The revenue from each coat is to be $100.

a) Find the total cost C of producing x coats.

b) Find the total revenue R from the sale of x coats.

c) Find the total profit P.

d) What profit or loss will the company realize if expected sales of 2000 coats occur?

e) Find the break-even point.

3.8 INCONSISTENT AND DEPENDENT EQUATIONS (Optional)

● INCONSISTENT EQUATIONS

If a system of equations has a solution, we say that it is *consistent*. If a system does not have a solution, we say that it is *inconsistent*.

Example 1   Determine whether this system is consistent or inconsistent:

$$x - 3y = 1$$
$$-2x + 6y = 5.$$

We attempt to find a solution:

$$2x - 6y = 2 \qquad \text{Multiplying the first equation by 2}$$
$$\underline{-2x + 6y = 5}$$
$$0 = 7. \qquad \text{Adding}$$

The last equation says that $0 \cdot x + 0 \cdot y = 7$. There are no numbers x and y for which this is true, so there is no solution. The system is inconsistent.

Whenever we obtain a statement such as $0 = 7$, which is clearly false, we know that the system we are trying to solve is inconsistent.

DO EXERCISE 1.

Example 2   Determine whether this system is consistent or inconsistent:

$$y + 3z = 4 \qquad (1)$$
$$-x - y + 2z = 0 \qquad (2)$$
$$x + 2y + z = 1. \qquad (3)$$

We attempt to find a solution. If we add equations (2) and (3), we get

$$y + 3z = 1.$$

If we multiply this by -1 and then add the result to equation (1) we get

$$0 = 3. \qquad \text{The system is inconsistent.}$$

DO EXERCISE 2.

●● DEPENDENT EQUATIONS

Consider this system:

$$2x + 3y = 1$$
$$4x + 6y = 2.$$

If we multiply the first equation by 2, we get the second equation, so the two equations have the same solutions (are equivalent). The system of two equations is equivalent to only one of them. We call them *dependent*.

OBJECTIVES

After finishing Section 3.8, you should be able to:

● Determine whether a system of two or three linear equations is consistent or inconsistent.

●● Determine whether a system of two or three linear equations is dependent or independent.

1. Determine whether these systems are consistent or inconsistent.

a) $3x - y = 2$
   $6x - 2y = 3$

b) $x + 4y = 2$
   $2x - y = 1$

2. Determine whether these systems are consistent or inconsistent.

a) $x + 2y + z = 1$
   $3x + 3y + z = 2$
   $2x + y = 2$

b) $x + z = 1$
   $y + z = 1$
   $x + y = 1$

ANSWERS ON PAGE A–13

3. Determine whether these systems are dependent or independent.

a) $3x - 2y = 1$
      $-6x + 4y = -2$

b) $x - y = 2$
     $x + y = 4$

4. Determine whether these systems are dependent or independent.

a) $x + y + 2z = 1$
     $x - y + z = 1$
     $2x + 3z = 2$

b) $x + y + 2z = 1$
     $x - y + z = 1$
     $x + 2y + z = 2$

> **If a system of n linear equations is equivalent to a system of fewer than n of them, we say that the system is *dependent*. If such is not the case, we call the system *independent*.**

Example 3   Determine whether this system is dependent or independent:

$$2x + 3y = 1$$
$$4x + 6y = 2.$$

We attempt to solve:

$$-4x - 6y = -2 \qquad \text{Multiplying the first equation by } -2$$
$$\underline{4x + 6y = 2}$$
$$0 = 0. \qquad \text{Adding}$$

The last equation says that $0 \cdot x + 0 \cdot y = 0$. This is true for all numbers x and y. The system is dependent.

Whenever we get an obviously true sentence like this, we know that the system we are trying to solve is dependent.

DO EXERCISE 3.

Example 4   Determine whether this system is dependent or independent:

$$x + 2y + z = 1$$
$$x - y + z = 1$$
$$2x + y + 2z = 2.$$

We attempt to solve:

$$x + 2y + z = 1$$
$$-3y = 0 \qquad \text{Multiplying the first equation by } -1 \text{ and adding it to the second}$$
$$-3y = 0 \qquad \text{Multiplying the first equation by } -2 \text{ and adding it to the third}$$

It is now obvious that our system of three equations is equivalent to a system of two equations. The system is dependent.

When we attempt to solve a system and obtain two equations that are identical, or obviously equivalent, we know that the system is dependent. Or, if we obtain a statement obviously true, such as $3 = 3$, we know that the system is dependent.

DO EXERCISE 4.

NAME

CLASS

ANSWERS

EXERCISE SET 3.8

▪●▪ Determine whether these systems are consistent or inconsistent.

1. $x + 2y = 6$
   $2x = 8 - 4y$

2. $y - 2x = 1$
   $2x - 3 = y$

3. $y - x = 4$
   $x + 2y = 2$

4. $y + x = 5$
   $y = x - 3$

5. $x - 3 = y$
   $2x - 2y = 6$

6. $3y = x - 2$
   $3x = 6 + 9y$

7. $x + z = 0$
   $x + y + 2z = 3$
   $y + z = 2$

8. $x + y = 0$
   $x + z = 1$
   $2x + y + z = 2$

9. $x + z = 0$
   $x + y = 1$
   $y + z = 1$

10. $2x + y = 1$
   $x + 2y + z = 0$
   $x + z = 1$

1. _____

2. _____

3. _____

4. _____

5. _____

6. _____

7. _____

8. _____

9. _____

10. _____

■ ■ Determine whether these systems are dependent or independent.

11. $x - 3 = y$
$2x - 2y = 6$

12. $3y = x - 2$
$3x = 6 + 9y$

13. $y - x = 4$
$x + 2y = 2$

14. $y + x = 5$
$y = x - 3$

15. $2x + 3y = 1$
$x + 1.5y = 0.5$

16. $15x + 6y = 20$
$7.5x - 10 = -3y$

17. $x + z = 0$
$x + y = 1$
$y + z = 1$

18. $2x + y = 1$
$x + 2y + z = 0$
$x + z = 1$

19. $x + y + z = 1$
$-x + 2y + z = 2$
$2x - y = -1$

20. $y + z = 1$
$x + y + z = 1$
$x + 2y + 2z = 2$

EXTENSION EXERCISES

SECTION 3.2

Solve.

1. $5x - 9y = 7$
   $7y - 3x = -5$

2. $a - 2b = 16$
   $b + 3 = 3a$

3. $3(a - b) = 15$
   $4a = b + 1$

4. $1.3x - 0.2y = 12$
   $0.4x + 17y = 89$

5. $x - \dfrac{1}{10}y = 100$

   $y - \dfrac{1}{10}x = -100$

6. $\dfrac{1}{8}x + \dfrac{3}{5}y = \dfrac{19}{2}$

   $-\dfrac{3}{10}x - \dfrac{7}{20}y = -1$

7. $\dfrac{x + y}{2} - \dfrac{x - y}{5} = 1$

   $\dfrac{x - y}{2} + \dfrac{x + y}{6} = -2$

8. Solve for x and y in terms of a and b.

   $5x + 2y = a$
   $x - y = b$

9. Determine a and b if $(-4, -3)$ is the solution of the system.

   $ax + by = -26$
   $bx - ay = 7$

10. The points $(0, -3)$ and $(-\frac{3}{2}, 6)$ are two of the solutions of the equation $px - qy = -1$. Find p and q.

SECTION 3.3

Solve.

11. Mike and Barbara Jensen have been mathematics professors at a state university. In total they have 46 years service. Two years ago Mr. Jensen had taught 2.5 times as many years as Barbara. How long has each taught at the university?

12. The ten's digit of a two-digit positive integer is 2 more than three times the unit's digit. If the digits are interchanged, the new number is 13 less than half the given number. Find the given integer.

13. The measure of one of two supplementary angles is 7° more than three times the other. Find the measure of the larger of the two angles.

14. A limited edition of a new hardback book published by a national historical society was offered for sale to only its membership. Orders were limited to not more than two books per member. The cost was one book for $12 or two books for $20. The society sold 880 books, and the total amount of money taken in was $9840. How many members ordered two books?

15. The numerator of a fraction is 12 more than the denominator. The sum of the numerator and the denominator is 5 more than three times the denominator. What is the reciprocal of the fraction?

16. Use a system of equations to solve the following problem. Two solutions of the equation $y = mx + b$ are $(1, 2)$ and $(-3, 4)$. Find m and b.

SECTION 3.4

Solve.

17. $2a - 3b = 2$

$7a + 4c = \dfrac{3}{4}$

$-3b + 2c = 1$

18. $\dfrac{2}{x} - \dfrac{1}{y} + \dfrac{3}{z} = -1$

$\dfrac{1}{x} - \dfrac{1}{y} - \dfrac{2}{z} = -5$

$\dfrac{4}{x} + \dfrac{3}{y} - \dfrac{1}{z} = 1$

$\left(\textit{Hint: } \text{Let } u = \dfrac{1}{x},\, v = \dfrac{1}{y},\, w = \dfrac{1}{z}. \right)$

19. $2x - 4y + 10z = 104$
$-5x + 3y - 6z = -63$
$-3x - 10y + 9z = 100$

20. $x - 2y = 0$
$3x + z = 1$
$5y - 2w = 4$
$3z + w = 1$

SECTION 3.5

Solve.

21. Sarah's age is the sum of the ages of Janice and Jason. Janice's age is 2 years more than the sum of the ages of Jason and Steve. Jason is four times as old as Steve. The sum of the ages of all four is 42. How old is Sarah?

22. Hal gives Tom as many raffle tickets as Tom already has and Gary as many as Gary already has. In like manner, Tom then gives Hal and Gary as many tickets as each then has. Similarly, Gary gives Hal and Tom as many tickets as each then has. If each finally has 40 tickets, with how many tickets does Tom begin?

23. For a musical production at a university, the costs of tickets for students, adults, and children were $4.00, $5.50, and $1.50, respectively. The total number of children's and adult tickets sold was 30 more than half the number of student tickets sold. The number of adult tickets sold was 5 more than four times the number of children's tickets. How many of each type of ticket was sold if the total receipts from ticket sales was $13,162.50?

24. Find a three-digit positive integer such that the sum of all three digits is 14, the ten's digit is 2 more than the unit's digit, and, if the digits are reversed, the number is unchanged.

SECTION 3.6

Solve.

25. $\begin{vmatrix} y & -2 \\ 4 & 3 \end{vmatrix} = 44$

26. $\begin{vmatrix} 2 & x & -1 \\ -1 & 3 & 2 \\ -2 & 1 & 1 \end{vmatrix} = -12$

27. $\begin{vmatrix} m+1 & -2 \\ m-2 & 1 \end{vmatrix} = 27$

28. Rewrite the expression $3a + 5b$ using a determinant. Answers may vary.

NAME   SCORE   ANSWERS

TEST OR REVIEW—CHAPTER 3

Review the material in the chapter. Then test yourself by doing these exercises. If you miss an item, restudy the section indicated to the left of the item.

[3.1, ● ●]   **1.** Solve graphically.

$$2x + y = 4$$
$$x - 3y = -5$$

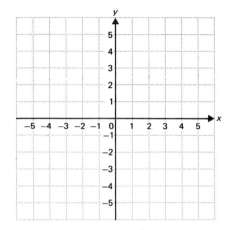

1. _____

[3.2, ● ●]   **2.** Solve.

$$3x + y = -1$$
$$2x - 3y = -8$$

2. _____

[3.3, ●]   **3.** The perimeter of a rectangular field is 710 meters. The length is 115 m greater than the width. Find the dimensions.

3. _____

[3.4, ● ●]   **4.** Solve.

$$2x + y + z = -3$$
$$2x - y + 2z = 0$$
$$3x - 2y + 2z = -3$$

4. _____

ANSWERS

[3.5, ●]   **5.** A student scores a total of 248 points on three tests. The total of the first two scores is 66 more than the score on the third test. The score on the third test is 6 more than that on the second test. Find the scores.

[3.6, ● ●]   **6.** Evaluate.

$$\begin{vmatrix} 1 & 0 & 2 \\ 3 & 1 & 5 \\ 1 & -2 & 2 \end{vmatrix}$$

5. _____

[3.7, ●]   **7.** Find the profit and the break-even point.

$$C = 15x + 90{,}000$$
$$R = 50x$$

6. _____

[3.8, ●]   **8.** Determine whether this system of equations is consistent or inconsistent.

$$2x + 2y = 0$$
$$4x + 4z = 4$$
$$2x + y + z = 2$$

7. _____

[3.8, ● ●]   **9.** Determine whether this system of equations is dependent or independent.

$$3x + 3z = 0$$
$$2x + 2y = 2$$
$$3y + 3z = 3$$

8. _____

9. _____

POLYNOMIALS AND FUNCTIONS

4

Polynomials arise in many kinds of applications. In baseball, for example, a polynomial can be used to calculate the "magic number," giving a team's progress toward winning the pennant. The number of games to be played in a season can also be calculated using a polynomial.

READINESS CHECK: SKILLS FOR CHAPTER 4

1. Find the additive inverse of -8.3.   **2.** Subtract: $\dfrac{3}{5} - \left(-\dfrac{4}{5}\right)$.

3. Factor: $10x + 15y - 5$.   **4.** Multiply: $3(y - 2)$.

5. Collect like terms: $3x + 2x$.   **6.** Rename without parentheses:

$-(-3y + 2)$.

7. Multiply and simplify: **8.** Simplify: $(x^3)^4$.

$(3y^2) \cdot (5y^4)$.

9. Solve: $(x - 3)(x + 5) = 0$.   **10.** Evaluate when $x = -2$, $y = 4$, and $z = 3$: $xy - xz$.

OBJECTIVES

After finishing Section 4.1, you should be able to:

● Rewrite a polynomial with plus signs between the terms and evaluate polynomials.

●● Identify the terms and coefficients of a polynomial.

●●● Determine the degree of each term of a polynomial and the degree of the polynomial.

⊞ Collect like terms.

⊞ Arrange a polynomial in ascending or descending powers of a given variable.

⊞⊞ Identify monomials, binomials, and trinomials.

1. Use the polynomial of Example 1 to determine how much it would cost to operate the automobile at 100 km/h.

4.1 SOME PROPERTIES OF POLYNOMIALS

● POLYNOMIALS

Expressions like these are called *polynomials in one variable:*

$-7x + 5, \qquad 2y^2 + 5y - 3.$

They show additions and subtractions. Each part to be added or subtracted is a number or a number times a variable to some whole-number power. A polynomial can also consist of just one of the parts. The following are polynomials:

$5x^2, \qquad -8a, \qquad 5, \qquad 3t, \qquad 0.$

Expressions like the following are called *polynomials in several variables:*

$5x - xy^2 + 7y + 2, \qquad 15x^3y^2, \qquad \pi r^2 + 2\pi rh.$

Evaluating polynomials is important in many applications.

Example 1   The cost, in cents per kilometer, of operating an automobile at speed s (in km/h) is approximated by the polynomial $0.002s^2 - 0.21s + 15$. How much does it cost to operate at 80 km/h?

We substitute:

$0.002(\boxed{80})^2 - 0.21(\boxed{80}) + 15 = 11.$

It costs 11¢ per kilometer.

DO EXERCISE 1.

Example 2   The *magic number* in baseball can be found by using the polynomial $G - P - L + 1$.

Here G is the number of games in the season, P is the number of games the leading team has played, and L is the number of games ahead in the loss column. Any combination of wins by the leading team and losses by the second-place team totaling the magic number results in the championship.

Find the magic number when there are 162 games in a season and team A has played 145 games and is in the lead by 5 games. The magic number is

$$G - P - L + 1 = 162 - 145 - 5 + 1, \quad \text{or} \quad 13.$$

Example 3   If there are n teams in a league and they play each other twice in a season, the total number of games played is given by the polynomial

$$n^2 - n.$$

If the teams play each other only once, the total number of games is given by the polynomial

$$\frac{1}{2}n^2 - \frac{1}{2}n.$$

Find the number of games played when there are 12 teams in a league and they play each other twice.

The total number of games is

$$n^2 - n = 12^2 - 12 = 144 - 12, \quad \text{or} \quad 132.$$

DO EXERCISES 2 AND 3.

Recall that we can subtract by adding an additive inverse. So any polynomial can be rewritten with only plus signs between the parts.

Examples

4. $4x - 3 = 4x + (-3)$

5. $7xy - 4x^2 = 7xy + (-4x^2)$

6. $-9x^3z - 7xyz^2 + y - 6x^3y^2 = -9x^3z + (-7xyz^2) + y + (-6x^3y^2)$

DO EXERCISES 4 AND 5.

●● TERMS AND COEFFICIENTS

When a polynomial is written with plus signs between the parts, the parts are called its *terms*. The numbers in the terms, other than the exponents, are called *coefficients*.

Example 7   Identify the terms and the coefficients in

$$-3x^2 - 4x^3z + 14x^4y^3 - 9.$$

We rewrite with plus signs:

$$-3x^2 + (-4x^3z) + 14x^4y^3 + (-9).$$

> *Caution!* Do not confuse terms (things to be added) with factors (things to be multiplied).

2. Find the magic number where $G = 162$, $P = 133$, and $L = 6$.

3. There are 15 teams in a league and they play each other once. Find the the total number of games.

Rewrite with plus signs between the parts.

4. $9xy - 5x^3 - y + 4$

5. $8pq - 7pqr^2 - 34xyz^6$

6. Identify the terms.

$$-92x^5 - 8x^4 + y^2 + 5$$

The terms are $-3x^2$, $-4x^3z$, $14x^4y^3$, and -9. The coefficients are -3, -4, 14, and -9.

DO EXERCISES 6 AND 7.

If a coefficient is 0, we usually do not write the term. When there is a zero coefficient in a polynomial with one variable, we say that there is a *missing* term.

Example 8 In $9x^5 - 2x^3 + 5x - 9$ there is no term with x^4. We say that the "x^4-term is missing." The x^2-term is also missing.

7. Identify the coefficient of each term.

$$5x^2y - 4xy^2 - 2x^3 + pq - y - 5$$

••• DEGREES

The *degree of a term* is the sum of the exponents of the variables. The *degree of a polynomial* is the same as that of its term of highest degree.

Example 9 Determine the degree of each term and the degree of the polynomial:

$$6x^2 + 8x^2y^3 - 17xy + 2y + 3.$$

8. Determine the degree of each term and the degree of the polynomial.

$$2y - 4 - 5x + 7x^2y^3z^2 + 5xy^2$$

Term	$6x^2$	$8x^2y^3$	$-17xy$	$2y$	3
Degree	2	5	2	1	0

You might think of $2y$ as $2y^1$ and 3 as $3y^0$.

The degree of the polynomial is 5.

DO EXERCISE 8.

▊▊ COLLECTING LIKE TERMS

Terms that have the same variables with the same exponents are called *like* terms (or *similar* terms). The following are examples of similar terms:

$$4x^2 \text{ and } -2x^2, \qquad 5x^2y^3 \text{ and } 9x^2y^3, \qquad 7y^6z^9 \text{ and } 23y^6z^9.$$

The following are examples of terms that are not similar:

$$x^2y \text{ and } xy^2, \qquad 5xyz^2 \text{ and } 13xyz.$$

We can often simplify expressions by collecting like terms. The process is based on the distributive laws. Thus we know that the simplified expression and the original expression will name the same number for all replacements. That is, the expressions are *equivalent*.

Collect like terms.

9. $3y - 4x + 6xy^2 - 2xy^2$

Examples Collect like terms.

10. $4x - 5x + 2xy + 3xy = (4 - 5)x + (2 + 3)xy$
$$= -1 \cdot x + 5xy, \quad \text{or} \quad -x + 5xy$$

10. $3xy^3 + 2x^3y + 5xy^3 - 3x^2y - 6x^2y$

11. $4x^3y + 5z - 4x^2 - 2x^3y + 5x^2 = 2x^3y + x^2 + 5z$

DO EXERCISES 9 AND 10.

::: ASCENDING AND DESCENDING ORDER

We usually arrange polynomials in one variable so that the exponents decrease (descending order) or so that they increase (ascending order).

Examples

12. Arrange $-2x + x^2 + 5x^7 + 6x^5$ in descending order:

$5x^7 + 6x^5 + x^2 - 2x$.

13. Arrange $9x^{10} - 2 + 4x - 8x^2$ in ascending order:

$-2 + 4x - 8x^2 + 9x^{10}$

We usually write polynomials in one variable in descending order, although it is not incorrect to use ascending or any other order.

DO EXERCISES 11 AND 12.

For polynomials in several variables we choose one of the variables and arrange the terms with respect to it.

Examples

14. Arrange $y^4 + 2 - 5x^2 + 3x^3y + 7xy$ in descending powers of x:

$3x^3y - 5x^2 + 7xy + y^4 + 2$.

15. Arrange $-17x^3y^2 + 4xy^3 + 13xy + 8$ in ascending powers of y:

$8 + 13xy - 17x^3y^2 + 4xy^3$.

DO EXERCISES 13 AND 14.

::: MONOMIALS, BINOMIALS, AND TRINOMIALS

Polynomials with just one term are called *monomials*. Polynomials with just two terms are called *binomials*. Those with just three terms are called *trinomials*.

Example 16

Monomials	*Binomials*	*Trinomials*
$5x^2$	$3x - 5$	$5x^2 - 6x + 1$
-1	$-2y^2 + 8t$	$3pq - 4q^2 - 10$

DO EXERCISES 15–18.

---

11. Arrange $3x^5 + x + 6x^3 - 5x^2 - 6x^7 - 7x^4$ in descending order.

12. Arrange $4y^2 + 9 - 7y^5 - 2y^3 - 5y^4$ in ascending order.

13. Arrange $-8xy^2 + 3xy + 7xy^4 - 2xy^3$ in descending powers of y.

14. Arrange $4x^2yz + 5xy^2 + 5x^3yz^2$ in ascending powers of x.

Tell whether each polynomial is a monomial, binomial, trinomial, or none of these.

15. $9y^4 - 8y^3 - 10y + 8$

16. $-8x^5$

17. $4x^3y - 5x^2y^2 + 8xy^3$

18. $P + Prt$

SOMETHING EXTRA

CALCULATOR CORNER: NUMBER PATTERNS

Do the following calculations. Look for a pattern in the answers. Then figure out what the next answer in the list should be. After that, check your answer by doing the actual calculation.

EXERCISES

1. $1 \cdot 9 + 1$
   $12 \cdot 9 + 2$
   $123 \cdot 9 + 3$
   $1234 \cdot 9 + 4$

2. $111,111 \div 7$
   $222,222 \div 14$
   $333,333 \div 21$
   $444,444 \div 28$

3. $9 \cdot 6$
   $99 \cdot 66$
   $999 \cdot 666$
   $9999 \cdot 6666$

4. $2 \cdot 999,999$
   $3 \cdot 999,999$
   $4 \cdot 999,999$
   $5 \cdot 999,999$

5. $9 \cdot 9 + 7$
   $98 \cdot 9 + 6$
   $987 \cdot 9 + 5$
   $9876 \cdot 9 + 4$

NAME   CLASS

EXERCISE SET 4.1

▪○▪ Rewrite each polynomial with plus signs and evaluate each when $x = -1$ and $y = 2$.

1. $8x^2 - 2x - 5$

2. $9y^3 - 3y^2 + 6$

3. $18xy - 8x^3 - y + 50$

4. $24y^3 - 9y^2 - xy - 21x$

▪○▪○ Identify the terms and coefficients.

5. $5x^3 + 7x^2 - 3x - 9$

6. $8y^3 - 9y^2 + 12y + 11$

7. $-3xyz + 7x^2y^2 - 5xy^2z + 4xyz^2$

8. $-9abc + 19a^2bc - 8ab^2c + 12abc^2$

▪▪▪ Determine the degree of each term and the degree of the polynomial.

9. $x^2 + 3y^5 - x^3y^4 - 7$

10. $y^3 + 2y^6 + x^2y - 9$

11. $x^5 + 3x^2y^4 - 5xy + 4x - 3$

12. $9y^6 + 2x^4y^4 - 8x^3y + 5x^2 - 9$

▪▪ ▪▪ Collect like terms.

13. $6x^2 - 7x^2 + 3x^2$

14. $-2y^2 - 7y^2 + 5y^2$

15. $5x - 4y - 2x + 5y$

16. $4a - 9b - 6a + 3b$

17. $5a + 7 - 4 + 2a - 6a + 3$

18. $9x + 12 - 8 - 7x + 5x + 10$

19. $3a^2b + 4b^2 - 9a^2b - 6b^2$

20. $5x^2y^2 + 4x^3 - 8x^2y^2 - 12x^3$

ANSWERS

1. _____

2. _____

3. _____

4. _____

5. _____

6. _____

7. _____

8. _____

9. _____

10. _____

11. _____

12. _____

13. _____

14. _____

15. _____

16. _____

17. _____

18. _____

19. _____

20. _____

21. $8x^2 - 3xy + 12y^2 + x^2 - y^2 + 5xy + 4y^2$

22. $a^2 - 2ab + b^2 + 9a^2 + 5ab - 4b^2 + a^2$

23. $4x^2y - 3y + 2xy^2 - 5x^2y + 7y + 7xy^2$

24. $3xy^2 + 4xy - 7xy^2 + 7xy + x^2y$

⦂⦂ Arrange in descending order.

25. $x - 3x^2 + 1 + x^3$ **26.** $x^2 - 8x^4 + 9 - x^3$

27. $a - a^3 + 5a^5 - 9 + 6a^2$ **28.** $y^2 - 9y^4 + 18 - y + 7y^5$

Arrange in ascending order.

29. $2y^3 - y + y^4 - 7 + 3y^2$ **30.** $5y^4 - 4y^3 + 8y - 11 - 2y^2$

31. $x^2 - x + 5x^5 - 9x^3 + 18x^4$ **32.** $-6x^3 + 9x - 42 - 3x^2 + 5x^4$

Arrange in descending powers of y.

33. $x^2y^2 + x^3y - xy^3 + 1$ **34.** $x^3y - x^2y^2 + xy^3 + 6$

Arrange in ascending powers of x.

35. $-9x^3y + 3xy^3 + x^2y^2 + 2x^4$ **36.** $5x^2y^2 - 9xy + 8x^3y^2 - 5x^4$

Collect like terms.

37. ▦ $0.00976x^2y^2 - 0.08054x^3y + 0.80149x^2y^2 + 0.00943x^3y$

38. ▦ $8{,}592{,}429xy^2z - 42{,}004x^2y^2 + 5{,}976{,}006x^2y^2 - 4{,}008{,}793xy^2z$

4.2 CALCULATIONS WITH POLYNOMIALS

■· ADDITION

The sum of two polynomials is a polynomial. To add polynomials we can write a plus sign between them and then collect like terms.

Example 1   Add: $-5x^3 + 3x - 5$ and $8x^3 + 4x^2 + 7$.

$(-5x^3 + 3x - 5) + (8x^3 + 4x^2 + 7)$
$= (-5 + 8)x^3 + 4x^2 + 3x + (-5 + 7)$
$= 3x^3 + 4x^2 + 3x + 2$

We still arrange the terms in descending order.

DO EXERCISE 1.

The use of columns is often helpful. To do this we write the polynomials one under the other, writing like terms under one another and leaving spaces for missing terms. Let us do the addition in Example 1 using columns.

$$
\begin{array}{cccc}
-5x^3 & & +3x & -5 \\
8x^3 & +4x^2 & & +7 \\
\hline
3x^3 & +4x^2 & +3x & +2
\end{array}
$$

Example 2   Add: $4ax^2 + 4bx - 5$ and $3ax^2 + 5bx + 8$.

$$
\begin{array}{l}
4ax^2 + 4bx - 5 \\
3ax^2 + 5bx + 8 \\
\hline
7ax^2 + 9bx + 3
\end{array}
$$

Although the use of columns is helpful for complicated examples, you should attempt to write only the answer when you can.

Example 3   Add.

$(13x^3y + 3x^2y - 5y) + (x^3y + 4x^2y - 3xy + 3y)$
$= 14x^3y + 7x^2y - 3xy - 2y$

DO EXERCISES 2 AND 3.

■·■ ADDITIVE INVERSES

If the sum of two polynomials is 0, they are called *additive inverses* of each other. For example, $(3x - 2) + (-3x + 2) = 0$, so the additive inverse of $(3x - 2)$ is $(-3x + 2)$. In symbols,

The additive
inverse of $3x - 2$ is $-3x + 2$
$-(3x - 2) = -3x + 2.$

OBJECTIVES

After finishing Section 4.2, you should be able to:
■· Add polynomials.
■·■ Rename the additive inverse of a polynomial without using parentheses.
■■■ Subtract polynomials.

1. Add.
$3x^3 + 4x^2 - 7x - 2$ and
$-7x^3 - 2x^2 + 3x + 4$

Add.

2. $7y^5 - 5$ and $3y^5 - 4y^2 + 10$

3. $5p^2q^4 - 2p^2q^2 - 3q$ and
$-6p^2q^2 + 3q + 5$

Rewrite each additive inverse without parentheses.

4. $-\left(4x^3 - 5x^2 + \dfrac{1}{4}x - 10\right)$

5. $-\left(8xy^2 - 4x^3y^2 - 9x - \dfrac{1}{5}\right)$

6. $-\left(-9y^5 - 8y^4 + \dfrac{1}{2}y^3 - y^2 + y - 1\right)$

Subtract.

7. $(6x^2 + 4) - (3x^2 - 1)$

8. $(9y^3 - 2y - 4) - (-5y^3 - 8)$

9. $(-3p^2 + 5p - 4) - (-4p^2 + 11p - 2)$

Subtract.

10. $(2y^5 - y^4 + 3y^3 - y^2 - y - 7) -$
$(-y^5 + 2y^4 - 2y^3 + y^2 - y - 4)$

11. $(4p^4q - 5p^3q^2 + p^2q^3 + 2q^4) -$
$(-5p^4q + 5p^3q^2 - 3p^2q^3 - 7q^4)$

12. $\left(\dfrac{3}{2}y^3 - \dfrac{1}{2}y^2 + 0.3\right) -$
$\left(\dfrac{1}{2}y^3 + \dfrac{1}{2}y^2 - \dfrac{4}{3}y + 0.2\right)$

ANSWERS ON PAGE A–14

Examples Rewrite each additive inverse without parentheses.

4. $\begin{aligned} -(5x^3 + 2) &= -1 \cdot (5x^3 + 2) \qquad \text{Replacing } - \text{ by } -1 \\ &= -1 \cdot 5x^3 + (-1) \cdot 2 \qquad \text{Using the distributive law} \\ &= -5x^3 + (-2) \\ &= -5x^3 - 2 \end{aligned}$

5. $\begin{aligned} -(9xy^2 - 4x^3y - 8x^4 + 7) &= (-1)9xy^2 + (-1)(-4x^3y) \\ &\qquad + (-1)(-8x^4) + (-1) \cdot 7 \\ &= -9xy^2 + 4x^3y + 8x^4 - 7 \end{aligned}$

Renaming an additive inverse like this amounts to "changing the sign" of every term inside the parentheses.

Example 6 Rewrite without parentheses.

$-(5xy^2 - 7x^3y - 8x + 4) = -5xy^2 + 7x^3y + 8x - 4$

DO EXERCISES 4–6.

●●● SUBTRACTION

To subtract one polynomial from another, we add the inverse of the subtrahend.

Example 7 Subtract.

$\begin{aligned} &(-9x^5 + 2x^2 + 4) - (2x^5 + 4x^3 - 3x^2) \\ &= (-9x^5 + 2x^2 + 4) + (-2x^5 - 4x^3 + 3x^2) \qquad \begin{array}{l}\text{Adding the inverse} \\ \text{of the subtrahend}\end{array} \\ &= -11x^5 - 4x^3 + 5x^2 + 4 \end{aligned}$

After some practice, you will find that you can skip some steps, by mentally "changing the sign of each term" and then collecting like terms. Eventually, all you will write is the answer.

DO EXERCISES 7–9.

We can also use columns for subtraction. We change signs mentally in the subtrahend and then add.

Example 8 Subtract.

$(4x^2y - 6x^3y^2 + x^2y^2) - (4x^2y + x^3y^2 + 3x^2y^3).$

$\begin{array}{l} 4x^2y - 6x^3y^2 \qquad\quad + x^2y^2 \\ \underline{4x^2y + \quad x^3y^2 + 3x^2y^3} \leftarrow \\ \qquad\quad - 7x^3y^2 - 3x^2y^3 + x^2y^2 \end{array}$

> Don't forget to change the sign of *every* term.

As with addition, you should avoid the use of columns as much as possible. After sufficient experience, you should write only the answer.

DO EXERCISES 10–12.

NAME   CLASS

ANSWERS

EXERCISE SET 4.2

● Add.

1. $3x^2 + 5y^2 + 6$ and
$2x^2 - 3y^2 - 1$

2. $9y^2 + 8y - 4$ and
$12y^2 - 5y + 8$

3. $2a + 3b - c$ and
$4a - 2b + 2c$

4. $5x - 4y + 2z$ and
$9x + 12y - 8z$

5. $a^2 - 3b^2 + 4c^2$ and
$-5a^2 + 2b^2 - c^2$

6. $x^2 - 5y^2 - 9z^2$ and
$-6x^2 + 9y^2 - 2z^2$

7. $x^2 + 2x - 3xy - 7$ and
$-3x^2 - x + 2xy + 6$

8. $3a^2 - 2b + ab + 6$ and
$-a^2 + 5b - 5ab - 2$

9. $7x^2y - 3xy^2 + 4xy$ and
$-2x^2y - xy^2 + xy$

10. $7ab - 3ac + 5bc$ and
$13ab - 15ac - 8bc$

11. $2r^2 + 12r - 11$ and
$6r^2 - 2r + 4$ and
$r^2 - r - 2$

12. $5x^2 + 19x - 23$ and
$-7x^2 - 11x + 12$ and
$-x^2 - 9x + 8$

1. _____

2. _____

3. _____

4. _____

5. _____

6. _____

7. _____

8. _____

9. _____

10. _____

11. _____

12. _____

ANSWERS

13. _____

14. _____

15. _____

16. _____

17. _____

18. _____

19. _____

20. _____

21. _____

22. _____

23. _____

24. _____

25. _____

26. _____

●● ● Rename each additive inverse without parentheses.

13. $-(5x^3 - 7x^2 + 3x - 6)$ **14.** $-(8y^4 - 18y^3 + 4y - 9)$

●●● Subtract.

15. $(8x - 4) - (-5x + 2)$

16. $(9y + 3) - (-4y - 2)$

17. $(-3x^2 + 2x + 9) - (x^2 + 5x - 4)$

18. $(-9y^2 + 4y + 8) - (4y^2 + 2y - 3)$

19. $(5a - 2b + c) - (3a + 2b - 2c)$

20. $(8x - 4y + z) - (4x + 6y - 3z)$

21. $(3x^2 - 2x - x^3) - (5x^2 - 8x - x^3)$

22. $(8y^2 - 3y - 4y^3) - (3y^2 - 9y - 7y^3)$

23. $(5a^2 + 4ab - 3b^2) - (9a^2 - 4ab + 2b^2)$

24. $(9y^2 - 14yz - 8z^2) - (12y^2 - 8yz + 4z^2)$

25. ▦ Add: $15.36x^2 - 4.039y^2 + 1.002x$ and
                  $12.58x^2 + 13.144y^2 + 7.148x$.

26. ▦ Subtract: $(8035x - 1735) - (-4111x + 3892)$.

4.3   MULTIPLYING POLYNOMIALS

⬛• MULTIPLYING MONOMIALS

To multiply monomials we multiply the coefficients. Then we multiply the variables. This process is based on the commutative and associative laws of multiplication.

Examples   Multiply and simplify.

1. $(-8x^4y^7)(5x^3y^2) = -8 \cdot 5 \cdot x^4 \cdot x^3 \cdot y^7 \cdot y^2$
$$= -40x^{4+3}y^{7+2} \qquad \text{Adding exponents}$$
$$= -40x^7y^9$$

2. $(3x^2yz^5)(-6x^5y^{10}z^2) = 3 \cdot (-6) \cdot x^2 \cdot x^5 \cdot y \cdot y^{10} \cdot z^5 \cdot z^2$
$$= -18x^7y^{11}z^7$$

You should try to work mentally, writing only the answer.

DO EXERCISES 1–3.

⬛•• MULTIPLYING MONOMIALS AND BINOMIALS

The distributive laws are the basis for multiplying polynomials other than monomials.

Example 3   Multiply: $2x$ and $3x - 5$.

$$\boxed{2x} \cdot (3x - 5) = \boxed{2x} \cdot (3x) - \boxed{2x} \cdot (5) \qquad \text{Using a distributive law}$$
$$= 6x^2 - 10x \qquad \text{Multiplying the monomials}$$

DO EXERCISES 4 AND 5.

Example 4   Multiply: $3y^2 + 4$ and $y - 2$.

$$(3y^2 + 4)\,\underset{(a)}{\boxed{(y-2)}} = 3y^2\,\underset{(a)}{\boxed{(y-2)}} + 4\,\underset{(b)}{\boxed{(y-2)}} \qquad \begin{array}{l}\text{Using a}\\\text{distributive law}\end{array}$$

We consider the two parts (a) and (b) separately.

a) $\boxed{3y^2}\,(y-2) = \boxed{3y^2}\,y - \boxed{3y^2}\,(2) \qquad \text{Using a distributive law}$
$$= 3y^3 - 6y^2 \qquad \text{Multiplying the monomials}$$

b) $\boxed{4}\,(y-2) = \boxed{4}\,y - \boxed{4}\,(2) \qquad \text{Using a distributive law}$
$$= 4y - 8 \qquad \text{Multiplying the monomials}$$

We replace the parts (a) and (b) in the original expression with their answers, and combine like terms if we can.

$$(3y^2 + 4)(y - 2) = (3y^3 - 6y^2) + (4y - 8)$$
$$= 3y^3 - 6y^2 + 4y - 8$$

DO EXERCISES 6 AND 7.

OBJECTIVES

After finishing Section 4.3, you should be able to:

⬛• Multiply monomials.

⬛•• Multiply a binomial by a monomial.

⬛••• Multiply any two polynomials.

Multiply and simplify.

1. $9y^2$ and $-2y$

2. $4x^3y$ and $6x^5y^2$

3. $-5xy^7z^4$ and $18x^3y^2z^8$

Multiply.

4. $-3y$ and $2y + 6$

5. $2xy$ and $4y^2 - 5$

6. Multiply.
$5x^2 - 4$ and $x + 3$

7. Multiply.
$2y + 3$ and $3y - 4$

Multiply.

8. $p - 3$ and $p^3 + 4p^2 - 5$

9. $2x^3 + 4x - 5$ and $x - 4$

10. Multiply.

$-4x^3 - 2x + 1$ and
$-2x^2 - 3x + 6$

●●● MULTIPLYING ANY TWO POLYNOMIALS

Example 5   Multiply: $p + 2$ and $p^4 - 2p^3 + 3$.

By the distributive laws we have:

$$(p + 2)(p^4 - 2p^3 + 3) = (p + 2)(p^4) - (p + 2)(2p^3)$$

$$+ (p + 2)(3)$$

$$= p \cdot (p^4) + 2(p^4) - p \cdot (2p^3)$$

$$- 2(2p^3) + p \cdot (3) + 2(3)$$

$$= p^5 + 2p^4 - 2p^4 - 4p^3 + 3p + 6$$

$$= p^5 - 4p^3 + 3p + 6$$

In the last step we combined like terms.

DO EXERCISES 8 AND 9.

Now we can see how to multiply any two polynomials.

> **To multiply two polynomials, multiply each term of one by every term of the other. Then add the results, combining like terms if possible.**

We can use columns for long multiplications. We multiply each term at the top by every term at the bottom, keeping like terms in columns. Then we add.

Example 6   Multiply: $5x^3 + x - 4$ and $-2x^2 + 3x + 6$.

$$
\begin{array}{llll}
5x^3 + & x - 4 & & \\
-2x^2 + 3x + 6 & & & \\
\hline
-10x^5 \qquad\qquad - & 2x^3 + & 8x^2 & \text{Multiplying by } -2x^2 \\
15x^4 \qquad\qquad + & 3x^2 - 12x & & \text{Multiplying by } 3x \\
30x^3 \qquad\qquad + & 6x - 24 & & \text{Multiplying by } 6 \\
\hline
-10x^5 + 15x^4 + 28x^3 + 11x^2 - & 6x - 24 & & \text{Adding}
\end{array}
$$

DO EXERCISE 10.

NAME
CLASS

EXERCISE SET 4.3

■● Multiply.

1. $2y^2$ and $5y$

2. $-3x^2$ and $2xy$

3. $5x$ and $-4x^2y$

4. $-3ab^2$ and $2a^2b^2$

5. $2x^3y^2$ and $-5x^2y^4$

6. $7a^2bc^4$ and $-8ab^3c^2$

●● Multiply.

7. $2x$ and $3 - x$

8. $4a$ and $a^2 - 5a$

9. $3ab$ and $a + b$

10. $2xy$ and $2x - 3y$

11. $5cd$ and $3c^2d - 5cd^2$

12. a^2 and $2a^2 - 5a^3$

●●● Multiply.

13. $2x + 3$ and $3x - 4$

14. $2a - 3b$ and $4a - b$

15. $s + 3t$ and $s - 3t$

16. $y + 4$ and $y - 4$

17. $x - y$ and $x - y$

18. $a + 2b$ and $a + 2b$

ANSWERS

1. _____

2. _____

3. _____

4. _____

5. _____

6. _____

7. _____

8. _____

9. _____

10. _____

11. _____

12. _____

13. _____

14. _____

15. _____

16. _____

17. _____

18. _____

ANSWERS

19. $y + 8x$ and $2y - 7x$ **20.** $x + y$ and $x - 2y$

19. _____

21. $a^2 - 2b^2$ and $a^2 - 3b^2$ **22.** $2m^2 - n^2$ and $3m^2 - 5n^2$

20. _____

23. $x - 4$ and $x^2 + 4x + 16$ **24.** $y + 3$ and $y^2 - 3y + 9$

21. _____

25. $x + y$ and $x^2 - xy + y^2$ **26.** $a - b$ and $a^2 + ab + b^2$

22. _____

23. _____

27. $a^2 + a - 1$ and   **28.** $x^2 - 2x + 1$ and
     $a^2 + 4a - 5$ $x^2 + x + 2$

24. _____

25. _____

29. $4a^2b - 2ab + 3b^2$ and   **30.** $2x^2 + y^2 - 2xy$ and
     $ab - 2b + a$ $x^2 - 2y^2 - xy$

26. _____

27. _____

28. _____

31. Express the area of this box   **32.** A box is to be made from a
as a polynomial. The box is   piece of cardboard 12 inches
rectangular with an open   square. Corners are cut out
top, and dimensions as   and the sides are folded up.
shown.   Express the volume of the
                                          box as a polynomial.

29. _____

30. _____

31. _____

32. _____

4.4 SPECIAL PRODUCTS OF POLYNOMIALS

We can multiply polynomials faster by learning some patterns.

◖•◗ PRODUCTS OF TWO BINOMIALS

To multiply any two polynomials we multiply each term of one by every term of the other. For two binomials we can think of it this way. Multiply the first terms, then the outside terms, then the inside terms, then the last terms. We abbreviate this FOIL

$$\text{F} \quad \text{O} \quad \text{I} \quad \text{L}$$
$$(A + B)(C + D) = AC + AD + BC + BD$$

Examples Multiply.

$$\text{F} \quad \text{O} \quad \text{I} \quad \text{L}$$

1. $(x + 5)(x - 8) = x^2 - 8x + 5x - 40$
$$= x^2 - 3x - 40 \quad \text{Combining like terms}$$

2. $(p - 3q)(2p - 5q) = 2p^2 - 5pq - 6pq + 15q^2$
$$= 2p^2 - 11pq + 15q^2 \quad \text{Combining like terms}$$

3. $(3xy + 2x)(x^2 + 2xy^2) = 3x^3y + 6x^2y^3 + 2x^3 + 4x^2y^2$

DO EXERCISES 1-3.

> Do this mentally if you can.

◖••◗ SQUARES OF BINOMIALS

Note the following:

$$(A + B)^2 = (A + B)(A + B)$$
$$= A^2 + AB + AB + B^2$$
$$= A^2 + 2AB + B^2$$

$$(A + B)^2 = A^2 + 2AB + B^2$$
$$(A - B)^2 = A^2 - 2AB + B^2$$

The square of a binomial is the square of the first expression, plus or minus twice the product of the expressions, plus the square of the second expression.

> Memorize this rule. Then, when you are squaring a binomial, say the words to yourself. That will help keep you from making errors.

OBJECTIVES

After finishing Section 4.4, you should be able to:

◖•◗ Multiply two binomials mentally.

◖••◗ Square a binomial mentally.

◖•••◗ Multiply, mentally, the product of the sum and the difference of the same two expressions, $(A + B)(A - B)$.

Multiply.

1. $(y - 4)(y + 10)$

2. $(p + 5q)(2p - 3q)$

3. $(x^2y + 2x)(xy^2 + y^2)$

Multiply.

4. $(a - b)^2$

Examples   Multiply.

$$(A - B)^2 = A^2 - 2AB + B^2$$

4. $(y - 5)^2 = y^2 - 2(5)(y) + 5^2$

> Note that these signs are the same.

$$= y^2 - 10y + 25$$

5. $(2x + 9)^2 = (2x)^2 + 2(2x)(9) + 9^2$
$$= 4x^2 + 36x + 81$$

Try to find such products mentally, and try to write only the answer.

5. $(x + 8)^2$

DO EXERCISES 4–6.

Examples   Multiply.

6. $(2x + 3y)^2 = (2x)^2 + 2(2x)(3y) + (3y)^2$
$$= 4x^2 + 12xy + 9y^2$$

> Remember to say the rule to yourself while multiplying.

7. $(3x^2 - 5xy^2)^2 = (3x^2)^2 - 2(3x^2)(5xy^2) + (5xy^2)^2$
$$= 9x^4 - 30x^3y^2 + 25x^2y^4$$

6. $(3x - 7)^2$

> A common error results from thinking that $(A + B)^2 = A^2 + B^2$, and thus leaving out the middle term. If you say the rule to yourself as you work, you will not make that mistake.

DO EXERCISES 7 AND 8.

Multiply.

7. $(4x - 5y)^2$

●●● PRODUCTS OF SUMS AND DIFFERENCES

Note the following:

$$\begin{array}{cccc} F & O & I & L \end{array}$$
$$(A + B)(A - B) = A^2 - AB + AB - B^2$$
$$= A^2 - B^2.$$

$$\boxed{(A + B)(A - B) = A^2 - B^2}$$

The product of the sum and difference of two expressions is the square of the first expression minus the square of the second.

8. $(2y^2 + 6x^2y)^2$

> Memorize this rule. Say it to yourself as you work.

Examples   Multiply.

$$(A + B)(A - B) = A^2 - B^2$$

8. $(y + 5)(y - 5) = y^2 - 5^2$
$$= y^2 - 25$$

9. $(3x - 2)(3x + 2) = (3x)^2 - 2^2$

$\qquad\qquad\qquad\quad - 9x^2 - 4$

> If you say the rule as you work, you will not need to write a middle step. You can simply write the answer.

DO EXERCISES 9 AND 10.

Examples   Multiply.

10. $(2xy^2 + 3x)(2xy^2 - 3x) = (2xy^2)^2 - (3x)^2$

$\qquad\qquad\qquad\qquad\qquad = 4x^2y^4 - 9x^2$

$$(\quad A \quad - B)(\quad A \quad + B) = \quad A^2 \quad - \quad B^2$$
$$\downarrow \qquad \downarrow \quad\quad \downarrow \qquad \downarrow \qquad\quad \downarrow \qquad\quad \downarrow$$

11. $(\boxed{5y + 4} - 3x)(\boxed{5y + 4} + 3x) = (\boxed{5y + 4})^2 - (3x)^2$

$\qquad\qquad\qquad\qquad\qquad\quad = 25y^2 + 40y + 16 - 9x^2$

> This multiplying could have been done by columns, but it is much faster using the present rule.

Example 12   Multiply: $(3xy^2 + 4y)(-3xy^2 + 4y)$.

$(3xy^2 + 4y)(-3xy^2 + 4y) = (4y + 3xy^2)(4y - 3xy^2)$

$\qquad\qquad\qquad\qquad\qquad = (4y)^2 - (3xy^2)^2$

$\qquad\qquad\qquad\qquad\qquad = 16y^2 - 9x^2y^4$

DO EXERCISES 11–13.

You should learn to multiply polynomials mentally, even when several kinds are mixed. First check to see what kinds of polynomials are to be multiplied. Then use the quickest method. In Exercise Set 4.4 the types are separated. In Exercise Set 4.4A they are mixed.

Multiply.

9. $(x + 8)(x - 8)$

10. $(4y - 7)(4y + 7)$

Multiply.

11. $(7x^2y + 2y)(7x^2y - 2y)$

12. $(\boxed{2x + 3} - 5y)(\boxed{2x + 3} + 5y)$

  a) Use column multiplication.

  b) Use the rule you have just learned, as in Example 11.

13. $(-2x^3y^2 + 5t)(2x^3y^2 + 5t)$

ANSWERS ON PAGE A–15

SOMETHING EXTRA
CALCULATOR CORNER

To evaluate polynomials with a calculator, it helps to first write what is called *nested form*, in which the variable occurs only to the first power in each occurrence. To find nested form, we factor out the variable successively.

Example   Find nested form and then evaluate for $x = 5.3$.

$$2x^4 + 4x^3 - x^2 + 7x - 9$$
$$x(2x^3 + 4x^2 - x + 7) - 9 \qquad \text{Factoring out an } x$$
$$x(x(2x^2 + 4x - 1) + 7) - 9 \qquad \text{Factoring out an } x \text{ again}$$
$$x(x(x(2x + 4) - 1) + 7) - 9 \qquad \text{Factoring out another } x$$

To evaluate when $x = 5.3$, start inside and work outward.

First enter 5.3, then calculate $2 \times 5.3 + 4$ (the innermost parentheses).

Next multiply the result by 5.3 and subtract 1.

Next multiply the result by 5.3 and add 7.

Finally, multiply the result by 5.3 and subtract 9.

The answer is 2173.6142.

EXERCISES

Find nested form. Then evaluate when (a) $x = 10.2$ and (b) $x = -1.4$. Round to four decimal places.

1. $x^5 - 3x^4 + x^3 + x^2 - x - 9$

2. $4x^4 - 12x^3 + 3x^2 - x + 8$

3. $5x^4 - 4x^3 + 7x^2 - 3x + 12$

4. $-4x^5 + 6x^3 - 3x^2 + 15$

NAME

CLASS

ANSWERS

EXERCISE SET 4.4

■ Multiply.

1. $(a + 2)(a + 3)$

2. $(x + 5)(x + 8)$

3. $(y + 3)(y - 2)$

4. $(y - 4)(y + 7)$

5. $\left(b - \dfrac{1}{3}\right)\left(b - \dfrac{1}{2}\right)$

6. $\left(x - \dfrac{1}{2}\right)\left(x - \dfrac{1}{4}\right)$

7. $(2x + 9)(x + 2)$

8. $(3b + 2)(2b - 5)$

9. $(2x - 3y)(2x + y)$

10. $(2a - 3b)(2a - b)$

1. _____

2. _____

3. _____

4. _____

5. _____

6. _____

7. _____

8. _____

9. _____

10. _____

ANSWERS

11. _____

12. _____

13. _____

14. _____

15. _____

16. _____

17. _____

18. _____

19. _____

20. _____

21. _____

22. _____

23. _____

24. _____

25. _____

26. _____

27. _____

28. _____

●● Multiply.

11. $(x + 3)^2$ **12.** $(y - 7)^2$ **13.** $\left(2a + \dfrac{1}{3}\right)^2$ **14.** $\left(3c - \dfrac{1}{2}\right)^2$

15. $(x - 2y)^2$ **16.** $(2s + 3t)^2$ **17.** $(2x^2 - 3y^2)^2$

18. $(3s^2 + 4t^2)^2$ **19.** $(a^2b^2 + 1)^2$ **20.** $(x^2y - xy^2)^2$

●●● Multiply.

21. $(c + 2)(c - 2)$ **22.** $(x - 3)(x + 3)$

23. $(2a + 1)(2a - 1)$ **24.** $(3 - 2x)(3 + 2x)$

25. $(3m - 2n)(3m + 2n)$ **26.** $(3x + 5y)(3x - 5y)$

27. $(x^2 + yz)(x^2 - yz)$ **28.** $(2a^2 + 5ab)(2a^2 - 5ab)$

NAME

CLASS

EXERCISE SET 4.4A

Multiply. (First determine the quickest way to multiply; then proceed.)

1. $(4x - y)(2x - y)$

2. $\left(4r - \dfrac{1}{3}\right)^2$

3. $(2c + 7d)^2$

4. $(3x^2 + 2y)(3x^2 - 2y)$

5. $(-2d + 3c)(2d + 3c)$

6. $(5c - 2d)(2c - 3d)$

7. $(x^2 + 2y)(2x^2 - y)$

8. $(a - 2b)(a - 4b)$

9. $(2m^2 - n^2)(m^2 + 3n^2)$

10. $(5y^2 + 7w^3)^2$

11. $(3x - 4y)^2$

12. $(a^3 - 3b^2)(a^3 + 3b^2)$

13. $(a^2 + b)(a^2 - b)$

14. $(3y - 2)(2y + 5)$

1. _____

2. _____

3. _____

4. _____

5. _____

6. _____

7. _____

8. _____

9. _____

10. _____

11. _____

12. _____

13. _____

14. _____

ANSWERS

15. _____

16. _____

17. _____

18. _____

19. _____

20. _____

21. _____

22. _____

23. _____

24. _____

25. _____

26. _____

27. _____

28. _____

29. _____

30. _____

15. $\left(\dfrac{1}{5} - x\right)\left(\dfrac{1}{5} + x\right)$

16. $(3a^2 + 2)^2$

17. $(2x^2 + 3y^2)(4x^2 - 5y^2)$

18. $(x^2 - 9)(x^2 + 9)$

19. $\left(\dfrac{1}{2}x - 6\right)^2$

20. $(0.2x - 0.4y)(0.2x + 0.4y)$

21. $(x + 1)(x - 1)(x^2 + 1)$
    Hint: First multiply $x + 1$
    and $x - 1$.

22. $(y - 2)(y + 2)(y^2 + 4)$
    Hint: First multiply $y - 2$
    and $y + 2$.

23. $(a + b)(a - b)(a^2 + b^2)$

24. $(2x - y)(2x + y)(4x^2 + y^2)$

25. $(a + b + 1)(a + b - 1)$

26. $(m + n + 2)(m + n - 2)$

27. $(2x + 3y + 4)(2x + 3y - 4)$

28. $(3a - 2b + c)(3a - 2b - c)$

The amount to which \$1000 will grow in 3 years, when interest is compounded annually, is given by the polynomial

$$1000r^3 + 3000r^2 + 3000r + 1000.$$

Here, r is the rate of interest.

29. Find the amount to which \$1000 will grow in 3 years at 12%.

30. Find the amount to which \$1000 will grow in 3 years at 15%.

4.5 FACTORING

The reverse of multiplication is factoring. To *factor* an expression means to write an equivalent expression that is a product.

▪ ▪ TERMS WITH COMMON FACTORS

Example 1　Factor out a common factor:　$4y^2 - 8$.

$$4y^2 - 8 = \boxed{4} \cdot y^2 - \boxed{4} \cdot 2 = \boxed{4} \cdot (y^2 - 2)$$

If there is more than one common factor, we usually choose the one with the largest coefficient and the greatest exponent. Try to write the answer directly.

Example 2　Factor:　$5x^4 - 20x^3$.

$$5x^4 - 20x^3 = \boxed{5x^3} \cdot x - \boxed{5x^3} \cdot 4 = \boxed{5x^3}\,(x - 4)$$

Example 3　Factor:　$15y^5 + 12y^4 - 27y^3 - 3y^2$.

$$15y^5 + 12y^4 - 27y^3 - 3y^2 = 3y^2(5y^3 + 4y^2 - 9y - 1)$$

> Always factor out the largest factor common to the terms.

DO EXERCISES 1–4.

▪ ▪ FACTORING BY GROUPING

Sometimes there is a common factor that is a binomial.

Example 4　Factor:　$(a - b)(x + 5) + (a - b)(x - y^2)$.

$$(\,\boxed{a - b}\,)(x + 5) + (\,\boxed{a - b}\,)(x - y^2) = (\,\boxed{a - b}\,)[(x + 5) + (x - y^2)]$$
$$= (a - b)(2x + 5 - y^2)$$

Example 5　Factor:　$y^2 + 3y + 4y + 12$.

$$y^2 + 3y + 4y + 12 = y(\,\boxed{y + 3}\,) + 4(\,\boxed{y + 3}\,) = (y + 4)(\,\boxed{y + 3}\,)$$

In Example 5 we factored two parts of the expression. Then we factored as in Example 4. We use trial and error to see which terms are to be grouped.

Example 6　Factor:　$4x^2 - 15 + 20x - 3x$.

$$4x^2 - 15 + 20x - 3x = 4x^2 + 20x - 3x - 15$$
$$= 4x(\,\boxed{x + 5}\,) - 3(\,\boxed{x + 5}\,)$$
$$= (4x - 3)(\,\boxed{x + 5}\,)$$

Example 7　Factor:　$ax^2 + ay + bx^2 + by$.

$$ax^2 + ay + bx^2 + by = a(\,\boxed{x^2 + y}\,) + b(\,\boxed{x^2 + y}\,)$$
$$= (a + b)(\,\boxed{x^2 + y}\,)$$

DO EXERCISES 5–8.

ANSWERS ON PAGE A–15

OBJECTIVES

After finishing Section 4.5, you should be able to:

▪　Factor polynomials whose terms have a common factor.

▪▪　Factor certain polynomials with four terms by grouping.

▪▪▪　Factor differences of two squares.

Factor out a common factor.

1.　$3x^2 - 6$

2.　$4x^5 - 8x^3$

3.　$9y^4 - 15y^3 + 3y^2$

4.　$6x^2y - 21x^3y^2 + 3x^2y^3$

Factor.

5.　$(p + q)(x + 2) + (p + q)(x + y)$

6.　$(y + 3)(y - 21) + (y + 3)(y + 10)$

7.　$x^2 + 5x - 4x - 20$

8.　$5y^2 + 2y - 10y - 4$

Factor.

9. $y^2 - 4$

10. $49x^4 - 25y^{10}$

Factor.

11. $25x^2y^2 - 4a^2$

12. $9x^2 - 16y^6$

13. $20x^2 - 5y^2$

14. $81x^4y^2 - 16y^2$

⦿⦿⦿ DIFFERENCES OF SQUARES

To factor a difference of squares, we use the result established in the last section, in reverse.

$$A^2 - B^2 = (A + B)(A - B)$$

To factor a difference of two squares, write the square root of the first expression *plus* the square root of the second, times the square root of the first *minus* the square root of the second.

Example 8 Factor: $x^2 - 9$.
$$x^2 - 9 = x^2 - 3^2 = (x + 3)(x - 3)$$

> You should memorize this rule, and then say it to yourself as you work.

Example 9 Factor: $25y^6 - 49x^2$.

$$A^2 \;-\; B^2 \;=\; (\;A\;+\;B\;)(\;A\;-\;B\;)$$
$$\downarrow \qquad \downarrow \qquad \downarrow \quad \downarrow \quad \downarrow \qquad \downarrow$$
$$25y^6 - 49x^2 = (5y^3 + 7x)(5y^3 - 7x)$$

DO EXERCISES 9 AND 10.

Example 10 Factor: $5 - 5x^2y^6$.

> *Always* look *first* for a factor common to the terms.

$$5 - 5x^2y^6 = 5(1 - x^2y^6)$$
$$= 5(1 + xy^3)(1 - xy^3) \quad \text{Factoring the difference of squares}$$

Example 11 Factor: $2x^4 - 8y^4$.

$$2x^4 - 8y^4 = 2(x^4 - 4y^4) \quad \text{Removing the common factor}$$
$$= 2(x^2 + 2y^2)(x^2 - 2y^2) \quad \text{Factoring the difference of squares}$$

Example 12 Factor: $16x^4y - 81y$.

$$16x^4y - 81y = y(16x^4 - 81) \quad \text{Removing the common factor}$$
$$= y(4x^2 + 9)(4x^2 - 9) \quad \text{Factoring the difference of squares}$$
$$= y(4x^2 + 9)(2x + 3)(2x - 3) \quad \text{Factoring } 4x^2 - 9, \text{ which is a difference of squares}$$

> *Always* continue to factor as long as you can. That way you will be factoring *completely*.

> *Caution!* A common error is trying to factor a *sum* of squares. $4x^2 + 9$ is a *sum* of squares and cannot be factored.

DO EXERCISES 11–14.

NAME

CLASS

EXERCISE SET 4.5

●

1. $4a^2 + 2a$

2. $6y^2 + 3y$

3. $3y^2 - 3y - 9$

4. $5x^2 - 5x + 15$

5. $6x^2 - 3x^4$

6. $8y^2 + 4y^4$

7. $4ab - 6ac + 12ad$

8. $8xy + 10xz - 14xw$

9. $10a^4 + 15a^2 - 25a - 30$

10. $12t^5 - 20t^4 + 8t^2 - 16$

●●

11. $a(b - 2) + c(b - 2)$

12. $a(x^2 - 3) - 2(x^2 - 3)$

13. $(x - 2)(x + 5) + (x - 2)(x + 8)$

14. $(m - 4)(m + 3) + (m - 4)(m - 3)$

15. $a^2(x - y) + a^2(x - y)$

16. $3x^2(x - 6) + 3x^2(x - 6)$

1. _____

2. _____

3. _____

4. _____

5. _____

6. _____

7. _____

8. _____

9. _____

10. _____

11. _____

12. _____

13. _____

14. _____

15. _____

16. _____

ANSWERS

17. _____

18. _____

19. _____

20. _____

21. _____

22. _____

23. _____

24. _____

25. _____

26. _____

27. _____

28. _____

29. _____

30. _____

31. _____

32. _____

33. _____

34. _____

35. _____

36. _____

17. $ac + ad + bc + bd$ **18.** $xy + xz + wy + wz$

19. $b^3 - b^2 + 2b - 2$ **20.** $y^3 - y^2 + 3y - 3$

21. $y^2 - 8y - y + 8$ **22.** $t^2 + 6t - 2t - 12$

●●● Factor. Remember to look first for a common factor.

23. $x^2 - 16$ **24.** $y^2 - 9$ **25.** $6x^2 - 6y^2$

26. $8x^2 - 8y^2$ **27.** $4xy^4 - 4xz^4$ **28.** $25ab^4 - 25az^4$

29. $9a^4 - 25a^2b^4$ **30.** $16x^6 - 121x^2y^4$ **31.** $\dfrac{1}{25} - x^2$

32. $\dfrac{1}{16} - y^2$ **33.** $0.04x^2 - 0.09y^2$ **34.** $0.01x^2 - 0.04y^2$

The number of diagonals of a polygon having n sides is given by the polynomial $\frac{1}{2}n^2 - \frac{3}{2}n$.

35. Find the number of diagonals of a polygon with 5 sides.

36. Find the number of diagonals of a polygon with 7 sides.

4.6 FACTORING TRINOMIALS

◼◦ TRINOMIAL SQUARES

Some trinomials are squares of binomials; for example,

$$x^2 + 6x + 9 = (x + 3)^2.$$

Trinomials like this can be factored. We must first learn to recognize trinomial squares.

> a) Two of the terms must be squares, such as A^2 and B^2.
> b) There must be no minus sign before A^2 or B^2.
> c) If we multiply A and B (the square roots of these expressions) and double the result, we get the remaining term, $2AB$, or its additive inverse, $-2AB$.

Examples　Determine whether these are trinomial squares.

1. $x^2 + 10x + 25$

　　a) Two terms are squares: x^2 and 25.
　　b) There is no minus sign before either x^2 or 25.
　　c) If we multiply the square roots, x and 5, and double, we get $10x$, the remaining term.

Thus the trinomial is a square.

2. $3x^2 + 4x + 16$

　　a) Only one term, 16, is a square ($3x^2$ is not a square). Therefore the trinomial is not a square.

DO EXERCISE 1.

To factor trinomial squares, we use these equations.

$$A^2 + 2AB + B^2 = (A + B)^2$$
$$A^2 - 2AB + B^2 = (A - B)^2$$

Example 3　Factor: $x^2 - 10x + 25$.

$x^2 - 10x + 25 = (x - 5)^2$　　We find the square terms and write their square roots with a minus sign between them.

Example 4　Factor: $16y^2 + 49 + 56y$.

$16y^2 + 49 + 56y = (4y + 7)^2$　　We find the square terms and write their square roots with a plus sign between them.

Example 5　Factor: $-20xy + 4y^2 + 25x^2$.

$-20xy + 4y^2 + 25x^2 = (2y - 5x)^2$

DO EXERCISES 2–5.

OBJECTIVES

After finishing Section 4.6, you should be able to:

◼◦ Factor trinomial squares.

◼◼◦ Factor trinomials of the type $x^2 + ax + b$.

◼◼◼ Factor trinomials of the type $ax^2 + bx + c$.

1. Which of the following are trinomial squares?

　a) $x^2 + 6x + 9$

　b) $x^2 - 8x + 16$

　c) $x^2 + 6x + 11$

　d) $4x^2 - 20x + 25$

　e) $16x^2 - 20x + 25$

　f) $5x^2 + 14x + 16$

　g) $x^2 + 8x - 16$

　h) $x^2 - 8x - 16$

Factor.

2. $x^2 + 14x + 49$

3. $9y^2 - 30y + 25$

4. $16x^2 + 72xy + 81y^2$

5. $16x^4 - 40x^2y^3 + 25y^6$

6. Factor.

$-8a^2 + 24ab - 18b^2$

Factor. Check by multiplying.

7. $x^2 + 5x - 14$

8. $y^2 - y - 2$

Example 6   Factor: $2x^2 - 12x + 18$.

$$2x^2 - 12x + 18 = 2(x^2 - 6x + 9) \qquad \text{Removing the common factor}$$

$$= 2(x - 3)^2 \qquad \text{Factoring the trinomial square}$$

> Always remember to look first for a common factor.

Example 7   Factor: $-4y^2 - 144y^8 + 48y^5$.

$$-4y^2 - 144y^8 + 48y^5 = -4y^2(1 + 36y^6 - 12y^3) \qquad \text{Removing the common factor}$$

$$= -4y^2(1 - 12y^3 + 36y^6) \qquad \text{Changing order}$$

$$= -4y^2(1 - 6y^3)^2 \qquad \text{Factoring the trinomial square}$$

DO EXERCISE 6.

⊙⊙ FACTORING TRINOMIALS OF THE TYPE $x^2 + ax + b$

Consider this product:

$$\begin{array}{cccc} \text{F} & \text{O} & \text{I} & \text{L} \\ (x + 3)(x + 5) = x^2 + 5x + 3x + 15 \end{array}$$

$$= x^2 + 8x + 15.$$

Note that the coefficient 8 is the sum of 3 and 5, and the 15 is the product of 3 and 5. In general, $(x + a)(x + b) = x^2 + (a + b)x + ab$.

To factor we can use the preceding equation in reverse, as follows:

$$x^2 + (a + b)x + ab = (x + a)(x + b).$$

Finding a and b is a trial-and-error process. Practice helps.

Example 8   Factor: $x^2 - 3x - 10$.

We look for pairs of integers whose product is -10 and whose sum is -3.

Pairs of factors	Sum of factors
$-2, \quad 5$	3
$2, -5$	-3
$10, -1$	9
$-10, \quad 1$	-9

Thus the desired integers are 2 and -5. Then

$$x^2 - 3x - 10 = (x + 2)(x - 5).$$

We can check by multiplying.

DO EXERCISES 7 AND 8.

●●● TRINOMIALS OF THE TYPE $ax^2 + bx + c$

Consider a multiplication.

$$
\begin{array}{cccccc}
 & \text{F} & \text{O} & \text{I} & \text{L} \\
(2x + 3)(5x + 4) = & 10x^2 + & 8x & + 15x & + 12 \\
 & 10x^2 + & & 23x & + 12 \\
 & \uparrow & & \uparrow & \uparrow \\
 & \text{F} & \text{O} & + \text{I} & \text{L} \\
 & 2 \cdot 5 & 2 \cdot 4 & + 3 \cdot 5 & 3 \cdot 4
\end{array}
$$

To factor $ax^2 + bx + c$ we look for two binomials

$$(\underline{\quad} x + \underline{\quad})(\underline{\quad} x + \underline{\quad})$$

where products of numbers in the blanks are as follows.

1. The numbers in the *first* blanks have product a.

2. The *outside* product and the *inside* product add up to b.

3. The numbers in the *last* blanks have product c.

Example 9   Factor: $5x^2 - 9x - 2$.

We first look for a common factor. There is none other than 1.

We look for numbers whose product is 5. These are 1, 5 and $-1, -5$.

We have these possibilities:

$$(x + \quad)(5x + \quad) \quad \text{or} \quad (-x + \quad)(-5x + \quad).$$

Now we look for numbers whose product is -2. These are 1, -2 and $-1, 2$.

We have these as some of the possibilities for factorization.	We multiply each. We must get $5x^2 - 9x - 2$.
a) $(x + 1)(5x - 2)$	**a)** $5x^2 + 3x - 2$
b) $(x - 2)(5x + 1)$	**b)** $5x^2 - 9x - 2$
c) $(-x + 1)(-5x - 2)$	**c)** $5x^2 - 3x - 2$
d) $(-x + 2)(-5x - 1)$	**d)** $5x^2 - 9x - 2$

We see that (b) and (d) are both factorizations. We prefer to have the first coefficients positive when that is possible. Thus the factorization we prefer is $(x - 2)(5x + 1)$.

DO EXERCISES 9 AND 10.

Example 10   Factor: $12x^2 + 34x + 14$.

We first look for a common factor. The number 2 is a common factor, so we factor it out: $2(6x^2 + 17x + 7)$. Now we consider $6x^2 + 17x + 7$. We look for numbers whose product is 6. These are 6, 1 and 2, 3. (From Example 9 we found that we needed to consider only positive factors of the first term.) We then have these possibilities:

$$(6x + \quad)(x + \quad) \quad \text{and} \quad (2x + \quad)(3x + \quad).$$

Factor.

9. $3x^2 + 5x + 2$

10. $4x^2 - 3 + 4x$

Factor.

11. $24y^2 - 46y + 10$

12. $16x^2 - 12 + 16x$

Factor.

13. $p^2q^2 + 7pq + 12$

14. $2x^4y^6 - 3x^2y^3 - 20$

Factor.

15. $3x^2 - 13x - 56$

16. $4x^2 + 37x + 9$

Next we look for pairs of numbers whose product is 7. They are

$$7, 1 \quad \text{and} \quad -7, -1. \qquad \text{Both positive or both negative}$$

By multiplying, we find that the answer is $2(2x + 1)(3x + 7)$.

DO EXERCISES 11 AND 12.

Example 11   Factor: $x^2y^2 + 5xy + 4$.

In this case, we can treat xy as if it were a single variable.

$$\begin{aligned} x^2y^2 + 5xy + 4 &= (xy)^2 + 5xy + 4 \\ &= (xy + 4)(xy + 1) \end{aligned}$$

DO EXERCISES 13 AND 14.

Optional

Another way to factor $ax^2 + bx + c$ is as follows:

a) First look for a common factor.

b) Multiply the first and last coefficients, a and c.

c) Try to factor the product ac so that the sum of the factors is b.

d) Write the middle term, bx, as a sum.

e) Factor by grouping.

Example 12   Factor: $2x^2 - 3x - 35$.

a) First look for a common factor. There is none (other than 1).

b) Multiply the first and last coefficients, 2 and -35:

$$2(-35) = -70.$$

c) Try to factor -70 so that the sum of the factors is -3.

Some pairs of factors	Sums of factors
$-2,\ \ 35$	33
$2, -35$	-33
$-14,\ \ \ 5$	-9
$7, -10$	-3

The desired factors are 7 and -10.

d) Write $-3x$ as a sum using the results of (c):

$$-3x = -10x + 7x.$$

e) Factor by grouping:

$$\begin{aligned} 2x^2 - 3x - 35 &= 2x^2 - 10x + 7x - 35 \\ &= (2x^2 - 10x) + (7x - 35) \\ &= 2x(x - 5) + 7(x - 5) \\ &= (2x + 7)(x - 5). \end{aligned}$$

DO EXERCISES 15 AND 16.

NAME   CLASS

EXERCISE SET 4.6

Factor. Remember to look first for a common factor.

■●■

1. $y^2 - 6y + 9$ **2.** $x^2 - 8x + 16$

3. $x^2 + 14x + 49$ **4.** $x^2 + 16x + 64$

5. $-18y^2 + y^3 + 81y$ **6.** $24a^2 + a^3 + 144a$

7. $12a^2 + 36a + 27$ **8.** $20y^2 + 100y + 125$

9. $2x^2 - 40x + 200$ **10.** $32x^2 + 48x + 18$

11. $0.25x^2 + 0.30x + 0.09$ **12.** $0.04x^2 - 0.28x + 0.49$

●●

13. $x^2 + 9x + 20$ **14.** $x^2 + 8x + 15$ **15.** $2y^2 - 16y + 32$

16. $2a^2 - 20a + 50$ **17.** $x^2 - 27 - 6x$ **18.** $t^2 - 15 - 2t$

ANSWERS

1. _____

2. _____

3. _____

4. _____

5. _____

6. _____

7. _____

8. _____

9. _____

10. _____

11. _____

12. _____

13. _____

14. _____

15. _____

16. _____

17. _____

18. _____

ANSWERS

19. _____

20. _____

21. _____

22. _____

23. _____

24. _____

25. _____

26. _____

27. _____

28. _____

29. _____

30. _____

31. _____

32. _____

33. _____

34. _____

35. _____

36. _____

37. _____

19. $32y + 4y^2 - y^3$ **20.** $56x + x^2 - x^3$ **21.** $15 + t^2 + 8t$

22. $27 + y^2 + 12y$ **23.** $x^4 + 11x^2 - 80$ **24.** $y^4 + 5y^2 - 84$

■■■

25. $3x^2 - 16x - 12$ **26.** $6x^2 - 5x - 25$ **27.** $6x^3 - 15x - x^2$

28. $10y^3 - 12y - 7y^2$ **29.** $3a^2 - 10a + 8$ **30.** $12a^2 - 7a + 1$

31. $35y^2 + 34y + 8$ **32.** $9a^2 + 18a + 8$ **33.** $4t + 10t^2 - 6$

34. $8x + 30x^2 - 6$ **35.** $8x^2 - 16 - 28x$ **36.** $18x^2 - 24 - 6x$

---

37. The area of a ring, as shown, is $\pi R^2 - \pi r^2$. Factoring this polynomial, we get $\pi(R + r)(R - r)$. Find the area of a ring, where $R = 10$ cm and $r = 4$ cm. Use 3.14 for π.

4.7 COMPLETING THE SQUARE

● FACTORING DIFFERENCES OF SQUARES

A difference of squares can have more than two terms. For example, one of the squares may be a trinomial. We can factor by a kind of grouping.

Example 1 Factor: $x^2 + 6x + 9 - y^2$.

$$x^2 + 6x + 9 - y^2 = (x^2 + 6x + 9) - y^2$$

Grouping as a trinomial minus y^2 to show a difference of squares

$$= (\boxed{x + 3})^2 - y^2$$
$$= (\boxed{x + 3} + y)(\boxed{x + 3} - y)$$

DO EXERCISES 1–4.

●● COMPLETING THE SQUARE

A trinomial such as $x^2 + 10x + 25$ may be the square of a binomial. This one is because $x^2 + 10x + 25 = (x + 5)^2$. Given the first two terms of a trinomial, we can find the third term that will make it a square as follows.

Example 2 What must be added to $x^2 + 12x$ to make it a trinomial square?

We take half the coefficient of x and square it.

$$x^2 + 12x$$

→ Half of 12 is 6, and $6^2 = 36$. We add 36.

$x^2 + 12x + 36$ is a trinomial square. It is equivalent to $(x + 6)^2$.

The process illustrated in Example 2 is called *completing the square*. Completing the square can be used in factoring and is used in a number of other ways.

Example 3 Complete the square on $x^2 - 8ax$.

We take half of $-8a$ (the coefficient of x) and get $-4a$.

We square $-4a$: $(-4a)^2 = 16a^2$.

We add the result to obtain $x^2 - 8ax + 16a^2$.

This is a trinomial square because it is equivalent to $(x - 4a)^2$.

Example 4 Complete the square on $y^2 + \frac{3}{4}y$.

We take half of $\frac{3}{4}$ (the coefficient of y) and get $\frac{1}{2} \cdot \frac{3}{4} = \frac{3}{8}$.

We square $\frac{3}{8}$: $\left(\frac{3}{8}\right)^2 = \frac{9}{64}$.

We add the result to obtain $y^2 + \frac{3}{4}y + \frac{9}{64}$.

DO EXERCISES 5–8.

OBJECTIVES

After finishing Section 4.7, you should be able to:

● Factor differences of squares having more than two terms.

●● Complete the square.

●●● Factor trinomials by completing the square.

Factor.

1. $x^2 + 2x + 1 - p^2$

2. $y^2 - 8y + 16 - 9m^2$

3. $x^2 + 8x + 16 - 100$

4. $64 - (x^2 + 8x + 16)$

Complete the square.

5. $x^2 + 14x$

6. $y^2 - 10by$

7. $x^2 - \frac{2}{5}x$

8. $x^2 + 4.2x$

Factor by completing the square.

9. $x^2 + 26x + 120$

10. $2x^2 - 48x + 126$

11. $x^2 + 8.4x + 13.64$

●●● FACTORING BY COMPLETING THE SQUARE

Completing the square can be used to factor trinomials.

Example 5   Factor $x^2 - 22x + 112$ by completing the square.

We first complete the square on $x^2 - 22x$. We take half of -22, which is -11.

We square -11: $(-11)^2 = 121$.

Now we add 0 to the original trinomial, naming it $121 - 121$.

$$x^2 - 22x + 112 = x^2 - 22x + 112 + (121 - 121) \quad \text{Adding } 121 - 121$$
$$= (x^2 - 22x + 121) + 112 - 121 \quad \text{Grouping}$$
$$= (x^2 - 22x + 121) - 9$$
$$= (x - 11)^2 - 3^2$$
$$= (x - 11 + 3)(x - 11 - 3), \text{ or } (x - 8)(x - 14)$$

Example 6   Factor $2x^2 + 80x + 768$ by completing the square.

$$2x^2 + 80x + 768 = 2(x^2 + 40x + 384) \quad \text{Removing the common factor 2}$$

To complete the square we take half of 40 to obtain 20.

We square 20 to get 400.

We add 0, naming it $400 - 400$.

$$2(x^2 + 40x + 384) = 2(x^2 + 40x + 400 - 400 + 384) \quad \text{Adding } 400 - 400$$
$$= 2(x^2 + 40x + 400 - 16) \quad \text{Grouping}$$
$$= 2[(x + 20)^2 - 4^2]$$
$$= 2(x + 20 + 4)(x + 20 - 4), \text{ or}$$
$$2(x + 24)(x + 16)$$

Example 7   Factor $x^2 - 6.2x + 8.61$ by completing the square.

To complete the square we take half of -6.2, obtaining -3.1.

We square -3.1: $(-3.1)^2 = 9.61$.

We then add 0, naming it $9.61 - 9.61$.

$$x^2 - 6.2x + 8.61 = x^2 - 6.2x + 8.61 + 9.61 - 9.61 \quad \text{Adding } 9.61 - 9.61$$
$$= (x^2 - 6.2x + 9.61) + 8.61 - 9.61 \quad \text{Grouping}$$
$$= (x - 3.1)^2 - 1$$
$$= (x - 3.1 + 1)(x - 3.1 - 1), \text{ or}$$
$$(x - 2.1)(x - 4.1)$$

DO EXERCISES 9–11.

ANSWERS ON PAGE A–16

NAME   CLASS

EXERCISE SET 4.7

Factor.

1. $a^2 + 2ab + b^2 - 9$ **2.** $x^2 - 2xy + y^2 - 25$

1. _____

2. _____

3. $r^2 - 2r + 1 - 4s^2$ **4.** $c^2 + 4cd + 4d^2 - 9p^2$

3. _____

4. _____

5. $2m^2 + 4mn + 2n^2 - 50b^2$ **6.** $12x^2 + 12x + 3 - 3y^2$

5. _____

6. _____

7. $9 - (a^2 + 2ab + b^2)$ **8.** $16 - (x^2 - 2xy + y^2)$

7. _____

8. _____

Complete the square.

9. $x^2 + 16x$ **10.** $y^2 - 24y$

9. _____

10. _____

11. $x^2 - 4.2x$ **12.** $y^2 + 3.6y$

11. _____

12. _____

13. $x^2 + \dfrac{2}{3}bx$ **14.** $y^2 - \dfrac{3}{4}ay$

13. _____

14. _____

ANSWERS

15. _____

16. _____

17. _____

18. _____

19. _____

20. _____

21. _____

22. _____

23. _____

24. _____

25. _____

26. _____

27. _____

28. _____

●●● Factor by completing the square.

15. $x^2 + 24x + 119$

16. $x^2 + 30x + 176$

17. $x^2 - 26x + 105$

18. $x^2 - 32x + 192$

19. $2x^2 + 56x + 342$

20. $3x^2 + 54x + 231$

21. $2x^2 - 32ax + 96a^2$

22. $3x^2 - 60bx + 192b^2$

23. $x^2 + 2.6x + 0.69$

24. $x^2 - 3.2x + 1.56$

25. $x^2 + \dfrac{3}{5}x + \dfrac{8}{100}$

26. $x^2 - \dfrac{7}{2}x + \dfrac{45}{16}$

_____

Factor by completing the square.

27. ▦ $x^2 + 4.482x - 7.403544$

28. ▦ $5.72x^2 + 35.464x - 1319.2608$

4.8　SUMS OR DIFFERENCES OF TWO CUBES

To succeed with this section, you will need to review and memorize the cubes of integers from -10 to 10.

DO EXERCISE 1.

 Next note the following:

$$(a + b)(a^2 - ab + b^2) = a(a^2 - ab + b^2) + b(a^2 - ab + b^2)$$
$$= u^3 - a^2b + ab^2 + a^2b - ab^2 + b^3$$
$$= a^3 + b^3$$

and

$$(a - b)(a^2 + ab + b^2) = a(a^2 + ab + b^2) - b(a^2 + ab + b^2)$$
$$= a^3 + a^2b + ab^2 - a^2b - ab^2 - b^3$$
$$= a^3 - b^3.$$

DO EXERCISES 2 AND 3.

The above equations (reversed) show how we can factor a sum or a difference of two cubes.

$$A^3 + B^3 = (A + B)(A^2 - AB + B^2)$$
$$A^3 - B^3 = (A - B)(A^2 + AB + B^2)$$

You will need to memorize these patterns and remember to use them as you factor.

Example 1　Factor:　$x^3 - 27$.

$$x^3 - 27 = x^3 - 3^3$$

In one set of parentheses we write the cube root of the first term, x. Then we write the cube root of the second term, -3. This gives us the expression $x - 3$.

$$(x - 3)(\quad\quad)$$

To get the next factor we think of $x - 3$ and do the following.

　　1. Square the first term: x^2.
　　2. Multiply the terms and then change the sign: $3x$.
　　3. Square the second term: 9.

$$(x - 3)(x^2 + 3x + 9)$$

Note:　We cannot factor $x^2 + 3x + 9$. (It is not a trinomial square.)

DO EXERCISES 4 AND 5.

OBJECTIVE

After finishing Section 4.8, you should be able to:

■ Factor sums and differences of two cubes.

1. Copy this table. Compute to fill it in. Then memorize it.

n	n^3
1	1
-1	-1
2	8
-2	-8
3	
-3	
4	
-4	
5	
-5	
6	
-6	
7	
-7	
8	
-8	
9	
-9	
10	1000
-10	-1000

Multiply.

2. $(x - 2)(x^2 + 2x + 4)$

3. $(a + 5)(a^2 - 5a + 25)$

Factor.

4. $x^3 - 8$

5. $64 - y^3$

Factor.

6. $27x^3 + y^3$

7. $8y^3 + z^3$

Factor.

8. $16x^7y + 54xy^7$

9. $729x^6 - 64y^6$

10. $x^3 - 0.027$

Example 2   Factor: $125x^3 + y^3$.

$$125x^3 + y^3 = (5x)^3 + y^3$$

In one set of parentheses we write the cube root of the first term, then a plus sign, and then the cube root of the second term.

$$(5x + y)(\qquad)$$

To get the next factor, we think of $5x + y$ and do the following.

1. Square the first term: $(5x)^2$, or $25x^2$.
2. Multiply the terms and then change the sign: $-5xy$.
3. Square the second term: y^2.

$$(5x + y)(25x^2 - 5xy + y^2)$$

DO EXERCISES 6 AND 7.

Example 3   Factor: $128y^7 - 250x^6y$.

We first look for a common factor.

$$2y(64y^6 - 125x^6) = 2y[(4y^2)^3 - (5x^2)^3]$$
$$= 2y(4y^2 - 5x^2)(16y^4 + 20x^2y^2 + 25x^4)$$

Example 4   Factor: $64a^6 - 729b^6$.

$$(8a^3 - 27b^3)(8a^3 + 27b^3) \qquad \text{Factoring a difference of squares}$$

Each factor is a sum or difference of cubes. We factor.

$$(2a - 3b)(4a^2 + 6ab + 9b^2)(2a + 3b)(4a^2 - 6ab + 9b^2)$$

Example 5   Factor: $y^3 + 0.008$.

We note that $0.008 = (0.2)^3$. Thus we have a sum of two cubes, and we factor as follows.

$$(y + 0.2)(y^2 - 0.2y + 0.04)$$

> Remember the following about factoring sums or differences of squares and cubes:
>
> | | |
> |---|---|
> | Sum of cubes: | $A^3 + B^3 = (A + B)(A^2 - AB + B^2)$ |
> | Difference of cubes: | $A^3 - B^3 = (A - B)(A^2 + AB + B^2)$ |
> | Difference of squares: | $A^2 - B^2 = (A + B)(A - B)$ |
> | Sum of squares: | $A^2 + B^2$ cannot be factored |

DO EXERCISES 8–10.

NAME   CLASS

EXERCISE SET 4.8

■ Factor.

1. $x^3 + 8$ **2.** $c^3 + 27$ **3.** $2y^3 - 128$

4. $3z^3 - 3$ **5.** $24a^3 + 3$ **6.** $54x^3 + 2$

7. $8 - 27b^3$ **8.** $64 - 125x^3$ **9.** $8x^3 + 27$

ANSWERS

1. _____

2. _____

3. _____

4. _____

5. _____

6. _____

7. _____

8. _____

9. _____

ANSWERS

10. _____

11. _____

12. _____

13. _____

14. _____

15. _____

16. _____

17. _____

18. _____

19. _____

20. _____

21. _____

22. _____

10. $27y^3 + 64$ **11.** $a^3 + \dfrac{1}{8}$ **12.** $b^3 + \dfrac{1}{27}$

$$\left[Hint: \dfrac{1}{8} = \left(\dfrac{1}{2}\right)^3.\right]$$

13. $rs^3 + 64r$ **14.** $ab^3 + 125a$ **15.** $5x^3 - 40z^3$

16. $2y^3 - 54z^3$ **17.** $x^3 + 0.001$ **18.** $y^3 + 0.125$

19. $64x^6 - 8t^6$ **20.** $125c^6 - 8d^6$

---

21. A 4-ft by 4-ft sandbox is placed on a square lawn x ft on a side. Express the area left over as a polynomial.

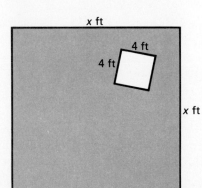

22. Express the colored area as a polynomial.

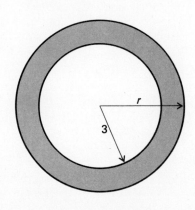

4.9 FACTORING: A GENERAL STRATEGY

▪● Here is a general strategy for factoring.

> **A. Always look first for a common factor.**
> **B. Then look at the number of terms.**
>
> > **Two terms:** Try factoring as a difference of squares first. Next, try factoring as a sum or difference of cubes. Do *not* try to factor a *sum* of squares.
> >
> > **Three terms:** Determine whether the trinomial is a square. If so, you know how to factor. If not, try trial and error.
> >
> > **Four or more terms:** Try factoring by grouping and removing a common binomial factor. Next, try grouping into a difference of squares, one of which is a trinomial.
>
> **C. Always** *factor completely.* **If a factor with more than one term can be factored, you should factor it.**

Example 1   Factor: $10a^2x - 40b^2x$.

A. We look first for a common factor:

$10x(a^2 - 4b^2)$.   Factoring out the largest common factor

B. The factor $a^2 - 4b^2$ has only two terms. It is a difference of squares. We factor it:

$10x(a + 2b)(a - 2b)$.   ⎫ It is a common error to forget to write the common factor in this step.

C. Have we factored completely? Yes, because no factor with more than one term can be factored further.

Example 2   Factor: $x^6 - y^6$.

A. We look for a common factor. There isn't one (other than 1 or -1).

B. There are only two terms. It is a difference of squares: $(x^3)^2 \quad (y^3)^2$. We factor it:

$(x^3 + y^3)(x^3 - y^3)$.

One factor is a sum of two cubes, and the other factor is a difference of two cubes. We factor them:

$(x + y)(x^2 - xy + y^2)(x - y)(x^2 + xy + y^2)$.

C. We have factored completely because no factors can be factored further.

In Example 2, had we thought of factoring first as a difference of two cubes, we would have had

$(x^2)^3 - (y^2)^3 = (x^2 - y^2)(x^4 + x^2y^2 + y^4)$

$= (x + y)(x - y)(x^4 + x^2y^2 + y^4).$

In this case, we might have missed some factors; $x^4 + x^2y^2 + y^4$ can be factored, but we might not have noticed.

DO EXERCISES 1–3.

After finishing Section 4.9, you should be able to:

▪● Factor polynomials using any of the methods considered in this chapter.

Factor completely.

1. $3y^3 - 12x^2y$

2. $7a^3 - 7$

3. $64x^6 - 729y^6$

Factor.

4. $3x - 6 - bx^2 + 2bx$

5. $5y^4 + 20x^6$

6. $6x^2 - 3x - 18$

7. $a^3 - ab^2 - a^2b + b^3$

8. $3x^2 + 18ax + 27a^2$

9. $2x^2 - 20x + 50 - 18b^2$

Example 3   Factor: $10x^6 + 40y^2$.

A. We remove the largest common factor: $10(x^6 + 4y^2)$.

B. In the parentheses there are two terms, a sum of squares, which cannot be factored.

Example 4   Factor: $2x^2 + 50a^2 - 20ax$.

A. We remove the largest common factor: $2(x^2 + 25a^2 - 10ax)$.

B. In the parentheses there are three terms. The trinomial is a square. We factor it: $2(x - 5a)^2$

C. No factor with more than one term can be factored further.

Example 5   Factor: $6x^2 - 20x - 16$.

A. We remove the largest common factor: $2(3x^2 - 10x - 8)$.

B. In the parentheses there are three terms. The trinomial is not a square. We factor by trial: $2(x - 4)(3x + 2)$.

C. We cannot factor further.

Example 6   Factor: $3x + 12 + ax^2 + 4ax$.

A. There is no common factor (other than 1 or -1).

B. There are four terms. We try grouping to remove a common binomial factor.

$\quad 3(x + 4) + ax(x + 4)$ Factoring two grouped binomials
$\quad (x + 4)(3 + ax)$ Removing the common binomial factor

C. No factor with more than one term can be factored further.

Example 7   Factor: $y^2 - 9a^2 + 12y + 36$.

A. There is no common factor (other than 1 or -1).

B. There are four terms. We try grouping to remove a common binomial factor, but that is not possible.
We try grouping as a difference of squares.

$\quad (y^2 + 12y + 36) - 9a^2$ Grouping
$\quad (y + 6 + 3a)(y + 6 - 3a)$ Factoring the difference of squares

C. No factor with more than one term can be factored further.

Example 8   Factor: $x^3 - xy^2 + x^2y - y^3$.

A. There is no common factor (other than 1 or -1).

B. There are four terms. We try grouping to remove a common binomial factor.

$\quad x(x^2 - y^2) + y(x^2 - y^2)$ Factoring two grouped binomials
$\quad (x^2 - y^2)(x + y)$ Removing the common binomial factor

C. A factor can be factored further.

$\quad (x + y)(x - y)(x + y)$ Factoring a difference of squares

No factor with more than one term can be factored further, so we have factored completely.

DO EXERCISES 4–9.

NAME   CLASS

EXERCISE SET 4.9

● Factor completely.

1. $x^2 - 144$ **2.** $y^2 - 81$ **3.** $2x^2 + 11x + 12$

4. $8a^2 + 18a - 5$ **5.** $3x^4 - 12$ **6.** $2xy^2 - 50x$

7. $a^2 + 25 + 10a$ **8.** $p^2 + 64 + 16p$ **9.** $2x^2 - 10x - 132$

10. $3y^2 - 15y - 252$ **11.** $9x^2 - 25y^2$ **12.** $16a^2 - 81b^2$

13. $4c^2 - 4cd + d^2$ **14.** $70b^2 - 3ab - a^2$

15. $-7x^2 + 2x^3 + 4x - 14$ **16.** $9m^2 + 3m^3 + 8m + 24$

17. $250x^3 - 128y^3$ **18.** $27a^3 - 343b^3$

1. _____

2. _____

3. _____

4. _____

5. _____

6. _____

7. _____

8. _____

9. _____

10. _____

11. _____

12. _____

13. _____

14. _____

15. _____

16. _____

17. _____

18. _____

ANSWERS

19. _____

20. _____

21. _____

22. _____

23. _____

24. _____

25. _____

26. _____

27. _____

28. _____

29. _____

30. _____

31. _____

32. _____

33. _____

34. _____

35. _____

36. _____

19. $8m^3 + m^6 - 20$ **20.** $-37x^2 + x^4 + 36$

21. $ac + cd - ab - bd$ **22.** $xw - yw + xz - yz$

23. $m^6 - 1$ **24.** $64t^6 - 1$

25. $x^2 + 6x - y^2 + 9$ **26.** $t^2 + 10t - p^2 + 25$

27. $36y^2 - 35 + 12y$ **28.** $2b - 28a^2b + 10ab$

29. $a^8 - b^8$ **30.** $2x^4 - 32$

31. $a^3b - 16ab^3$ **32.** $x^3y - 25xy^3$

33. $2x^3 + 6x^2 - 8x - 24$ **34.** $3x^3 + 6x^2 - 27x - 54$

35. $16x^3 + 54y^3$ **36.** $250a^3 + 54b^3$

4.10 SOLVING EQUATIONS BY FACTORING

■ In Chapter 2 we solved equations using the *principle of zero products*. To use this principle in solving equations we make sure that there is a 0 on one side; then we factor the other side and set each factor equal to 0. (See p. 55.)

Example 1 Solve: $x^2 - 3x - 28 = 0$.

We first factor the polynomial. Then we use the principle of zero products.

$$x^2 - 3x - 28 = 0$$
$$(x - 7)(x + 4) = 0 \quad \text{Factoring}$$

The expression $x^2 - 3x - 28$ and $(x - 7)(x + 4)$ name the same number for any replacement. Hence the equations have the same solutions.

$$x - 7 = 0 \quad \text{or} \quad x + 4 = 0 \quad \text{Using the principle of zero products}$$
$$x = 7 \quad \text{or} \quad x = -4$$

Check:

For 7:
$$\begin{array}{c|c} x^2 - 3x - 28 = 0 \\ \hline 7^2 - 3(7) - 28 & 0 \\ 49 - 21 - 28 & \\ 0 & \end{array}$$

For -4:
$$\begin{array}{c|c} x^2 - 3x - 28 = 0 \\ \hline (-4)^2 - 3(-4) - 28 & 0 \\ 16 + 12 - 28 & \\ & 0 \end{array}$$

The solutions are 7 and -4.

When we use the principle of zero products, a check is not necessary except to detect errors.

Example 2 Solve: $7y + 3y^2 = -2$.
$$3y^2 + 7y + 2 = 0 \quad \text{Adding 2 to get 0 on one side and arranging in descending order}$$
$$(3y + 1)(y + 2) = 0 \quad \text{Factoring}$$
$$3y + 1 = 0 \quad \text{or} \quad y + 2 = 0 \quad \text{Using the principle of zero products}$$
$$3y = -1 \quad \text{or} \quad y = -2$$
$$y = -\tfrac{1}{3} \quad \text{or} \quad y = -2$$

The solutions are $-\tfrac{1}{3}$ and -2.

Important: There *must* be a 0 on one side of the equation. It is a common error to forget this.

DO EXERCISES 1 AND 2.

Example 3 Solve: $5b^2 - 10b = 0$.
$$5b(b - 2) = 0 \quad \text{Factoring}$$
$$5b = 0 \quad \text{or} \quad b - 2 = 0 \quad \text{Using the principle of zero products}$$
$$b = 0 \quad \text{or} \quad b = 2$$

The solutions are 0 and 2.

DO EXERCISE 3.

OBJECTIVES

After finishing Section 4.10, you should be able to:

■ Solve equations by factoring and using the principle of zero products.

■■ Solve applied problems by translating to equations and solving the equations.

Solve.

1. $x^2 + 8 = 6x$

2. $5y + 2y^2 = 3$

Solve.

3. $8b^2 - 16b = 0$

4. The square of a number minus twice the number is 48. Find the number.

5. The width of a rectangle is 5 cm less than the length. The area is 24 cm². Find the dimensions.

●● SOLVING PROBLEMS

To solve problems, we first translate the problem situation to an equation and then solve. We check to see whether the solution(s) of the equation(s) satisfies the conditions of the problem.

Example 4 The square of a number minus the number is 20. Find the number.

$$\underbrace{\text{The square of a number}}_{x^2} \;\underbrace{\text{minus}}_{-}\; \underbrace{\text{the number}}_{x} \;\underbrace{\text{is}}_{=}\; \underbrace{20}_{20} \qquad \text{Translating}$$

We solve the equation.

$$x^2 - x = 20$$
$$x^2 - x - 20 = 0 \qquad \text{Adding} -20 \text{ to get } 0 \text{ on one side}$$
$$(x - 5)(x + 4) = 0 \qquad \text{Factoring}$$
$$x - 5 = 0 \quad \text{or} \quad x + 4 = 0 \qquad \text{Using the principle of zero products}$$
$$x = 5 \quad \text{or} \quad x = -4$$

The numbers 5 and −4 both check. They are the answers to the problem.

DO EXERCISE 4.

Example 5 The width of a rectangle is 2 m less than the length. The area is 15 m². Find the dimensions.

We first make a drawing and label the length and width.

w, **or**
$l - 2$

l

Area is length times width, so we have the translation

$$l \cdot (l - 2) = 15.$$

> A common error here is to write $l = 0$ or $l - 2 = 0$, using the principle of zero products. That is incorrect because there is not a 0 on the other side of the equation.

We solve the equation.

$$l \cdot (l - 2) = 15$$
$$l^2 - 2l = 15 \qquad \text{Multiplying}$$
$$l^2 - 2l - 15 = 0 \qquad \text{Adding} -15 \text{ to get } 0 \text{ on one side}$$
$$(l - 5)(l + 3) = 0 \qquad \text{Factoring}$$
$$l - 5 = 0 \quad \text{or} \quad l + 3 = 0 \qquad \text{Using the principle of zero products}$$
$$l = 5 \quad \text{or} \quad l = -3$$

The solutions of the equation are 5 and −3. Now we check in the problem. The length of a rectangle cannot be negative. Thus the length is 5 m. Since the width is 2 m less than the length, the width is 3 m.

DO EXERCISE 5.

EXERCISE SET 4.10

■● Solve.

1. $x^2 + 3x - 28 = 0$

2. $y^2 - 4y - 45 = 0$

3. $y^2 - 8y + 16 = 0$

4. $r^2 - 2r + 1 = 0$

5. $9x + x^2 + 20 = 0$

6. $8y + y^2 + 15 = 0$

7. $x^2 + 8x = 0$

8. $t^2 + 9t = 0$

9. $z^2 = 36$

10. $y^2 = 81$

11. $x^2 + 14x + 45 = 0$

12. $y^2 + 12y + 32 = 0$

13. $p^2 - 11p = -28$

14. $x^2 - 14x = -45$

15. $2x^2 - 15x = -7$

16. $x^2 - 9x = -8$

17. $14 = x(x - 5)$

18. $24 = x(x - 5)$

ANSWERS
1.
2.
3.
4.
5.
6.
7.
8.
9.
10.
11.
12.
13.
14.
15.
16.
17.
18.

ANSWERS

● ● Solve.

19. Four times the square of a number is 21 more than eight times the number. What is the number?

20. The square of a number plus the number is 132. What is the number?

19. _____

20. _____

21. The length of the top of a table is 5 in. more than the width. Find the length and width if the area is 84 in^2.

22. Sam Sylow has a garden 25 feet longer than it is wide. The garden has an area of 7500 ft^2. What are its dimensions?

21. _____

22. _____

23. The sum of the squares of two consecutive odd positive integers is 202. Find the integers.

24. The sum of the squares of two consecutive odd positive integers is 394. Find the integers.

23. _____

24. _____

25. The square of a number plus the number is 156. What is the number?

26. Four times the square of a number is 45 more than eight times the number. What is the number?

27. The length of the top of a work bench is 4 cm greater than the width. The area is 96 cm². Find the length and the width.

28. A flower bed is to be 3 m longer than it is wide. The flower bed will have an area of 108 m². What will its dimensions be?

29. The perimeter of a square is 4 more than its area. Find the length of a side.

30. The area of a square is 12 more than its perimeter. Find the length of a side.

25. _____

26. _____

27. _____

28. _____

29. _____

30. _____

ANSWERS

31. The base of a triangle is 9 cm greater than the height. The area is 56 cm². Find the height and base.

32. The base of a triangle is 5 cm less than the height. The area is 18 cm². Find the height and base.

31. _____

33. Three consecutive even integers are such that the square of the first plus the square of the third is 136. Find the three integers.

34. Three consecutive even integers are such that the square of the third is 76 more than the square of the second. Find the three integers.

32. _____

33. _____

34. _____

35. A rectangular piece of tin is twice as long as it is wide. Squares 2 cm on a side are cut out of each corner and the ends are turned up to make a box whose volume is 480 cm³. What are the dimensions of the piece of tin?

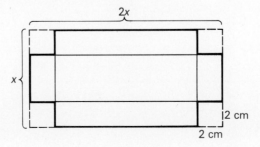

35. _____

4.11 EQUATIONS AND FORMULAS

▮● EQUATIONS WITH LETTERS FOR CONSTANTS

If a letter is used to represent one specific number (although it may be unknown) we call it a constant. Usually, but not always, we use letters at the beginning of the alphabet (a, b, c) for constants and letters near the end (x, y, z) for variables.

To solve an equation with unknown constants, we treat them as if they were known. The laws of numbers apply in any case.

Example 1 Solve for y: $ay + c^2 = 5$.

$$ay = 5 - c^2 \qquad \text{Adding } -c^2$$

$$y = \frac{5 - c^2}{a} \qquad \text{Multiplying by } \frac{1}{a}$$

In Example 1, we treated all constants—a, c^2, and 5—in the same way.

DO EXERCISE 1.

Usually when we solve an equation, we try to get the variable alone on one side. If the variable appears in several terms, we get those terms on one side and factor out the variable.

Examples Solve for the variable.

2. $5cdx + 7 = -3cdx$

$\quad 5cdx + 3cdx = -7 \qquad$ Adding -7 and $3cdx$ to get all the x-terms on one side

$\quad (5cd + 3cd)x = -7 \qquad$ Factoring out an x

$\qquad\qquad 8cdx = -7$

$\qquad\qquad\quad x = \dfrac{-7}{8cd} \qquad$ Multiplying by $\dfrac{1}{8cd}$

3. $(x + b)(x - b) = x(x - 2)$

$\quad x^2 - b^2 = x^2 - 2x \qquad$ Multiplying to eliminate parentheses

$\qquad -b^2 = -2x \qquad$ Adding $-x^2$

$\qquad\quad b^2 = 2x \qquad$ Multiplying by -1

$\qquad\quad \dfrac{b^2}{2} = x \qquad$ Multiplying by $\dfrac{1}{2}$

DO EXERCISES 2–4.

▮▮● FORMULAS

Formulas are important in business, science, etc. It is important to be able to solve a formula for a given letter. It is impossible to merely look at a formula and tell which letters are constants. In fact, a letter can be a constant in one situation and a variable in another.

OBJECTIVES

After finishing Section 4.11, you should be able to:

▮● Solve simple equations in which letters are used as constants.

▮▮● Solve a formula for a given letter.

1. Solve for x.

$\quad cx + a^2 = 4$

2. Solve for y.

$\quad 3aby + 6 = -2aby$

3. Solve for x.

$\quad 3(7 + 2x) = 30 + 7(x - 1)$

4. Solve for x.

$\quad (x + a)(x - b) = x^2 + 5$

ANSWERS ON PAGE A–17

5. Solve $F = \dfrac{kmM}{d^2}$ for M.

Examples

4. The formula

$$N = \frac{kP_1P_2}{d^2}$$

is used to find the number N of telephone calls between two cities of populations P_1 and P_2 at a distance d from each other. Solve it for P_1.

Note that P_1 and P_2 are different variables (or constants). So, don't forget to write the subscripts, 1 and 2. We first clear of fractions.

$$d^2N = kP_1P_2 \qquad \text{Multiplying by } d^2$$

Now there is no fractional expression in the equation. Next,

$$\frac{d^2N}{kP_2} = P_1. \qquad \text{Multiplying by } \frac{1}{kP_2} \text{ to get } P_1 \text{ alone}$$

6. Solve $L = \dfrac{t}{5}(r - w)$ for t.

5. Solve $P = \dfrac{k}{4}(a + b)$ for k.

$$4P = k(a + b) \qquad \text{Multiplying by } 4$$

$$\frac{4P}{a + b} = k \qquad \text{Multiplying by } \frac{1}{a + b}$$

DO EXERCISES 5 AND 6.

Example 6 Solve the following formula for R^2:

$$V = \frac{1}{3}\pi h(R^2 + r^2).$$

Note carefully that R and r are different. Be careful how you write them.

$$3V = \pi h(R^2 + r^2) \qquad \text{Multiplying by } 3 \leftarrow$$

7. Solve $M = \dfrac{m}{2}(v^2 + V^2)$ for v^2.

$$\frac{3V}{\pi h} = R^2 + r^2 \qquad \text{Multiplying by } \frac{1}{\pi h}$$

$$\frac{3V}{\pi h} - r^2 = R^2 \qquad \text{Adding } -r^2$$

In solving, you could multiply $(R^2 + r^2)$ by πh before using the addition principle. If so, you would get

$$\frac{3V - \pi h r^2}{\pi h} = R^2.$$

That answer is also correct, but it is not as simply arrived at as the one we obtained. There are also other correct answers.

DO EXERCISE 7.

ANSWERS ON PAGE A–17

NAME   CLASS   ANSWERS

EXERCISE SET 4.11

■ Solve for x, y, or z.

1. $15 - ay - b = 3y$ **2.** $bx - a = b - ax$ **3.** $ab - bz = cz - ac$

4. $bx - a + cx = a + b$ **5.** $10(a + x) = 8(a - x)$ **6.** $y(b + c) - a = d$

7. $3(2x - 4) + 2(x - 3) = 5(x + 3)$ **8.** $4y(2 + b) - b(3y + 1) = 5b$

9. $a^2 - ax = b^2 - bx$ **10.** $a^2 = b^2 - b^3y + a^2by$

1. _____

2. _____

3. _____

4. _____

5. _____

6. _____

7. _____

8. _____

9. _____

10. _____

11. Solve $V = \pi r^2 h$ for r^2.
(A volume formula)

12. Solve $V = \frac{1}{3} Bh$ for B.

(A volume formula)

11. _____

12. _____

13. Solve $s = \frac{1}{2} gt^2$ for t^2.

(A motion formula)

14. Solve $F = \frac{mv^2}{r}$ for m.

(A motion formula)

13. _____

14. _____

15. Solve $A = \pi r^2 + 2\pi rh$ for h.
(An area formula)

16. Solve $A = \pi rs + \pi r^2$ for s.
(An area formula)

15. _____

16. _____

17. Solve $A = P(1 + rt)$ for r.
(An interest formula)

18. Solve $V = \frac{1}{3}\pi h^2 (3R - h)$ for R.

(A volume formula)

17. _____

18. _____

4.12 FUNCTIONS

●   A *function* is a certain kind of correspondence, from one set to another.

Examples

1. To each person   there corresponds   his or her age.
   in a math class

2. To each item in   there corresponds   its price.
   a store

3. To each positive   there corresponds   its positive square root.
   real number

In each example, the first set is called the *domain*. The second set is called the *range*. Given a member of the domain, there is *just one* member of the range to which it corresponds. This kind of correspondence is called a *function*.

A *function* **is a correspondence between a first set, called the** *domain*, **and a second set, called the** *range*, **such that to any member of the domain, there corresponds** *exactly one* **member of the range.**

Examples   Which of the following are functions?

Domain	*Correspondence*	*Range*
4. A family	Each person's weight	A set of positive numbers

This correspondence is a function, because each person has *only one* weight.

5. The natural numbers	Square of each number	A set of natural numbers

This correspondence is a function, because each natural number has *only one* square.

6. The set of positive real numbers	Square roots of each number	The set of all real numbers

This correspondence is not a function, because every positive real number has two square roots.*

DO EXERCISES 1 AND 2.

---

* In a function, two different members of the domain can correspond to the same member of the range. In Example 4, for instance, two different family members could have the same weight.

1. Which of the following are functions?

Domain	*Corre-spondence*	*Range*
a) A family	Each person's age	A set of positive numbers
b) A math class	Each person's final grade	A set of letters
c) The people in a city	Each person's brothers	A set of males

2. In this correspondence, which is the domain and which is the range?

   The items in a store correspond to their prices.

Find these function values.

3. $f(x) = 2x + 5$; find $f(6)$.

4. $P(x) = 2x^3 - x + 1$; find $P(2)$.

5. $r(x) = 2x^2 - x - 1$; find $r(-1)$.

▣▣ NOTATION FOR FUNCTIONS

To understand function notation, it helps to think of a *function machine*. Think of putting a member of the domain (an *input*) into the machine. The machine knows the correspondence and gives you a member of the range (the *output*).

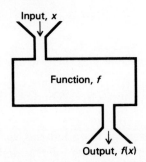

The function has been named f. We call the input x, and its output is named $f(x)$. This is read "f of x," or "f at x," or "the value of f at x." Note that $f(x)$ does *not* mean "f times x."

Some functions can be described by formulas or equations. For example, $f(x) = 2x + 3$ describes the function that takes an input x, multiplies it by 2, and then adds 3.

$$\underset{\text{Double}}{\underbrace{}}\ f(x)\ =\ \overset{\overset{\text{Input}}{\downarrow}}{2x}\ \underset{\text{Add 3}}{\underbrace{+\ 3}}$$

To find the output $f(4)$, we take the input, 4, double it, and add 3 to get 11. That is, we substitute 4 into the formula for $f(x)$:

$$f(4) = 2 \cdot 4 + 3$$
$$= 11.$$

Examples Find these function values.

7. $f(x) = 3x + 2$; find $f(5)$.

  $f(5) = 3 \cdot 5 + 2 = 17$

8. $g(x) = 5x^2 - 4$; find $g(3)$.

  $g(3) = 5(3)^2 - 4 = 41$

9. $A(x) = 3x^2 + 2x$; find $A(-2)$.

  $A(-2) = 3(-2)^2 + 2(-2) = 8$

> Remember, $f(x)$ does *not* mean f times x. Rather, $f(x)$ is the output when the input is x.

DO EXERCISES 3–5.

POLYNOMIAL FUNCTIONS

Any function that can be described by a polynomial such as

$$f(x) = 5x^7 + 3x^5 - 4x^2 - 5,$$

where the right-hand side is a polynomial, is called a *polynomial function*.

◼◼◼ GRAPHS OF FUNCTIONS

Consider the function $f(x) = 3x + 2$. If you think of this as $y = 3x + 2$, you already know how to draw a graph. The members of the domain are shown along the horizontal axis, and the members of the range along the vertical axis. Given an x (an input), the output, which is y or $f(x)$, is found by going up to the line and then over to the $f(x)$ axis.

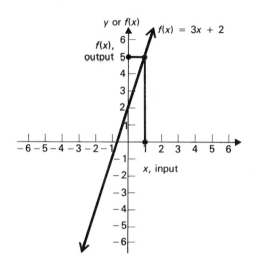

Some graphs are graphs of functions and some are not. If a vertical line drawn anywhere on the graph intersects the curve in more than one point, then there will be more than one number in the range corresponding to a member of the domain. The graph is not the graph of a function.

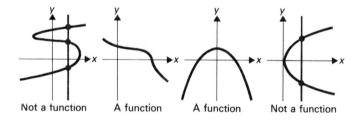

The vertical line test: **If it is possible for a vertical line to intersect a graph more than once, the graph is not the graph of a function.**

DO EXERCISES 6–9.

Which of the following are graphs of functions?

6.

7.

8.

9.

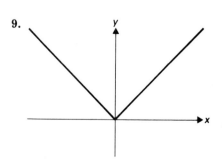

SOMETHING EXTRA
POLYNOMIALS AND MULTIPLICATION

Some of the special products for polynomials can be used in arithmetic.

Example 1   Find 24^2.

$$24^2 = (25 - 1)^2$$
$$= 25^2 - 2 \cdot 25 + 1^2 \qquad \text{Using } (A - B)^2 = A^2 - 2AB + B^2$$
$$= 625 - 50 + 1$$
$$= 576$$

Example 2   Find $48 \cdot 52$.

$$48 \cdot 52 = (50 - 2)(50 + 2)$$
$$= 50^2 - 2^2 \qquad \text{Using } (A - B)(A + B) = A^2 - B^2$$
$$= 2500 - 4$$
$$= 2496$$

Many such calculations can be done mentally.

EXERCISES

Calculate, using a special product.

1. 23^2 **2.** 26^2 **3.** 97^2 **4.** 102^2

5. $23 \cdot 27$ **6.** $98 \cdot 102$ **7.** 101^2 **8.** 27^2

NAME	CLASS

EXERCISE SET 4.12

▪● Which of the following are functions?

Domain	*Correspondence*	*Range*
1. A math class	Each person's seat number	A set of numbers
2. A set of numbers	Square each number and then add 4	A set of numbers

●● Given the function g described by $g(x) = -2x - 4$, find:

3. $g(-3)$ **4.** $g\left(-\dfrac{1}{2}\right)$ **5.** $g\left(\dfrac{3}{2}\right)$ **6.** $g(1.2)$

Given the function h described by $h(x) = 3x^2$, find:

7. $h(-1)$ **8.** $h(1)$ **9.** $h(2)$ **10.** $h(-2)$

A polynomial function P is described by $P(x) = x^3 - x$. Find:

11. $P(2)$ **12.** $P(-2)$

A polynomial function Q is described by $Q(x) = x^4 - 2x^3 + x^2 - x + 2$. Find:

13. $Q(2)$ **14.** $Q(-2)$

The function A described by $A(s) = \dfrac{s^2}{4}\sqrt{3}$ gives the area of an equilateral triangle with side s. Find:

15. $A(4)$ **16.** $A(6)$

15. _____

The function V described by $V(r) = \dfrac{4}{3}\pi r^3$ gives the volume of a sphere with radius r. Find:

16. _____

17. $V(3)$ **18.** $V(6)$

●●● Which of the following are graphs of functions?

17. _____

19.

20.

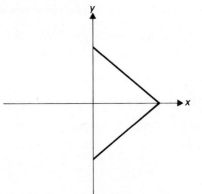

18. _____

19. _____

21.

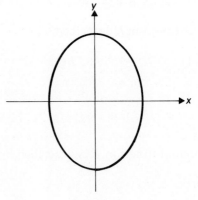

20. _____

22.

21. _____

22. _____

Which of the following are functions?

	Domain	Correspondence	Range
23.	A set of pairs of numbers (a, b)	Take the larger of a and b.	A set of numbers
24.	A set of ordered pairs of numbers (a, b)	Subtract a from b. Then take the larger of a and $b - a$.	A set of numbers

23. _____

24. _____

EXTENSION EXERCISES

SECTION 4.1

Evaluate each polynomial for the given value of its variables.

1. $3 - xy - x^2 - y^2$; $x = \dfrac{1}{2}$, $y = -\dfrac{1}{2}$

2. $\dfrac{1}{10,000}x^3 + \dfrac{1}{1000}x^2 - \dfrac{1}{100}x + \dfrac{1}{10}$; $x = 10$

3. $0.008z^4 - 0.002z^6$; $z = 0.1$

4. $a^2 + ab - 12b^2$; $a = 3t$, $b = t$

SECTION 4.2

Simplify.

5. $\dfrac{2}{3}(18a^2 - 33ab + 3b^2) - \dfrac{3}{4}(20a^2 + 44ab + 56b^2)$

6. $5(3 - x) + (5 - x - x^2) - 5(3 - 2x - x^3)$

7. $(0.06u^2 + 0.003v^2) - (v^2 - 0.6u^2) - (0.003u^2 - 6v^2)$

8. $\left(\dfrac{1}{2} - \dfrac{1}{3}y + \dfrac{1}{4}y^2\right) + \left(\dfrac{1}{2}y + \dfrac{1}{3}y^2 - \dfrac{1}{4}y^3\right) - \left(\dfrac{1}{2}y^2 - \dfrac{1}{3}y^3 - \dfrac{1}{4}y^4\right)$

SECTIONS 4.3–4.4

Multiply. Assume that variables in exponents represent positive integers.

9. $(6y)^2\left(-\dfrac{1}{3}x^2y^3\right)^3$

10. $[(a^{2n})^{2n}]^4$

11. $(-r^6s^2)^3\left(-\dfrac{r^2}{6}\right)^2(9s^4)^2$

12. $(z^{n^2})^{n^3}(z^{4n^3})^{n^2}$

13. $\left(-\dfrac{3}{8}xy^2\right)\left(-\dfrac{7}{9}x^3y\right)\left(-\dfrac{8}{7}x^2y\right)^2$

14. $(a^xb^{2y})\left(\dfrac{1}{2}a^{3x}b\right)^2$

15. $(-8s^3t)(2s^5 - 3s^3t^4 + st^7 - t^{10})$

16. $y^3z^n(y^{3n}z^3 - 4yz^{2n})$

17. $[(2x - 1)^2 - 1]^2$

18. $[x + y + 1][x^2 - x(y + 1) + (y + 1)^2]$

19. $[(a + b)(a - b)][5 - (a + b)][5 + (a + b)]$

20. $(y - 1)^6(y + 1)^6$

21. $(r^2 + s^2)^2(r^2 + 2rs + s^2)(r^2 - 2rs + s^2)$

22. $\left(3x^5 - \dfrac{5}{11}\right)^2$

23. $(a - b + c - d)(a + b + c + d)$

24. $\left(\dfrac{2}{3}x + \dfrac{1}{3}y + 1\right)\left(\dfrac{2}{3}x - \dfrac{1}{3}y - 1\right)$

25. $[2(y - 3) - 6(x + 4)][5(y - 3) - 4(x + 4)]$

26. $\left(x - \dfrac{1}{7}\right)\left(x^2 + \dfrac{1}{7}x + \dfrac{1}{49}\right)$

27. $(4x^2 + 2xy + y^2)(4x^2 - 2xy + y^2)$

28. $(x^2 - 7x + 12)(x^2 + 7x + 12)$

29. $(x^a + y^b)(x^a - y^b)(x^{2a} + y^{2b})$

30. $\left[1 + \dfrac{1}{5}(x + 1)^2\right]^2$

31. $[a - (b - 1)][(b - 1)^2 + a(b - 1) + a^2]$

32. $(x - 1)(x^2 + x + 1)(x^3 + 1)$

33. $\left[\left(\frac{1}{3}x^3 - \frac{2}{3}y^2\right)\left(\frac{1}{3}x^3 + \frac{2}{3}y^2\right)\right]^2$

34. $10(0.1x^4y - 0.01xy^4)^2$

35. $(x^{a-b})^{a+b}$

36. $(M^{x+y})^{x+y}$

SECTIONS 4.5–4.9

Factor. Assume that variables in exponents represent positive integers.

37. $20x^2 - 4x - 39$

38. $30y^4 - 97xy^2 + 60x^2$

39. $72a^2 + 284a - 160$

40. $3x^2y^2z + 25xyz^2 + 28z^3$

41. $7x^3 - \frac{7}{8}$

42. $-16 + 17(5 - y^2) - (5 - y^2)^2$

43. $[(c - d)^3 - d^3]^2$

44. $(x - p)^2 - p^2$

45. $x^4 - 50x^2 + 49$

46. $(y - 1)^4 - (y - 1)^2$

47. $s^6 - 729t^6$

48. $x^6 - 2x^5 + x^4 - x^2 + 2x - 1$

49. $27x^{6s} + 64y^{3t}$

50. $5c^{100} - 80d^{100}$

51. $4x^2 + 4xy + y^2 - r^2 + 6rs - 9s^2$

52. $(y + x)^3 - x^3$

53. $c^4d^4 - a^{16}$

54. $(1 - x)^3 + (x - 1)^6$

55. $c^{2w+1} + 2c^{w+1} + c$

56. $24x^{2a} - 6$

57. $y^9 - y$

58. $1 - \dfrac{x^{27}}{1000}$

59. $3a^2 + 3b^2 - 3c^2 - 3d^2 + 6ab - 6cd$

60. $3(x + 1)^2 + 9(x + 1) - 12$

61. $(a + 2)^3 - (a - 2)^3$

62. $8(a - 3)^2 - 64(a - 3) + 128$

63. Let $\left(x + \dfrac{2}{x}\right)^2 = 6$. Find $x^3 + \dfrac{8}{x^3}$.

SECTION 4.10

Solve.

64. The sum of two numbers is 17, and the sum of their squares is 205. Find the numbers.

65. The top and base of a fish tank are rectangles in which the length is 10 inches more than the width. If the depth and the width of the tank total 50 inches and the combined area of the top and base is 400 square inches less than the total area of the four sides, what are the dimensions of the tank?

66. A rectangular swimming pool with dimensions 11 meters and 8 meters is built in a rectangular backyard. The area of the backyard is 1120 square meters. If the strip of yard surrounding the pool is of uniform width, how wide is the strip?

67. The hypotenuse of a right triangle is 3 cm more than one of its legs and 6 cm more than the other leg. What is the area of the triangle?

SECTION 4.11

Solve for x.

68. $(x + 6)(x - 3) + b = (x - 5)(x - 2) - 3b$

69. $x^2 - 3(x - b) + c = (x - b)^2$

SECTION 4.12

70. Given the functions f, g, and h described by $f(x) = (x + 1)^3$, $g(x) = (1 - x)^2$, and $h(x) = -2x + 1$, find:
   a) $3h(x) - g(x) + f(x)$
   b) $f(x) \cdot g(x)$
   c) $[g(x) - h(x)]^3$
   d) $3h(10) - g(10) + f(10)$
   e) $f(-1) \cdot g(-1)$
   f) $[g(2) - h(2)]^3$
   g) $h(5) - 2g(-3)$
   h) $f(0) \cdot g(-1) \cdot h(1)$
   i) $[g(\frac{1}{2}) - h(\frac{1}{2})]^2$

NAME   SCORE

TEST OR REVIEW—CHAPTER 4

Review the material in the chapter. Then test yourself by doing these
exercises. If you miss an item, restudy the section indicated to the
left of the item.

1. _____

[4.1, ⠿]   **1.** Given the polynomial $3xy^3 - 4x^2y + 5x^5y^4 - 2x^4y$:

[4.1, ⠿]   a) Arrange in descending powers of x.
                                                              2. _____

                     b) What is the degree of the polynomial?

                                                              3. _____

[4.1, ⠿]   **2.** Collect like terms: $5xy - 2xy^2 - 2xy + 5xy^2$.

[4.2, ⠄]   **3.** Add: $3x^2y^4 - 2x^2y^2 - 5x$ and
                                                              4. _____
                        $-3xy^2 + 2x^2y + 4$.

[4.2, ⠿]   **4.** Subtract: $(6y^2 - 2y - 5y^3) - (4y^2 - 7y - 6y^3)$.

                                                              _____

[4.3, ⠿]   **5.** Multiply: $4x^3 - 2x + 3$ and $-2x^2 - 3x + 5$.

                                                              6. _____

[4.4, ⠿]   **6.** Multiply: $(3x - 4y)^2$.

                                                              7. _____

[4.4, ⠿]   **7.** Multiply: $(3x + 5)(3x - 5)$.

[4.5, ⠄]   **8.** Factor: $6a^2b - 21a^3b^2 + 3a^2b^3$.
                                                              8. _____

[4.5, ⠿]   **9.** Factor: $y^2 + 5y - 4y - 20$.

                                                              9. _____

[4.5, ⠿]   **10.** Factor: $20a^2 - 5b^2$.

                                                              10. _____

ANSWERS

11. _____

12. _____

13. _____

14. _____

15. _____

16. _____

17. _____

18. _____

19. _____

20. _____

[4.6, ▪]   **11.** Factor: $9x^2 - 30x + 25$.

[4.6, ▪▪▪]   **12.** Factor: $24x^2 - 46x + 10$.

[4.7, ▪]   **13.** Factor: $y^2 + 8y + 16 - 100t^2$.

[4.7, ▪▪]   **14.** Complete the square: $x^2 - 5x$.

[4.7, ▪▪]   **15.** Factor by completing the square:
$x^2 + 4.4x + 0.84$

[4.8, ▪]
[4.5, ▪]   **16.** Factor: $16a^7b + 54ab^7$.
[4.9, ▪]

[4.10, ▪]   **17.** Solve: $2x^2 = 3 - 5x$.

[4.10, ▪▪]   **18.** A photograph is 3 cm longer than it is wide. Its area is 40 cm². Find its length and width.

[4.11, ▪▪]   **19.** Solve for r: $F = \dfrac{mv^2}{r}$.

[4.12, ▪▪]   **20.** Given the function f described by $f(x) = x^2 - 3x + 5$, find $f(-2)$.

FRACTIONAL
EXPRESSIONS
AND EQUATIONS

5

A boat goes more slowly upstream than downstream. Certain problems relating the speed of a boat in a current can be solved using fractional equations.

OBJECTIVES

After finishing Section 5.1, you should be able to:

◻◻ Multiply fractional expressions.

◻◻◻ Multiply a fractional expression by 1, using an expression like A/A.

◻◻◻ Simplify fractional expressions.

◼◼ Multiply fractional expressions and simplify.

◼◼◼ Divide fractional expressions and simplify.

5.1 MULTIPLYING AND DIVIDING

The following are fractional expressions. They are quotients of polynomials.

$$\frac{7}{8}, \qquad \frac{8}{y + 5}, \qquad \frac{x^2 + 7xy - 4}{x^3 - y^3}$$

Fractional expressions indicate division.

$$\frac{7}{8} \quad \text{means} \quad 7 \div 8 \quad \text{and} \quad \frac{2x + y}{x - y} \quad \text{means} \quad (2x + y) \div (x - y).$$

Certain substitutions are not sensible in fractional expressions. Since division by 0 is not defined, any number that makes a denominator 0 is not a sensible replacement. In $8/(y + 5)$, -5 is not a sensible replacement.

Most of the calculations we do with fractional expressions are very much like calculations with fractional notation in arithmetic.

◦ MULTIPLYING

> **To multiply two fractional expressions, multiply numerators and multiply denominators.**

Example 1 Multiply.

$$\frac{x+3}{y-4}\cdot\frac{x^3}{y+5}=\frac{(x+3)x^3}{(y-4)(y+5)}$$ Multiplying numerators and multiplying denominators

> In algebra we do the same thing as in arithmetic. For example,
> $$\frac{2}{3}\cdot\frac{5}{7}=\frac{2\cdot5}{3\cdot7},$$
> multiplying numerators and denominators. We could multiply out to get $\frac{10}{21}$, but we shall not do it yet for fractional expressions because we will learn to simplify first.

DO EXERCISES 1 AND 2.

◦◦ MULTIPLYING BY 1

Any number multiplied by 1 is that same number. Any fractional expression with the same numerator and denominator names the number 1.

$$\frac{y+5}{y+5},\quad\frac{4x^2-5}{4x^2-5},\quad\frac{-1}{-1}$$ All name the number 1 for all sensible replacements

We can multiply by 1 to get equivalent expressions. For example, let us multiply $(x+y)/5$ by 1.

$$\frac{x+y}{5}\cdot\boxed{\frac{x-y}{x-y}}=\frac{(x+y)(x-y)}{5(x-y)}$$ Multiplying by $\frac{x-y}{x-y}$, which is 1

We know that $\dfrac{x+y}{5}$ and $\dfrac{(x+y)(x-y)}{5(x-y)}$ are equivalent. That means they will name the same number for all replacements that do not make a denominator 0.

Examples Multiply to obtain equivalent expressions.

2. $\dfrac{x^2+3}{x-1}\cdot\boxed{\dfrac{x+1}{x+1}}=\dfrac{(x^2+3)(x+1)}{(x-1)(x+1)}$

3. $\boxed{\dfrac{-1}{-1}}\cdot\dfrac{x-4}{x-y}=\dfrac{-1\cdot(x-4)}{-1\cdot(x-y)}$

DO EXERCISES 3–5.

Multiply.

1. $\dfrac{x-2}{5}\cdot\dfrac{x+2}{x+4}$

2. $\dfrac{x+y}{x+3}\cdot\dfrac{x+y}{x-3}$

Multiply.

3. $\dfrac{3x+2y}{5x+4y}\cdot\dfrac{x}{x}$

4. $\dfrac{2x^2-y}{3x+4}\cdot\dfrac{3x+2}{3x+2}$

5. $\dfrac{-1}{-1}\cdot\dfrac{2a-5}{a-b}$

Simplify by removing factors of 1.

6. $\dfrac{7x^2}{x}$

7. $\dfrac{6a + 9}{3}$

●●● SIMPLIFYING FRACTIONAL EXPRESSIONS

We can simplify fractional expressions by "removing" factors that are equal to 1. We first factor numerator and denominator, then factor the fractional expression, so that a factor is equal to 1.

Examples   Simplify by removing factors equal to 1.

4. $\dfrac{5x^2}{x} = \dfrac{5x \cdot x}{1 \cdot x}$ Factoring numerator and denominator

$\quad = \dfrac{5x}{1} \cdot \boxed{\dfrac{x}{x}}$ Factoring the fractional expression

$\quad = 5x \cdot 1 \qquad \dfrac{x}{x} = 1$ for all sensible replacements

$\quad = 5x$ "Removing" a factor of 1

In this example we supplied a 1 in the denominator. This can always be done, but it is not necessary.

5. $\dfrac{4a + 8}{2} = \dfrac{2(2a + 4)}{2 \cdot 1}$ Factoring numerator and denominator

$\quad = \boxed{\dfrac{2}{2}} \cdot \dfrac{2a + 4}{1}$ Factoring the fractional expression

$\quad = \dfrac{2a + 4}{1}$ "Removing" a factor of 1

$\quad = 2a + 4$ ⟶ A common error here is to get $2a + 8$. This happens because of taking an incorrect short-cut. You should go through the *entire* process of removing a factor of 1.

DO EXERCISES 6 AND 7.

6. $\dfrac{2x^2 + 4x}{6x^2 + 2x} = \dfrac{2x(x + 2)}{2x(3x + 1)}$ Factoring numerator and denominator

$\quad = \boxed{\dfrac{2x}{2x}} \cdot \dfrac{x + 2}{3x + 1}$ Factoring the fractional expression

$\quad = \dfrac{x + 2}{3x + 1}$ "Removing" a factor of 1

7. $\dfrac{x^2 - 1}{2x^2 - x - 1} = \dfrac{(x - 1)(x + 1)}{(2x + 1)(x - 1)}$ Factoring numerator and denominator

$\quad = \boxed{\dfrac{x - 1}{x - 1}} \cdot \dfrac{x + 1}{2x + 1}$ Factoring the fractional expression

$\quad = \dfrac{x + 1}{2x + 1}$ "Removing" a factor of 1

8. $\dfrac{9x^2 + 6xy - 3y^2}{12x^2 - 12y^2} = \dfrac{3(x + y)(3x - y)}{12(x + y)(x - y)}$ Factoring numerator and denominator

$= \dfrac{3(x + y)}{3(x + y)} \cdot \dfrac{3x - y}{4(x - y)}$ Factoring the fractional expression

$= \dfrac{3x - y}{4(x - y)}$ "Removing" a factor of 1

After removing all possible factors of 1, we usually multiply out the numerator and the denominator. The final answer is then

$\dfrac{3x - y}{4x - 4y}.$

DO EXERCISES 8 AND 9.

▐▐ MULTIPLYING AND SIMPLIFYING

After multiplying you should usually simplify, if possible.

Examples Multiply. Then simplify by removing factors of 1.

9. $\dfrac{x + 2}{x - 3} \cdot \dfrac{x^2 - 4}{x^2 + x - 2} = \dfrac{(x + 2)(x^2 - 4)}{(x - 3)(x^2 + x - 2)}$ Multiplying numerators and also denominators

$= \dfrac{(x + 2)(x - 2)(x + 2)}{(x - 3)(x + 2)(x - 1)}$ Factoring numerators and denominators

$= \dfrac{x + 2}{x + 2} \cdot \dfrac{(x + 2)(x - 2)}{(x - 3)(x - 1)}$ Factoring the fractional expression

$= \dfrac{(x + 2)(x - 2)}{(x - 3)(x - 1)}$ "Removing" a factor of 1

$= \dfrac{x^2 - 4}{x^2 - 4x + 3}$ Multiplying out numerator and denominator

10. $\dfrac{a^3 - b^3}{a^2 - b^2} \cdot \dfrac{a^2 + 2ab + b^2}{a^2 + ab + b^2} = \dfrac{(a^3 - b^3)(a^2 + 2ab + b^2)}{(a^2 - b^2)(a^2 + ab + b^2)}$

$= \dfrac{(a - b)(a^2 + ab + b^2)(a + b)(a + b)}{(a - b)(a + b)(a^2 + ab + b^2)}$

$= \dfrac{(a - b)(a^2 + ab + b^2)(a + b)}{(a - b)(a^2 + ab + b^2)(a + b)} \cdot \dfrac{a + b}{1}$

$= a + b$

DO EXERCISES 10 AND 11.

Simplify by removing factors of 1.

8. $\dfrac{6x^2 + 4x}{4x^2 + 8x}$

9. $\dfrac{2y^2 + 6y + 4}{y^2 - 1}$

Multiply and simplify.

10. $\dfrac{(x - y)^3}{x + y} \cdot \dfrac{3x + 3y}{x^2 - y^2}$

11. $\dfrac{a^3 + b^3}{a^2 - b^2} \cdot \dfrac{a^2 - 2ab + b^2}{a^2 - ab + b^2}$

Find the reciprocal.

12. $\dfrac{x + 3}{x - 5}$

13. $x + 7$

14. $\dfrac{1}{y^3 - 9}$

Divide. Simplify by removing a factor of 1 if possible. Then multiply out the numerator and denominator.

15. $\dfrac{x^2 + 7x + 10}{2x - 4} \div \dfrac{x^2 - 3x - 10}{x - 2}$

16. $\dfrac{a^2 - b^2}{ab} \div \dfrac{a^2 - 2ab + b^2}{2a^2b^2}$

⠛ DIVIDING AND SIMPLIFYING

Two expressions are reciprocals of each other if their product is 1. To find the reciprocal of a fractional expression we interchange numerator and denominator.

The reciprocal of $\dfrac{x + 2y}{x + y - 1}$ is $\dfrac{x + y - 1}{x + 2y}$.

The reciprocal of $y - 8$ is $\dfrac{1}{y - 8}$.

DO EXERCISES 12–14.

> **We divide with fractional expressions the way we did in arithmetic. We multiply by the reciprocal of the divisor.**

> We sometimes say "invert the divisor and multiply."

Examples Divide. Simplify by removing a factor of 1 if possible. Then multiply out the numerator and denominator.

11. $\dfrac{x - 2}{x + 1} \div \dfrac{x + 5}{x - 3} = \dfrac{x - 2}{x + 1} \cdot \dfrac{x - 3}{x + 5}$ Multiplying by the reciprocal

$= \dfrac{x^2 - 5x + 6}{x^2 + 6x + 5}$ Multiplying numerators and denominators

12. $\dfrac{a^2 - 1}{a - 1} \div \dfrac{a^2 - 2a + 1}{a + 1}$

$= \dfrac{a^2 - 1}{a - 1} \cdot \dfrac{a + 1}{a^2 - 2a + 1}$ Multiplying by the reciprocal

$= \dfrac{(a^2 - 1)(a + 1)}{(a - 1)(a^2 - 2a + 1)}$ Multiplying numerators and denominators

$= \dfrac{(a + 1)(a - 1)(a + 1)}{(a - 1)(a - 1)(a - 1)}$ Factoring numerator and denominator

$= \dfrac{a - 1}{a - 1} \cdot \dfrac{(a + 1)(a + 1)}{(a - 1)(a - 1)}$ Factoring the fractional expression

$= \dfrac{(a + 1)(a + 1)}{(a - 1)(a - 1)}$ "Removing" a factor of 1

$= \dfrac{a^2 + 2a + 1}{a^2 - 2a + 1}$ Multiplying out numerator and denominator

DO EXERCISES 15 AND 16.

EXERCISE SET 5.1

■ • , ■ • • Multiply to obtain equivalent expressions. Do not simplify.

1. $\dfrac{3x}{3x} \cdot \dfrac{x + 1}{x + 3}$

2. $\dfrac{4 - y^2}{6 - y} \cdot \dfrac{-1}{-1}$

3. $\dfrac{t - 3}{t + 2} \cdot \dfrac{t + 3}{t + 3}$

4. $\dfrac{p - 4}{p - 5} \cdot \dfrac{p + 5}{p + 5}$

■ • • Simplify by removing factors of 1.

5. $\dfrac{9y^2}{15y}$

6. $\dfrac{6x^3}{18x^2}$

7. $\dfrac{2a - 6}{2}$

8. $\dfrac{3a - 6}{3}$

9. $\dfrac{4y - 12}{4y + 12}$

10. $\dfrac{8x + 16}{8x - 16}$

11. $\dfrac{t^2 - 16}{t^2 - 8t + 16}$

12. $\dfrac{p^2 - 25}{p^2 + 10p + 25}$

1. _____

2. _____

3. _____

4. _____

5. _____

6. _____

7. _____

8. _____

9. _____

10. _____

11. _____

12. _____

■■ ■■ Multiply. Simplify by removing a factor of 1 if possible. Then multiply out the numerator and denominator.

13. $\dfrac{x^2 - 16}{x^2} \cdot \dfrac{x^2 - 4x}{x^2 - x - 12}$

14. $\dfrac{y^2 + 10y + 25}{y^2 - 9} \cdot \dfrac{y + 3}{y + 5}$

13. _____

14. _____

15. $\dfrac{y^2 - 16}{2y + 6} \cdot \dfrac{y + 3}{y - 4}$

16. $\dfrac{m^2 - n^2}{4m + 4n} \cdot \dfrac{m + n}{m - n}$

15. _____

16. _____

17. $\dfrac{x^2 - 2x - 35}{2x^3 - 3x^2} \cdot \dfrac{4x^3 - 9x}{7x - 49}$

18. $\dfrac{y^2 - 10y + 9}{y^2 - 1} \cdot \dfrac{y + 4}{y^2 - 5y - 36}$

17. _____

18. _____

19. $\dfrac{c^3 + 8}{c^2 - 4} \cdot \dfrac{c^2 - 4c + 4}{c^2 - 2c + 4}$

20. $\dfrac{x^3 - 27}{x^2 - 9} \cdot \dfrac{x^2 - 6x + 9}{x^2 + 3x + 9}$

19. _____

20. _____

21. $\dfrac{x^2 - y^2}{x^3 - y^3} \cdot \dfrac{x^2 + xy + y^2}{x^2 + 2xy + y^2}$

22. $\dfrac{4x^2 - 9y^2}{8x^3 - 27y^3} \cdot \dfrac{4x^2 + 6xy + 9y^2}{4x^2 + 12xy + 9y^2}$

21. _____

22. _____

Divide. Simplify by removing a factor of 1 if possible. Then multiply out the numerator and denominator.

23. $\dfrac{3y + 15}{y} \div \dfrac{y + 5}{y}$

24. $\dfrac{6x + 12}{x} \div \dfrac{x + 2}{x^3}$

23. _____

24. _____

25. $\dfrac{y^2 - 9}{y} \div \dfrac{y + 3}{y + 2}$

26. $\dfrac{x^2 - 4}{x} \div \dfrac{x - 2}{x + 4}$

25. _____

26. _____

27. $\dfrac{4a^2 - 1}{a^2 - 4} \div \dfrac{2a - 1}{a - 2}$

28. $\dfrac{25x^2 - 4}{x^2 - 9} \div \dfrac{5x - 2}{x + 3}$

27. _____

28. _____

29. $\dfrac{x^2 - 16}{x^2 - 10x + 25} \div \dfrac{3x - 12}{x^2 - 3x - 10}$

30. $\dfrac{y^2 - 36}{y^2 - 8y + 16} \div \dfrac{3y - 18}{y^2 - y + 12}$

29. _____

30. _____

31. $\dfrac{y^3 + 3y}{y^2 - 9} \div \dfrac{y^2 + 5y - 14}{y^2 + 4y - 21}$

32. $\dfrac{a^3 + 4a}{a^2 - 16} \div \dfrac{a^2 + 8a + 15}{a^2 + a - 20}$

31. _____

32. _____

33. $\dfrac{x^3 - 64}{x^3 + 64} \div \dfrac{x^2 - 16}{x^2 - 4x + 16}$

34. $\dfrac{8y^3 + 27}{64y^3 - 1} \div \dfrac{4y^2 - 9}{16y^2 + 4y + 1}$

33. _____

Perform the indicated operations and simplify.

34. _____

35. $\left[\dfrac{r^2 - 4s^2}{r + 2s} \div (r + 2s)\right] \cdot \dfrac{2s}{r - 2s}$

36. $\left[\dfrac{d^2 - d}{d^2 - 6d + 8} \cdot \dfrac{d - 2}{d^2 + 5d}\right] \div \dfrac{5d}{d^2 - 9d + 20}$

35. _____

36. _____

Multiply or divide.

37. $\boxed{\boxplus}$ $\dfrac{834x}{y - 427.2} \cdot \dfrac{26.3x}{y + 427.2}$

38. $\boxed{\boxplus}$ $\dfrac{527}{x + 93.87} \div \dfrac{x - 93.87}{468}$

37. _____

38. _____

5.2 LEAST COMMON MULTIPLES

◼◦ FINDING LCM'S BY FACTORING

To add fractional expressions when denominators are different, we first find a common denominator. Let us review the procedure in arithmetic first. To do the addition

$$\frac{5}{42} + \frac{7}{12}$$

we first find a common denominator. We look for a number that is a multiple of both 42 and 12 (a common multiple). Usually we try to get the smallest such number, or the *least common multiple* (LCM). To find the LCM we factor both numbers completely (into primes).

$$42 = 2 \cdot 3 \cdot 7 \longleftarrow \boxed{\text{Any multiple of 42 has these factors.}}$$
$$12 = 2 \cdot 2 \cdot 3 \longleftarrow$$
$$\boxed{\text{Any multiple of 12 has these factors.}}$$

The LCM is the number that has 2 as a factor twice, 3 as a factor once, and 7 as a factor once. The LCM is $2 \cdot 2 \cdot 3 \cdot 7$, or 84.

> **To obtain the LCM we use each factor the greatest number of times it occurs in any one factorization.**

Example 1 Find the LCM of 18 and 24.

$$\left. \begin{array}{l} 18 = \boxed{3 \cdot 3} \cdot 2 \\ 24 = \boxed{2 \cdot 2 \cdot 2} \cdot 3 \end{array} \right\} \quad \text{The LCM is } \boxed{3 \cdot 3} \cdot \boxed{2 \cdot 2 \cdot 2}, \text{ or } 72.$$

DO EXERCISES 1 AND 2.

◼◼ ADDING

Now let us return to adding $\frac{5}{42}$ and $\frac{7}{12}$.

$$\frac{5}{42} + \frac{7}{12} = \frac{5}{2 \cdot 3 \cdot 7} + \frac{7}{2 \cdot 2 \cdot 3} \qquad \text{Factoring denominators}$$

The LCM is $2 \cdot 2 \cdot 3 \cdot 7$. To get this LCM in the first denominator we need a 2. In the second denominator we need a 7. We multiply by 1, as follows:

$$\frac{5}{2 \cdot 3 \cdot 7} \cdot \boxed{\frac{2}{2}} + \frac{7}{2 \cdot 2 \cdot 3} \cdot \boxed{\frac{7}{7}} = \frac{10}{2 \cdot 2 \cdot 3 \cdot 7} + \frac{49}{2 \cdot 2 \cdot 3 \cdot 7}$$

$$= \frac{59}{2 \cdot 2 \cdot 3 \cdot 7} = \frac{59}{84}$$

Multiplying the first fraction by $\frac{2}{2}$ gave us a denominator that is the LCM. Multiplying the second fraction by $\frac{7}{7}$ also gave us a denominator that is the LCM. When we have a common denominator, we add the numerators.

DO EXERCISES 3 AND 4.

ANSWERS ON PAGE A–18

OBJECTIVES

After finishing Section 5.2, you should be able to:

◼◦ Find the LCM of several numbers by factoring.

◼◼ Add numbers (fractional notation) by first finding a common denominator.

◼◼◼ Find the LCM of several algebraic expressions by factoring.

Find the LCM by factoring.

1. 18, 30

2. 12, 18, 24

Add, first finding the LCM of the denominators.

3. $\dfrac{5}{12} + \dfrac{11}{30}$

4. $\dfrac{7}{12} + \dfrac{13}{18} + \dfrac{1}{24}$

Find the LCM.

5. $a^2b^2, \quad 5a^3b$

6. $y^2 + 7y + 12, \quad y^2 + 8y + 16, \quad y + 4$

7. $x^2 - 9, \quad x^3 - x^2 - 6x, \quad 2x^2$

8. $a^2 - b^2, \quad 2b - 2a$

○○○ LCM'S OF ALGEBRAIC EXPRESSIONS

Examples

2. Find the LCM of $12xy^2$ and $15x^3y$.

We factor each expression completely.

$12xy^2 = \boxed{2 \cdot 2 \cdot 3} \cdot x \cdot \boxed{y \cdot y}$

$15x^3y = 3 \cdot \boxed{5 \cdot x \cdot x \cdot x} \cdot y$

Remember: To find the LCM, use each factor the greatest number of times it occurs in any one factorization.

$\text{LCM} = \boxed{2 \cdot 2 \cdot 3} \cdot \boxed{5 \cdot x \cdot x \cdot x} \cdot \boxed{y \cdot y} = 60x^3y^2$

3. Find the LCM of $x^2 + 2x + 1, 5x^2 - 5x,$ and $x^2 - 1$.

$x^2 + 2x + 1 = \boxed{(x + 1)(x + 1)}$

$5x^2 - 5x = \boxed{5x(x - 1)}$ Factoring

$x^2 - 1 = (x + 1)(x - 1)$

$\text{LCM} = 5x(x + 1)(x + 1)(x - 1)$

4. Find the LCM of $x^2 - y^2, x^3 + y^3,$ and $x^2 + 2xy + y^2$.

$x^2 - y^2 = \boxed{(x - y)}\ (x + y)$

$x^3 + y^3 = (x + y)\ \boxed{(x^2 - xy + y^2)}$

$x^2 + 2xy + y^2 = \boxed{(x + y)(x + y)}$

$\text{LCM} = (x - y)(x + y)(x + y)(x^2 - xy + y^2)$

The additive inverse of an LCM is also an LCM. For example if $(x + 2)(x - 3)$ is an LCM, then $-(x + 2)(x - 3)$ is also an LCM. We can name the latter $(x + 2)(-1)(x - 3)$, or $(x + 2)(3 - x)$. If, when finding LCMs, factors that are additive inverses occur, we do not use them both. For example if $(a - b)$ occurs in one factorization and $(b - a)$ occurs in another, we do not use them both since these are additive inverses.

Example 5 Find the LCM of $x^2 - y^2$ and $3y - 3x$.

$x^2 - y^2 = \boxed{(x + y)}\ (x - y) \longleftarrow$ We can use $(x - y)$ or $(y - x)$ but we do not use both.

$3y - 3x = \boxed{3(y - x)} \longleftarrow$

$\text{LCM} = 3(x + y)(x - y), \quad \text{or} \quad 3(x + y)(y - x)$

DO EXERCISES 5–8.

NAME

CLASS

ANSWERS

EXERCISE SET 5.2

■ ● ■ Find the LCM by factoring.

1. 12, 18

2. 15, 20

3. 18, 48

4. 45, 54

5. 24, 36

6. 30, 75

7. 9, 15, 5

8. 27, 35, 63

■ ● ● ■ Add, first finding the LCM of the denominators.

9. $\dfrac{5}{6} + \dfrac{4}{15}$

10. $\dfrac{5}{36} + \dfrac{5}{24}$

11. $\dfrac{7}{12} + \dfrac{11}{18}$

12. $\dfrac{11}{30} + \dfrac{19}{75}$

13. $\dfrac{3}{4} + \dfrac{2}{5} + \dfrac{1}{6}$

14. $\dfrac{5}{8} + \dfrac{7}{12} + \dfrac{11}{40}$

1. _____

2. _____

3. _____

4. _____

5. _____

6. _____

7. _____

8. _____

9. _____

10. _____

11. _____

12. _____

13. _____

14. _____

••• Find the LCM.

15. $12x^2y$, $4xy$

16. $18r^2s$, $12rs^3$

15. _____

16. _____

17. $y^2 - 9$, $3y + 9$

18. $a^2 - b^2$, $ab + b^2$

17. _____

19. $15ab^2$, $3ab$, $10a^3b$

20. $6x^2y^2$, $9x^3y$, $15y^3$

18. _____

19. _____

21. $5y - 15$, $y^2 - 6y + 9$

22. $x^2 + 10x + 25$, $x^2 + 2x - 15$

20. _____

21. _____

23. $x^2 - 4$, $2 - x$

24. $y^2 - 9$, $3 - y$

22. _____

25. $2r^2 - 5r - 12$, $3r^2 - 13r + 4$, $r^2 - 16$

23. _____

24. _____

26. $2x^2 - 5x - 3$, $2x^2 - x - 1$, $x^2 - 6x + 9$

25. _____

26. _____

5.3   ADDITION AND SUBTRACTION

■ ● ■   WHEN DENOMINATORS ARE THE SAME

When denominators are the same, we add or subtract the numerators and keep the same denominator.

Example 1   Add.

$$\frac{3 + x}{x} + \frac{4}{x} = \frac{7 + x}{x}$$ ← This expression does *not* simplify to 7.

Example 1 shows that $\frac{3 + x}{x} + \frac{4}{x}$ and $\frac{7 + x}{x}$ are equivalent expressions.

They name the same number for all replacements except 0.

Example 2   Add.

$$\frac{4x^2 - 5xy}{x^2 - y^2} + \frac{2xy - y^2}{x^2 - y^2} = \frac{4x^2 - 3xy - y^2}{x^2 - y^2}$$ Adding numerators

$$= \frac{(4x + y)(x - y)}{(x + y)(x - y)}$$ Factoring numerator and denominator

$$= \frac{x - y}{x - y} \cdot \frac{4x + y}{x + y}$$ Factoring the fractional expression

$$= \frac{4x + y}{x + y}$$ "Removing" a factor of 1

DO EXERCISES 1 AND 2.

Example 3   Subtract.

$$\frac{4x + 5}{x + 3} - \frac{x - 2}{x + 3} = \frac{4x + 5 - (x - 2)}{x + 3}$$ ← Subtracting numerators

$$= \frac{4x + 5 - x + 2}{x + 3}$$ A common error: Forgetting these parentheses. If you forget them, you will be subtracting only *part* of the numerator, $x - 2$.

$$= \frac{3x + 7}{x + 3}$$

DO EXERCISES 3 AND 4.

■ ● ■   WHEN DENOMINATORS ARE ADDITIVE INVERSES OF EACH OTHER

When one denominator is the additive inverse of the other, we multiply one expression by $-1/-1$. This gives a common denominator.

OBJECTIVES

After finishing Section 5.3, you should be able to:

■ ● ■ Add and subtract fractional expressions having the same denominator.

■ ● ■ Add and subtract fractional expressions whose denominators are additive inverses of each other.

■ ● ● ● Add and subtract fractional expressions having different denominators.

■ ■ Simplify combined additions and subtractions of fractional expressions.

Add.

1. $\dfrac{5 + y}{y} + \dfrac{7}{y}$

2. $\dfrac{2x^2 + 5x - 9}{x - 5} + \dfrac{x^2 - 19x + 4}{x - 5}$

Subtract.

3. $\dfrac{a}{b + 2} - \dfrac{b}{b + 2}$

4. $\dfrac{4y + 7}{x^2 + y^2} - \dfrac{3y - 5}{x^2 + y^2}$

ANSWERS ON PAGE A–19

Add.

5. $\dfrac{b}{3b} + \dfrac{b^3}{-3b}$

6. $\dfrac{3x^2 + 4}{x - 5} + \dfrac{x^2 - 7}{5 - x}$

Subtract.

7. $\dfrac{3}{4y} - \dfrac{7x}{-4y}$

8. $\dfrac{4x^2}{2x - y} - \dfrac{7x^2}{y - 2x}$

Example 4 Add.

$$\frac{a}{2a} + \frac{a^3}{-2a} = \frac{a}{2a} + \boxed{\frac{-1}{-1}} \cdot \frac{a^3}{-2a} \qquad \text{Multiplying by } \frac{-1}{-1}$$

This is equal to 1 (not -1).

$$= \frac{a}{2a} + \frac{-a^3}{2a}$$

$$= \frac{a - a^3}{2a} \qquad \text{Adding numerators}$$

$$= \frac{a(1 - a^2)}{2a} \qquad \text{Factoring}$$

$$= \boxed{\frac{a}{a}} \cdot \frac{1 - a^2}{2} \qquad \text{Factoring}$$

$$= \frac{1 - a^2}{2} \qquad \text{"Removing" a factor of 1}$$

DO EXERCISES 5 AND 6.

Example 5 Subtract.

$$\frac{5x}{x - 2y} - \frac{3y - 7}{2y - x} = \frac{5x}{x - 2y} - \boxed{\frac{-1}{-1}} \cdot \frac{3y - 7}{2y - x}$$

$$= \frac{5x}{x - 2y} - \frac{7 - 3y}{x - 2y} \qquad \text{Remember: } -1(2y - x) = -(2y - x) = (x - 2y).$$

$$= \frac{5x - (7 - 3y)}{x - 2y} \qquad \text{Subtracting numerators}$$

Don't forget these parentheses.

$$= \frac{5x - 7 + 3y}{x - 2y}$$

DO EXERCISES 7 AND 8.

⬤⬤⬤ WHEN DENOMINATORS ARE DIFFERENT

When denominators are different, but not additive inverses of each other, we first find a common denominator, using the LCM, and then add or subtract numerators.

Example 6 Add: $\dfrac{2a}{5} + \dfrac{3b}{2a}$.

First find the LCM of the denominators.

$$\begin{matrix} 5 \\ 2a \end{matrix} \qquad \text{The LCM is } 5 \cdot 2a \quad \text{or} \quad 10a.$$

Now we multiply each expression by 1. We choose whatever symbol for 1 will give us the LCM in each denominator. In this case we use $2a/2a$ and $5/5$.

$$\frac{2a}{5} \cdot \boxed{\frac{2a}{2a}} + \frac{3b}{2a} \cdot \boxed{\frac{5}{5}} = \frac{4a^2}{10a} + \frac{15b}{10a} = \frac{4a^2 + 15b}{10a}$$

Multiplying by $2a/2a$ in the first term gave us a denominator of $10a$. Multiplying by $\frac{5}{5}$ in the second term also gave us a denominator of $10a$.

DO EXERCISE 9.

Example 7 Add: $\dfrac{3x^2 + 3xy}{x^2 - y^2} + \dfrac{2 - 3x}{x - y}$.

We first find the LCM of the denominators.

$$\left.\begin{array}{l} x^2 - y^2 = (x + y)(x - y) \\[4pt] x - y = x - y \end{array}\right\} \quad \text{The LCM is } (x + y)(x - y)$$

We now multiply by 1 to get the LCM in the second expression. Then we add and simplify if possible.

$$\frac{3x^2 + 3xy}{(x + y)(x - y)} + \frac{2 - 3x}{x - y} \cdot \boxed{\frac{x + y}{x + y}} \qquad \begin{array}{l}\text{Multiplying by 1 to}\\ \text{get LCM}\end{array}$$

$$= \frac{3x^2 + 3xy}{(x + y)(x - y)} + \frac{(2 - 3x)(x + y)}{(x - y)(x + y)}$$

$$= \frac{3x^2 + 3xy}{(x + y)(x - y)} + \frac{2x + 2y - 3x^2 - 3xy}{(x - y)(x + y)} \qquad \begin{array}{l}\text{Multiplying out}\\ \text{in numerator}\end{array}$$

$$= \frac{3x^2 + 3xy + 2x + 2y - 3x^2 - 3xy}{(x + y)(x - y)} \qquad \text{Adding numerators}$$

$$= \frac{2x + 2y}{(x + y)(x - y)} = \frac{2(x + y)}{(x + y)(x - y)} \qquad \begin{array}{l}\text{Combining like terms}\\ \text{and then factoring}\\ \text{numerator}\end{array}$$

$$= \boxed{\frac{x + y}{x + y}} \cdot \frac{2}{x - y} = \frac{2}{x - y} \qquad \begin{array}{l}\text{Simplifying by ``removing'' a}\\ \text{factor of 1}\end{array}$$

DO EXERCISES 10 AND 11.

Example 8 Subtract.

$$\frac{2y + 1}{y - 7y + 6} - \frac{y + 3}{y - 5y - 6}$$

$$= \frac{2y + 1}{(y - 6)(y\quad 1)} - \frac{y + 3}{(y - 6)(y + 1)} \qquad \begin{array}{l}\text{The LCM is}\\ (y - 6)(y - 1)(y + 1).\end{array}$$

$$= \frac{2y + 1}{(y - 6)(y - 1)} \cdot \boxed{\frac{y + 1}{y + 1}} - \frac{y + 3}{(y - 6)(y + 1)} \cdot \boxed{\frac{y - 1}{y - 1}}$$

$$= \frac{(2y + 1)(y + 1) - (y + 3)(y - 1)}{(y - 6)(y - 1)(y + 1)}$$

$$= \frac{2y^2 + 3y + 1 - (y^2 + 2y - 3)}{(y - 6)(y - 1)(y + 1)}$$

$$= \frac{2y^2 + 3y + 1 - y^2 - 2y + 3}{(y - 6)(y - 1)(y + 1)} = \frac{y^2 + y + 4}{(y - 6)(y - 1)(y + 1)}$$

If a denominator has 3 or more factors, we usually do not multiply it out.

DO EXERCISES 12 AND 13.

Add.

9. $\dfrac{3x}{7} + \dfrac{4y}{3x}$

10. $\dfrac{2xy - 2x^2}{x^2 - y^2} + \dfrac{2x + 3}{x + y}$

11. $\dfrac{1}{3y} + \dfrac{4y}{y^2 - 1} + \dfrac{7}{y - 1}$

Subtract.

12. $\dfrac{a}{a + 3} - \dfrac{a - 4}{a}$

13. $\dfrac{4y - 5}{y^2 - 7y + 12} - \dfrac{y + 7}{y^2 + 2y - 15}$

Do this calculation.

14. $\dfrac{8x}{x^2 - 1} + \dfrac{2}{1 - x} - \dfrac{4}{x + 1}$

⠿ COMBINED ADDITIONS AND SUBTRACTIONS

Example 9 Do this calculation.

$$\frac{2x}{x^2 - 4} + \frac{5}{2 - x} - \frac{1}{2 + x}$$

$$= \frac{2x}{(x - 2)(x + 2)} + \frac{5}{2 - x} - \frac{1}{2 + x}$$

$$= \frac{2x}{(x - 2)(x + 2)} + \boxed{\frac{-1}{-1}} \cdot \frac{5}{(2 - x)} - \frac{1}{x + 2}$$

$$= \frac{2x}{(x - 2)(x + 2)} + \frac{-5}{x - 2} - \frac{1}{x + 2} \qquad \begin{array}{l}\text{The LCM is} \\ (x - 2)(x + 2).\end{array}$$

$$= \frac{2x}{(x - 2)(x + 2)} + \frac{-5}{x - 2} \cdot \boxed{\frac{x + 2}{x + 2}} - \frac{1}{x + 2} \cdot \boxed{\frac{x - 2}{x - 2}}$$

$$= \frac{2x - 5(x + 2) - (x - 2)}{(x - 2)(x + 2)}$$

$$= \frac{2x - 5x - 10 - x + 2}{(x - 2)(x + 2)}$$

$$= \frac{-4x - 8}{(x - 2)(x + 2)}$$

$$= \frac{-4(x + 2)}{(x - 2)(x + 2)}$$

$$= \frac{-4}{x - 2} \cdot \boxed{\frac{x + 2}{x + 2}}$$

$$= \frac{-4}{x - 2}$$

DO EXERCISE 14.

SOMETHING EXTRA

CALCULATOR CORNER: NUMBER PATTERNS

Do the following calculations. Look for a pattern in the answers. Then figure out what the next answer in the list should be. After that, check your answer by doing the actual calculation.

EXERCISES

1. 3^2	**2.** 6^2	**3.** $37 \cdot 3$	**4.** $37 \cdot 3$
33^2	66^2	$37 \cdot 6$	$37 \cdot 33$
333^2	666^2	$37 \cdot 9$	$37 \cdot 333$
3333^2	6666^2	$37 \cdot 12$	$37 \cdot 3333$

EXERCISE SET 5.3

● Add or subtract. Simplify by removing a factor of 1 when possible.

1. $\dfrac{a-3b}{a+b} + \dfrac{a+5b}{a+b}$

2. $\dfrac{x-5y}{x+y} + \dfrac{x+7y}{x+y}$

3. $\dfrac{4y+3}{y-2} - \dfrac{y-2}{y-2}$

4. $\dfrac{3t+2}{t-4} - \dfrac{t-4}{t-4}$

● ● Add or subtract. Simplify by removing a factor of 1 when possible.

5. $\dfrac{a^2}{a-b} + \dfrac{b^2}{b-a}$

6. $\dfrac{r^2}{r-s} + \dfrac{s^2}{s-r}$

7. $\dfrac{3}{x} - \dfrac{8}{-x}$

8. $\dfrac{2}{a} - \dfrac{5}{-a}$

9. $\dfrac{2x-10}{x^2-25} - \dfrac{5-x}{25-x^2}$

10. $\dfrac{y-9}{y^2-16} - \dfrac{7-y}{16-y^2}$

1. _____

2. _____

3. _____

4. _____

5. _____

6. _____

7. _____

8. _____

9. _____

10. _____

●●● Add or subtract. Simplify by removing a factor of 1 when possible. If a denominator has three or more factors (other than monomials), leave it factored.

11. $\dfrac{y-2}{y+4} + \dfrac{y+3}{y-5}$

12. $\dfrac{x-2}{x+3} + \dfrac{x+2}{x-4}$

11. _____

12. _____

13. $\dfrac{4xy}{x^2-y^2} + \dfrac{x-y}{x+y}$

14. $\dfrac{5ab}{a^2-b^2} + \dfrac{a+b}{a-b}$

13. _____

14. _____

15. $\dfrac{9x+2}{3x^2-2x-8} + \dfrac{7}{3x^2+x-4}$

16. $\dfrac{3y+2}{2y^2-y-10} + \dfrac{8}{2y^2-7y+5}$

15. _____

16. _____

17. $\dfrac{4}{x+1} + \dfrac{x+2}{x^2-1} + \dfrac{3}{x-1}$

18. $\dfrac{-2}{y+2} + \dfrac{5}{y-2} + \dfrac{y+3}{y^2-4}$

17. _____

18. _____

19. $\dfrac{x-1}{3x+15} - \dfrac{x+3}{5x+25}$

20. $\dfrac{y-2}{4y+8} - \dfrac{y+6}{5y+10}$

19. _____

20. _____

21. $\dfrac{5ab}{a^2 - b^2} - \dfrac{a - b}{a + b}$

22. $\dfrac{6xy}{x^2 - y^2} - \dfrac{x + y}{x - y}$

23. $\dfrac{3y}{y^2 - 7y + 10} - \dfrac{2y}{y^2 - 8y + 15}$

24. $\dfrac{5x}{x^2 - 6x + 8} - \dfrac{3x}{x^2 - x - 12}$

25. $\dfrac{y}{y^2 - y - 20} + \dfrac{2}{y + 4}$

26. $\dfrac{6}{y^2 + 6y + 9} + \dfrac{5}{y^2 - 9}$

27. $\dfrac{3y + 2}{y^2 + 5y - 24} + \dfrac{7}{y^2 + 4y - 32}$

28. $\dfrac{3y + 2}{y^2 - 7y + 10} + \dfrac{2y}{y^2 - 8y + 15}$

29. $\dfrac{3x - 1}{x^2 + 2x - 3} - \dfrac{x + 4}{x^2 - 9}$

30. $\dfrac{3p - 2}{p^2 + 2p - 24} - \dfrac{p - 3}{p^2 - 16}$

21. _____

22. _____

23. _____

24. _____

25. _____

26. _____

27. _____

28. _____

29. _____

30. _____

ANSWERS

■■ Do these calculations.

31. $\dfrac{1}{x+1} - \dfrac{x}{x-2} + \dfrac{x^2+2}{x^2-x-2}$

32. $\dfrac{2}{y+3} - \dfrac{y}{y-1} + \dfrac{y^2+2}{y^2+2y-3}$

31. _____

32. _____

33. $\dfrac{x-1}{x-2} - \dfrac{x+1}{x+2} + \dfrac{x-6}{x^2-4}$

34. $\dfrac{y-3}{y-4} - \dfrac{y+2}{y+4} + \dfrac{y-7}{y^2-16}$

33. _____

34. _____

35. $\dfrac{4x}{x^2-1} + \dfrac{3x}{1-x} - \dfrac{4}{x-1}$

36. $\dfrac{5y}{1-2y} - \dfrac{2y}{2y+1} + \dfrac{3}{4y^2-1}$

35. _____

36. _____

5.4 COMPLEX FRACTIONAL EXPRESSIONS

⬛•

A *complex fractional expression* is one that has a fractional expression in its numerator or its denominator, or both.

These are complex fractional expressions:

$$\frac{x}{x - \frac{1}{3}}, \quad \frac{2x - \frac{4x}{3y}}{\frac{5x^2 + 2x}{6y^2}}, \quad \frac{\frac{1}{a} + \frac{1}{b}}{\frac{1}{a} - \frac{1}{b}}, \quad \frac{\frac{5}{x}}{\frac{x}{y}}.$$

To simplify a complex fractional expression:

1. **Add or subtract, as necessary, to get a single fractional expression in the numerator.**
2. **Add or subtract, as necessary, to get a single fractional expression in the denominator.**
3. **Divide the numerator by the denominator.**
4. **If possible, simplify by removing a factor of 1.**

Example 1   Simplify: $\dfrac{x + \frac{1}{5}}{x - \frac{1}{3}}$.

$$\frac{x + \frac{1}{5}}{x - \frac{1}{3}} = \frac{x \cdot \boxed{\frac{5}{5}} + \frac{1}{5}}{x - \frac{1}{3}} = \frac{\frac{5x + 1}{5}}{x - \frac{1}{3}}$$ Finding LCM in numerator, multiplying by 1 and adding

$$= \frac{\frac{5x + 1}{5}}{x \cdot \boxed{\frac{3}{3}} - \frac{1}{3}} = \frac{\frac{5x + 1}{5}}{\frac{3x - 1}{3}}$$ Finding LCM in denominator, multiplying by 1, and subtracting

$$= \frac{5x + 1}{5} \cdot \frac{3}{3x - 1}$$ Multiplying by the reciprocal of the denominator

$$= \frac{15x + 3}{15x - 5}$$ Multiplying numerators and denominators

Simplifying by removing a factor of 1 is not possible in this case.

If you feel more comfortable doing so, you can always write denominators of 1 where there are no denominators. You could start out by writing

$$\frac{\frac{x}{1} + \frac{1}{5}}{\frac{x}{1} - \frac{1}{3}}.$$

Simplify.

1. $\dfrac{y + \dfrac{1}{2}}{y - \dfrac{1}{7}}$

2. $\dfrac{1 - \dfrac{1}{x}}{1 - \dfrac{1}{x^2}}$

Simplify.

3. $\dfrac{\dfrac{1}{a} + \dfrac{1}{b}}{\dfrac{1}{a} - \dfrac{1}{b}}$

4. $\dfrac{\dfrac{1}{a} - \dfrac{1}{b}}{\dfrac{1}{a^3} - \dfrac{1}{b^3}}$

Example 2 Simplify.

$$\frac{1 + \dfrac{1}{x}}{1 - \dfrac{1}{x^2}} = \frac{\dfrac{x}{x} + \dfrac{1}{x}}{\dfrac{x^2}{x^2} - \dfrac{1}{x^2}} \qquad \text{Finding the LCM and multiplying by 1}$$

$$= \frac{\dfrac{x + 1}{x}}{\dfrac{x^2 - 1}{x^2}} \qquad \begin{array}{l}\text{Adding in the numerator and} \\ \text{subtracting in the denominator}\end{array}$$

$$= \frac{x + 1}{x} \cdot \frac{x^2}{x^2 - 1} \qquad \begin{array}{l}\text{Multiplying by the reciprocal} \\ \text{of the denominator}\end{array}$$

$$= \frac{(x + 1) \cdot x^2}{x(x^2 - 1)}$$

$$= \frac{x(x + 1)}{x(x + 1)} \cdot \frac{x}{x - 1} \qquad \begin{array}{l}\text{Factoring the fractional} \\ \text{expression}\end{array}$$

$$= \frac{x}{x - 1} \qquad \text{Removing a factor of 1}$$

Example 2 shows that the expressions $\dfrac{1 + \dfrac{1}{x}}{1 - \dfrac{1}{x^2}}$ and $\dfrac{x}{x - 1}$ are equivalent.

They name the same number for all replacements except 0, 1, and -1.

DO EXERCISES 1 AND 2.

Example 3 Simplify.

$$\frac{\dfrac{1}{a} + \dfrac{1}{b}}{\dfrac{1}{a^3} + \dfrac{1}{b^3}} = \frac{\dfrac{1}{a} \cdot \dfrac{b}{b} + \dfrac{1}{b} \cdot \dfrac{a}{a}}{\dfrac{1}{a^3} \cdot \dfrac{b^3}{b^3} + \dfrac{1}{b^3} \cdot \dfrac{a^3}{a^3}} = \frac{\dfrac{b}{ab} + \dfrac{a}{ab}}{\dfrac{b^3}{a^3b^3} + \dfrac{a^3}{a^3b^3}}$$

$$= \frac{\dfrac{b + a}{ab}}{\dfrac{b^3 + a^3}{a^3b^3}} \qquad \text{Adding in the numerator and denominator}$$

$$= \frac{b + a}{ab} \cdot \frac{a^3b^3}{b^3 + a^3} \qquad \begin{array}{l}\text{Multiplying by the reciprocal} \\ \text{of the denominator}\end{array}$$

$$= \frac{(b + a)a^3b^3}{ab(b^3 + a^3)} = \frac{(b + a) \cdot ab \cdot a^2b^2}{ab(b + a)(b^2 - ab + a^2)}$$

$$= \frac{(b + a)ab}{(b + a)ab} \cdot \frac{a^2b^2}{b^2 - ab + a^2} = \frac{a^2b^2}{b^2 - ab + a^2}$$

DO EXERCISES 3 AND 4.

NAME   CLASS   ANSWERS

EXERCISE SET 5.4

● Simplify.

1. $\dfrac{\dfrac{1}{x} + 4}{\dfrac{1}{x} - 3}$

2. $\dfrac{\dfrac{1}{y} + 7}{\dfrac{1}{y} - 5}$

3. $\dfrac{x - \dfrac{1}{x}}{x + \dfrac{1}{x}}$

4. $\dfrac{y + \dfrac{1}{y}}{y - \dfrac{1}{y}}$

5. $\dfrac{\dfrac{3}{x} + \dfrac{4}{y}}{\dfrac{4}{x} - \dfrac{3}{y}}$

6. $\dfrac{\dfrac{2}{y} + \dfrac{5}{z}}{\dfrac{1}{y} - \dfrac{4}{z}}$

7. $\dfrac{\dfrac{x^2 - y^2}{xy}}{\dfrac{x - y}{y}}$

8. $\dfrac{\dfrac{a^2 - b^2}{ab}}{\dfrac{a - b}{b}}$

1. _____

2. _____

3. _____

4. _____

5. _____

6. _____

7. _____

8. _____

ANSWERS

9. _____

10. _____

11. _____

12. _____

13. _____

14. _____

15. _____

16. _____

9. $\dfrac{a - \dfrac{3a}{b}}{b - \dfrac{b}{a}}$

10. $\dfrac{1 - \dfrac{2}{3x}}{x - \dfrac{4}{9x}}$

11. $\dfrac{\dfrac{1}{a} + \dfrac{1}{b}}{\dfrac{a^2 - b^2}{ab}}$

12. $\dfrac{\dfrac{1}{x} + \dfrac{1}{y}}{\dfrac{x^2 - y^2}{xy}}$

13. $\dfrac{\dfrac{1}{x + h} - \dfrac{1}{x}}{h}$ ←

It may help you to write h as $\dfrac{h}{1}$.

14. → $\dfrac{\dfrac{1}{a - h} - \dfrac{1}{a}}{h}$

15. $\dfrac{\dfrac{y^2 - y - 6}{y^2 - 5y - 14}}{\dfrac{y^2 + 6y + 5}{y^2 - 6y - 7}}$

16. $\dfrac{\dfrac{x^2 - x - 12}{x^2 - 2x - 15}}{\dfrac{x^2 + 8x + 12}{x^2 - 5x - 14}}$

5.5 DIVISION OF POLYNOMIALS

■• DIVISOR A MONOMIAL

Remember that fractional expressions indicate division. Division by a monomial can be done by first writing a fractional expression.

Example 1   Divide $12x^3 + 8x^2 + x + 4$ by $4x$.

$$\frac{12x^3 + 8x^2 + x + 4}{4x} \quad \text{Writing a fractional expression}$$

$$= \frac{12x^3}{4x} + \frac{8x^2}{4x} + \frac{x}{4x} + \frac{4}{4x} \quad \text{Doing the reverse of adding*}$$

$$= 3x^2 + 2x + \frac{1}{4} + \frac{1}{x} \quad \text{Doing the four indicated divisions}$$

DO EXERCISE 1.

Example 2   Divide: $(8x^4 - 3x^3 + 5x^2) \div x^2$.

$$\frac{8x^4 - 3x^3 + 5x^2}{x^2} = \frac{8x^4}{x^2} - \frac{3x^3}{x^2} + \frac{5x^2}{x^2}$$

$$= 8x^2 - 3x + 5$$

You should try to write only the answer.

> **To divide a polynomial by a monomial we can divide each term by the monomial.**

DO EXERCISES 2 AND 3.

■•• DIVISOR NOT A MONOMIAL

When the divisor is not a monomial, we use a procedure very much like long division in arithmetic.

Example 3   Divide $x^2 + 5x + 8$ by $x + 3$.

$$\begin{array}{r} x \phantom{+ 3\overline{)x^2 + 5x}} \\ x + 3\overline{)x^2 + 5x + 8} \\ x^2 + 3x \\ \hline 2x \end{array}$$

Divide first term of dividend by first term of divisor: $x^2/x = x$.
Multiply x above by divisor.
Subtract.

The subtraction we do here is $(x^2 + 5x) - (x^2 + 3x)$. Remember, to subtract, add the inverse (change the sign of every term, then add).

* To see this, do the addition

$$\frac{12x^3}{4x} + \frac{8x^2}{4x} + \frac{x}{4x} + \frac{4}{4x}$$

and you will get the expression

$$\frac{12x^3 + 8x^2 + x + 4}{4x}.$$

OBJECTIVES

After finishing Section 5.5, you should be able to:
■• Divide a polynomial by a monomial.
■•• Divide a polynomial by a divisor that is not a monomial, and if there is a remainder, express the result in two ways.

Divide.

1. $\dfrac{x^3 + 16x^2 + 6x}{2x}$

Divide.

2. $(15y^5 - 6y^4 + 18y^3) \div 3y^2$

3. $(x^4 + 10x^3 + 16x^2) \div 2x^2$

Divide and check.

4. $x - 2\overline{)x^2 + 3x - 10}$

We now "bring down" the other terms of the dividend—in this case, 8.

$$
\begin{array}{r}
x \;\; + 2 \longleftarrow \text{Divide first term by first term}: 2x/x = 2. \\
x + 3\overline{)x^2 + 5x + 8} \\
\underline{x^2 + 3x} \\
2x + 8 \longleftarrow \text{The 8 has been ``brought down.''} \\
\underline{2x + 6} \longleftarrow \text{Multiply 2 by divisor.} \\
2 \longleftarrow \text{Subtract}: (2x + 8) - (2x + 6).
\end{array}
$$

Answer: Quotient $x + 2$, remainder 2.

To check, we multiply quotient by divisor and add the remainder, to see if we get the dividend.

$$
\begin{array}{ccc}
\text{Divisor} & \text{Quotient} & \text{Remainder} \\
\overbrace{(x + 3)} & \overbrace{(x + 2)} + & \overbrace{2}
\end{array}
\begin{array}{l}
= (x^2 + 5x + 6) + 2 \\
= \underline{x^2 + 5x + 8} \\
\qquad \text{Dividend}
\end{array}
$$

The answer checks.

DO EXERCISE 4.

Always remember the following.

1. Arrange polynomials in descending order.

2. If there are missing terms in the dividend, either write them with 0 coefficients or leave space for them.

Example 4 Divide: $(125y^3 - 8) \div (5y - 2)$.

a)
$$
\begin{array}{r}
25y^2 + 10y \;\; + \;\; 4 \\
5y - 2\overline{)125y^3 + 0y^2 + 0y - 8} \\
\underline{125y^3 - 50y^2} \\
50y^2 + 0y - 8 \\
\underline{50y^2 - 20y} \\
20y - 8 \\
\underline{20y - 8} \\
0
\end{array}
$$

When there are missing terms, we can write them in as in (a), or leave space for them as in (b).

> This subtraction is $125y^3 - (125y^3 - 50y^2)$. We get $50y^2$.

Divide and check.

5. $(9y^4 + 14y^2 - 8) \div (3y + 2)$

b)
$$
\begin{array}{r}
25y^2 + 10y \;\; + \;\; 4 \\
5y - 2\overline{)125y^3 - 8} \\
\underline{125y^3 - 50y^2} \\
50y^2 - 8 \\
50y^2 - 20y \\
\underline{20y - 8} \\
\underline{20y - 8} \\
0
\end{array}
$$

> This subtraction is $50y^2 - (50y^2 - 20y)$. We get $20y$.

DO EXERCISE 5.

Example 5   Divide: $(x^4 - 9x^2 - 5) \div (x - 2)$.

$$
\begin{array}{r}
x^3 + 2x^2 - 5x - 10 \\
x - 2 \overline{)x^4 \quad\quad - 9x^2 \quad\quad\quad - 5} \\
\underline{x^4 - 2x^3} \\
2x^3 - 9x^2 \quad\quad\quad - 5 \\
\underline{2x^3 - 4x^2} \\
-5x^2 \quad\quad\quad - 5 \\
\underline{-5x^2 + 10x} \\
-10x - 5 \\
\underline{-10x + 20} \\
-25
\end{array}
$$

The first subtraction is $x^4 - (x^4 - 2x^3)$.

The second subtraction is $(2x^3 - 9x^2) - (2x^3 - 4x^2)$.

The answer is $x^3 + 2x^2 - 5x - 10$, with R $= -25$, or

$$x^3 + 2x^2 - 5x - 10 + \frac{-25}{x - 2}.$$

DO EXERCISES 6 AND 7.

When dividing, we may "come out even" (have a remainder of 0), or we may not. If not, how long should we keep working? We continue until the degree of the remainder is less than the degree of the divisor, as in the next example. The answer can be given by writing the quotient and remainder, or by using a fractional expression, as you saw in Example 5 and as you will see in the next example.

Example 6   Divide: $(x^3 + 9x^2 - 5) \div (x^2 - 1)$.

$$
\begin{array}{r}
x + 9 \\
x^2 - 1 \overline{)x^3 + 9x^2 + 0x - 5} \\
\underline{x^3 \quad\quad - x} \\
9x^2 + x - 5 \\
\underline{9x^2 \quad\quad - 9} \\
x + 4
\end{array}
$$

Again we have a missing term, so we can write it in.

The degree of the remainder is less than the degree of the divisor, so we are finished.

The answer is $x + 9$, with R $= x + 4$, or

$$x + 9 + \frac{x + 4}{x^2 - 1}.$$

This expression is the remainder over the divisor.

DO EXERCISE 8.

Divide and check.

6. $(y^3 - 11y^2 + 6) \div (y - 3)$

7. $(x^3 + 9x^2 - 5) \div (x - 1)$

Divide and check.

8. $(y^3 - 11y^2 + 6) \div (y^2 - 3)$

SOMETHING EXTRA

AN APPLICATION: DEPRECIATION—STRAIGHT-LINE METHOD (PART II)*

On p. 94 we considered a copying machine depreciated over an 8-year period. The cost was $5200 and the salvage value was $1300, so the total depreciation in 8 years was $3900. Each year the depreciation was $\frac{1}{8}$, or 12.5%, of $3900, or $487.50. Following is a *depreciation schedule*. Note how it shows the *book value* at the end of each year.

Year	Rate of depreciation	Annual depreciation	Total depreciation	Book value, V
0				$5200.00
1	$\frac{1}{8}$ or 12.5%	$487.50	$487.50	4712.50
2	same	same	975.00	4225.00
3	same	same	1462.50	3737.50
4	same	same	1950.00	3250.00
5	same	same	2437.50	2762.50
6	same	same	2925.00	2275.00
7	same	same	3412.50	1787.50
8	same	same	3900.00	1300.00

We can develop a formula for V, the book value. We choose notation as follows: C = the original cost, S = the salvage value, N = the number of years of useful life, and n = the number of years passed since purchase. The total depreciation is, of course, $C - S$. The depreciation each of the N years is $(C - S)/N$. After n years have passed, the total depreciation will be

$$n\left(\frac{C - S}{N}\right).$$

Thus the book value V will be given by

$$V = C - n\left(\frac{C - S}{N}\right).$$

Example Find the book value in the above situation after 7 years.

$$V = 5200 - 7\left(\frac{5200 - 1300}{8}\right) = \$1787.50.$$

EXERCISE

An item costs $9400 and will have a salvage value of $2200 after 10 years. Construct a depreciation schedule.

---

* For more on depreciation, see pp. 94, 390, and 404.

NAME

CLASS

ANSWERS

EXERCISE SET 5.5

⬤ Divide.

1. $\dfrac{30x^8 - 15x^6 + 40x^4}{5x^4}$

2. $\dfrac{24y^6 + 18y^5 - 36y^2}{6y^2}$

3. $(9y^4 - 18y^3 + 27y^2) \div 9y$

4. $(24a^3 + 28a^2 - 20a) \div 2a$

⬤⬤ Divide.

5. $(x^2 + 10x + 21) \div (x + 3)$

6. $(y^2 - 8y + 16) \div (y - 4)$

7. $(a^2 - 8a - 16) \div (a + 4)$

8. $(y^2 - 10y - 25) \div (y - 5)$

1. _____

2. _____

3. _____

4. _____

5. _____

6. _____

7. _____

8. _____

ANSWERS

9. $(x^2 + 7x + 14) \div (x + 5)$

10. $(t^2 - 7t - 9) \div (t - 3)$

9. _____

10. _____

11. $(4y^3 + 6y^2 + 14) \div (2y + 4)$

12. $(6x^3 - x^2 - 10) \div (3x + 4)$

11. _____

13. $(10y^3 + 6y^2 - 9y + 10) \div (5y - 2)$

14. $(6x^3 - 11x^2 + 11x - 2) \div (2x - 3)$

12. _____

13. _____

15. $(2x^4 - x^3 - 5x^2 + x - 6) \div (x^2 + 2)$

16. $(3x^4 + 2x^3 - 11x^2 - 2x + 5) \div (x^2 - 2)$

14. _____

15. _____

16. _____

5.6 SYNTHETIC DIVISION

● To divide a polynomial by a binomial of the type $x - a$, we can streamline the usual procedure, by a process called *synthetic division*.

Compare the following. In **A** we perform a division. In **B** we do not write the variables.

A.
$$
\begin{array}{r}
4x^2 + 5x + 11 \\
x - 2\overline{)4x^3 - 3x^2 + x + 7} \\
\underline{4x^3 - 8x^2} \\
5x^2 + x \\
\underline{5x^2 - 10x} \\
11x + 7 \\
\underline{11x - 22} \\
29
\end{array}
$$

B.
$$
\begin{array}{r}
4 + 5 + 11 \\
1 - 2\overline{)4 - 3 + 1 + 7} \\
\underline{4 - 8} \\
5 + 1 \\
\underline{5 - 10} \\
11 + 7 \\
\underline{11 - 22} \\
29
\end{array}
$$

In **B** there is still some duplication of writing. Also, since we can subtract by adding an inverse, we can use 2 instead of -2 and then add instead of subtracting.

C. *Synthetic Division*

a) $\underline{2|\,4 - 3 + 1 + 7}$

Write the 2 of $x - 2$ and the coefficients of the dividend.

4 Bring down the first coefficient.

b) $\underline{2|\,4 - 3 + 1 + 7}$

Multiply 4 by 2, to get 8. Add 8 and -3.

$4 \quad 5$

c) $\underline{2|\,4 - 3 + 1 + 7}$

$8 \quad 10$

Multiply 5 by 2 to get 10. Add 10 and 1.

$4 \quad 5 \quad 11$

d) $\underline{2|\,4 - 3 + 1 + 7}$

$8 \quad 10 \quad 22$

Multiply 11 by 2 to get 22. Add 22 and 7.

$4 \quad 5 \quad 11 \,|\, 29$

Quotient   Remainder

The last number, 29, is the remainder. The other numbers are the coefficients of the quotient, with that of the term of highest degree first, as follows.

$$4 \quad 5 \quad 11 \,|\, 29$$

Zero-degree coefficient

First-degree coefficient

Second-degree coefficient

The quotient is $4x^2 + 5x + 11$. The remainder is 29.

> It is important to remember that for this method to work, the divisor must be of the form $x - a$, that is, a variable minus a constant. The coefficient of the variable must be 1.

Use synthetic division to find the quotient and remainder.

1. $(2x^3 - 4x^2 + 8x - 8) \div (x - 3)$

Use synthetic division to find the quotient and remainder.

2. $(x^3 - 2x^2 + 5x - 4) \div (x + 2)$

3. $(y^3 + 1) \div (y + 1)$

Example 1   Use synthetic division to divide:

$$(x^3 + 6x^2 - x - 30) \div (x - 2).$$

$$\underline{2\,|}\;\begin{array}{rrrr}1 & 6 & -1 & -30 \\ & 2 & 16 & 30 \\ \hline 1 & 8 & 15| & 0\end{array}$$

The quotient is $x^2 + 8x + 15$. The remainder is 0.

DO EXERCISE 1.

When there are missing terms, be sure to write 0's.

Examples   Use synthetic division to find the quotient and remainder.

2. $(2x^3 + 7x^2 - 5) \div (x + 3)$

There is no x-term, so we must write a 0 for the coefficient. Note that $x + 3 = x - (-3)$.

$$\underline{-3\,|}\;\begin{array}{rrrr}2 & 7 & 0 & -5 \\ & -6 & -3 & 9 \\ \hline 2 & 1 & -3| & 4\end{array}$$

The quotient is $2x^2 + x - 3$. The remainder is 4.

3. $(x^3 + 4x^2 - x - 4) \div (x + 4)$

$$\underline{-4\,|}\;\begin{array}{rrrr}1 & 4 & -1 & -4 \\ & -4 & 0 & 4 \\ \hline 1 & 0 & -1| & 0\end{array}$$

The quotient is $x^2 - 1$. The remainder is 0.

4. $(x^4 - 1) \div (x - 1)$

$$\underline{1\,|}\;\begin{array}{rrrrr}1 & 0 & 0 & 0 & -1 \\ & 1 & 1 & 1 & 1 \\ \hline 1 & 1 & 1 & 1| & 0\end{array}$$

The quotient is $x^3 + x^2 + x + 1$. The remainder is 0.

5. $(8x^5 - 6x^3 + x - 8) \div (x + 2)$

$$\underline{-2\,|}\;\begin{array}{rrrrrr}8 & 0 & -6 & 0 & 1 & -8 \\ & -16 & 32 & -52 & 104 & -210 \\ \hline 8 & -16 & 26 & -52 & 105| & -218\end{array}$$

The quotient is $8x^4 - 16x^3 + 26x^2 - 52x + 105$. The remainder is -218.

DO EXERCISES 2 AND 3.

NAME   CLASS   ANSWERS

EXERCISE SET 5.6

■ Use synthetic division to find the quotient and remainder.

1. $(x^3 - 2x^2 + 2x - 5) \div (x - 1)$ **2.** $(x^3 - 2x^2 + 2x - 5) \div (x + 1)$

1. _____

2. _____

3. $(a^2 + 11a - 19) \div (a + 4)$ **4.** $(a^2 + 11a - 19) \div (a - 4)$

3. _____

5. $(x^3 - 7x^2 - 13x + 3) \div (x - 2)$

4. _____

6. $(x^3 - 7x^2 - 13x + 3) \div (x + 2)$

5. _____

7. $(3x^3 + 7x^2 - 4x + 3) \div (x + 3)$

6. _____

8. $(3x^3 + 7x^2 - 4x + 3) \div (x - 3)$

7. _____

8. _____

ANSWERS

9. _____

10. _____

11. _____

12. _____

13. _____

14. _____

15. _____

16. _____

17. _____

18. _____

9. $(y^3 - 3y + 10) \div (y - 2)$

10. $(x^3 - 2x^2 + 8) \div (x + 2)$

11. $(3x^4 - 25x^2 - 18) \div (x - 3)$

12. $(6y^4 + 15y^3 + 28y + 6) \div (y + 3)$

13. $(x^3 - 27) \div (x - 3)$ **14.** $(y^3 + 27) \div (y + 3)$

15. $(y^4 - 1) \div (y - 1)$ **16.** $(x^5 - 32) \div (x - 2)$

_____

Divide, using synthetic division.

17. $\boxplus$ $(3.41x^4 - 24.25x^2 - 13.47) \div (x - 2.41)$

18. $\boxplus$ $(5.032x^5 + 11.414x^3 + 217.3) \div (x + 17.07)$

5.7 SOLVING FRACTIONAL EQUATIONS

● A fractional equation is an equation that contains one or more fractional expressions. These are fractional equations:

$$\frac{2}{3} + \frac{5}{6} = \frac{1}{x}, \qquad \frac{x-1}{x-5} = \frac{4}{x-5}.$$

> **To solve a fractional equation we multiply on both sides by the LCM of all the denominators. This is called** *clearing of fractions.*

Example 1   Solve: $\dfrac{2}{3} - \dfrac{5}{6} = \dfrac{1}{x}.$

The LCM of all denominators is $6x$, or $2 \cdot 3 \cdot x$. We multiply on both sides of the equation by the LCM.

$$(2 \cdot 3 \cdot x) \cdot \left(\frac{2}{3} - \frac{5}{6}\right) = (2 \cdot 3 \cdot x) \cdot \frac{1}{x} \qquad \text{Multiplying by LCM}$$

$$2 \cdot 3 \cdot x \cdot \frac{2}{3} - 2 \cdot 3 \cdot x \cdot \frac{5}{6} = 2 \cdot 3 \cdot x \cdot \frac{1}{x} \qquad \begin{array}{l}\text{Multiplying to remove} \\ \text{parentheses}\end{array}$$

$$\frac{2 \cdot 3 \cdot x \cdot 2}{3} - \frac{2 \cdot 3 \cdot x \cdot 5}{6} = \frac{2 \cdot 3 \cdot x}{x}$$

$$4x - 5x = 6 \qquad \text{Simplifying}$$

$$-x = 6$$

$$-1 \cdot x = 6$$

$$x = -6$$

> When clearing of fractions, be sure to multiply *every* term in the equation by the LCM.

Check:

$$\frac{2}{3} - \frac{5}{6} = \frac{1}{x}$$

$$\begin{array}{c|c} \dfrac{2}{3} - \dfrac{5}{6} & \dfrac{1}{-6} \\ \hline \dfrac{4}{6} - \dfrac{5}{6} & -\dfrac{1}{6} \\ -\dfrac{1}{6} & \end{array}$$

Note that when we *clear of fractions* all the denominators disappear. Thus we have an equation without fractional expressions, which we know how to solve.

> *Note:* We have introduced a new use of the LCM here. Before, you used the LCM in adding or subtracting fractional expressions. *Now* we have equations. There are = signs. We clear of fractions by multiplying on both sides of the equation by the LCM. This makes the denominators disappear. *Do not* make the mistake of trying to "clear of fractions" when you do not have an equation!

DO EXERCISE 1.

Solve.

1. $\dfrac{2}{3} + \dfrac{5}{6} = \dfrac{1}{x}$

Solve.

2. $\dfrac{x - 7}{x - 9} = \dfrac{2}{x - 9}$

3. $\dfrac{x^2}{x + 3} = \dfrac{9}{x + 3}$

Example 2   Solve: $\dfrac{x + 1}{2} - \dfrac{x - 3}{3} = 3$.

The LCM of all the denominators is $2 \cdot 3$, or 6. We multiply on both sides of the equation by the LCM.

$$\boxed{2 \cdot 3}\left(\dfrac{x + 1}{2} - \dfrac{x - 3}{3}\right) = \boxed{2 \cdot 3} \cdot 3 \qquad \begin{array}{l}\text{Multiplying on} \\ \text{both sides by the} \\ \text{LCM}\end{array}$$

$$\boxed{2 \cdot 3} \cdot \dfrac{x + 1}{2} - \boxed{2 \cdot 3} \cdot \dfrac{x - 3}{3} = \boxed{2 \cdot 3} \cdot 3 \qquad \begin{array}{l}\text{Multiplying to} \\ \text{remove parentheses}\end{array}$$

$$\dfrac{2 \cdot 3(x + 1)}{2} - \dfrac{2 \cdot 3(x - 3)}{3} = 18$$

$$3(x + 1) - 2(x - 3) = 18$$

$$\left.\begin{array}{r}3x + 3 - 2x + 6 = 18 \\ x + 9 = 18\end{array}\right\} \quad \begin{array}{l}\text{Multiplying and} \\ \text{collecting like terms}\end{array}$$

$$x = 9$$

Check:
$$\dfrac{x + 1}{2} - \dfrac{x - 3}{3} = 3$$

$$\begin{array}{c|c}\dfrac{9 + 1}{2} - \dfrac{9 - 3}{3} & 3 \\[2mm] 5 - 2 & \\ 3 & \end{array}$$

When we multiply by an expression with a variable, we may not get equivalent equations. Thus we must *always* check possible solutions in the original equation.

Example 3   Solve: $\dfrac{x - 1}{x - 5} = \dfrac{4}{x - 5}$.

The LCM of the denominators is $x - 5$. We multiply by $x - 5$.

$$\boxed{(x - 5)} \cdot \dfrac{x - 1}{x - 5} = \boxed{(x - 5)} \cdot \dfrac{4}{x - 5}$$

$$x - 1 = 4$$

$$x = 5$$

Check:
$$\dfrac{x - 1}{x - 5} = \dfrac{4}{x - 5}$$

$$\begin{array}{c|c}\dfrac{5 - 1}{5 - 5} & \dfrac{4}{5 - 5} \\[2mm] \dfrac{4}{0} & \dfrac{4}{0}\end{array}$$

5 is not a solution of the original equation because it results in division by 0. In fact, the equation has no solution.

DO EXERCISES 2 AND 3.

Solve,

4. $\dfrac{x^2}{x-3} = \dfrac{9}{x-3}$

When in solving an equation we multiply by an expression containing a variable, we may get an equation having solutions that are not solutions of the original equation. In such a case you *must* check.

Example 4   Solve: $\dfrac{x^2}{x-2} = \dfrac{4}{x-2}$.

The LCM of the denominators is $x-2$. We multiply by $x-2$.

$$(x-2) \cdot \dfrac{x^2}{x-2} = (x-2) \cdot \dfrac{4}{x-2}$$

$$x^2 = 4$$

$$x^2 - 4 = 0$$

$$(x+2)(x-2) = 0$$

$$x = -2 \quad \text{or} \quad x = 2 \qquad \text{Using the principle of zero products}$$

Check: For 2: $\dfrac{x^2}{x-2} = \dfrac{4}{x-2}$ For -2: $\dfrac{x^2}{x-2} = \dfrac{4}{x-2}$

$$\dfrac{2^2}{2-2} \;\bigg|\; \dfrac{4}{2-2} \qquad\qquad \dfrac{(-2)^2}{-2-2} \;\bigg|\; \dfrac{4}{-2-2}$$

$$\dfrac{4}{-4} \;\bigg|\; \dfrac{4}{-4}$$

The number -2 is a solution, but 2 is not (it results in division by 0).

5. $x - \dfrac{12}{x} = 1$

Example 5   Solve: $x + \dfrac{6}{x} = 5$.

The LCM of the denominators is x. We multiply on both sides by x.

$$x\left(x + \dfrac{6}{x}\right) = 5 \cdot x \qquad \text{Multiplying on both sides by } x$$

$$x^2 + x \cdot \dfrac{6}{x} = 5x$$

$$x^2 + 6 = 5x \qquad \text{Simplifying}$$

$$x^2 - 5x + 6 = 0 \qquad \text{Getting 0 on one side}$$

$$(x-3)(x-2) = 0 \qquad \text{Factoring}$$

$$x = 3 \quad \text{or} \quad x = 2 \qquad \text{Using the principle of zero products}$$

Check: For 3: $x + \dfrac{6}{x} = 5$ For 2: $x + \dfrac{6}{x} = 5$

$$3 + \dfrac{6}{3} \;\bigg|\; 5 \qquad\qquad 2 + \dfrac{6}{2} \;\bigg|\; 5$$

$$3 + 2 \;\bigg|\; \qquad\qquad 2 + 3 \;\bigg|\;$$

$$5 \;\bigg|\; \qquad\qquad\qquad 5 \;\bigg|\;$$

The solutions are 2 and 3.

DO EXERCISES 4 AND 5.

Solve.

6. $\dfrac{2}{x-1} = \dfrac{3}{x+2}$

Example 6 Solve: $\dfrac{2}{x-1} = \dfrac{3}{x+1}$.

The LCM of the denominators is $(x-1)(x+1)$.

$$(x-1)(x+1) \cdot \dfrac{2}{(x-1)} = (x-1)(x+1) \cdot \dfrac{3}{x+1} \qquad \text{Multiplying}$$

$$2(x+1) = 3(x-1) \qquad \text{Simplifying}$$

$$2x + 2 = 3x - 3$$

$$5 = x$$

Check:

$$\dfrac{2}{x-1} = \dfrac{3}{x+1}$$

$\dfrac{2}{5-1}$	$\dfrac{3}{5+1}$
$\dfrac{2}{4}$	$\dfrac{3}{6}$
$\dfrac{1}{2}$	$\dfrac{1}{2}$

The number 5 is the solution.

Example 7 Solve: $\dfrac{2}{x+5} + \dfrac{1}{x-5} = \dfrac{16}{x^2-25}$.

The LCM is $(x+5)(x-5)$. We multiply by $(x+5)(x-5)$.

$$(x+5)(x-5) \cdot \left[\dfrac{2}{x+5} + \dfrac{1}{x-5} \right] = (x+5)(x-5) \cdot \dfrac{16}{x^2-25}$$

$$(x+5)(x-5) \cdot \dfrac{2}{x+5} + (x+5)(x-5) \cdot \dfrac{1}{x-5}$$

$$= (x+5)(x-5) \cdot \dfrac{16}{x^2-25}$$

$$2(x-5) + (x+5) = 16$$

$$2x - 10 + x + 5 = 16$$

$$3x - 5 = 16$$

$$3x = 21$$

$$x = 7$$

7. $\dfrac{2}{x^2-9} + \dfrac{5}{x-3} = \dfrac{3}{x+3}$

Check:

$$\dfrac{2}{x+5} + \dfrac{1}{x-5} = \dfrac{16}{x^2-25}$$

$\dfrac{2}{7+5} + \dfrac{1}{7-5}$	$\dfrac{16}{7^2-25}$
$\dfrac{2}{12} + \dfrac{1}{2}$	$\dfrac{16}{49-25}$
$\dfrac{8}{12}$	$\dfrac{16}{24}$
$\dfrac{2}{3}$	$\dfrac{2}{3}$

The solution is 7.

DO EXERCISES 6 AND 7.

EXERCISE SET 5.7

● Solve.

1. $\dfrac{2}{5} + \dfrac{7}{8} = \dfrac{y}{20}$

2. $\dfrac{4}{5} + \dfrac{1}{3} = \dfrac{t}{9}$

3. $\dfrac{1}{3} - \dfrac{5}{6} = \dfrac{1}{x}$

4. $\dfrac{5}{8} - \dfrac{2}{5} = \dfrac{1}{y}$

5. $\dfrac{x}{3} - \dfrac{x}{4} = 12$

6. $\dfrac{y}{5} - \dfrac{y}{3} = 15$

7. $y + \dfrac{5}{y} = -6$

8. $x + \dfrac{4}{x} = -5$

1. _____

2. _____

3. _____

4. _____

5. _____

6. _____

7. _____

8. _____

ANSWERS

9. $\dfrac{4}{z} + \dfrac{2}{z} = 3$

10. $\dfrac{4}{3y} - \dfrac{3}{y} = \dfrac{10}{3}$

9. _____

10. _____

11. $\dfrac{x-3}{x+2} = \dfrac{1}{5}$

12. $\dfrac{y-5}{y+1} = \dfrac{3}{5}$

11. _____

12. _____

13. $\dfrac{3}{y+1} = \dfrac{2}{y-3}$

14. $\dfrac{4}{x-1} = \dfrac{3}{x+2}$

13. _____

14. _____

15. $\dfrac{y-1}{y-3} = \dfrac{2}{y-3}$

16. $\dfrac{x-2}{x-4} = \dfrac{2}{x-4}$

15. _____

16. _____

17. $\dfrac{x+1}{x}=\dfrac{3}{2}$　　　**18.** $\dfrac{y+2}{y}=\dfrac{5}{3}$　　　**19.** $\dfrac{2}{x}-\dfrac{3}{x}+\dfrac{4}{x}=5$　　　**20.** $\dfrac{4}{y}-\dfrac{6}{y}+\dfrac{8}{y}=8$

17.＿＿＿＿＿＿＿

18.＿＿＿＿＿＿＿

19.＿＿＿＿＿＿＿

21. $\dfrac{1}{2}-\dfrac{4}{9x}=\dfrac{4}{9}-\dfrac{1}{6x}$　　　**22.** $-\dfrac{1}{3}-\dfrac{5}{4y}=\dfrac{3}{4}-\dfrac{1}{6y}$　　　**23.** $\dfrac{60}{x}-\dfrac{60}{x-5}=\dfrac{2}{x}$

20.＿＿＿＿＿＿＿

21.＿＿＿＿＿＿＿

22.＿＿＿＿＿＿＿

24. $\dfrac{50}{y}-\dfrac{50}{y-2}=\dfrac{4}{y}$　　　**25.** $\dfrac{7}{5x-2}=\dfrac{5}{4x}$　　　**26.** $\dfrac{5}{y+4}=\dfrac{3}{y-2}$

23.＿＿＿＿＿＿＿

24.＿＿＿＿＿＿＿

25.＿＿＿＿＿＿＿

26.＿＿＿＿＿＿＿

27. $\dfrac{x}{x-2} + \dfrac{x}{x^2-4} = \dfrac{x+3}{x+2}$

28. $\dfrac{3}{y-2} + \dfrac{2y}{4-y^2} = \dfrac{5}{y+2}$

27. _____

28. _____

29. $\dfrac{a}{2a-6} - \dfrac{3}{a^2-6a+9} = \dfrac{a-2}{3a-9}$

30. $\dfrac{2}{x+4} + \dfrac{2x-1}{x^2+2x-8} = \dfrac{1}{x-2}$

29. _____

30. _____

31. $\dfrac{2x+3}{x-1} = \dfrac{10}{x^2-1} + \dfrac{2x-3}{x+1}$

32. $\dfrac{y}{y+1} + \dfrac{3y+5}{y^2+4y+3} = \dfrac{2}{y+3}$

31. _____

32. _____

5.8 SOLVING PROBLEMS

● WORK PROBLEMS

Suppose a machine can do a certain job in 5 hours. Then in 1 hour it can do $\frac{1}{5}$ of the job. In 3 hours it can do $\frac{3}{5}$ of the job, and so on. This reasoning gives us a principle for solving certain problems.

> **If a job can be done in t hours (or days), then $1/t$ of it can be done in 1 hour (or day).**

> Don't forget the principles that hold in general for solving problems. You should review them. Go to Section 2.3, p. 59.

Example 1 Lon Moore can mow a lawn in 4 hours. Penny Push can mow the same lawn in 5 hours. How long would it take both of them, working together, to mow the lawn?

> Use some common sense here. Don't just start manipulating numbers. The answer is going to be less than 4 hours because Lon alone can do the job in 4 hours.

Lon can mow the lawn in 4 hours, so he can mow $\frac{1}{4}$ of it in 1 hour. Penny can mow the lawn in 5 hours, so she can mow $\frac{1}{5}$ of it in 1 hour. Thus they can mow $\frac{1}{4} + \frac{1}{5}$ of it in 1 hour working together.

Let t represent the time it takes them, working together. Then they mow $1/t$ of it in 1 hour. We now have an equation:

$$\frac{1}{4} + \frac{1}{5} = \frac{1}{t}. \qquad \text{This is the translation to mathematical language.}$$

We solve the equation:

$$20t\left(\frac{1}{4} + \frac{1}{5}\right) = 20t \cdot \frac{1}{t}$$

$$\frac{20t}{4} + \frac{20t}{5} = \frac{20t}{t} \qquad \left.\begin{array}{l} \text{Multiplying on both sides by the} \\ \text{LCM of denominators and clearing} \\ \text{of fractions} \end{array}\right.$$

$$5t + 4t = 20$$

$$9t = 20$$

$$t = \frac{20}{9}, \quad \text{or} \quad 2\frac{2}{9}.$$

Now, of course, we check in the original problem to see if $\frac{20}{9}$ is the right answer.

Lon does $\frac{1}{4}$ of the lawn in 1 hour. In $\frac{20}{9}$ hr he does $\frac{1}{4} \cdot \frac{20}{9}$, or $\frac{5}{9}$ of it. Penny does $\frac{1}{5}$ of the job in 1 hour. In $\frac{20}{9}$ hr she does $\frac{1}{5} \cdot \frac{20}{9}$, or $\frac{4}{9}$ of it. Altogether they do $\frac{5}{9} + \frac{4}{9}$ of the job, or all of it. The number checks, so the answer is that it takes them $\frac{20}{9}$ hr to do the job working together.

DO EXERCISE 1.

ANSWER ON PAGE A–20

OBJECTIVES

After finishing Section 5.8, you should be able to:

● Solve work problems.

●● Solve motion problems.

1. Fred Huntinpeck does a certain typing job in 6 hours. Fran Ikfingers can do the same job in 4 hours. How long would it take them to do the same amount of typing working together?

2. Two pipes carry water to the same tank. Pipe A, working alone, can fill the tank three times as fast as pipe B. Together the pipes can fill the tank in 24 hours. Find the time each would take to fill the tank alone.

Example 2   At a factory, smokestack A pollutes the air twice as fast as smokestack B. When the stacks operate together they yield a certain amount of pollution in 15 hours. Find the time it would take each to yield that same amount of pollution operating alone.

Let x represent the amount of time it takes A to yield the pollution. Then $2x$ is the time it takes B to yield the same amount of pollution. Then

$\dfrac{1}{x}$ is the fraction of the pollution by A in 1 hour,

and

$\dfrac{1}{2x}$ is the fraction of the pollution by B in 1 hour.

> Common sense will help with this translation, too. The slower smokestack will take *longer* to generate pollution than the faster stack. Since B is slower, the time required is longer. That way we get $2x$ for the time required for B.

Together, the stacks yield $1/x + 1/2x$ of the pollution in 1 hour. They also yield $\frac{1}{15}$ of it in one hour. We now have an equation:

$$\frac{1}{x} + \frac{1}{2x} = \frac{1}{15}.\qquad \text{This is the translation.}$$

We solve the equation:

$$30x\left(\frac{1}{x} + \frac{1}{2x}\right) = 30x \cdot \frac{1}{15}\qquad \begin{array}{l}\text{Multiplying on both sides by the}\\ \text{LCM of denominators and clearing}\\ \text{of fractions}\end{array}$$

$$\frac{30x}{x} + \frac{30x}{2x} = \frac{30x}{15}$$

$$30 + 15 = 2x$$

$$45 = 2x$$

$$x = \frac{45}{2}, \quad \text{or} \quad 22\frac{1}{2}.$$

Thus if the numbers check, the answer will be A takes x, or $\frac{45}{2}$, hours and B takes $2x$, or 45, hours. Part of the check is obvious already. B will take twice as long, so A pollutes twice as fast.

Now if A does $1/x$ of the total pollution in 1 hour, or $1/\frac{45}{2}$, which is $\frac{2}{45}$ of the total pollution, then in $\frac{45}{2}$ hours it will do $\frac{2}{45} \cdot \frac{45}{2}$, or all of it. This checks.

If B does $1/2x$, or $\frac{1}{45}$, of the total pollution in 1 hour, it will do $\frac{1}{45} \cdot 45$ of it in 45 hours. This is the entire amount. This checks.

DO EXERCISE 2.

⦿⦿ MOTION PROBLEMS

Problems dealing with speed, distance, and time are called *motion* problems. To translate them we use the definition of speed.

$$\textbf{Speed} = \frac{\textbf{Distance}}{\textbf{Time}}$$

$$r = \frac{d}{t}, \quad \textbf{or} \quad d = rt.$$

⟨ From the equation $r = d/t$ we can easily get $d = rt$ or $t = d/r$. ⟩

Example 3   A train leaves Sioux City traveling east at 30 mph. Two hours later another train leaves Sioux City in the same direction on a parallel track at 45 mph. How far from Sioux City will the faster train catch the slower one?

We first make a drawing.

From the drawing we see that the distances are the same. Let's call the distance d. We don't know the times. Let t represent the time for the faster train. Then the time for the slower one will be $t + 2$. We can organize the information in a table.

	Distance	Speed	Time
Slow train	d	30	$t + 2$
Fast train	d	45	t

Using $d = rt$ in each row of the table, we get an equation. Thus we get a system of two equations:

   $d = 30(t + 2)$ and $d = 45t$.   This is the translation.

We solve:

   $30(t + 2) = 45t$ Using substitution

   $30t + 60 = 45t$

   $60 = 15t$

   $t = 4.$

Thus the time for the faster train should be 4 hours and the time for the slower train 6 hours. At 45 mph the faster train would go $45 \cdot 4$, or 180 miles in 4 hours. At 30 mph the slower train would go $30 \cdot 6$, or 180 miles in 6 hours. We have a check, and the answer is 180 miles.

DO EXERCISE 3.

3. A train leaves Gigville traveling east at 35 mph. One hour later a faster train leaves Gigville, also traveling east on a parallel track at 40 mph. How far from Gigville will the faster train catch the slower one?

ANSWER ON PAGE A–20

4. A boat travels 246 miles downstream in the same time that it takes to travel 180 miles upstream. The speed of the current in the stream is 5.5 mph. Find the speed of the boat in still water.

Example 4 An airplane flies 1062 miles with the wind. In the same amount of time it can fly 738 miles against the wind. The speed of the plane in still air is 200 mph. Find the speed of the wind.

We first make a drawing. We let r represent the speed of the wind, and organize the facts in a chart.

	Distance	Speed	Time
With wind	1062	$200 + r$	t
Against wind	738	$200 - r$	t

Using $t = d/r$ with each row of the table, we get a system of equations:

$$t = \frac{1062}{200 + r} \quad \text{and} \quad t = \frac{738}{200 - r}. \qquad \text{This is the translation.}$$

We solve:

$$\frac{1062}{200 + r} = \frac{738}{200 - r} \qquad \text{Using substitution}$$

$$(200 + r)(200 - r)\left(\frac{1062}{200 + r}\right) = (200 + r)(200 - r)\left(\frac{738}{200 - r}\right)$$

Multiplying by LCM to clear of fractions

$$(200 - r)1062 = (200 + r)738$$
$$212{,}400 - 1062r = 147{,}600 + 738r$$
$$64{,}800 = 1800r$$
$$r = 36.$$

Thus, if it checks, the speed of the wind is 36 mph. With the wind, the speed of the plane would be 236 mph. Dividing the distance, 1062 miles, by the speed, 236 mph, we get 4.5 hours for the time. Against the wind, the speed of the plane would be 164 mph. Dividing the distance, 738 miles, by the speed, 164 mph, we get 4.5 hours. The answer checks, so the speed of the wind is 36 mph.

DO EXERCISE 4.

ANSWER ON PAGE A–20

NAME

CLASS

EXERCISE SET 5.8

▪ Solve.

1. Sam Strong, an experienced shipping clerk, can fill a certain order in 5 hours. Willy Weak, a new clerk, needs 9 hours to do the same job. Working together, how long would it take them to fill the order?

2. Paul Putty can paint a room in 4 hours. Sally Spackle can paint the same room in 3 hours. Working together, how long would it take them to paint the room?

1. _____

3. A swimming pool can be filled in 12 hours if water enters through a pipe alone, or in 30 hours if water enters through a hose alone. If water is entering through both the pipe and the hose, how long will it take to fill the pool?

4. A tank can be filled in 18 hours by pipe A alone and in 22 hours by pipe B alone. How long would it take to fill the tank if both pipes were working?

2. _____

3. _____

4. _____

5. Bill Dozer can clear a lot in 5.5 hours. His partner can do the same job in 7.5 hours. How long would it take them to clear the lot working together?

> *Hint:* You may find that multiplying on both sides of your equation by $\frac{1}{10}$ will clear of decimals.

6. One of Ty Psetter's printing presses can print an order of booklets in 4.5 hours. Another press can do the same job in 5.5 hours. How long would it take if both presses are used?

5. _____

6. _____

7. A can paint the neighbor's house 4 times as fast as B. The year they worked together it took them 8 days. How long would it take each to paint the house alone?

8. A can deliver papers 3 times as fast as B. If they work together, it takes them 1 hour. How long would it take each to deliver the papers alone?

7. _____

8. _____

▪▫ Solve.

9. The speed of a stream is 3 mph. A boat travels 4 mi upstream in the same time it takes to travel 10 mi downstream. What is the speed of the boat in still water?

10. The speed of a stream is 4 mph. A boat travels 6 mi upstream in the same time it takes to travel 12 mi downstream. What is the speed of the boat in still water?

9. _____

11. A train leaves a station and travels north at 75 km/h. Two hours later a second train leaves on a parallel track traveling north at 125 km/h. How far from the station will the second train overtake the first train?

12. A private airplane leaves an airport and flies due east at 180 km/h. Two hours later a jet leaves the same airport and flies due east at 900 km/h. How far from the airport will the jet overtake the private plane?

10. _____

11. _____

12. _____

13. The speed of train A is 12 mph slower than the speed of train B. Train A travels 230 mi in the same time it takes train B to travel 290 mi. Find the speed of each train.

14. The speed of a passenger train is 14 mph faster than the speed of a freight train. The passenger train travels 400 mi in the same time it takes the freight train to travel 330 mi. Find the speed of each train.

13. _____

14. _____

15. Suzie Skiff has a boat that can move at a speed of 15 km/h in still water. She rides 140 km downstream in a river in the same time it takes to ride 35 km upstream. What is the speed of the river?

16. A paddleboat can move at a speed of 2 km/h in still water. The boat is paddled 4 km downstream in a river in the same time it takes to go 1 km upstream. What is the speed of the river?

15. _____

16. _____

17. _____

17. At what time after 4:00 will the minute hand and the hour hand of a clock first be in the same position?

18. At what time after 10:30 will the hands of a clock first be perpendicular?

18. _____

5.9 FORMULAS

● As we mentioned earlier, solving formulas is important in business, technology, science, and engineering.

Example 1   The formula $A = 9R/I$ tells how to calculate a pitcher's earned run average (ERA). In this formula, A is the ERA, R is the number of earned runs, and I is the number of innings pitched. Solve this formula for I.

We multiply by the LCM, which is I.

$$I \cdot A = I \cdot \frac{9R}{I} \qquad I \cdot A = 9R$$

Next we multiply by $\frac{1}{A}$.

$$I \cdot A \cdot \frac{1}{A} = 9R \cdot \frac{1}{A} \qquad I = \frac{9R}{A}$$

> In solving formulas we proceed as if we were solving equations with numbers. In this case we have a fractional equation, so the first thing we do is clear of fractions.

From the formula solved for I, you can calculate the number of innings a pitcher pitched if you know the earned run average and the number of earned runs given up.

DO EXERCISE 1.

Example 2   The formula

$$\frac{1}{R} = \frac{1}{r_1} + \frac{1}{r_2}$$

gives the resistance R of two resistors r_1 and r_2 connected in parallel.* Solve it for r_1.

We multiply by the LCM, which is Rr_1r_2:

$$Rr_1r_2 \cdot \frac{1}{R} = Rr_1r_2 \cdot \left[\frac{1}{r_1} + \frac{1}{r_2} \right] \qquad \text{Multiplying by LCM}$$

$$Rr_1r_2 \cdot \frac{1}{R} = Rr_1r_2 \cdot \frac{1}{r_1} + Rr_1r_2 \cdot \frac{1}{r_2} \qquad \begin{array}{l}\text{Multiplying to}\\ \text{remove parentheses}\end{array}$$

$$\left. \begin{array}{l} \dfrac{R}{R} \cdot r_1r_2 = \dfrac{r_1}{r_1} \cdot Rr_2 + \dfrac{r_2}{r_2} \cdot Rr_1 \\[2mm] r_1r_2 = Rr_2 + Rr_1. \end{array} \right\} \quad \begin{array}{l}\text{Simplifying by}\\ \text{removing factors}\\ \text{of 1}\end{array}$$

One might be tempted at this point to multiply by $1/r_2$ to get r_1 alone on the left, BUT note that there is an r_1 on the right. We must get all the terms involving r_1 on the *same side* of the equation.

$$r_1r_2 - Rr_1 = Rr_2 \qquad \text{Adding } -Rr_1$$
$$r_1(r_2 - R) = Rr_2 \qquad \text{Factoring out } r_1$$

$$r_1 = \frac{Rr_2}{r_2 - R} \qquad \text{Multiplying by } \frac{1}{r_2 - R} \text{ to get } r_1 \text{ alone}$$

DO EXERCISE 2.

---

* Note that R, r_1, and r_2 are all different variables. It is common to use subscripts, as in r_1 and r_2, to distinguish variables. In many applications, this helps to keep ideas straight.

OBJECTIVE

After finishing Section 5.9, you should be able to:

● Solve a formula containing fractional expressions for a given letter.

1. The formula

$$\frac{PV}{T} = k$$

relates the pressure, volume, and temperature of a gas. Solve it for T. (*Hint:* Begin by clearing of fractions.)

2. The formula

$$\frac{1}{t} = \frac{1}{u} + \frac{1}{v}$$

gives the total time t for some work to be done by two workers, whose individual times are u and v. Solve it for t. (*Hint:* Begin by clearing of fractions. Then multiply to remove parentheses. Get all terms with t alone on one side. Then factor out t.)

ANSWERS ON PAGE A–20

3. Solve $V = \dfrac{A}{n(T - t)}$ for T.

To solve a formula we do the same things we would do to solve an equation with numbers. For fractional formulas the procedure is as follows.

> **To solve a fractional formula:**
>
> 1. **Clear of fractions.**
> 2. **Multiply if necessary to remove parentheses.**
> 3. **Get all terms with the unknown alone on one side of the equation.**
> 4. **Factor out the unknown.**
> 5. **Use the multiplication principle to get the unknown alone on one side.**

Example 3   Solve the formula $I = \dfrac{pT}{M + pn}$ for p.

We first clear of fractions by multiplying by the LCM, $M + pn$.

$$I(M + pn) = (M + pn)\frac{pT}{M + pn}$$

$$IM + Ipn = pT$$

We must now get all terms containing p alone on one side.

$$IM = pT - Ipn \qquad \text{Adding } -Ipn$$

$$IM = p(T - In) \qquad \text{Factoring out the unknown } p$$

$$\frac{IM}{T - In} = p \qquad \text{Multiplying by } \frac{1}{T - In}$$

We now have p alone on one side, and p does *not* appear on the other side, so we have solved the formula for p.

DO EXERCISE 3.

SOMETHING EXTRA

A CHALLENGE

This seemingly simple problem is not as easy as it looks. Can you solve it?

> A woman drives to work at 45 mph and arrives one minute early.
> She drives to work at 40 mph and arrives one minute late.
> How far does she live from work?

NAME   CLASS   ANSWERS

EXERCISE SET 5.9

▪•

1. Solve $\dfrac{W_1}{W_2} = \dfrac{d_1}{d_2}$ for d_1.

2. Solve $\dfrac{W_1}{W_2} = \dfrac{d_1}{d_2}$ for W_2.

1. _____

2. _____

3. Solve $s = \dfrac{(v_1 + v_2)t}{2}$ for t.

4. Solve $s = \dfrac{(v_1 + v_2)t}{2}$ for v_1.

3. _____

4. _____

5. Solve $\dfrac{1}{R} = \dfrac{1}{r_1} + \dfrac{1}{r_2}$ for r_2.

6. Solve $\dfrac{1}{R} = \dfrac{1}{r_1} + \dfrac{1}{r_2}$ for R.

5. _____

6. _____

7. Solve $R = \dfrac{gs}{g + s}$ for s.

8. Solve $I = \dfrac{2V}{R + 2r}$ for r.

7. _____

8. _____

9. Solve $\dfrac{1}{p} + \dfrac{1}{q} = \dfrac{1}{f}$ for p.

10. Solve $\dfrac{1}{p} + \dfrac{1}{q} = \dfrac{1}{f}$ for f.

9. _____

10. _____

11. Solve $I = \dfrac{nE}{R + nr}$ for r.

12. Solve $I = \dfrac{nE}{R + nr}$ for n.

11. _____

12. _____

13. Solve $S = \dfrac{H}{m(t_1 - t_2)}$ for H.

14. Solve $S = \dfrac{H}{m(t_1 - t_2)}$ for t_1.

13. _____

14. _____

15. Solve $\dfrac{E}{e} = \dfrac{R + r}{r}$ for e.

16. Solve $\dfrac{E}{e} = \dfrac{R + r}{r}$ for r.

15. _____

16. _____

5.10   VARIATION

● DIRECT VARIATION

A plumber earns $9 per hour. In 1 hour, $9 is earned. In 2 hours, $18 is earned. In 3 hours, $27 is earned, and so on. This gives rise to a set of ordered pairs of numbers, all having the same ratio:

(1, 9),   (2, 18),   (3, 27),   (4, 36),   and so on.

The ratio of earnings to time is $\frac{9}{1}$ in every case.

Whenever a situation gives rise to pairs of numbers in which the ratio is constant, we say that there is *direct variation*. Here the earnings *vary directly* as the time:

$$\frac{E}{t} = 9 \text{ (a constant)},   \text{or}   E = 9t.$$

Note that the constant, 9, is positive. Note also that when one variable increases, so does the other. When one decreases, so does the other.

> **Whenever a situation gives rise to a relation among variables $y = kx$, where k is a positive constant, we say that there is *direct variation*, or that y varies directly as x. The number k is called the *variation constant*.**

Example 1   Find the variation constant and an equation of variation, where y varies directly as x, and where $y = 32$ when $x = 2$.

We know that (2, 32) is a solution of $y = kx$. Thus,

$32 = k \cdot 2$ Substituting

$\frac{32}{2} = k,$ or $k = 16.$ Solving for k

The variation constant is 16.

The equation of variation is $y = 16x$.

DO EXERCISES 1 AND 2.

PROPORTIONS

Suppose that y varies directly as x. Then we have $y = kx$. Suppose also that (x_1, y_1) and (x_2, y_2) are solutions of the equation. Then, substituting, we have

$y_1 = kx_1$ and $y_2 = kx_2.$

Dividing, we get

$$\frac{y_1}{y_2} = \frac{kx_1}{kx_2} = \frac{k}{k} \cdot \frac{x_1}{x_2} = \frac{x_1}{x_2}.$$

The equation

$$\frac{y_1}{y_2} = \frac{x_1}{x_2}$$

is called a *proportion*. Such equations are helpful in solving problems.

OBJECTIVES

After finishing Section 5.10, you should be able to:

● Given a description of direct variation, find an equation of variation.

●● Solve problems involving direct variation.

●●● Given a description of inverse variation, write an equation of variation.

●●●● Solve problems involving inverse variation.

1. Find the variation constant and an equation of variation where y varies directly as x and $y = 8$ when $x = 20$.

2. Find the variation constant and an equation of variation where y varies directly as x and $y = 5.6$ when $x = 8$.

Solve two ways.

3. *Ohm's law* states that the voltage V in an electric circuit varies directly as the number of amperes I of electric current in the circuit. If the voltage is 10 volts when the current is 3 amperes, what is the voltage when the current is 15 amperes?

4. *An ecology problem.* The amount of garbage G produced in the United States varies directly as the number of people N who produce the garbage. It is known that 50 tons of garbage is produced by 200 people in 1 year. The population of San Francisco is 705,000. How much garbage is produced by San Francisco in 1 year?

⬤ ⬤   DIRECT-VARIATION PROBLEMS

Example 2   The number of centimeters W of water, produced from melting snow, varies directly as S, the number of centimeters of snow. Meteorologists have found that 150 cm of snow will melt to 16.8 cm of water. How many cm of water will 200 cm of snow melt to?

Method 1.   First find the variation constant using the data and then find an equation of variation.

$$W = kS$$
$$16.8 = k \cdot 150 \qquad \text{Substituting}$$
$$\frac{16.8}{150} = k \qquad \text{Solving for } k$$
$$0.112 = k \qquad \text{This is the variation constant.}$$

The equation of variation is $W = 0.112S$.

Next, use the equation to find how many cm of water will result from melting 200 cm of snow.

$$W = 0.112S$$
$$W = 0.112(200) \qquad \text{Substituting}$$
$$W = 22.4$$

Thus 200 cm of snow will melt to 22.4 cm of water.

Method 2.   Use a proportion and solve for W without finding the variation constant.

$$\frac{W_1}{W_2} = \frac{S_1}{S_2} \qquad \text{Writing the proportion}$$
$$\frac{16.8}{W_2} = \frac{150}{200} \qquad \text{Substituting} \leftarrow \boxed{\text{Think of (16.8, 150) as } (W_1, S_1). \text{ For a second solution we have } (W_2, 200).}$$
$$\frac{200}{150} \cdot 16.8 = W_2 \qquad \text{Solving}$$
$$22.4 = W_2$$

DO EXERCISES 3 AND 4.

⬤⬤⬤   INVERSE VARIATION

A bus is traveling a distance of 20 mi. At a speed of 20 mph, it will take 1 hour. At 40 mph it will take $\frac{1}{2}$ hour. At 60 mph it will take $\frac{1}{3}$ hour, and so on. This gives rise to a set of pairs of numbers, all having the same product:

$$(20, 1), \quad \left(40, \frac{1}{2}\right), \quad \left(60, \frac{1}{3}\right), \quad \left(80, \frac{1}{4}\right), \quad \text{and so on.}$$

Whenever a situation gives rise to pairs of numbers whose product is constant, we say that there is *inverse variation*. Here the time *varies inversely* as the speed:

$$rt = 20 \text{ (a constant)} \quad \text{or} \quad t = \frac{20}{r}.$$

Note that the constant, 20, is positive. Note also that when one variable increases, the other decreases.

> **Whenever a situation gives rise to a relation among variables** $y = k/x$, **where** k **is a positive constant, we say that there is** *inverse variation*, **or that** y *varies inversely as* x. **The number** k **is called the** *variation constant*.

Example 3   Find the variation constant and then an equation of variation where y varies inversely as x, and $y = 32$ when $x = 0.2$.

We know that $(0.2, 32)$ is a solution of $y = k/x$. We substitute

$$y = \frac{k}{x}$$

$$32 = \frac{k}{0.2} \qquad \text{Substituting}$$

$$(0.2)32 = k$$

$$6.4 = k. \qquad \text{Solving}$$

The variation constant is 6.4.

The equation of variation is $y = \dfrac{6.4}{x}$.

DO EXERCISE 5.

PROPORTIONS

Suppose that y varies inversely as x. Then we have $y = k/x$. Suppose also that (x_1, y_1) and (x_2, y_2) are solutions of the equation.

We substitute into $y = k/x$. Then we have

$$y_1 = \frac{k}{x_1} \quad \text{and} \quad y_2 = \frac{k}{x_2}.$$

Dividing, we get

$$\frac{y_1}{y_2} = \frac{\dfrac{k}{x_1}}{\dfrac{k}{x_2}} = \frac{k}{x_1} \cdot \frac{x_2}{k} = \frac{k}{k} \cdot \frac{x_2}{x_1} = \frac{x_2}{x_1}.$$

The equation

$$\frac{y_1}{y_2} = \frac{x_2}{x_1}$$

is called a *proportion*. Compare this with proportions obtained in direct variation. Note how the x-values are interchanged.

5. Find the variation constant and an equation of variation where y varies inversely as x and $y = 0.012$ when $x = 50$

ANSWER ON PAGE A–20

Solve two ways.

6. The time t required to drive a fixed distance varies inversely as the speed r. It takes 5 hours at 60 km/h to drive a fixed distance. How long would it take to drive that same distance at 40 km/h?

▪▪ INVERSE-VARIATION PROBLEMS

Example 4   The time t required to do a certain job varies inversely as the number of people P who work on the job (assuming that all do the same amount of work). It takes 4 hours for 12 people to erect some football bleachers. How long would it take 3 people to do the same job?

Method 1.   First find the variation constant using the data and then find an equation of variation:

$$t = \frac{k}{P}$$

$$4 = \frac{k}{12} \qquad \text{Substituting}$$

$$48 = k. \qquad \text{Solving for } k, \text{ the variation constant}$$

The equation of variation is $t = \dfrac{48}{P}$.

Next, use the equation to find the time it would take 3 people to do the job:

$$t = \frac{48}{P}$$

$$t = \frac{48}{3} \qquad \text{Substituting}$$

$$t = 16.$$

It would take 16 hours.

Method 2.   Use a proportion and solve for t without finding the variation constant.

> We think of (4, 12) as (t_1, P_1). For a second solution we have $(t_2, 3)$.

$$\frac{t_1}{t_2} = \frac{P_2}{P_1} \qquad \text{Writing the proportion}$$

$$\frac{4}{t_2} = \frac{3}{12} \qquad \text{Substituting}$$

$$4 \cdot 12 = 3t_2 \qquad \text{Solving}$$

$$16 = t_2$$

DO EXERCISE 6.

NAME   CLASS   ANSWERS

EXERCISE SET 5.10

▪ • Find the variation constant and an equation of variation where y varies directly as x and the following are true.

1. $y = 24$ when $x = 3$ **2.** $y = 5$ when $x = 12$ **3.** $y = 3.6$ when $x = 1$

1. _____

2. _____

4. $y = 2$ when $x = 5$ **5.** $y = 0.8$ when $x = 0.5$ **6.** $y = 0.6$ when $x = 0.4$
   (*Hint:* After you have substituted, clear of decimals.)

3. _____

4. _____

▪ • ▪ Solve.

7. The electric current I, in amperes, in a circuit varies directly as the voltage V. When 12 volts are applied, the current is 4 amperes. What is the current when 18 volts are applied?

8. Hooke's law states that the distance d that a spring is stretched by a hanging object varies directly as the weight w of the object. If the distance is 40 cm when the weight is 3 kg, what is the distance when the weight is 5 kg?

5. _____

6. _____

7. _____

9. The number N of plastic straws produced by a machine varies directly as the amount of time t the machine is operating. If the machine produces 20,000 straws in 8 hours, how many straws can it produce in 50 hours?

10. The number N of aluminum cans used each year varies directly as the number of people using the cans. If 250 people use 60,000 cans in one year, how many cans are used each year in Dallas, which has a population of 850,000?

8. _____

9. _____

10. _____

ANSWERS

●●● Find the variation constant and an equation of variation where y varies inversely as x and the following are true.

11. $y = 6$ when $x = 10$ **12.** $y = 16$ when $x = 4$ **13.** $y = 12$ when $x = 3$

11. _____

12. _____

14. $y = 9$ when $x = 5$ **15.** $y = 0.4$ when $x = 0.8$ **16.** $y = 1.5$ when $x = 0.3$
                                    (*Hint:* After you have
                                    substituted, clear of
                                    decimals.)

13. _____

14. _____   ⠿ Solve.

17. The current I in an electrical conductor varies inversely as the resistance R of the conductor. If the current is $\frac{1}{2}$ ampere when the resistance is 240 ohms, what is the current when the resistance is 540 ohms?

18. The time t required to empty a tank varies inversely as the rate r of pumping. If a pump can empty a tank in 45 minutes at the rate of 600 kL per minute, how long will it take the pump to empty the same tank at the rate of 1000 kL per minute?

15. _____

16. _____

17. _____

19. The weight W that a horizontal beam can support varies inversely as the length L of the beam. Suppose an 8-meter beam can support 1200 kilograms. How many kilograms can a 14-meter beam support?

20. The wavelength W of a radio wave varies inversely as its frequency F. A wave with a frequency of 1200 kilohertz has a length of 300 meters. What is the length of a wave with a frequency of 800 kilohertz?

18. _____

19. _____

20. _____

EXTENSION EXERCISES

SECTION 5.1

Simplify.

1. $\dfrac{x(x + 1) - 2(x + 3)}{(x + 1)(x + 2)(x + 3)}$

2. $\dfrac{2x - 5(x + 2) - (x - 2)}{x^2 - 4}$

3. $\dfrac{m^2 - t^2}{m^2 + t^2 + m + t + 2mt}$

4. $\dfrac{a^3 - 2a^2 + 2a - 4}{a^3 - 2a^2 - 3a + 6}$

5. $\dfrac{x^3 + x^2 - y^3 - y^2}{x^2 - 2xy + y^2}$

6. $\dfrac{u^6 + v^6 + 2u^3v^3}{u^3 - v^3 + u^2v - uv^2}$

7. $\dfrac{x^5 - x^3 + x^2 - 1 - (x^3 - 1)(x + 1)^2}{(x^2 - 1)^2}$

SECTION 5.2

Find the LCM.

8. 18, 42, 82, 120, 300, 700

9. $x^8 - x^4, x^5 - x^2, x^5 - x^3, x^5 + x^2$

10. The LCM of two expressions is $8a^4b^7$. One of the expressions is $2a^3b^7$. List all possibilities for the other expression.

SECTION 5.3

Do these calculations.

11. $2x^{-2} + 3x^{-2}y^{-2} - 7xy^{-1}$

12. $2(a^2 - b^2)^{-1} + 2(a^2 + b^2)^{-1} + 4a^2(a^4 - b^4)^{-1}$

13. $4(y - 1)(2y - 5)^{-1} + 5(2y + 3)(5 - 2y)^{-1} + (y - 4)(2y - 5)^{-1}$

SECTION 5.4

Simplify.

14. $\dfrac{5x^{-1} - 5y^{-1} + 10x^{-1}y^{-1}}{6x^{-1} - 6y^{-1} + 12x^{-1}y^{-1}}$

15. $\dfrac{\dfrac{4a}{2a^2 - a - 1} - \dfrac{4}{a - 1}}{\dfrac{1}{a - 1} + \dfrac{6}{2a + 1}}$

16. $2 + \dfrac{2}{2 + \dfrac{2}{2 + \dfrac{2}{2 + \dfrac{2}{x}}}}$

17. $\left[\dfrac{\dfrac{x + 3}{x - 3} + 1}{\dfrac{x + 3}{x - 3} - 1}\right]^4$

18. $(a^2 - ab + b^2)^{-1}(a^2b^{-1} + b^2a^{-1})(a^{-2} - b^{-2})(a^{-2} + 2a^{-1}b^{-1} + b^{-2})^{-1}$

Find the reciprocal of each of the following and simplify.

19. $x^2 - \dfrac{1}{x}$

20. $\dfrac{1 - \dfrac{1}{a}}{a - 1}$

21. $\dfrac{a^3 + b^3}{a + b}$

SECTION 5.5

Divide.

22. $(2x^6 + 5x^4 - x^3 + 1) \div (x^2 + x + 1)$

23. $(2y^5 + y^3 - 2y - 3) \div (y^2 - 3y + 1)$

24. $(x^4 - x^3y + x^2y^2 + 2x^2y - 2xy^2 + 2y^3) \div (x^2 - xy + y^2)$

25. $(4a^3b + 5a^2b^2 + a^4 + 2ab^3) \div (a^2 + 2b^2 + 3ab)$ **26.** $(x^4 - y^4) \div (x - y)$

27. $(a^7 + b^7) \div (a + b)$ **28.** Find k so that when $x^3 - kx^2 + 3x + 7k$ is divided by $x + 2$ the remainder will be 0.

SECTION 5.7

Solve.

29. $\left(\dfrac{1}{1+x} + \dfrac{x}{1-x}\right) \div \left(\dfrac{x}{1+x} - \dfrac{1}{1-x}\right) = -1$ **30.** $\dfrac{x+3}{x+2} - \dfrac{x+4}{x+3} = \dfrac{x+5}{x+4} - \dfrac{x+6}{x+5}$

SECTION 5.8

31. A boat travels 96 km downstream in 4 hr. It travels 28 km upstream in 7 hr. Find the speed of the boat and the speed of the stream.

32. An airplane carries enough fuel for 6 hours of flight time, and its speed in still air is 240 mph. It leaves an airport against a wind of 40 mph and returns to the same airport with a wind of 40 mph. How far can it fly under those conditions without refueling?

33. A motor boat travels 3 times as fast as the current. A trip up the river and back takes 10 hours, and the total distance of the trip is 100 km. Find the speed of the current.

34. An employee drives to work at 50 mph and arrives one minute late. The employee drives to work at 60 mph and arrives 5 minutes early. How far does the employee live from work?

Average speed is defined as *total distance divided by total time.*

35. A driver went 200 km. For the first half of the trip the driver traveled at a speed of 40 km/h. For the second half of the trip the driver traveled at a speed of 60 km/h. What was the average speed for the entire trip? (It is *not* 50 km/h.)

36. For half of a trip a driver travels at 40 mph. What speed would the driver have to travel for the last half of the trip so that the average speed for the entire trip would be 45 mph?

37. Three trucks A, B, and C, working together, can move a load of sand in t hours. When working alone, it takes A, 1 extra hour to move the sand; B, 6 extra hours; and C, t extra hours. Find t.

SECTION 5.10

38. A loan of $9000 earns simple interest of $1665. Find an equation of variation that describes the simple interest I in terms of the loan principal P.

39. A 96-fluid-ounce container of fabric softener costs $2.99. A 128-fluid-ounce container of the same softener costs $3.99. Which is the better value? Find a way to use proportions to solve this problem.

40. During the summer Steve sold 4 sweepers for every 3 sweepers sold by his sister, Stephanie. If together they sold 98 sweepers, how many did each sell?

41. In each of the following equations state whether y varies directly as x, inversely as x, or neither directly nor inversely as x.

a) $7xy = 14$ b) $x - 2y = 12$ c) $-2x + 3y = 0$ d) $x = \dfrac{3}{4}y$ e) $\dfrac{x}{y} = 2$

NAME   SCORE   ANSWERS

TEST OR REVIEW—CHAPTER 5

Review the material in the chapter. Then test yourself by doing these exercises. If you miss
an item, restudy the section indicated to the left of the item.

1. _____

[5.1, ⁘]   **1.** Multiply and simplify: $\dfrac{2x^2 + 20x + 50}{x^2 - 4} \cdot \dfrac{x + 2}{x + 5}$.

2. _____

[5.1, ⁘]   **2.** Divide and simplify: $\dfrac{y^2 - 16}{2y + 6} \div \dfrac{y - 4}{y + 3}$.

[5.2, ⁚⁚]   **3.** Find the LCM: $x^2 + x - 6, x^2 + 8x + 15$.

3. _____

[5.3, ••]   **4.** Add and simplify: $\dfrac{x^2}{x - y} + \dfrac{y^2}{y - x}$.

4. _____

[5.3, ⁘]   **5.** Calculate and simplify: $\dfrac{1}{x + 1} - \dfrac{x + 2}{x^2 - 1} + \dfrac{3}{x - 1}$.

5. _____

[5.4, •]   **6.** Simplify: $\dfrac{1 - \dfrac{1}{x^2}}{1 - \dfrac{1}{x}}$.

6. _____

[5.5, ••]   **7.** Divide: $(x^2 - 10x - 25) \div (x - 3)$.

7. _____

[5.6, •]   **8.** Divide. Use synthetic division.

$(6y^4 + 15y^3 - 28y + 6) \div (y + 2)$

8. _____

ANSWERS

[5.7, ▮ ▮]   **9.** Solve: $\dfrac{2}{x-1} = \dfrac{3}{x+3}$.

9. _____

[5.8, ▮ ▮]   **10.** Two pipes are filling a tank. One of them can fill the tank alone in 6 hours. The other can fill the tank alone in 4 hours. How long will it take them together?

10. _____

[5.8, ▮ ▮]   **11.** A boat can go upstream a certain distance in 6 hours. It can go that same distance downstream in 2 hours and 15 minutes. The speed of the stream is 5 km/h. What is the speed of the boat in still water?

11. _____

[5.9, ▮ ▮]   **12.** Solve $m = \dfrac{2t}{r+t}$ for t.

12. _____

[5.10, ▮ ▮]   **13.** Income varies directly as the time worked. If a job pays $275 for 40 hours, what does it pay for 72 hours?

13. _____

[5.10, ▮ ▮]   **14.** The time required to drive a certain distance varies inversely as the speed. If you can make a trip in $3\frac{1}{2}$ hours at 45 mph, how long will it take you at 60 mph?

14. _____

EXPONENTS, POWERS, AND ROOTS

6

In cold weather we feel colder if there is wind than if there is not. Wind chill can be computed using a formula involving square root.

1. Simplify: $|-8|$.

2. Multiply: $(3x^2y^{-4})(4x^3y^2)$.

3. Divide: $\dfrac{32x^3y^{10}}{4x^4y^{-3}}$.

4. Simplify: $(4x^2y^{-3})^4$.

5. Factor: $4x^2 - 49$.

6. Factor: $2x^2 - 26x + 72$.

7. Factor: $9y^2 - 12y + 4$.

8. Multiply: $(2x - 3)(2x + 3)$.

9. Multiply: $(2a + b)(a - 3b)$.

10. Solve: $3x(3x - 5) = 0$.

11. Solve:

$$2(8 - 3x) = 3 - 5(x - 1).$$

OBJECTIVES

After finishing Section 6.1, you should be able to:

[•] Simplify expressions with integer exponents.

[••] Convert between scientific notation and ordinary decimal notation.

[•••] Multiply and divide using scientific notation.

[⁞⁞] Use scientific notation in estimating.

6.1 SCIENTIFIC NOTATION

[•] INTEGER EXPONENTS

Let us recall from Chapter 1 some of the properties of integer exponents.

$a^m a^n = a^{m+n}$ Add exponents when multiplying.

$\dfrac{a^m}{a^n} = a^{m-n}$ Subtract exponents when dividing.

$(a^m)^n = a^{mn}$ Multiply exponents when raising a power to a power.

$a^{-n} = \dfrac{1}{a^n}$,   hence $\dfrac{1}{a^{-n}} = a^n$. a^n and a^{-n} are reciprocals.

$a^1 = a$

$a^0 = 1$ if $a \neq 0$

Examples   Simplify.

1. $(4xy)^1 = 4xy$

2. $(-8x^3y^7)^0 = 1$, assuming neither x nor y is 0

3. $4x^2x^{-5} = 4x^{2-5} = 4x^{-3}$, or $\dfrac{4}{x^3}$

4. $(-8x^5y^7)(4x^3y^{-2}) = -8 \cdot 4x^{5+3}y^{7-2} = -32x^8y^5$

5. $\dfrac{-10x^7}{2x^6} = \dfrac{-10}{2}x^{7-6} = -5x$

6. $\dfrac{42x^8y^3}{21x^4y^8} = \dfrac{42}{21}x^{8-4}y^{3-8} = 2x^4y^{-5}$, or $\dfrac{2x^4}{y^5}$

7. $(x^{-3})^5 = x^{-15}$, or $\dfrac{1}{x^{15}}$

8. $\dfrac{-8^6y^4z^{-3}}{2^{-1}y^{-2}z^2} = \dfrac{-(2^3)^6y^4z^{-3}}{2^{-1}y^{-2}z^2} = -2^{18}2^{-(-1)}y^4y^{-(-2)}z^{-3}z^{-2}$

$$= \dfrac{-2^{19}y^6}{z^5}$$

9. $(5x^{-2}y^4)^3 = 5^3x^{-6}y^{12} = 125x^{-6}y^{12}, \text{ or } \dfrac{125y^{12}}{x^6}$

DO EXERCISES 1–10.

The following illustrates a new property of exponents.

$$\left(\dfrac{x}{y}\right)^3 = \dfrac{x}{y} \cdot \dfrac{x}{y} \cdot \dfrac{x}{y} = \dfrac{x \cdot x \cdot x}{y \cdot y \cdot y} = \dfrac{x^3}{y^3}$$

> **For any real numbers a and b, where $b \neq 0$, and any integer n,**
> $$\left(\dfrac{a}{b}\right)^n = \dfrac{a^n}{b^n}.$$

Examples Simplify.

To raise a fractional expression to a power we can raise both numerator and denominator to that power.

10. $\left(\dfrac{4x^2}{y}\right)^3 = \dfrac{(4x^2)^3}{y^3} = \dfrac{4^3(x^2)^3}{y^3} = \dfrac{64x^6}{y^3}$

11. $\left(\dfrac{x^{-5}}{y^4}\right)^{-3} = \dfrac{(x^{-5})^{-3}}{(y^4)^{-3}} = \dfrac{x^{15}}{y^{-12}}, \text{ or } x^{15}y^{12}$

12. $\left(\dfrac{2x^{-2}}{y^3}\right)^4 = \dfrac{(2x^{-2})^4}{(y^3)^4} = \dfrac{2^4(x^{-2})^4}{(y^3)^4} = \dfrac{16x^{-8}}{y^{12}}, \text{ or } \dfrac{16}{x^8y^{12}}$

13. $\left(\dfrac{4x^3y^{-1}}{3z^3}\right)^2 = \dfrac{(4x^3y^{-1})^2}{(3z^3)^2} = \dfrac{4^2x^6y^{-2}}{3^2z^6}, \text{ or } \dfrac{16}{9}x^6y^{-2}z^{-6}$

DO EXERCISES 11–13.

●● SCIENTIFIC NOTATION

The following are examples of scientific notation, which is very useful when naming or calculating with very large or very small numbers. It is also very helpful when estimating.

6.4×10^{13} means 64000000000000; 4.6×10^{-6} means 0.0000046.

> *Scientific notation* **for a number consists of exponential notation for a power of 10, and, if needed, decimal notation for a number a between 1 and 10 and a multiplication sign. The following are both scientific notation: $a \times 10^b$; 10^b.**

We can convert to scientific notation by multiplying by 1, choosing a name like $10^b \cdot 10^{-b}$ for the number 1.

Simplify.

1. $(-9ab^2)^1$

2. $(-9x^2y)^0$

3. $-\dfrac{1}{2}y^{-2} \cdot 4y^{-7}$

4. $(7z^{10})(-4y^{-3}z^{-8})$

5. $30a^4 \cdot 6a^{-9}$

6. $(-5a^2b^{-3})(4a^{-3}b^6)$

7. $\dfrac{-24x^{-3}y^5}{6x^5y^7}$

8. $\dfrac{10a^3b^2c^5}{15a^5b^7c^2}$

9. $(y^{-4})^6$

10. $(8a^2b^{-4})^3$

Simplify.

11. $\left(\dfrac{a}{b}\right)^4$

12. $\left(\dfrac{x^2y^{-3}}{3a}\right)^4$

13. $\left(\dfrac{y^{-2}}{x^3}\right)^{-4}$

ANSWERS ON PAGE A–21

14. Convert 460,000,000,000 to scientific notation.

15. The distance from the earth to the sun is about 93,000,000 miles. Write scientific notation for this number.

16. Convert 0.00000001235 to scientific notation.

17. The mass of a hydrogen atom is about 0.0000000000000000000000017 grams. Write scientific notation for this number.

Convert to decimal notation.

18. 7.893×10^{11}

19. 5.67×10^{-5}

Multiply and write scientific notation for the answer.

20. $(9.1 \times 10^{-17})(8.2 \times 10^3)$

21. $(1.12 \times 10^{-8})(5 \times 10^{-7})$

Example 14   Light travels about 9,460,000,000,000 kilometers in one year. Write scientific notation for the number.

We want to move the decimal point 12 places, between the 9 and the 4, so we choose $10^{-12} \times 10^{12}$ as a name for 1. Then we multiply.

$9{,}460{,}000{,}000{,}000 \times 10^{-12} \times 10^{12}$ Multiplying by 1

$= 9.46 \times 10^{12}$ The 10^{-12} moved the decimal point 12 places to the left and we have scientific notation.

DO EXERCISES 14 AND 15.

Example 15   Write scientific notation for 0.0000000000156.

We want to move the decimal point 11 places. We choose $10^{11} \times 10^{-11}$ as a name for 1, and then multiply.

$0.0000000000156 \times 10^{11} \times 10^{-11}$ Multiplying by 1

$= 1.56 \times 10^{-11}$ The 10^{11} moved the decimal point 11 places to the right and we have scientific notation.

You should try to make conversions to scientific notation mentally as much as possible.

DO EXERCISES 16 AND 17.

Examples   Convert mentally to decimal notation.

16. $7.893 \times 10^5 = 789{,}300$ Moving decimal point 5 places to the right

17. $4.7 \times 10^{-8} = 0.000000047$ Moving decimal point 8 places to the left

> A positive exponent indicates a large number. A negative exponent indicates a small number.

DO EXERCISES 18 AND 19.

●●● MULTIPLYING AND DIVIDING

Multiplying and dividing in scientific notation is easy because we can use the properties of exponents.

Example 18   Multiply and write scientific notation for the answer: $(3.1 \times 10^5)(4.5 \times 10^{-3})$.

We apply the commutative and associative laws to get

$(3.1 \times 4.5)(10^5 \times 10^{-3}) = 13.95 \times 10^2.$

To find scientific notation for the result, we convert 13.95 to scientific notation and then simplify.

$13.95 \times 10^2 = (1.395 \times 10^1) \times 10^2 = 1.395 \times 10^3$

DO EXERCISES 20 AND 21.

Example 19   Divide and write scientific notation for the answer: $\dfrac{6.4 \times 10^{-7}}{8.0 \times 10^{6}}$.

$$\frac{6.4 \times 10^{-7}}{8.0 \times 10^{6}} = \frac{6.4}{8.0} \times \frac{10^{-7}}{10^{6}} \qquad \text{Factoring}$$

This shows two divisions.

$$= 0.8 \times 10^{-13} \qquad \text{Doing the divisions separately}$$

$$= (8.0 \times 10^{-1}) \times 10^{-13} \qquad \text{Converting 0.8 to scientific notation}$$

$$= 8.0 \times 10^{-14}$$

The answer 0.8×10^{-13} is not incorrect, except that we have not converted it to scientific notation.

DO EXERCISES 22 AND 23.

Divide and write scientific notation for the answer.

22. $\dfrac{4.2 \times 10^{5}}{2.1 \times 10^{2}}$

▣▣ ESTIMATING AND APPROXIMATING

Scientific notation is helpful in estimating or approximating.

Example 20   Estimate: $\dfrac{780,000,000 \times 0.00071}{0.000005}$.

We first convert to scientific notation:

$$\frac{(7.8 \times 10^{8})(7.1 \times 10^{-4})}{5 \times 10^{-6}}.$$

Next, we group as follows:

$$\frac{7.8 \times 7.1}{5} \times \frac{(10^{8})(10^{-4})}{10^{-6}}.$$

We round the numbers that are between 1 and 10 and calculate:

$$\frac{8 \times 7}{5} = \frac{56}{5} \approx \frac{55}{5} = 11.$$

This answer will vary, depending on how you round.

Next, we consider the powers of 10.

$$\frac{10^{8} \times 10^{-4}}{10^{-6}} = \frac{10^{4}}{10^{-6}} = 10^{10}$$

With some practice you will be able to estimate mentally.

We now have an approximation of 11×10^{10}, or 1.1×10^{11}.

DO EXERCISE 24.

23. $\dfrac{1.1 \times 10^{-4}}{2.0 \times 10^{-7}}$

24. Estimate. Answers will vary depending on how you round.

$$\frac{830,000,000 \times 0.0000012}{3,100,000}$$

Some examples using scientific notation

The distance from the planet Pluto to the sun is 3.664×10^{9} miles.

The mass of an electron is 9.11×10^{-28} grams.

The population of the United States was 2.04×10^{8}.

The wavelength of a certain red light is 6.6×10^{-5} cm.

The gross national product (GNP) one year was \$$9.32 \times 10^{11}$.

An electron has a charge of 4.8×10^{-10} electrostatic units.

ANSWERS ON PAGE A–21

SOMETHING EXTRA

AN APPLICATION: COMPOUND INTEREST

Compound interest can be computed from a formula. The amount of money A in an account is given by

$$A = P\left(1 + \frac{r}{n}\right)^{nt},$$

where P is the principal, r is the annual interest rate, and n is the number of times per year interest is compounded. Of course, t is the time in years.

Example   Suppose $1000 is invested at 8% compounded quarterly. At the end of two years, how much will be in the account?

$P = 1000$, $r = 8\%$ or 0.08, $n = 4$, because interest is compounded quarterly, and $t = 2$.

We substitute into the formula:

$$A = 1000\left(1 + \frac{0.08}{4}\right)^{4 \cdot 2}$$

$$= 1000(1.02)^8$$

$$= 1171.66.$$

The amount is $1171.66.

EXERCISES

A calculator with an exponential key $\boxed{a^b}$ will be useful for some of these.

1. Suppose $1000 is invested at 8% compounded semi-annually. How much will be in the account at the end of two years?

2. Suppose $1000 is invested at 8% compounded monthly. How much will be in the account at the end of two years?

3. Suppose $1000 is invested at 8%. How much will be in the account at the end of one year if interest is compounded daily?

4. Suppose $1000 is invested at 8%. How much will be in the account at the end of one year if interest is compounded hourly?

NAME   CLASS

EXERCISE SET 6.1

▪●▪   Simplify.

1. $(-4m^2n^3)^0$ **2.** $(-9x^3y^5)^0$ **3.** $(x^2 + y)^1$

4. $(y^3 + z)^1$ **5.** $(-3r^2s^{-3})(5r^{-6}s^5)$ **6.** $(8s^{-4}t^5)(-2s^8t^{-7})$

7. $(-2m^{-2}n^3)^3$ **8.** $(-3x^{-3}y^5)^2$ **9.** $\dfrac{5x^{-4}}{25x^{-10}}$

10. $\dfrac{40z^{-8}}{5z^{-6}}$ **11.** $(-2a^2b^3)(6a^{-4}b^{-1})$ **12.** $(-5c^3d^{-5})(7d^4c^{-1})$

13. $\dfrac{a^2b^{-3}}{a^4b^{-2}}$ **14.** $\dfrac{x^3y^{-5}}{x^5y^{-7}}$ **15.** $\dfrac{-6^5y^4z^{-5}}{2^{-2}y^{-2}z^3}$

16. $\dfrac{9^{-2}x^{-4}y}{3^{-3}x^{-3}y^2}$ **17.** $(2x^{-4}y^{-2})^{-3}$ **18.** $(4a^{-4}b^{-5})^{-3}$

19. $\left(\dfrac{3a^{-2}b}{c^2}\right)^3$ **20.** $\left(\dfrac{2x^2y^{-2}}{3a^2}\right)^3$

1. _____

2. _____

3. _____

4. _____

5. _____

6. _____

7. _____

8. _____

9. _____

10. _____

11. _____

12. _____

13. _____

14. _____

15. _____

16. _____

17. _____

18. _____

19. _____

20. _____

ANSWERS

21. _____

22. _____

23. _____

24. _____

25. _____

26. _____

27. _____

28. _____

29. _____

30. _____

31. _____

32. _____

33. _____

34. _____

35. _____

36. _____

37. _____

38. _____

39. _____

40. _____

41. _____

42. _____

●● Convert to scientific notation.

21. 47,000,000,000

22. 2,600,000,000,000

23. 863,000,000,000,000,000

24. 957,000,000,000,000,000

25. 0.000000016

26. 0.000000263

27. 0.00000000007

28. 0.00000000009

Convert to decimal notation.

29. 6.73×10^8

30. 9.24×10^7

31. 8.923×10^{-10}

32. 7.034×10^{-2}

●●● Multiply and write scientific notation for the answer.

33. $(2.3 \times 10^6)(4.2 \times 10^{-11})$

34. $(6.5 \times 10^3)(5.2 \times 10^{-8})$

35. $(2.34 \times 10^{-8})(5.7 \times 10^{-4})$

36. $(3.26 \times 10^{-6})(8.2 \times 10^9)$

Divide and write scientific notation for the answer.

37. $\dfrac{8.5 \times 10^8}{3.4 \times 10^5}$

38. $\dfrac{5.1 \times 10^6}{3.4 \times 10^3}$

39. $\dfrac{4.0 \times 10^{-6}}{8.0 \times 10^{-3}}$

40. $\dfrac{7.5 \times 10^{-9}}{2.5 \times 10^{-4}}$

●● Estimate.

41. $\dfrac{(6.1 \times 10^4)(7.2 \times 10^{-6})}{9.8 \times 10^{-4}}$

42. $\dfrac{(8.05 \times 10^{-11})(5.9 \times 10^7)}{3.1 \times 10^{14}}$

6.2 RADICAL EXPRESSIONS

�⚬ SQUARE ROOTS

A *square root* of a number a is a number c whose second power is a, that is, $c^2 = a$.

> 5 is a square root of 25 because $5^2 = 5 \cdot 5 = 25$.
>
> -5 is a square root of 25 because $(-5)^2 = (-5)(-5) = 25$.

-4 does not have a real square root because there is no real number b such that $b^2 = -4$.

> **Every positive real number has two real-number square roots. The number 0 has just one square root, 0 itself. Negative numbers do not have real number square roots.***

Example 1   Find the two square roots of 64.

The square roots are 8 and -8 because $8^2 = 64$ and $(-8)^2 = 64$.

DO EXERCISES 1–3.

> **The *principal* square root of a nonnegative number is its non-negative square root. The symbol $\sqrt{a}$ represents the principal square root of a. To name the negative square root of a we write $-\sqrt{a}$.**

Examples   Simplify.

2. $\sqrt{25} = 5$ ⟨ Remember: $\sqrt{}$ indicates the principal (nonnegative) square root.

3. $\sqrt{\dfrac{25}{64}} = \dfrac{5}{8}$ **4.** $-\sqrt{64} = -8$

5. $\sqrt{0} = 0$ **6.** $\sqrt{0.0049} = 0.07$

7. $\sqrt{-49}$ Does not exist. Negative numbers do not have real number square roots.

DO EXERCISES 4–10.

> **The symbol $\sqrt{}$ is called a *radical*. An expression written with a radical is called a *radical expression*. The expression written under the radical is called the *radicand*.**

These are radical expressions:

$$\sqrt{5}, \qquad \sqrt{a}, \qquad -\sqrt{5x}, \qquad \frac{\sqrt{y^2 + 7}}{\sqrt{x}}.$$

---

* Later, we will see that there is a number system other than the real-number system, in which negative numbers do have square roots.

Find the square roots of each number.

1. 9

2. 36

3. 121

Simplify.

4. $\sqrt{1}$

5. $\sqrt{36}$

6. $\sqrt{\dfrac{81}{100}}$

7. $\sqrt{0.0064}$

Find the following.

8. (a) $\sqrt{16}$ (b) $-\sqrt{16}$ (c) $\sqrt{-16}$

9. (a) $\sqrt{49}$ (b) $-\sqrt{49}$ (c) $\sqrt{-49}$

10. (a) $\sqrt{144}$ (b) $-\sqrt{144}$ (c) $\sqrt{-144}$

ANSWERS ON PAGE A–22

Identify the radicand in each
expression.

11. $\sqrt{28 + x}$

12. $\sqrt{\dfrac{y}{y + 3}}$

Find the following. Assume that letters
can represent any real numbers.

13. $\sqrt{y^2}$

14. $\sqrt{(-24)^2}$

15. $\sqrt{(5y)^2}$

16. $\sqrt{16y^2}$

17. $\sqrt{(x + 7)^2}$

18. $\sqrt{4(x - 2)^2}$

19. $\sqrt{49(y + 5)^2}$

20. $\sqrt{x^2 - 6x + 9}$

Example 8 Identify the radicand in $\sqrt{x^2 - 9}$.

The radicand is $x^2 - 9$.

DO EXERCISES 11 AND 12.

●● FINDING $\sqrt{a^2}$

In the expression $\sqrt{a^2}$, the radicand is a perfect square.

Suppose $a = 5$. Then we have $\sqrt{5^2}$, which is $\sqrt{25}$, or 5.
Suppose $a = -5$. Then we have $\sqrt{(-5)^2}$, which is $\sqrt{25}$, or 5.
Suppose $a = 0$. Then we have $\sqrt{0^2}$, which is $\sqrt{0}$, or 0.

The symbol $\sqrt{a^2}$ does not represent a negative number. It represents the principal square root of a^2. Note that if a represents a positive number or 0, then $\sqrt{a^2}$ represents a. If a is negative, then $\sqrt{a^2}$ represents the additive inverse of a. In all cases the radical expression represents the absolute value of a.

> For any real number a, $\sqrt{a^2} = |a|$. The principal (nonnegative) square root of a^2 is the absolute value of a.

Examples Find the following. Assume that letters can represent any real numbers.

9. $\sqrt{(-16)^2} = |-16|$, or 16

10. $\sqrt{(3b)^2} = |3b|$, or $3|b|$ ←

> $|3b|$ can be simplified to $3|b|$ because the absolute value of any product is the product of the absolute values. That is, $|a \cdot b| = |a| \cdot |b|$. In this case, $|3b| = |3| \cdot |b|$, or $3|b|$.

11. $\sqrt{(x - 1)^2} = |x - 1|$

12. $\sqrt{x^2 + 8x + 16} = \sqrt{(x + 4)^2}$
$= |x + 4|$ ←

> $|x + 4|$ does *not* mean the same as $|x| + 4$.

DO EXERCISES 13–20.

●●● CUBE ROOTS

The number c is the cube root of a if its third power is a, that is, $c^3 = a$. For example,

2 is the cube root of 8 because $2^3 = 2 \cdot 2 \cdot 2 = 8$;
-4 is the cube root of -64 because $(-4)^3 = (-4)(-4)(-4) = -64$.

We used the word "the" with cube roots because of the following.

> Every real number has exactly one cube root in the system of real numbers.

The symbol $\sqrt[3]{a}$ represents the cube root of a.

Examples Find the following.

13. $\sqrt[3]{8} = 2$ **14.** $\sqrt[3]{-27} = -3$ **15.** $\sqrt[3]{-\dfrac{216}{125}} = -\dfrac{6}{5}$

16. $\sqrt[3]{x^3} = x$ **17.** $\sqrt[3]{-8y^3} = -2y$

No absolute value signs are needed when finding cube roots because a real number has just one cube root. The cube root of a positive number is positive. The cube root of a negative number is negative.

DO EXERCISES 21–24.

⚃ ODD AND EVEN kth ROOTS

The 5th root of a number a is the number c for which $c^5 = a$. There are also 7th roots, 9th roots, and so on. Whenever the number k in $\sqrt[k]{}$ is an odd number, we say that we are taking an *odd* root. The number k is called the *index*. When the index is 2 we do not write it.

Every number has just one odd root. If the number is positive, the root is positive. If the number is negative, the root is negative.

Examples Find the following.

18. $\sqrt[5]{32} = 2$ **19.** $\sqrt[5]{-32} = -2$

20. $-\sqrt[5]{32} = -2$ **21.** $-\sqrt[5]{-32} = -(-2) = 2$

22. $\sqrt[7]{x^7} = x$ **23.** $\sqrt[9]{(x-1)^9} = x - 1$

Absolute value signs are never needed when finding odd roots.

DO EXERCISES 25 AND 26.

When the index k in $\sqrt[k]{}$ is an even number, we say that we are taking an *even* root. Every positive real number has two kth roots when k is even. One of those roots is positive and one is negative. Negative real numbers do not have kth roots when k is even. When finding even kth roots, absolute value signs are sometimes necessary, as they are with square roots.

Examples Find the following. Assume that letters can represent any real number.

24. $\sqrt[4]{16} = 2$ **25.** $-\sqrt[4]{16} = -2$

26. $\sqrt[4]{-16}$ Does not exist   **27.** $\sqrt[4]{81x^4} = 3|x|$

28. $\sqrt[6]{(y+7)^6} = |y + 7|$

> **For any real number a:**
>
> a) $\sqrt[k]{a^k} = |a|$ **when k is even. We use absolute value when k is even unless a is nonnegative.**
>
> b) $\sqrt[k]{a^k} = a$ **when k is odd. We do not use absolute value when k is odd.**

DO EXERCISES 27–30.

Find the following.

21. $\sqrt[3]{-64}$

22. $\sqrt[3]{27y^3}$

23. $\sqrt[3]{8(x+2)^3}$

24. $\sqrt[3]{-\dfrac{343}{64}}$

Find the following. Assume that letters can represent any real number.

25. $\sqrt[5]{-32x^5}$

26. $\sqrt[7]{(3x+2)^7}$

Find the following.

27. (a) $\sqrt[4]{81}$ (b) $-\sqrt[4]{81}$ (c) $\sqrt[4]{-81}$

28. $\sqrt[4]{16(x-2)^4}$

29. $\sqrt[6]{x^6}$

30. $\sqrt[8]{(x+3)^8}$

SOMETHING EXTRA
AMAZE YOUR FRIENDS

Ask a friend to pick any number; then:

Double it,
Add 4,
Divide by 2,
Add the original number,
Divide by 2.

Then ask the result. Subtract 1 from the result, and this will always be the original number. You can then tell the person what the original number was.

How does this work? Let's call the original number x, and then write down, algebraically, the sequence of calculations above.

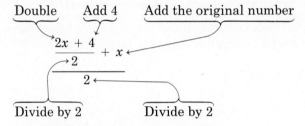

Now this expression can be simplified.

$$\frac{\dfrac{2x + 4}{2} + x}{2} = \frac{\dfrac{2(x + 2)}{2} + x}{2} \qquad \text{Factoring } 2x + 4$$

$$= \frac{x + 2 + x}{2} \qquad \text{Removing a factor of 1}$$

$$= \frac{2x + 2}{2}$$

$$= \frac{2(x + 1)}{2} \qquad \text{Factoring } 2x + 2$$

$$= x + 1 \qquad \text{Removing a factor of 1}$$

Thus the two expressions, $x + 1$ and the original one, name the same number for all sensible replacements. Your friend does the long calculation, and *you* know that adding 1 gives the same answer.

Any expression that can be simplified like this can be used as the basis of a number trick.

EXERCISE

Make up your own number trick and use it to amaze your friends.

NAME   CLASS

EXERCISE SET 6.2

⬤ Find the square roots of each number.

1. 16   **2.** 225   **3.** 144   **4.** 9   **5.** 400   **6.** 81

Find the following.

7. $-\sqrt{\dfrac{49}{36}}$ **8.** $-\sqrt{\dfrac{361}{9}}$ **9.** $\sqrt{196}$ **10.** $\sqrt{441}$

11. $-\sqrt{\dfrac{16}{81}}$ **12.** $-\sqrt{\dfrac{81}{144}}$ **13.** $\sqrt{0.09}$ **14.** $\sqrt{0.36}$

15. $-\sqrt{0.0049}$ **16.** $\sqrt{0.0144}$

Identify the radicand in each expression.

17. $5\sqrt{p^2 + 4}$ **18.** $-7\sqrt{y^2 - 8}$ **19.** $x^2 y^2 \sqrt{\dfrac{x}{y + 4}}$ **20.** $a^2 b^3 \sqrt{\dfrac{a}{a^2 - b}}$

⬤⬤ Find the following. Assume that letters can represent any real number.

21. $\sqrt{16x^2}$ **22.** $\sqrt{25t^2}$ **23.** $\sqrt{(-7c)^2}$

24. $\sqrt{(-6b)^2}$ **25.** $\sqrt{(a + 1)^2}$ **26.** $\sqrt{(5 - b)^2}$

27. $\sqrt{x^2 - 4x + 4}$ **28.** $\sqrt{y^2 + 16y + 64}$

29. $\sqrt{4x^2 + 28x + 49}$ **30.** $\sqrt{9x^2 - 30x + 25}$

ANSWERS

1. _____
2. _____
3. _____
4. _____
5. _____
6. _____
7. _____
8. _____
9. _____
10. _____
11. _____
12. _____
13. _____
14. _____
15. _____
16. _____
17. _____
18. _____
19. _____
20. _____
21. _____
22. _____
23. _____
24. _____
25. _____
26. _____
27. _____
28. _____
29. _____
30. _____

ANSWERS

31. _____

32. _____

33. _____

34. _____

35. _____

36. _____

37. _____

38. _____

39. _____

40. _____

41. _____

42. _____

43. _____

44. _____

45. _____

46. _____

47. _____

48. _____

49. _____

50. _____

51. _____

52. _____

53. _____

54. _____

55. _____

56. _____

57. a) _____

b) _____

c) _____

d) _____

●●● Simplify.

31. $\sqrt[3]{27}$

32. $-\sqrt[3]{64}$

33. $\sqrt[3]{-64x^3}$

34. $\sqrt[3]{-125y^3}$

35. $\sqrt[3]{-216}$

36. $-\sqrt[3]{-1000}$

37. $\sqrt[3]{0.343(x+1)^3}$

38. $\sqrt[3]{0.000008(y-2)^3}$

⦂⦂ Find the following. Assume that letters can represent any real number.

39. $\sqrt[4]{625}$

40. $-\sqrt[4]{256}$

41. $\sqrt[5]{-1}$

42. $-\sqrt[5]{-32}$

43. $\sqrt[5]{-\dfrac{32}{243}}$

44. $\sqrt[5]{-\dfrac{1}{32}}$

45. $\sqrt[6]{x^6}$

46. $\sqrt[8]{y^8}$

47. $\sqrt[4]{(5a)^4}$

48. $\sqrt[4]{(7b)^4}$

49. $\sqrt[10]{(-6)^{10}}$

50. $\sqrt[12]{(-10)^{12}}$

51. $\sqrt[414]{(a+b)^{414}}$

52. $\sqrt[1976]{(2a+b)^{1976}}$

53. $\sqrt[7]{y^7}$

54. $\sqrt[3]{(-6)^3}$

55. $\sqrt[5]{(x-2)^5}$

56. $\sqrt[9]{(2xy)^9}$

57. *Parking.* A parking lot has attendants to park the cars. The number N of temporary stalls needed for waiting cars before attendants can get to them is given by the formula $N = 2.5\sqrt{A}$, where A is the number of arrivals in peak hours. Find the number of spaces needed when the average number of arrivals in peak hours is (a) 25; (b) 36; (c) 49; (d) 64.

6.3 MULTIPLYING AND SIMPLIFYING WITH RADICALS

● MULTIPLYING

Notice that $\sqrt{4}\sqrt{25} = 2 \cdot 5 = 10$.

Also $\sqrt{4 \cdot 25} = \sqrt{100} = 10$.

Likewise, $\sqrt[3]{27}\sqrt[3]{8} = 3 \cdot 2 = 6$ and $\sqrt[3]{27 \cdot 8} = \sqrt[3]{216} = 6$.

These examples suggest the following.

> **For any nonnegative real numbers a and b and any index k,**
> $\sqrt[k]{a} \cdot \sqrt[k]{b} = \sqrt[k]{a \cdot b}$.   **(To multiply, we multiply the radicands.)**

Examples   Multiply.

> Note that the index k must be the same throughout.

1. $\sqrt{3} \cdot \sqrt{5} = \sqrt{3 \cdot 5} = \sqrt{15}$

2. $\sqrt{x + 2}\sqrt{x - 2} = \sqrt{(x + 2)(x - 2)} = \sqrt{x^2 - 4}$

3. $\sqrt[3]{4}\sqrt[3]{5} = \sqrt[3]{4 \cdot 5} = \sqrt[3]{20}$

4. $\sqrt[4]{\dfrac{y}{5}}\sqrt[4]{\dfrac{7}{x}} = \sqrt[4]{\dfrac{y}{5} \cdot \dfrac{7}{x}} = \sqrt[4]{\dfrac{7y}{5x}}$

> A common error is to omit the index in the answer.

DO EXERCISES 1–4.

●● SIMPLIFYING BY FACTORING

Reversing the above statement we have $\sqrt[k]{a \cdot b} = \sqrt[k]{a} \cdot \sqrt[k]{b}$. This shows a way to factor and thus simplify radical expressions. Consider $\sqrt{20}$. The number 20 has the factor 4, which is a perfect square. Therefore,

$$\sqrt{20} = \sqrt{4 \cdot 5} \qquad \text{Factoring the radicand (4 is a perfect square)}$$
$$\phantom{\sqrt{20}} = \sqrt{4} \cdot \sqrt{5} \qquad \text{Factoring into two radicals}$$
$$\phantom{\sqrt{20}} = 2\sqrt{5}. \qquad \text{Taking the square root of 4}$$

> **To simplify a radical expression by factoring, look for factors of the radicand that are perfect kth powers (where k is the index). Then take the kth root of those factors.**

Examples   Simplify by factoring.

5. $\sqrt{50} = \sqrt{25 \cdot 2} = \sqrt{25} \cdot \sqrt{2} = 5\sqrt{2}$

> This factor is a perfect square.

6. $\sqrt[3]{32} = \sqrt[3]{8 \cdot 4} = \sqrt[3]{8} \cdot \sqrt[3]{4} = 2\sqrt[3]{4}$

> This factor is a perfect cube (third power).

DO EXERCISES 5 AND 6.

ANSWERS ON PAGE A–22

OBJECTIVES

After finishing Section 6.3, you should be able to:

● Multiply with radical notation.

●● Simplify radical expressions by factoring, assuming that all expressions represent non-negative numbers.

●●● Multiply and simplify radical expressions.

∷ Approximate expressions using a square root table.

∷∷ Identify rational and irrational numbers.

Multiply.

1. $\sqrt{19}\sqrt{7}$

2. $\sqrt{x + 2y}\sqrt{x - 2y}$

3. $\sqrt[4]{403}\sqrt[4]{7}$

4. $\sqrt[3]{8x}\sqrt[3]{x^4 + 5}$

Simplify by factoring.

5. $\sqrt{32}$

6. $\sqrt[3]{80}$

Simplify by factoring. Assume that all expressions represent nonnegative numbers. Hence no absolute value signs will be needed.

7. $\sqrt{300}$

8. $\sqrt{36y^2}$

9. $\sqrt{3x^2 + 12x + 12}$

10. $\sqrt{12ab^3c^2}$

11. $\sqrt[3]{16}$

12. $\sqrt[3]{81x^4y^8}$

13. $\sqrt[3]{(a + b)^4}$

Multiply and then simplify by factoring. Assume that all expressions under radicals represent nonnegative numbers.

14. $\sqrt{3}\sqrt{6}$

15. $\sqrt{18y}\sqrt{14y}$

16. $\sqrt[3]{3x^2y}\sqrt[3]{36x}$

17. $\sqrt{7a}\sqrt{21b}$

18. $\sqrt[3]{2(y + 5)}\sqrt[3]{4(y + 5)^4}$

In many situations, expressions under radicals never represent negative numbers. In such cases, absolute value notation is not necessary.

Examples Simplify by factoring. Assume that all expressions under radicals represent nonnegative numbers.

7. $\sqrt{5x^2} = \sqrt{x^2 \cdot 5}$ Factoring the radicand
$\phantom{\sqrt{5x^2}} = \sqrt{x^2} \cdot \sqrt{5}$ Factoring into two radicals
$\phantom{\sqrt{5x^2}} = x \cdot \sqrt{5}$ Taking the square root of x^2

8. $\sqrt{2x^2 - 4x + 2} = \sqrt{2(x - 1)^2}$
$\phantom{\sqrt{2x^2 - 4x + 2}} = \sqrt{(x - 1)^2} \cdot \sqrt{2}$
$\phantom{\sqrt{2x^2 - 4x + 2}} = (x - 1) \cdot \sqrt{2}$

> Absolute value is not needed because expressions are not negative.

9. $\sqrt{216x^5y^3} = \sqrt{36 \cdot 6 \cdot x^4 \cdot x \cdot y^2 \cdot y}$
$\phantom{\sqrt{216x^5y^3}} = \sqrt{36 \cdot x^4 \cdot y^2 \cdot 6 \cdot x \cdot y}$ Factoring the radicand
$\phantom{\sqrt{216x^5y^3}} = \sqrt{36}\sqrt{x^4}\sqrt{y^2}\sqrt{6xy}$ Factoring into several radicals
$\phantom{\sqrt{216x^5y^3}} = 6x^2y\sqrt{6xy}$ Taking square roots

Note: Had we not seen that $216 = 36 \cdot 6$, where 36 is the largest square factor of 216, we could have found the prime factorization $2 \cdot 2 \cdot 2 \cdot 3 \cdot 3 \cdot 3$. Each pair of factors makes a square, so

$$\sqrt{2 \cdot 2 \cdot 2 \cdot 3 \cdot 3 \cdot 3} = \sqrt{2^2 \cdot 3^2 \cdot 2 \cdot 3} = 2 \cdot 3\sqrt{2 \cdot 3}.$$

DO EXERCISES 7–13.

●●● MULTIPLYING AND SIMPLIFYING

Sometimes after we multiply we can then simplify by factoring.

Examples Multiply and then simplify by factoring. Assume that all expressions under radicals represent nonnegative numbers.

10. $\sqrt{15}\sqrt{6} = \sqrt{15 \cdot 6} = \sqrt{90} = \sqrt{9 \cdot 10} = 3\sqrt{10}$

11. $3\sqrt[3]{25} \cdot 2\sqrt[3]{5} = 6 \cdot \sqrt[3]{25 \cdot 5}$ Multiplying radicands
$\phantom{3\sqrt[3]{25} \cdot 2\sqrt[3]{5}} = 6 \cdot \sqrt[3]{125}$
$\phantom{3\sqrt[3]{25} \cdot 2\sqrt[3]{5}} = 6 \cdot 5,$ or 30   Taking the cube root of 125

12. $\sqrt[3]{18y^3}\sqrt[3]{4x^2} = \sqrt[3]{18y^3 \cdot 4x^2} = \sqrt[3]{72y^3x^2}$ Multiplying radicands
$\phantom{\sqrt[3]{18y^3}\sqrt[3]{4x^2}} = \sqrt[3]{8y^3 \cdot 9x^2}$ Factoring the radicand
$\phantom{\sqrt[3]{18y^3}\sqrt[3]{4x^2}} = \sqrt[3]{8y^3}\sqrt[3]{9x^2}$ Factoring into two radicals
$\phantom{\sqrt[3]{18y^3}\sqrt[3]{4x^2}} = 2y\sqrt[3]{9x^2}$ Taking the cube root

DO EXERCISES 14–18.

⦂⦂ APPROXIMATING SQUARE ROOTS

Table 1 in the back of the book contains approximate square roots for the natural numbers 1 through 100. If a radicand is not listed in the table we can factor the radical expression, find exact or approximate square roots of the factors, and then find the product of those square roots.

Example 13   Approximate to the nearest tenth.

$$\sqrt{275} = \sqrt{25 \cdot 11} \qquad \text{Factoring the radicand}$$
$$= \sqrt{25} \cdot \sqrt{11} \qquad \text{Factoring into two radicals}$$
$$\approx 5 \times 3.317 \approx 16.6 \qquad \text{From Table 1, } \sqrt{11} \approx 3.317$$

DO EXERCISES 19 AND 20.

Example 14   Approximate to the nearest tenth.

$$\frac{16 - \sqrt{640}}{4} = \frac{16 - \sqrt{64 \cdot 10}}{4} \qquad \text{Factoring the radicand}$$

$$= \frac{16 - \sqrt{64} \cdot \sqrt{10}}{4} \qquad \text{Factoring into two radicals}$$

$$= \frac{16 - 8\sqrt{10}}{4} \qquad \longleftarrow \boxed{\text{A common error here is to ``cancel'' and get } 4 - 8\sqrt{10}. \text{ Always remember to } \textit{remove a factor of 1.}}$$

$$= \frac{4(4 - 2\sqrt{10})}{4} = 4 - 2\sqrt{10} \qquad \text{Removing a factor of 1}$$

$$\approx 4 - 2 \times 3.162 \qquad \text{Using Table 1}$$

$$\approx -2.324, \text{ or } -2.3 \text{ to nearest tenth}$$

DO EXERCISES 21 AND 22.

⦂⦂ RATIONAL AND IRRATIONAL NUMBERS

The real numbers consist of the rational numbers and the irrational numbers. We can describe these kinds of numbers in several ways.

> **Rational numbers are:**
>
> 1. **Those numbers that can be named with fractional notation** a/b, **where** a **and** b **are integers. (Definition from p. 2)**
> 2. **Those numbers for which decimal notation either ends or repeats.**

Examples

15. $\dfrac{5}{16} = 0.3125$ Ending (terminating) decimal

16. $\dfrac{8}{7} = 1.142857142857\ldots = 1.\overline{142857}$ Repeating decimal

17. $\dfrac{3}{11} = 0.2727\ldots = 0.\overline{27}$ Repeating decimal

---

Simplify and approximate to the nearest tenth.

19. $\sqrt{160}$

20. $\sqrt{341}$
(*Hint:* $341 = 11 \cdot 31$.)

Approximate to the nearest tenth.

21. $\dfrac{18 - \sqrt{810}}{3}$

22. $\dfrac{6 + \sqrt{450}}{3}$

Which of the following are rational?
Which are irrational?

23. $\dfrac{-4}{5}$

24. $\dfrac{59}{37}$

25. 7.42

26. 0.47474747...
(Numeral repeats)

27. 2.57340046631...
(Numeral does not repeat)

28. $\sqrt{7}$

29. $\sqrt{16}$

Irrational numbers are:

1. **Those real numbers that are not rational.**
2. **Those real numbers that cannot be named with fractional notation** a/b, **where** a **and** b **are integers.**
3. **Those real numbers for which decimal notation does not end and does not repeat.**

There are many irrational numbers. For example, $\sqrt{2}$ is irrational. We can find rational numbers a/b for which $(a/b) \cdot (a/b)$ is close to 2, but we cannot find such a number a/b for which $(a/b) \cdot (a/b)$ is *exactly* 2.

Unless a whole number is a perfect square its square root is irrational. For example, $\sqrt{9}$ and $\sqrt{25}$ are rational, but all of the following are irrational:

$$\sqrt{3}, \qquad -\sqrt{14}, \qquad \sqrt{45}.$$

There are also many irrational numbers that we do not obtain by taking square roots. The number π is an example.* Decimal notation for π does not end and does not repeat.

Examples All of these are irrational.

18. $\pi = 3.1415926535...$ Numeral does not repeat.

19. 1.101001000100001000001...   Numeral does not repeat.

20. $\sqrt{2} = 1.41421356...$ Numeral does not repeat.

In Example 19 there is a pattern, but it is not a repeating pattern.

DO EXERCISES 23–29.

SOMETHING EXTRA

AN APPLICATION: WIND CHILL

In cold weather we feel colder if there is wind than if there is not. *Wind chill temperature* is the temperature at which, without wind, we would feel as cold as in an actual situation with wind. Here is a formula for finding wind chill temperature:

$$T_w = 91.4 - \frac{(10.45 + 6.68\sqrt{v} - 0.447v)(457 - 5T)}{110}.$$

Here T is the actual temperature as given by a thermometer, in degrees Fahrenheit, and v is the wind speed in mph.

EXERCISES

You will find a calculator very helpful. If it does not take square roots, you can get them from Table 1. Find the wind chill temperature in each case. Round to the nearest degree.

1. $T = 30°F$, $v = 25$ mph

2. $T = 10°F$, $v = 25$ mph

3. $T = 20°F$, $v = 20$ mph

4. $T = 20°F$, $v = 40$ mph

5. $T = -10°F$, $v = 30$ mph

6. $T = -30°F$, $v = 30$ mph

* 3.14 and $\frac{22}{7}$ are only rational approximations to π.

NAME

CLASS

ANSWERS

EXERCISE SET 6.3

● Multiply.

1. $\sqrt{3}\sqrt{2}$

2. $\sqrt{5}\sqrt{7}$

3. $\sqrt[3]{2}\sqrt[3]{5}$

4. $\sqrt[3]{7}\sqrt[3]{2}$

5. $\sqrt[4]{8}\sqrt[4]{9}$

6. $\sqrt[4]{6}\sqrt[4]{3}$

7. $\sqrt{3a}\sqrt{10b}$

8. $\sqrt{2x}\sqrt{13y}$

9. $\sqrt[5]{9t^2}\sqrt[5]{2t}$

10. $\sqrt[5]{8y^3}\sqrt[5]{10y}$

11. $\sqrt{x-a}\sqrt{x+a}$

12. $\sqrt{y-b}\sqrt{y+b}$

●● Simplify by factoring. Assume that all expressions represent nonnegative numbers.

13. $\sqrt{8}$

14. $\sqrt{18}$

15. $\sqrt{24}$

16. $\sqrt{20}$

17. $\sqrt{40}$

18. $\sqrt{90}$

19. $\sqrt{180x^4}$

20. $\sqrt{175y^6}$

21. $\sqrt[3]{54x^8}$

22. $\sqrt[3]{40y^3}$

●●● Multiply and simplify by factoring. Assume that all expressions represent non-negative numbers.

23. $\sqrt{15}\sqrt{6}$

24. $\sqrt{2}\sqrt{32}$

25. $\sqrt[3]{3}\sqrt[3]{18}$

26. $\sqrt[3]{2}\sqrt[3]{20}$

1. _____

2. _____

3. _____

4. _____

5. _____

6. _____

7. _____

8. _____

9. _____

10. _____

11. _____

12. _____

13. _____

14. _____

15. _____

16. _____

17. _____

18. _____

19. _____

20. _____

21. _____

22. _____

23. _____

24. _____

25. _____

26. _____

ANSWERS

27. _____

28. _____

29. _____

30. _____

31. _____

32. _____

33. _____

34. _____

35. _____

36. _____

37. _____

38. _____

39. _____

40. _____

41. _____

42. _____

43. _____

44. _____

45. _____

46. _____

47. _____

48. _____

49. a) _____

b) _____

c) _____

27. $\sqrt{45}\sqrt{60}$

28. $\sqrt{24}\sqrt{75}$

29. $\sqrt{5b^3}\sqrt{10c^4}$

30. $\sqrt{2x^3y}\sqrt{12xy}$

31. $\sqrt[3]{y^4}\sqrt[3]{16y^5}$

32. $\sqrt[3]{5^2t^4}\sqrt[3]{5^4t^6}$

33. $\sqrt[3]{(b+3)^4}\sqrt[3]{(b+3)^2}$

34. $\sqrt[3]{(x+y)^3}\sqrt[3]{(x+y)^5}$

35. $\sqrt{12a^3b}\sqrt{8a^4b^2}$

36. $\sqrt{18x^2y^3}\sqrt{6xy^4}$

⠿ Approximate to the nearest tenth.

37. $\sqrt{180}$

38. $\sqrt{124}$

39. $\dfrac{8+\sqrt{480}}{4}$

40. $\dfrac{12-\sqrt{450}}{3}$

⠿ Which of the following are irrational numbers? Which are rational?

41. $-\dfrac{5}{6}$

42. 8.93

43. $8.23\overline{23}$

44. $\sqrt{5}$

45. $\sqrt{36}$

46. $-\sqrt{15}$

47. $2.101001\ldots$
(Numeral does not repeat.)

48. $3.1415926\ldots$
(Numeral does not repeat.)

49. *Speed of a skidding car.* After an accident, police can estimate the speed that a car was traveling by measuring its skid marks. The formula

$$r = 2\sqrt{5L}$$

can be used, where r is the speed in mph, and L is the length of the skid marks in feet. Estimate the speed of a car that left skid marks (a) 20 ft long; (b) 70 ft long; (c) 90 ft long.

6.4   DIVIDING AND SIMPLIFYING WITH RADICALS

• ROOTS OF QUOTIENTS

Notice that $\sqrt[3]{\dfrac{27}{8}} = \dfrac{3}{2}$ and that $\dfrac{\sqrt[3]{27}}{\sqrt[3]{8}} = \dfrac{3}{2}$. This example suggests that we can take the root of a quotient by taking the root of the numerator and denominator separately. This is true.

> **Rule A: For any nonnegative number a and any positive number b, and any index k,**
>
> $$\sqrt[k]{\dfrac{a}{b}} = \dfrac{\sqrt[k]{a}}{\sqrt[k]{b}}.$$ **(We can take the kth root of the numerator and denominator separately.)**

Examples   Simplify by taking roots of numerator and denominator. Assume that all expressions under radicals represent positive numbers.

1. $\sqrt[3]{\dfrac{27}{125}} = \dfrac{\sqrt[3]{27}}{\sqrt[3]{125}} = \dfrac{3}{5}$ ┌ We take the cube root of the numerator and denominator. ┐

2. $\sqrt{\dfrac{25}{y^2}} = \dfrac{\sqrt{25}}{\sqrt{y^2}} = \dfrac{5}{y}$ ┌ We take the square root of the numerator and denominator. ┐

3. $\sqrt{\dfrac{16x^3}{y^4}} = \dfrac{\sqrt{16x^3}}{\sqrt{y^4}} = \dfrac{\sqrt{16x^2 \cdot x}}{\sqrt{y^4}} = \dfrac{4x\sqrt{x}}{y^2}$

4. $\sqrt[3]{\dfrac{27y^5}{343x^3}} = \dfrac{\sqrt[3]{27y^5}}{\sqrt[3]{343x^3}} = \dfrac{\sqrt[3]{27y^3 \cdot y^2}}{\sqrt[3]{343x^3}} = \dfrac{\sqrt[3]{27y^3} \cdot \sqrt[3]{y^2}}{\sqrt[3]{343x^3}} = \dfrac{3y\sqrt[3]{y^2}}{7x}$

DO EXERCISES 1–3.

┌ We are assuming that no expression represents 0 or a negative number. Thus we need not be concerned about zero denominators. ┐

•• DIVIDING RADICAL EXPRESSIONS

Reversing Rule A shows a way to divide radical expressions as follows.

> **Rule B: $\dfrac{\sqrt[k]{a}}{\sqrt[k]{b}} = \sqrt[k]{\dfrac{a}{b}}$ To divide, we divide the radicands. After doing this we can sometimes simplify by taking roots.**

Examples   Divide. Then simplify by taking roots if possible. Assume that all expressions under radicals represent positive numbers.

5. $\dfrac{\sqrt{80}}{\sqrt{5}} = \sqrt{\dfrac{80}{5}} = \sqrt{16} = 4$ ┌ We divide the radicands. ┐

6. $\dfrac{3\sqrt{2}}{5\sqrt{3}} = \dfrac{3}{5} \cdot \dfrac{\sqrt{2}}{\sqrt{3}} = \dfrac{3}{5} \cdot \sqrt{\dfrac{2}{3}}$

7. $\dfrac{5\sqrt[3]{32}}{\sqrt[3]{2}} = 5\sqrt[3]{\dfrac{32}{2}} = 5\sqrt[3]{16} = 5\sqrt[3]{8 \cdot 2} = 5\sqrt[3]{8}\sqrt[3]{2} = 5 \cdot 2\sqrt[3]{2} = 10\sqrt[3]{2}$

ANSWERS ON PAGE A–22

OBJECTIVES

After finishing Section 6.4, you should be able to:

• Simplify radical expressions having a quotient for a radicand.

•• Divide and simplify radical expressions.

••• Calculate combinations of roots and powers.

Simplify by taking roots of numerator and denominator. Assume that all expressions under radicals represent positive numbers.

1. $\sqrt{\dfrac{25}{36}}$

2. $\sqrt{\dfrac{x^2}{100}}$

3. $\sqrt[3]{\dfrac{54x^5}{125}}$

Divide. Then simplify by taking roots if possible. Assume that all expressions under radicals represent positive numbers.

4. $\dfrac{\sqrt{75}}{\sqrt{3}}$

5. $\dfrac{14\sqrt{128xy}}{2\sqrt{2}}$

6. $\dfrac{\sqrt{50a^3}}{\sqrt{2a}}$

7. $\dfrac{4\sqrt[3]{250}}{7\sqrt[3]{2}}$

8. $\dfrac{\sqrt[3]{8a^3b}}{\sqrt[3]{27b^{-2}}}$

Calculate as shown. Then use Rule C to calculate another way.

9. $(\sqrt[3]{125})^2$

10. $(\sqrt{6y})^3$

8. $\dfrac{\sqrt{72xy}}{2\sqrt{2}} = \dfrac{1}{2}\dfrac{\sqrt{72xy}}{\sqrt{2}} = \dfrac{1}{2}\sqrt{\dfrac{72xy}{2}} = \dfrac{1}{2}\sqrt{36xy} = \dfrac{1}{2}\sqrt{36}\sqrt{xy}$

$= \dfrac{1}{2}\cdot 6\sqrt{xy} = 3\sqrt{xy}$

9. $\dfrac{\sqrt[4]{33a^5b^3}}{\sqrt[4]{2b^{-1}}} = \sqrt[4]{\dfrac{33a^5b^3}{2b^{-1}}} = \sqrt[4]{\dfrac{33}{2}a^5b^4} = \sqrt[4]{a^4b^4}\sqrt[4]{\dfrac{33}{2}a} = ab\sqrt[4]{\dfrac{33}{2}a}$

DO EXERCISES 4–8.

●●● POWERS AND ROOTS COMBINED

Consider the following.

$\sqrt[3]{8^2} = \sqrt[3]{64} = 4$

$(\sqrt[3]{8})^2 = (2)^2 = 4$

This suggests another important property of radical expressions.

> **Rule C:** For any nonnegative number a, any index k, and any natural number m,
> $\sqrt[k]{a^m} = (\sqrt[k]{a})^m$. **(We can raise to a power and then take a root, or we can take a root and then raise to a power.)**

In some cases one way of calculating is easier than the other.

Examples Calculate as shown. Then use Rule C to calculate another way. Assume that all expressions under radicals are positive.

10. a) $\sqrt[3]{27^2} = \sqrt[3]{729} = 9$ Using Table 1

b) $(\sqrt[3]{27})^2 = (3)^2 = 9$

11. a) $\sqrt[3]{2^6} = \sqrt[3]{64} = 4$

b) $(\sqrt[3]{2})^6 = \sqrt[3]{2}\sqrt[3]{2}\sqrt[3]{2}\sqrt[3]{2}\sqrt[3]{2}\sqrt[3]{2} = 2\cdot 2 = 4$

12. a) $(\sqrt{5x})^3 = \sqrt{5x}\sqrt{5x}\sqrt{5x} = \sqrt{5^3x^3} = \sqrt{5^2x^2}\sqrt{5x} = 5x\sqrt{5x}$

b) $\sqrt{(5x)^3} = \sqrt{5^3x^3} = \sqrt{5^2x^2}\sqrt{5x} = 5x\sqrt{5x}$

DO EXERCISES 9 AND 10.

NAME   **CLASS**   ANSWERS

EXERCISE SET 6.4

● Simplify by taking roots of numerator and denominator. Assume that all expressions under radicals represent positive numbers.

2. _____

1. $\sqrt{\dfrac{16}{25}}$ **2.** $\sqrt{\dfrac{100}{81}}$ **3.** $\sqrt[3]{\dfrac{64}{27}}$ **4.** $\sqrt[3]{\dfrac{343}{512}}$ **5.** $\sqrt{\dfrac{49}{y^2}}$

3. _____

4. _____

5. _____

6. _____

6. $\sqrt{\dfrac{121}{x^2}}$ **7.** $\sqrt{\dfrac{25y^3}{x^4}}$ **8.** $\sqrt{\dfrac{36a^5}{b^6}}$ **9.** $\sqrt[3]{\dfrac{8x^5}{27y^3}}$ **10.** $\sqrt[3]{\dfrac{64x^7}{216y^6}}$

7. _____

8. _____

9. _____

● ● Divide. Then simplify by taking roots if possible. Assume that all expressions under radicals represent positive numbers.

10. _____

11. $\dfrac{\sqrt{21a}}{\sqrt{3a}}$ **12.** $\dfrac{\sqrt{28y}}{\sqrt{4y}}$ **13.** $\dfrac{\sqrt[3]{54}}{\sqrt[3]{2}}$ **14.** $\dfrac{\sqrt[3]{40}}{\sqrt[3]{5}}$

11. _____

12. _____

13. _____

14. _____

15. $\dfrac{\sqrt{40xy^3}}{\sqrt{8x}}$ **16.** $\dfrac{\sqrt{56ab^3}}{\sqrt{7a}}$ **17.** $\dfrac{\sqrt[3]{96a^4b^2}}{\sqrt[3]{12a^2b}}$ **18.** $\dfrac{\sqrt[3]{189x^5y^7}}{\sqrt[3]{7x^2y^2}}$

15. _____

16. _____

17. _____

18. _____

19. $\dfrac{\sqrt{72xy}}{2\sqrt{2}}$

20. $\dfrac{\sqrt{75ab}}{3\sqrt{3}}$

21. $\dfrac{\sqrt{x^3 - y^3}}{\sqrt{x - y}}$

22. $\dfrac{\sqrt{r^3 + s^3}}{\sqrt{r + s}}$

Hint: To divide $x^3 - y^3$ by $x - y$, factor $x^3 - y^3$. Then simplify the fractional expression by removing a factor of 1.

●●● Calculate as shown. Then use Rule C to calculate another way. Assume that all expressions under radicals represent positive numbers.

23. $\sqrt{(6a)^3}$

24. $\sqrt{(7y)^3}$

25. $(\sqrt[3]{16b^2})^2$

26. $(\sqrt[3]{25r^2})^2$

27. $\sqrt{(18a^2b)^3}$

28. $\sqrt{(12x^2y)^3}$

29. $(\sqrt[3]{12c^2d})^2$

30. $(\sqrt[3]{9x^2y})^2$

31. *Pendulums.* The *period* of a pendulum is the time it takes to complete one cycle, swinging to and fro. If a pendulum consists of a ball on a string, the period T is given by the formula

$$T = 2\pi\sqrt{\dfrac{L}{980}},$$

where T is in seconds and L is the length of the pendulum in centimeters. Find the period of a pendulum of length (a) 65 cm; (b) 98 cm; (c) 120 cm. Use 3.14 for π.

6.5 ADDITION AND SUBTRACTION INVOLVING RADICALS

■ Any two real numbers can be added. For instance, the sum of 7 and $\sqrt{3}$ can be expressed as

$$7 + \sqrt{3}.$$

We cannot simplify this name for the sum. However, when we have *like radicals* (radicals having the same index and radicand), we can use the distributive laws to simplify by collecting like radical terms.

Examples   Add or subtract. Simplify by collecting like radical terms if possible.

1. $6\sqrt{7} + 4\sqrt{7} = (6 + 4)\sqrt{7}$ Using the distributive law (factoring out $\sqrt{7}$)

$$= 10\sqrt{7}$$

2. $8\sqrt[3]{2} - 7x\sqrt[3]{2} + 5\sqrt[3]{2} = (8 - 7x + 5)\sqrt[3]{2}$ Factoring out $\sqrt[3]{2}$

$$= (13 - 7x)\sqrt[3]{2}$$

3. $6\sqrt[5]{4x} + 4\sqrt[5]{4x} - \sqrt[3]{4x} = (6 + 4)\sqrt[5]{4x} - \sqrt[3]{4x}$

$$= 10\sqrt[5]{4x} - \sqrt[3]{4x}$$

DO EXERCISES 1 AND 2.

Sometimes we need to factor in order to have terms with like radicals.

Examples   Add or subtract. Simplify by collecting like radical terms if possible.

4. $3\sqrt{8} - 5\sqrt{2} = 3\sqrt{4 \cdot 2} - 5\sqrt{2}$ Factoring 8

$$= 3\sqrt{4} \cdot \sqrt{2} - 5\sqrt{2}$$ Factoring $\sqrt{4 \cdot 2}$ into two radicals

$$= 3 \cdot 2\sqrt{2} - 5\sqrt{2}$$ Taking the square root of 4

$$= 6\sqrt{2} - 5\sqrt{2}$$

$$= (6 - 5)\sqrt{2}$$

$$= \sqrt{2}$$ Collecting like radical terms

5. $5\sqrt{2} - 4\sqrt{3}$ No simplification possible

6. $5\sqrt[3]{16y^4} + 7\sqrt[3]{2y} = 5\sqrt[3]{8y^3 \cdot 2y} + 7\sqrt[3]{2y}$ ⎫   Factoring the first radical

$$= 5\sqrt[3]{8y^3} \cdot \sqrt[3]{2y} + 7\sqrt[3]{2y}$$ ⎬

$$= 5 \cdot 2y \cdot \sqrt[3]{2y} + 7\sqrt[3]{2y}$$ Taking cube root

$$= 10y\sqrt[3]{2y} + 7\sqrt[3]{2y}$$

$$= (10y + 7)\sqrt[3]{2y}$$ Collecting like radical terms

DO EXERCISES 3-5.

ANSWERS ON PAGE A-22

OBJECTIVE

After finishing Section 6.5, you should be able to:

■ Add or subtract with radical notation and simplify.

Add or subtract. Simplify by collecting like radical terms if possible.

1. $5\sqrt{2} + 8\sqrt{2}$

2. $7\sqrt[4]{5x} + 3\sqrt[4]{5x} - \sqrt{7}$

Add or subtract. Simplify by collecting like radical terms if possible.

3. $7\sqrt{45} - 2\sqrt{5}$

4. $3\sqrt[3]{y^5} + 4\sqrt[3]{y^2} + \sqrt[3]{8y^6}$

5. $\sqrt{25x - 25} - \sqrt{9x - 9}$

SOMETHING EXTRA

CALCULATOR CORNER: CALCULATING SQUARE ROOTS

If your calculator does not have a square root key, you can still calculate square roots to any desired accuracy, and you can do it rather rapidly. The idea is to divide and average repeatedly.

Example Find the square root of 28 to five decimal places.

a) First we make a guess. Any positive number will do. Let's try 4.

b) Then divide 28 by 4. We get 7.

If we had guessed the square root exactly, the quotient would have been the same as the guess. The quotient is larger than the guess, so we know the guess was too small. To get a better approximation, we *average* 4 and 7.

c) Average 4 and 7: $\dfrac{4 + 7}{2} = 5.5$.

d) Let's use 5.5 the way we did our guess. Divide 28 by 5.5:

$$\frac{28}{5.5} = 5.09090909 \text{ to 8 decimal places.}$$

e) Now average the quotient and divisor:

$$\frac{5.5 + 5.09090909}{2} = 5.29545454.$$

f) Repeat the process and divide 28 by 5.29545454:

$$\frac{28}{5.29545454} = 5.28755365.$$

We know that the actual square root is between the quotient and the divisor (this is true at every step).

Now we average: $\dfrac{5.29545454 + 5.28755365}{2} = 5.29150409.$

g) We repeat:

$$\frac{28}{5.29150 \boxed{4} 09} = 5.29150 \boxed{1} 15.$$

We now have the square root to five decimal places. The sixth place is 1, 2, 3, or 4. We have the answer correct to 5 decimal places even if we round to five places.

h) The answer correct to five decimal places is 5.29150.

EXERCISES

Find these square roots to five decimal places without using the square root key.

1. $\sqrt{57}$ **2.** $\sqrt{88}$ **3.** $\sqrt{119}$ **4.** $\sqrt{324}$

NAME

CLASS

ANSWERS

EXERCISE SET 6.5

● Add or subtract. Simplify by collecting like radical terms if possible, assuming that all expressions under radicals represent nonnegative numbers.

1. $6\sqrt{3} + 2\sqrt{3}$

2. $8\sqrt{5} + 9\sqrt{5}$

3. $9\sqrt[3]{5} - 6\sqrt[3]{5}$

4. $14\sqrt[5]{2} - 6\sqrt[5]{2}$

5. $4\sqrt[3]{y} + 9\sqrt[3]{y}$

6. $6\sqrt[4]{t} - 3\sqrt[4]{t}$

7. $8\sqrt{2} - 6\sqrt{2} + 5\sqrt{2}$

8. $2\sqrt{6} + 8\sqrt{6} - 3\sqrt{6}$

9. $4\sqrt[3]{3} - \sqrt{5} + 2\sqrt[3]{3} + \sqrt{5}$

10. $5\sqrt{7} - 8\sqrt[4]{11} + \sqrt{7} + 9\sqrt[4]{11}$

11. $8\sqrt{27} - 3\sqrt{3}$

12. $9\sqrt{50} - 4\sqrt{2}$

13. $8\sqrt{45} + 7\sqrt{20}$

1. _____

2. _____

3. _____

4. _____

5. _____

6. _____

7. _____

8. _____

9. _____

10. _____

11. _____

12. _____

13. _____

14. _____

15. _____

16. _____

17. _____

18. _____

19. _____

20. _____

21. _____

22. _____

23. _____

24. _____

25. _____

26. _____

27. _____

28. _____

29. _____

14. $9\sqrt{12} + 16\sqrt{27}$

15. $18\sqrt{72} + 2\sqrt{98}$

16. $12\sqrt{45} - 8\sqrt{80}$

17. $3\sqrt[3]{16} + \sqrt[3]{54}$

18. $\sqrt[3]{27} - 5\sqrt[3]{8}$

19. $2\sqrt{128} - \sqrt{18} + 4\sqrt{32}$

20. $5\sqrt{50} - 2\sqrt{18} + 9\sqrt{32}$

21. $\sqrt{5a} + 2\sqrt{45a^3}$

22. $4\sqrt{3x^3} - \sqrt{12x}$

23. $\sqrt[3]{24x} - \sqrt[3]{3x^4}$

24. $\sqrt[3]{54x} - \sqrt[3]{2x^4}$

25. $\sqrt{8y - 8} + \sqrt{2y - 2}$

26. $\sqrt{12t + 12} + \sqrt{3t + 3}$

27. $\sqrt{x^3 - x^2} + \sqrt{9x - 9}$

28. $\sqrt{4x - 4} - \sqrt{x^3 - x^2}$

29. Find as many pairs of numbers, a and b, as you can for which

$$\sqrt{a} + \sqrt{b} = \sqrt{a + b}.$$

Find several pairs of numbers for which the above equation is false.

6.6 MORE ABOUT MULTIPLICATION WITH RADICALS

⬛• To multiply expressions in which some factors contain more than one term, we use the procedures for multiplying polynomials.

Examples   Multiply.

1. $\sqrt{3}\,(x - \sqrt{5}) = \boxed{\sqrt{3}} \cdot x - \boxed{\sqrt{3}} \cdot \sqrt{5}$ Using the distributive law

$\qquad\qquad\quad = x\sqrt{3} - \sqrt{15}$ Multiplying radicals

2. $\sqrt[3]{y}\,(\sqrt[3]{y^2} + \sqrt[3]{2}) = \boxed{\sqrt[3]{y}} \cdot \sqrt[3]{y^2} + \boxed{\sqrt[3]{y}} \cdot \sqrt[3]{2}$ Using the distributive law

$\qquad\qquad\qquad = \sqrt[3]{y^3} + \sqrt[3]{2y}$ Multiplying radicals

$\qquad\qquad\qquad = y + \sqrt[3]{2y}$ Simplifying $\sqrt[3]{y^3}$

DO EXERCISES 1 AND 2.

Example 3   Multiply.

$$\begin{aligned}
&\qquad\qquad\quad\; F \qquad\quad O \qquad\; I \qquad\;\; L\\
(4\sqrt{3} + \sqrt{2})(\sqrt{3} - 5\sqrt{2}) &= 4(\sqrt{3})^2 - 20\sqrt{3}\cdot\sqrt{2} + \sqrt{2}\cdot\sqrt{3} - 5(\sqrt{2})^2\\
&= 4\cdot 3 - 20\sqrt{6} + \sqrt{6} - 5\cdot 2\\
&= 12 - 20\sqrt{6} + \sqrt{6} - 10\\
&= 2 - 19\sqrt{6}
\end{aligned}$$

Example 4   Multiply. Assume that all expressions under radicals represent positive numbers.

$$\begin{aligned}
(\sqrt{a} + \sqrt{3})(\sqrt{b} + \sqrt{3}) &= \sqrt{a}\sqrt{b} + \sqrt{a}\sqrt{3} + \sqrt{3}\sqrt{b} + \sqrt{3}\sqrt{3}\\
&= \sqrt{ab} + \sqrt{3a} + \sqrt{3b} + 3
\end{aligned}$$

DO EXERCISES 3 AND 4.

Example 5   Multiply.

$$(\sqrt{5} + \sqrt{7})(\sqrt{5} - \sqrt{7}) = (\sqrt{5})^2 - (\sqrt{7})^2 \qquad \text{This is now a difference of two squares.}$$

$$\qquad\qquad\qquad\qquad\quad = 5 - 7$$
$$\qquad\qquad\qquad\qquad\quad = -2$$

DO EXERCISE 5.

Example 6   Multiply.

$$(\sqrt{3} + x)^2 = (\sqrt{3})^2 + 2x\sqrt{3} + x^2 \qquad \text{Squaring a binomial}$$
$$\qquad\qquad = 3 + 2x\sqrt{3} + x^2$$

DO EXERCISES 6 AND 7.

OBJECTIVE

After finishing Section 6.6, you should be able to:

⬛• Multiply expressions involving radicals, where some of the expressions contain more than one term.

Multiply.

1. $\sqrt{2}(5\sqrt{3} + 3\sqrt{7})$

2. $\sqrt[3]{a^2}(\sqrt[3]{3a} - \sqrt[3]{2})$

Multiply.

3. $(\sqrt{3} - 5\sqrt{2})(2\sqrt{3} + \sqrt{2})$

4. $(\sqrt{a} + 2\sqrt{3})(3\sqrt{b} - 4\sqrt{3})$

Multiply.

5. $(\sqrt{2} + \sqrt{5})(\sqrt{2} - \sqrt{5})$

Multiply.

6. $(2\sqrt{5} - y)^2$

7. $(3\sqrt{6} + 2)^2$

SOMETHING EXTRA
CALCULATOR CORNER: MONTHLY MORTGAGE PAYMENTS

You are about to buy a house. The cost is $50,000 and you have a down payment of $8000. The bank is going to lend you $42,000 at 14% over a period of 20 years. What will your monthly payment be? You can use the following formula:

$$M = P\left[\frac{\dfrac{r}{12}\left(1 + \dfrac{r}{12}\right)^n}{\left(1 + \dfrac{r}{12}\right)^n - 1}\right],$$

where M = monthly payment, n = total number of monthly payments, r = interest rate, and P = principal (the amount of the loan).

We will find your monthly payment in the above situation. You will need a calculator with a power key $\boxed{a^b}$. The total number of monthly payments is $20 \cdot 12$, or 240. The interest rate is 0.14. The principal is 42,000. We substitute.

$$M = 42,000\left[\frac{\dfrac{0.14}{12}\left(1 + \dfrac{0.14}{12}\right)^{240}}{\left(1 + \dfrac{0.14}{12}\right)^{240} - 1}\right]$$

$$= 522.28$$

Your monthly payment is $522.28

EXERCISES

1. The cost of a house is $47,000. The down payment is $9000. The interest rate is 13%. The loan period is 25 years. What is the monthly payment?

2. The cost of a house is $63,000. The down payment is $12,000 and the interest rate is 14.5%. The loan period is 25 years. What is the monthly payment?

3. A rule of thumb to use in deciding whether you can afford a house is not to spend more than 25% of your monthly salary for a mortgage payment. If you make $20,000 per year, can you afford the house in Exercise 1?

4. See Exercise 3. If you are making $38,000 per year can you afford the house in Exercise 2?

NAME   CLASS   ANSWERS

EXERCISE SET 6.6

▪️ Multiply.

1. $\sqrt{6}(2 - 3\sqrt{6})$ **2.** $\sqrt{3}(4 + \sqrt{3})$ **3.** $\sqrt{2}(\sqrt{3} - \sqrt{5})$ **4.** $\sqrt{5}(\sqrt{5} - \sqrt{2})$

5. $\sqrt{3}(2\sqrt{5} - 3\sqrt{4})$ **6.** $\sqrt{2}(3\sqrt{10} - 2\sqrt{2})$ **7.** $\sqrt[3]{2}(\sqrt[3]{4} - 2\sqrt[3]{32})$

8. $\sqrt[3]{3}(\sqrt[3]{9} - 4\sqrt[3]{21})$ **9.** $\sqrt[3]{a}(\sqrt[3]{2a^2} + \sqrt[3]{16a^2})$ **10.** $\sqrt[3]{x}(\sqrt[3]{3x^2} - \sqrt[3]{81x^2})$

11. $(\sqrt{3} - \sqrt{2})(\sqrt{3} + \sqrt{2})$ **12.** $(\sqrt{5} + \sqrt{6})(\sqrt{5} - \sqrt{6})$

13. $(\sqrt{8} + 2\sqrt{5})(\sqrt{8} - 2\sqrt{5})$ **14.** $(\sqrt{18} + 3\sqrt{7})(\sqrt{18} - 3\sqrt{7})$

1. _____

2. _____

3. _____

4. _____

5. _____

6. _____

7. _____

8. _____

9. _____

10. _____

11. _____

12. _____

13. _____

14. _____

For the following exercises assume that all expressions under radicals represent positive numbers.

15. _____

15. $(\sqrt{a} + \sqrt{b})(\sqrt{a} - \sqrt{b})$ **16.** $(\sqrt{x} - \sqrt{y})(\sqrt{x} + \sqrt{y})$ **17.** $(3 - \sqrt{5})(2 + \sqrt{5})$

16. _____

17. _____

18. $(2 + \sqrt{6})(4 - \sqrt{6})$ **19.** $(\sqrt{3} + 1)(2\sqrt{3} + 1)$ **20.** $(4\sqrt{3} + 5)(\sqrt{3} - 2)$

18. _____

19. _____

21. $(2\sqrt{7} - 4\sqrt{2})(3\sqrt{7} + 6\sqrt{2})$ **22.** $(4\sqrt{5} + 3\sqrt{3})(3\sqrt{5} - 4\sqrt{3})$

20. _____

21. _____

23. $(\sqrt{a} + \sqrt{2})(\sqrt{a} + \sqrt{3})$ **24.** $(2 - \sqrt{x})(1 - \sqrt{x})$

22. _____

23. _____

24. _____

25. $(2\sqrt[3]{3} + \sqrt[3]{2})(\sqrt[3]{3} - 2\sqrt[3]{2})$ **26.** $(3\sqrt[4]{7} + \sqrt[4]{6})(2\sqrt[4]{9} - 3\sqrt[4]{6})$

25. _____

26. _____

27. $(2 + \sqrt{3})^2$ **28.** $(\sqrt{5} + 1)^2$

27. _____

28. _____

29. a) _____

29. ▦ a) Calculate $(3 + \sqrt{2})^2$ by first finding $\sqrt{2}$, then adding 3, and then squaring.

   b) Find $(3 + \sqrt{2})^2$ by expanding, to get $9 + 6\sqrt{2} + 2$, and then evaluating. Compare

b) _____
   your answers.

6.7 RATIONALIZING NUMERATORS OR DENOMINATORS

● RATIONALIZING DENOMINATORS

Let us consider several ways of finding $\sqrt{\dfrac{1}{2}}$.

1. $\sqrt{\dfrac{1}{2}} = \dfrac{\sqrt{1}}{\sqrt{2}} = \dfrac{1}{\sqrt{2}}$

We find an approximation to $\sqrt{2}$, in a table or by computing, to be 1.414. Then

$$\frac{1}{\sqrt{2}} \approx \frac{1}{1.414}.$$

By long division we can then find decimal notation for the answer. The division is rather lengthy.

2. $\sqrt{\dfrac{1}{2}} = \sqrt{\dfrac{1}{2} \cdot \boxed{\dfrac{2}{2}}}$ Multiplying by $\dfrac{2}{2}$

$$= \sqrt{\frac{2}{4}} = \frac{\sqrt{2}}{\sqrt{4}} = \frac{\sqrt{2}}{2} \longleftarrow \boxed{\frac{\sqrt{2}}{2} \text{ is often written } \frac{1}{2}\sqrt{2}}$$

We find an approximation to $\sqrt{2}$ and then compute.

$$\frac{\sqrt{2}}{2} \approx \frac{1.414}{2} = 0.707$$

This time the division is easy.

3. $\sqrt{\dfrac{1}{2}} = \sqrt{0.5}$ Finding decimal notation for $\dfrac{1}{2}$

We can look up the square root of 0.5 in a table or compute it with a calculator. Before the days of calculators Method (3) was not very practical, so we had a choice between Methods (1) and (2). Method (2) is much easier, hence it was preferred. The procedure used in Method (2) involves finding a denominator that has no radicals in it. That procedure is called *rationalizing the denominator*. Today, rationalizing the denominator is not as important as it once was because we have calculators. However, the procedure is still important in some other ways. It is more or less standard to write radical expressions with the denominator rationalized.

Example 1   Rationalize the denominator: $\sqrt[3]{\dfrac{7}{9}}$

$$\sqrt[3]{\frac{7}{9}} = \sqrt[3]{\frac{7}{3 \cdot 3} \cdot \boxed{\frac{3}{3}}} \qquad \text{Multiplying by } \frac{3}{3} \text{ to make the}$$
$$\text{denominator a perfect cube}$$

$$= \sqrt[3]{\frac{21}{3 \cdot 3 \cdot 3}} = \frac{\sqrt[3]{21}}{\sqrt[3]{3^3}} = \frac{\sqrt[3]{21}}{3}$$

DO EXERCISES 1–3.

OBJECTIVES

After finishing Section 6.7, you should be able to:

●   Rationalize the denominator of a radical expression.

●●   Rationalize the numerator of a radical expression.

●●●   Rationalize denominators or numerators having two terms.

Rationalize the denominator.

1. $\sqrt{\dfrac{2}{3}}$

2. $\sqrt{\dfrac{10}{7}}$

3. $\sqrt[3]{\dfrac{3}{6}}$

Rationalize the denominator. Assume that all expressions under radicals represent positive numbers.

4. $\sqrt{\dfrac{4a}{3b}}$

5. $\dfrac{\sqrt{4x^5}}{\sqrt{3y^3}}$

Rationalize the denominator.

6. $\dfrac{\sqrt[3]{7}}{\sqrt[3]{2}}$

7. $\sqrt[3]{\dfrac{3x^5}{2y}}$

ANSWERS ON PAGE A–23

The idea in rationalizing a denominator is to multiply by 1 in order to make the denominator a perfect power. In Example 1 we multiplied by 1 under the radical. We can also multiply by 1 outside the radical, as in Example 2.

Example 2   Rationalize the denominator: $\sqrt{\dfrac{2a}{5b}}$. Assume that all expressions under radicals represent positive numbers.

$$\sqrt{\dfrac{2a}{5b}} = \dfrac{\sqrt{2a}}{\sqrt{5b}} \qquad \text{Converting to a quotient of radicals}$$

$$= \dfrac{\sqrt{2a}}{\sqrt{5b}} \cdot \dfrac{\sqrt{5b}}{\sqrt{5b}} \qquad \text{Multiplying by 1}$$

$$= \dfrac{\sqrt{10ab}}{\sqrt{25b^2}} \qquad \text{The radicand in the denominator is a perfect square.}$$

$$= \dfrac{\sqrt{10ab}}{5b} \qquad \text{Taking the square root of } 25b^2$$

DO EXERCISES 4 AND 5.

Example 3   Rationalize the denominator: $\dfrac{\sqrt[3]{a}}{\sqrt[3]{9x}}$.

$$\dfrac{\sqrt[3]{a}}{\sqrt[3]{9x}} = \dfrac{\sqrt[3]{a}}{\sqrt[3]{9x}} \cdot \dfrac{\sqrt[3]{3x^2}}{\sqrt[3]{3x^2}} \qquad \text{Multiplying by 1}$$

To choose the symbol for 1, we look at the radicand $9x$. This is $3 \cdot 3 \cdot x$. To make it a cube we need another 3 and two more x's. Thus we multiply by $\sqrt[3]{3x^2}/\sqrt[3]{3x^2}$. We have

$$\dfrac{\sqrt[3]{3ax^2}}{\sqrt[3]{27x^3}}, \quad \text{which simplifies to} \quad \dfrac{\sqrt[3]{3ax^2}}{3x}.$$

DO EXERCISES 6 AND 7.

◗◗   RATIONALIZING NUMERATORS

Sometimes in advanced mathematics it is necessary to rationalize a numerator. We use a similar procedure.

Example 4   Rationalize the numerator: $\dfrac{\sqrt{7}}{\sqrt{5}}$.

$$\dfrac{\sqrt{7}}{\sqrt{5}} = \dfrac{\sqrt{7}}{\sqrt{5}} \cdot \dfrac{\sqrt{7}}{\sqrt{7}} \qquad \text{Multiplying by 1}$$

$$= \dfrac{\sqrt{49}}{\sqrt{35}} \qquad \text{The radicand in the numerator is a perfect square.}$$

$$= \dfrac{7}{\sqrt{35}}$$

Example 5   Rationalize the numerator: $\dfrac{\sqrt[3]{5}}{\sqrt[3]{3}}$.

$$\frac{\sqrt[3]{5}}{\sqrt[3]{3}} = \frac{\sqrt[3]{5}}{\sqrt[3]{3}} \cdot \frac{\sqrt[3]{5^2}}{\sqrt[3]{5^2}} \qquad \text{Multiplying by 1}$$

$$= \frac{\sqrt[3]{5^3}}{\sqrt[3]{3 \cdot 5^2}}$$

$$= \frac{\sqrt[3]{5^3}}{\sqrt[3]{75}}$$

$$= \frac{5}{\sqrt[3]{75}}$$

DO EXERCISES 8 AND 9.

●●● RATIONALIZING WHEN THERE ARE TWO TERMS

When the denominator to be rationalized has two terms, choose a symbol for 1 as illustrated below.

Example 6   Rationalize the denominator: $\dfrac{4 + \sqrt{2}}{\sqrt{5} - \sqrt{2}}$.

$$\frac{4 + \sqrt{2}}{\sqrt{5} - \sqrt{2}} = \frac{4 + \sqrt{2}}{\sqrt{5} - \sqrt{2}} \cdot \frac{\sqrt{5} + \sqrt{2}}{\sqrt{5} + \sqrt{2}} \qquad \text{Multiplying by 1}$$

$$= \frac{(4 + \sqrt{2})(\sqrt{5} + \sqrt{2})}{(\sqrt{5} - \sqrt{2})(\sqrt{5} + \sqrt{2})} \qquad \begin{array}{l}\text{Multiplying numerators}\\\text{and denominators}\end{array}$$

$$= \frac{4\sqrt{5} + 4\sqrt{2} + \sqrt{2}\sqrt{5} + (\sqrt{2})^2}{(\sqrt{5})^2 - (\sqrt{2})^2} \qquad \text{Using FOIL}$$

$$= \frac{4\sqrt{5} + 4\sqrt{2} + \sqrt{10} + 2}{5 - 2} \qquad \begin{array}{l}\text{Squaring in the}\\\text{denominator}\end{array}$$

$$= \frac{4\sqrt{5} + 4\sqrt{2} + \sqrt{10} + 2}{3}$$

Note that the denominator in this example was $\sqrt{5} \quad \sqrt{2}$. We chose a symbol for 1 that had $\sqrt{5} + \sqrt{2}$ in the numerator and denominator. If the denominator had been $\sqrt{5} + \sqrt{2}$ we would have chosen $\dfrac{5 - \sqrt{2}}{5 - \sqrt{2}}$ for 1.

Examples   What symbol for 1 would you use to rationalize the denominator?

Expression	*Symbol for 1*
7. $\dfrac{3}{x + \sqrt{7}}$	$\dfrac{x - \sqrt{7}}{x - \sqrt{7}}$
8. $\dfrac{\sqrt{7} + 4}{3 - 2\sqrt{5}}$	$\dfrac{3 + 2\sqrt{5}}{3 + 2\sqrt{5}}$

DO EXERCISES 10 AND 11.

Rationalize the numerator.

8. $\dfrac{\sqrt{11}}{\sqrt{6}}$

9. $\dfrac{\sqrt[3]{7}}{\sqrt[3]{4}}$

What symbol for 1 would you use to rationalize the denominator?

10. $\dfrac{\sqrt{5} + 1}{\sqrt{3} - y}$

11. $\dfrac{1}{\sqrt{2} + \sqrt{3}}$

12. Rationalize the denominator.

$$\frac{5}{1 - \sqrt{2}}$$

Example 9   Rationalize the denominator: $\dfrac{4}{\sqrt{3} + x}$.

$$\frac{4}{\sqrt{3} + x} = \frac{4}{\sqrt{3} + x} \cdot \frac{\sqrt{3} - x}{\sqrt{3} - x}$$

$$= \frac{4(\sqrt{3} - x)}{(\sqrt{3} + x)(\sqrt{3} - x)}$$

$$= \frac{4(\sqrt{3} - x)}{(\sqrt{3})^2 - x^2}$$

$$= \frac{4(\sqrt{3} - x)}{3 - x^2}$$

$$= \frac{4\sqrt{3} - 4x}{3 - x^2}$$

DO EXERCISE 12.

We can also rationalize numerators. Of course we choose our symbol for 1 a little differently.

13. What symbol for 1 would you use to rationalize the numerator of

$$\frac{5 - \sqrt{7}}{\sqrt{6} - 3} ?$$

Examples   What symbol for 1 would you use to rationalize the numerator?

	Expression	*Symbol for 1*
10.	$\dfrac{\sqrt{5} + \sqrt{2}}{1 - \sqrt{3}}$	$\dfrac{\sqrt{5} - \sqrt{2}}{\sqrt{5} - \sqrt{2}}$
11.	$\dfrac{x - \sqrt{7}}{3}$	$\dfrac{x + \sqrt{7}}{x + \sqrt{7}}$
12.	$\dfrac{3 - 2\sqrt{5}}{\sqrt{7} + 4}$	$\dfrac{3 + 2\sqrt{5}}{3 + 2\sqrt{5}}$

DO EXERCISE 13.

14. Rationalize the numerator.

$$\frac{3 + \sqrt{5}}{\sqrt{2} - \sqrt{6}}$$

Example 13   Rationalize the numerator: $\dfrac{4 + \sqrt{2}}{\sqrt{5} - \sqrt{2}}$.

$$\frac{4 + \sqrt{2}}{\sqrt{5} - \sqrt{2}} = \frac{4 + \sqrt{2}}{\sqrt{5} - \sqrt{2}} \cdot \frac{4 - \sqrt{2}}{4 - \sqrt{2}}$$

$$= \frac{16 - (\sqrt{2})^2}{4\sqrt{5} - \sqrt{5}\sqrt{2} - 4\sqrt{2} + (\sqrt{2})^2}$$

$$= \frac{14}{4\sqrt{5} - \sqrt{10} - 4\sqrt{2} + 2}$$

DO EXERCISE 14.

NAME   CLASS   ANSWERS

EXERCISE SET 6.7

■● Rationalize the denominator.

1. $\sqrt{\dfrac{6}{5}}$ 2. $\sqrt{\dfrac{11}{6}}$ 3. $\sqrt{\dfrac{10}{7}}$ 4. $\sqrt{\dfrac{22}{3}}$ 5. $\dfrac{6\sqrt{5}}{5\sqrt{3}}$

6. $\dfrac{2\sqrt{3}}{5\sqrt{2}}$ 7. $\sqrt[3]{\dfrac{16}{9}}$ 8. $\sqrt[3]{\dfrac{3}{9}}$ 9. $\dfrac{\sqrt[3]{3a}}{\sqrt[3]{5c}}$ 10. $\dfrac{\sqrt[3]{7x}}{\sqrt[3]{3y}}$

11. $\dfrac{\sqrt[3]{2y^4}}{\sqrt[3]{6x^4}}$ 12. $\dfrac{\sqrt[3]{3a^4}}{\sqrt[3]{7b^2}}$ 13. $\dfrac{1}{\sqrt[3]{xy}}$ 14. $\dfrac{1}{\sqrt[3]{ub}}$

1. _____

2. _____

3. _____

4. _____

5. _____

6. _____

7. _____

8. _____

9. _____

10. _____

11. _____

12. _____

13. _____

14. _____

ANSWERS

● ◐ Rationalize the numerator.

15. _____

16. _____

17. _____

18. _____

19. _____

15. $\dfrac{\sqrt{7}}{\sqrt{3x}}$ 16. $\dfrac{\sqrt{6}}{\sqrt{5x}}$ 17. $\sqrt{\dfrac{14}{21}}$ 18. $\sqrt{\dfrac{12}{15}}$ 19. $\dfrac{4\sqrt{13}}{3\sqrt{7}}$

20. _____

21. _____

22. _____

23. _____

20. $\dfrac{5\sqrt{21}}{2\sqrt{6}}$ 21. $\dfrac{\sqrt[3]{7}}{\sqrt[3]{2}}$ 22. $\dfrac{\sqrt[3]{5}}{\sqrt[3]{4}}$ 23. $\sqrt{\dfrac{7x}{3y}}$ 24. $\sqrt{\dfrac{6a}{2b}}$

24. _____

25. _____

26. _____

27. _____

25. $\dfrac{\sqrt[3]{5y^4}}{\sqrt[3]{6x^5}}$ 26. $\dfrac{\sqrt[3]{3a^5}}{\sqrt[3]{7b^2}}$ 27. $\dfrac{\sqrt{ab}}{3}$ 28. $\dfrac{\sqrt{xy}}{5}$

28. _____

●●● Rationalize the denominator. Assume that all expressions under radicals represent positive numbers.

29. $\dfrac{5}{8 - \sqrt{6}}$ **30.** $\dfrac{7}{9 + \sqrt{10}}$ **31.** $\dfrac{-4\sqrt{7}}{\sqrt{5} - \sqrt{3}}$ **32.** $\dfrac{-3\sqrt{2}}{\sqrt{3} - \sqrt{5}}$

29. _____

30. _____

31. _____

33. $\dfrac{\sqrt{5} - 2\sqrt{6}}{\sqrt{3} - 4\sqrt{5}}$ **34.** $\dfrac{\sqrt{6} - 3\sqrt{5}}{\sqrt{3} - 2\sqrt{7}}$ **35.** $\dfrac{\sqrt{x} - \sqrt{y}}{\sqrt{x} + \sqrt{y}}$ **36.** $\dfrac{\sqrt{a} + \sqrt{b}}{\sqrt{a} - \sqrt{b}}$

32. _____

33. _____

34. _____

37. $\dfrac{5\sqrt{3} - 3\sqrt{2}}{3\sqrt{2} - 2\sqrt{3}}$ **38.** $\dfrac{7\sqrt{2} + 4\sqrt{3}}{4\sqrt{3} - 3\sqrt{2}}$

35. _____

36. _____

37. _____

38. _____

Rationalize the numerator. Assume that all expressions under radicals represent positive numbers.

39. $\dfrac{\sqrt{3}+5}{8}$

40. $\dfrac{3-\sqrt{2}}{5}$

41. $\dfrac{\sqrt{3}-5}{\sqrt{2}+5}$

42. $\dfrac{\sqrt{6}-3}{\sqrt{3}+7}$

43. $\dfrac{\sqrt{x}-\sqrt{y}}{\sqrt{x}+\sqrt{y}}$

44. $\dfrac{\sqrt{x}+\sqrt{y}}{\sqrt{x}-\sqrt{y}}$

45. $\dfrac{4\sqrt{6}-5\sqrt{3}}{2\sqrt{3}+7\sqrt{6}}$

46. $\dfrac{8\sqrt{2}+5\sqrt{3}}{5\sqrt{3}-7\sqrt{2}}$

Pythagorean formula in three dimensions. The length d of a diagonal of a rectangular box is given by $d = \sqrt{a^2 + b^2 + c^2}$, where a, b, and c are the lengths of the sides.

47. ▦ Find the length of a diagonal of a box whose sides have lengths 2, 4, and 5.

48. ▦ Will a baton 57 cm long fit inside an attache case that is 12.5 cm by 45 cm by 31.5 cm?

6.8   RATIONAL NUMBERS AS EXPONENTS

■ FRACTIONAL EXPONENTS

Expressions like $a^{1/2}$, $5^{-1/4}$, and $(2y)^{4/5}$ have not yet been defined. We shall define such expressions in such a way that the usual properties of exponents hold.

Consider $a^{1/2} \cdot a^{1/2}$. If we still want to multiply by adding exponents it must follow that $a^{1/2} \cdot a^{1/2} = a^{1/2+1/2}$ or a^1. Thus we should define $a^{1/2}$ to be a square root of a. Similarly, $a^{1/3} \cdot a^{1/3} \cdot a^{1/3} = a^{1/3+1/3+1/3}$ or a^1, so $a^{1/3}$ should be defined to mean $\sqrt[3]{a}$.

> **For any nonnegative number a and any index n, $a^{1/n}$ means $\sqrt[n]{a}$ (the nonnegative nth root of a).**

Whenever we use fractional exponents we assume that the bases are nonnegative.

Examples   Rewrite without fractional exponents.

1. $x^{1/2} = \sqrt{x}$ **2.** $27^{1/3} = \sqrt[3]{27}$, or 3

3. $(abc)^{1/5} = \sqrt[5]{abc}$

DO EXERCISES 1–5.

Examples   Rewrite with fractional exponents.

4. $\sqrt[5]{7xy} = (7xy)^{1/5}$

5. $\sqrt[7]{\dfrac{x^3 y}{9}} = \left(\dfrac{x^3 y}{9}\right)^{1/7}$

DO EXERCISES 6–8.

How should we define $a^{2/3}$? If the usual properties of exponents are to hold, we have $a^{2/3} = (a^{1/3})^2$, or $(\sqrt[3]{a})^2$, or $\sqrt[3]{a^2}$. We define this accordingly.

> **For any natural numbers m and n ($n \neq 1$) and any nonnegative number a,**
>
> $$a^{m/n} \quad \text{means} \quad \sqrt[n]{a^m}, \text{ or } (\sqrt[n]{a})^m.$$

Examples   Rewrite without fractional exponents.

6. $(27)^{2/3} = \sqrt[3]{27^2}$, or $(\sqrt[3]{27})^2$
$\qquad\quad = 3^2$, or 9

7. $4^{3/2} = \sqrt[2]{4^3}$, or $(\sqrt[2]{4})^3$
$\qquad\; = 2^3$, or 8.

DO EXERCISES 9–11.

OBJECTIVES

After finishing Section 6.8, you should be able to:

■ Write expressions with or without fractional exponents.

■■ Write expressions without negative exponents.

■■■ Use the laws of exponents with fractional exponents.

■■ Use fractional exponents to simplify radical expressions.

Rewrite without fractional exponents.

1. $y^{1/4}$

2. $(3a)^{1/2}$

3. $16^{1/4}$

4. $(125)^{1/3}$

5. $(a^3 b^2 c)^{1/5}$

Rewrite with fractional exponents.

6. $\sqrt[3]{19}$

7. $\sqrt{abc}$

8. $\sqrt[5]{\dfrac{x^2 y}{16}}$

Rewrite without fractional exponents.

9. $x^{3/2}$

10. $8^{2/3}$

11. $4^{5/2}$

ANSWERS ON PAGE A–23

Rewrite with fractional exponents.

12. $(\sqrt[3]{7abc})^4$

13. $\sqrt[5]{6^7}$

Rewrite with positive exponents.

14. $16^{-1/4}$

15. $(3xy)^{-7/8}$

Use the laws of exponents to simplify.

16. $7^{1/3} \cdot 7^{3/5}$

17. $\dfrac{5^{7/6}}{5^{5/6}}$

18. $(9^{3/5})^{2/3}$

Examples   Rewrite with fractional exponents.

8. $\sqrt[3]{9^4} = (\sqrt[3]{9})^4 = 9^{4/3}$ **9.** $(\sqrt[4]{7xy})^5 = (7xy)^{5/4}$

DO EXERCISES 12 AND 13.

■■ **NEGATIVE FRACTIONAL EXPONENTS**

Negative fractional exponents have a meaning similar to that of negative integer exponents.

> **For any rational number m/n and any positive real number a,**
>
> $$a^{-m/n} \quad \text{means} \quad \frac{1}{a^{m/n}},$$
>
> **that is, $a^{m/n}$ and $a^{-m/n}$ are reciprocals.**

Examples   Rewrite with positive exponents.

10. $4^{-1/2} = \dfrac{1}{4^{1/2}}$ $4^{-1/2}$ is the reciprocal of $4^{1/2}$

Since $4^{1/2} = \sqrt{4} = 2$, the answer simplifies to $\frac{1}{2}$.

> Don't make the mistake of thinking that a negative exponent means that the entire expression is negative.

11. $(5xy)^{-4/5} = \dfrac{1}{(5xy)^{4/5}}$ $(5xy)^{-4/5}$ is the reciprocal of $(5xy)^{4/5}$

DO EXERCISES 14 AND 15.

■■■ **LAWS OF EXPONENTS**

The same laws hold for rational number exponents as for integer exponents. We list them for review.

> **For any real number a and any rational exponents m and n,**
>
> **1.** $a^m \cdot a^n = a^{m+n}$ **In multiplying, we can add exponents if the bases are the same.**
>
> **2.** $\dfrac{a^m}{a^n} = a^{m-n}$ **In dividing, we can subtract exponents if the bases are the same.**
>
> **3.** $(a^m)^n = a^{m \cdot n}$ **To raise a power to a power, we can multiply the exponents.**

Examples   Use the laws of exponents to simplify.

12. $3^{1/5} \cdot 3^{3/5} = 3^{1/5 + 3/5} = 3^{4/5}$ Adding exponents

13. $\dfrac{7^{1/4}}{7^{1/2}} = 7^{1/4 - 1/2} = 7^{1/4 - 2/4} = 7^{-1/4}$ Subtracting exponents

14. $(7.2^{2/3})^{3/4} = 7.2^{2/3 \cdot 3/4} = 7.2^{6/12}$ Multiplying exponents
$$= 7.2^{1/2}$$

DO EXERCISES 16–18.

🎱 SIMPLIFYING RADICAL EXPRESSIONS

Fractional exponents can be used to simplify some radical expressions. The procedure is as follows.

> 1. Convert radical expressions to exponential expressions.
> 2. Use arithmetic and the laws of exponents to simplify.
> 3. Convert back to radical notation when appropriate.
>
> *Important:* This works only when all expressions under radicals are nonnegative since fractional exponents are not defined otherwise. No absolute value signs will be needed.

Examples   Use fractional exponents to simplify.

15. $\sqrt[6]{x^3} = x^{3/6}$ Converting to an exponential expression

   $\quad\quad\ = x^{1/2}$ Using arithmetic to simplify the exponent

   $\quad\quad\ = \sqrt{x}$ Converting back to radical notation

16. $\sqrt[6]{4} = 4^{1/6}$ Converting to exponential notation

   $\quad\quad = (2^2)^{1/6}$ Renaming 4 as 2^2

   $\quad\quad = 2^{2/6}$ Using the third law of exponents

   $\quad\quad = 2^{1/3}$ Using arithmetic to simplify the exponent

   $\quad\quad = \sqrt[3]{2}$ Converting back to radical notation

DO EXERCISES 19–21.

Example 17   Use fractional exponents to simplify.

   $\sqrt[8]{a^2b^4} = (a^2b^4)^{1/8}$ Converting to exponential notation

   $\quad\quad\quad = a^{2/8} \cdot b^{4/8}$ Using the third law of exponents

   $\quad\quad\quad = a^{1/4} \cdot b^{2/4}$ Using arithmetic to simplify the exponents

   $\quad\quad\quad = (ab^2)^{1/4}$ Using the third law of exponents (in reverse)

   $\quad\quad\quad = \sqrt[4]{ab^2}$ Converting back to radical notation

DO EXERCISES 22–24.

We can use properties of fractional exponents to write a single radical expression for a product or quotient.

Examples   Use fractional exponents to write a single radical expression.

18. $\sqrt[3]{5} \cdot \sqrt{2} = 5^{1/3} \cdot 2^{1/2}$ Converting to exponential notation

   $\quad\quad\quad\quad = 5^{2/6} \cdot 2^{3/6}$ Rewriting so that exponents have a common denominator

   $\quad\quad\quad\quad = (5^2 \cdot 2^3)^{1/6}$ Using the third law of exponents

   $\quad\quad\quad\quad = \sqrt[6]{5^2 \cdot 2^3}$ Converting back to radical notation

   $\quad\quad\quad\quad = \sqrt[6]{200}$ Multiplying under the radical

Use fractional exponents to simplify.

19. $\sqrt[4]{a^2}$

20. $\sqrt[4]{x^4}$

21. $\sqrt[6]{8}$

Use fractional exponents to simplify.

22. $\sqrt[5]{a^5b^{10}}$

23. $\sqrt[4]{x^4y^{12}}$

24. $\sqrt[12]{x^3y^6}$

ANSWERS ON PAGE A–23

Use fractional exponents to write a single radical expression.

25. $\sqrt[4]{7} \cdot \sqrt{3}$

26. $\dfrac{\sqrt[4]{(a-b)^5}}{(a-b)}$

Write a single radical expression.

27. $x^{2/3}y^{1/2}z^{5/6}$

28. $\dfrac{a^{1/2}b^{3/8}}{a^{1/4}b^{1/8}}$

19.
$$\sqrt{x-2} \cdot \sqrt[4]{3y} = (x-2)^{1/2}(3y)^{1/4} \quad \text{Converting to exponential notation}$$
$$= (x-2)^{2/4}(3y)^{1/4} \quad \text{Writing exponents with a common denominator}$$
$$= [(x-2)^2(3y)]^{1/4} \quad \text{Using the third law of exponents (in reverse)}$$
$$= \sqrt[4]{(x^2-4x+4)\cdot 3y} \quad \text{Converting back to radical notation}$$
$$= \sqrt[4]{3x^2y - 12xy + 12y} \quad \text{Multiplying under the radical}$$

20.
$$\frac{\sqrt[4]{(x+y)^3}}{\sqrt{x+y}} = \frac{(x+y)^{3/4}}{(x+y)^{1/2}} \quad \text{Converting to exponential notation}$$
$$= (x+y)^{3/4-1/2} \quad \text{Using the second law of exponents}$$
$$= (x+y)^{1/4} \quad \text{Subtracting exponents}$$
$$= \sqrt[4]{x+y} \quad \text{Converting back to radical notation}$$

DO EXERCISES 25 AND 26.

Example 21 Write a single radical expression.
$$a^{1/2}b^{-1/2}c^{5/6} = a^{3/6}b^{-3/6}c^{5/6} \quad \text{Rewriting exponents with a common denominator}$$
$$= (a^3b^{-3}c^5)^{1/6} \quad \text{Using the third law of exponents}$$
$$= \sqrt[6]{a^3b^{-3}c^5} \quad \text{Converting to radical notation}$$

DO EXERCISES 27 AND 28.

We have now seen several different methods of simplifying radical expressions. We list them.

Some Ways to Simplify Radical Expressions:

1. *Simplifying by factoring.* **We factor the radicand, looking for factors that are perfect powers.**
   Example: $\sqrt[3]{16} = \sqrt[3]{8}\,\sqrt[3]{2} = 2\sqrt[3]{2}$

2. *Rationalizing denominators.* **Radical expressions are usually considered simpler if there are no radicals in the denominator.**
   Example: $\dfrac{1}{\sqrt{2}} = \dfrac{\sqrt{2}}{2}$

3. *Collecting like radical terms.*
   Example: $\sqrt{8} + 3\sqrt{2} = \sqrt{4}\cdot\sqrt{2} + 3\sqrt{2} = 2\sqrt{2} + 3\sqrt{2} = 5\sqrt{2}$

4. *Using fractional exponents to simplify.* **We convert to exponential notation and then use arithmetic and the laws of exponents to simplify the exponents. Then we convert back to radical notation.** *Caution:* **This works only when there are only nonnegative expressions under radical signs.**

NAME   CLASS

EXERCISE SET 6.8

◼•◼ Rewrite without fractional exponents.

1. $x^{1/4}$ **2.** $y^{1/5}$ **3.** $(8)^{1/3}$ **4.** $(16)^{1/2}$

5. $(a^2b^2)^{1/5}$ **6.** $(x^3y^3)^{1/4}$ **7.** $16^{3/4}$ **8.** $4^{7/2}$

Rewrite with fractional exponents.

9. $\sqrt[3]{20}$ **10.** $\sqrt[3]{19}$ **11.** $\sqrt[5]{xy^2z}$ **12.** $\sqrt[7]{x^3y^2z^2}$

13. $(\sqrt{3mn})^3$ **14.** $(\sqrt[3]{7xy})^4$ **15.** $(\sqrt[7]{8x^2y})^5$ **16.** $(\sqrt[6]{2a^5b})^7$

◼•◼•◼ Rewrite with positive exponents.

17. $x^{-1/3}$ **18.** $y^{-1/4}$ **19.** $\dfrac{1}{x^{-2/3}}$ **20.** $\dfrac{1}{x^{-5/6}}$

◼•◼•◼ Use the laws of exponents to simplify.

21. $5^{3/4} \cdot 5^{1/8}$ **22.** $11^{2/3} \cdot 11^{1/2}$ **23.** $\dfrac{7^{5/8}}{7^{3/8}}$ **24.** $\dfrac{9^{9/11}}{9^{7/11}}$

25. $\dfrac{8.3^{3/4}}{8.3^{2/5}}$ **26.** $\dfrac{3.9^{3/5}}{3.9^{1/4}}$ **27.** $(10^{3/8})^{2/5}$ **28.** $(5^{5/9})^{3/7}$

◼◼ Use fractional exponents to simplify.

29. $\sqrt[6]{a^4}$ **30.** $\sqrt[6]{y^2}$ **31.** $\sqrt[3]{8y^6}$ **32.** $\sqrt{x^4y^6}$

33. $\sqrt[4]{32}$ **34.** $\sqrt[8]{81}$ **35.** $\sqrt[6]{4x^2}$ **36.** $\sqrt[4]{16x^4y^2}$

ANSWERS
1.
2.
3.
4.
5.
6.
7.
8.
9.
10.
11.
12.
13.
14.
15.
16.
17.
18.
19.
20.
21.
22.
23.
24.
25.
26.
27.
28.
29.
30.
31.
32.
33.
34.
35.
36.

37. $\sqrt[5]{32c^{10}d^{15}}$ **38.** $\sqrt[4]{16x^{12}y^{16}}$ **39.** $\sqrt[6]{\dfrac{m^{12}n^{24}}{64}}$ **40.** $\sqrt[5]{\dfrac{x^{15}y^{20}}{32}}$

41. $\sqrt[8]{r^4s^2}$ **42.** $\sqrt[12]{64t^6s^6}$ **43.** $\sqrt[3]{27a^3b^9}$ **44.** $\sqrt[4]{81x^8y^8}$

Use fractional exponents to write a single radical expression.

45. $\sqrt[3]{5} \cdot \sqrt{2}$ **46.** $\sqrt[3]{7} \cdot \sqrt[4]{5}$ **47.** $\sqrt{x}\sqrt[3]{2x}$ **48.** $\sqrt[3]{y}\sqrt[5]{3y}$

49. $\sqrt{x}\sqrt[3]{x-2}$ **50.** $\sqrt[4]{3x}\sqrt{y+4}$ **51.** $\dfrac{\sqrt[3]{(a+b)^2}}{\sqrt{(a+b)}}$ **52.** $\dfrac{\sqrt[3]{(x+y)^2}}{\sqrt[4]{(x+y)^3}}$

53. $a^{2/3} \cdot b^{3/4}$ **54.** $x^{1/3} \cdot y^{1/4} \cdot z^{1/6}$ **55.** $\dfrac{s^{7/12} \cdot t^{5/6}}{s^{1/3} \cdot t^{-1/6}}$ **56.** $\dfrac{x^{8/15} \cdot y^{4/5}}{x^{1/3} \cdot y^{-1/5}}$

6.9 SOLVING RADICAL EQUATIONS

◼•◼ THE PRINCIPLE OF POWERS

These are radical equations:

$$\sqrt[3]{2x} + 1 = 5, \qquad \sqrt{x} + \sqrt{4x - 2} = 7.$$

A radical equation has variables in one or more radicands. To solve such equations we need a new principle. Suppose an equation $a = b$ is true. If we square both sides, we get another true equation: $a^2 = b^2$. This can be generalized.

> *The Principle of Powers:* **If an equation $a = b$ is true, then $a^n = b^n$ is true for any natural number n.**

If an equation $a^n = b^n$ is true, it *may not* be true that $a = b$. For example, $3^2 = (-3)^2$ is true, but $3 = -3$ is not true.

Example 1   Solve: $\sqrt{x} - 3 = 4$.

$$\sqrt{x} = 7 \qquad \text{Adding to isolate the radical}$$
$$x = 7^2, \text{ or } 49 \qquad \text{Using the principle of powers}$$

Check:
$$\begin{array}{c|c} \sqrt{x} - 3 = 4 & \\ \hline \sqrt{49} - 3 & 4 \\ 7 - 3 & \\ 4 & \end{array}$$

The solution is 49.

The principle of powers does not always give equivalent equations.

Example 2   Solve: $\sqrt{x} = -3$.

We might observe at the outset that this equation has no solution because the principal square root of a number is never negative. Let us continue as above, for comparison.

$$x = (-3)^2, \text{ or } 9 \qquad \text{Principle of powers (squaring)}$$

Check:
$$\begin{array}{c|c} \sqrt{x} = -3 & \\ \hline \sqrt{9} & -3 \\ 3 & \end{array}$$

The number 9 does not check. Hence the equation has no solution.

> **In solving radical equations, possible solutions found using the principle of powers *must* be checked!**

DO EXERCISES 1 AND 2.

OBJECTIVES

After finishing Section 6.9, you should be able to:

◼•◼ Solve radical equations with one radical term.

◼•◼ Solve radical equations with two radical terms.

Solve.

1. $\sqrt{x} - 7 = 3$

2. $\sqrt{x} = -2$

3. Solve.

$$x - 1 = \sqrt{x + 5}$$

To solve an equation with a radical term, we first isolate the radical term on one side of the equation. Then we use the principle of powers.

Example 3 Solve: $x = \sqrt{x + 7} + 5$.

$$x - 5 = \sqrt{x + 7} \qquad \text{Adding } -5 \text{ to isolate the radical term}$$

$$(x - 5)^2 = (\sqrt{x + 7})^2 \qquad \text{Principle of powers: squaring both sides}$$

$$x^2 - 10x + 25 = x + 7$$

$$x^2 - 11x + 18 = 0$$

$$(x - 9)(x - 2) = 0 \qquad \text{Factoring}$$

$$x = 9 \quad \text{or} \quad x = 2 \qquad \text{Using the principle of zero products}$$

The possible solutions are 9 and 2. Let us check.

For 9:
$$\begin{array}{c|c} x = \sqrt{x + 7} + 5 \\ \hline 9 & \sqrt{9 + 7} + 5 \\ & 9 \end{array}$$

For 2:
$$\begin{array}{c|c} x = \sqrt{x + 7} + 5 \\ \hline 2 & \sqrt{2 + 7} + 5 \\ & 8 \end{array}$$

Since 9 checks but 2 does not, the solution is 9.

Example 4 Solve: $\sqrt[3]{2x + 1} + 5 = 0$.

$$\sqrt[3]{2x + 1} = -5 \qquad \text{Adding } -5 \text{; this isolates the radical term}$$

$$2x + 1 = (-5)^3, \text{ or } -125 \qquad \text{Principle of powers; raising to the third power}$$

$$2x = -126 \qquad \text{Adding } -1$$

$$x = -63$$

4. Solve.

$$\sqrt[4]{x - 1} - 2 = 0$$

Check:
$$\begin{array}{c|c} \sqrt[3]{2x + 1} + 5 = 0 \\ \hline \sqrt[3]{2 \cdot (-63) + 1} + 5 & 0 \\ \sqrt[3]{-125} + 5 & \\ -5 + 5 & \\ 0 & \end{array}$$

The solution is -63.

DO EXERCISES 3 AND 4.

⬤⬤⬤ EQUATIONS WITH TWO RADICAL TERMS

A general strategy for solving equations with two radical terms is as follows.

1. **Isolate one of the radical terms.**
2. **Use the principle of powers.**
3. **If a radical remains, perform steps 1 and 2 again.**
4. **Check possible solutions.**

Example 5   Solve: $\sqrt{x - 3} + \sqrt{x + 5} = 4$.

$$\sqrt{x - 3} = 4 - \sqrt{x + 5} \qquad \text{Adding } -\sqrt{x + 5}\text{; this isolates} \atop \text{one of the radical terms}$$

$$(\sqrt{x - 3})^2 = (4 - \sqrt{x + 5})^2 \qquad \text{Principle of powers;} \atop \text{squaring both sides}$$

> Here we are squaring a binomial. We square 4, then find twice the product of 4 and $\sqrt{x + 5}$, and then the square of $\sqrt{x + 5}$. (See the rule on p. 173.)

$$x - 3 = 16 - 8\sqrt{x + 5} + (x + 5)$$
$$-24 = -8\sqrt{x + 5} \qquad \text{Isolating the remaining radical term}$$
$$3 = \sqrt{x + 5}$$
$$3^2 = x + 5 \qquad \text{Squaring}$$
$$4 = x$$

The number 4 checks and is the solution.

> A common error in solving equations like $\sqrt{x - 3} + \sqrt{x + 5} = 4$ is to square the left side, obtaining $(x - 3) + (x + 5)$. That is wrong because the square of a sum is *not* the sum of the squares.
> *Example:* $\sqrt{9} + \sqrt{16} = 7$, but $9 + 16 = 25$, not 7^2.

Example 6   Solve: $\sqrt{2x - 5} = 1 + \sqrt{x - 3}$.

$$(\sqrt{2x - 5})^2 = (1 + \sqrt{x - 3})^2 \qquad \text{One radical is already} \atop \text{isolated; we square both sides}$$
$$2x - 5 = 1 + 2\sqrt{x - 3} + (x - 3)$$
$$x - 3 = 2\sqrt{x - 3} \qquad \text{Isolating the remaining} \atop \text{radical term}$$
$$(x - 3)^2 = (2\sqrt{x - 3})^2 \qquad \text{Squaring both sides}$$
$$x^2 - 6x + 9 = 4(x - 3)$$
$$x^2 - 6x + 9 = 4x - 12$$
$$x^2 - 10x + 21 = 0$$
$$(x - 7)(x - 3) = 0 \qquad \text{Factoring}$$
$$x = 7 \quad \text{or} \quad x = 3 \qquad \text{Using the principle} \atop \text{of zero products}$$

The numbers 7 and 3 check and are the solutions.

DO EXERCISES 5 AND 6.

Solve.

5. $\sqrt{x} - \sqrt{x - 5} = 1$

6. $\sqrt{3x + 1} = 1 + \sqrt{x + 4}$

SOMETHING EXTRA

AN APPLICATION: ROAD PAVEMENT SIGNS

In a psychological study it was determined that the proper length L of the letters of a word painted on pavement is given by

$$L = \frac{0.00252d^{2.27}}{h},$$

where d is the distance of a car from the lettering and h is the height of the eye above the road. All units are in meters.

According to the study and this formula, if a person h meters above the road is to read a message d meters away, that message will be the most readable if the length of the letters is L.

EXERCISES

Find L, given the values of d and h. You will need a calculator with an exponential key $\boxed{a^b}$.

1. $h = 1$ m, $d = 60$ m
2. $h = 2.4$ m, $d = 80$ m
3. $h = 0.9906$ m, $d = 75$ m
4. $h = 1.1$ m, $d = 100$ m

NAME

CLASS

ANSWERS

EXERCISE SET 6.9

■● Solve.

1. $\sqrt{y + 1} - 5 = 8$

2. $\sqrt{x - 2} - 7 = -4$

3. $\sqrt{3y + 1} = 9$

4. $\sqrt{2y + 1} = 13$

5. $\sqrt[3]{x} = -3$

6. $\sqrt[3]{y} = -4$

7. $\sqrt{x + 2} = -4$

8. $\sqrt{y - 3} = -2$

9. $\sqrt[3]{6x + 9} + 8 = 5$

10. $\sqrt[3]{3y + 6} + 2 = 3$

1. _____
2. _____
3. _____
4. _____
5. _____
6. _____
7. _____
8. _____
9. _____
10. _____

■ ■ Solve.

11. $\sqrt{3y + 1} = \sqrt{2y + 6}$

12. $\sqrt{5x - 3} = \sqrt{2x + 3}$

11. _____

12. _____

13. $\sqrt{y - 5} + \sqrt{y} = 5$

14. $\sqrt{x - 9} + \sqrt{x} = 1$

13. _____

14. _____

15. $3 + \sqrt{z - 6} = \sqrt{z + 9}$

16. $\sqrt{4x - 3} = 2 + \sqrt{2x - 5}$

15. _____

16. _____

17. $\sqrt{20 - x} + 8 = \sqrt{9 - x} + 11$

18. $4 + \sqrt{10 - x} = 6 + \sqrt{4 - x}$

17. _____

18. _____

19. $\sqrt{4y + 1} - \sqrt{y - 2} = 3$

20. $\sqrt{y + 15} - \sqrt{2y + 7} = 1$

19. _____

20. _____

21. Solve: $\dfrac{x + \sqrt{x + 1}}{x - \sqrt{x + 1}} = \dfrac{5}{11}$.

21. _____

22. Refer to Exercise 31 on p. 312. Calculate the length of a pendulum that will beat seconds (having a period of two seconds). Make a pendulum like this and check it with a stopwatch.

22. _____

6.10 THE COMPLEX NUMBERS

● IMAGINARY AND COMPLEX NUMBERS

Negative numbers do not have square roots in the system of real numbers. *Imaginary** numbers were invented so that negative numbers would have square roots. We begin by making up a number that is a square root of -1. We call this new number i. Thus $i^2 = -1$, or $i = \sqrt{-1}$. All other imaginary numbers can be expressed as a product of i and a real number.

Examples   Express in terms of i.

1. $\sqrt{-5} = \sqrt{-1 \cdot 5} = \sqrt{-1} \cdot \sqrt{5} = i\sqrt{5}$ or $\sqrt{5}i$

2. $-\sqrt{-49} = -\sqrt{-1 \cdot 49} = -\sqrt{-1} \cdot \sqrt{49} = -i\sqrt{49} = -7i$

> This procedure works, even though the general rule for factoring radicals does not hold for negative radicands.

DO EXERCISES 1–4.

> **An imaginary number is a number that can be named bi, where b is some real number.**

To form the system of *complex numbers* we take the imaginary numbers and the real numbers, and all possible sums of real and imaginary numbers. These are complex numbers:

$$3 - 4i, \quad -\sqrt{46} + 2i, \quad 14, \quad i\sqrt{2}.$$

> **A complex number is any number that can be named $a + bi$, where a and b are any real numbers. (Note that a or b or both can be 0.)**

●● ADDITION AND SUBTRACTION

The complex numbers follow the commutative and associative laws of addition. Thus we can add and subtract them as we do binomials in real numbers.

Examples   Add or subtract.

3. $(8 + 6i) + (3 + 2i) = (8 + 3) + (6 + 2)i$ Collecting the real parts and the imaginary parts

   $= 11 + 8i$

4. $(3 + 2i) - (5 - 2i) = (3 - 5) + [2 - (-2)]i$

   $= -2 + 4i$

DO EXERCISES 5–8.

---

* Don't let the name *imaginary* fool you. The imaginary numbers are very important in such fields as engineering and the physical sciences.

OBJECTIVES

After finishing Section 6.10, you should be able to:

● Express imaginary numbers as bi, where b is a real number, and complex numbers as $a + bi$, where a and b are real numbers.

●● Add and subtract complex numbers.

●●● Multiply complex numbers.

⦂⦂ Find conjugates of complex numbers and divide complex numbers.

⦂⦂ Determine whether a given complex number is a solution of an equation.

Express in terms of i.

1. $\sqrt{-7}$

2. $\sqrt{-25}$

3. $-\sqrt{-11}$

4. $-\sqrt{-36}$

Add or subtract.

5. $(7 + 4i) + (8 - 7i)$

6. $(-5 - 6i) + (-7 + 12i)$

7. $(8 + 3i) - (5 + 8i)$

8. $(5 - 4i) - (-7 + 3i)$

ANSWERS ON PAGE A–24

Multiply.

9. $-3i(4 - 3i)$

10. $5i(-5 + 7i)$

11. $(1 + 3i)(1 + 5i)$

12. $(3 - 2i)(1 + 4i)$

Simplify.

13. i^5
   [*Hint:* $i^5 = (i^2)(i^2)i$.]

14. $(2i)^3$

15. $(3 + 2i)^2$

16. $(2 - 5i)^2$

●●● MULTIPLYING

The complex numbers also obey the commutative and associative laws of multiplication and the distributive laws. Thus we multiply them in much the same way as we do binomials in real numbers. We must remember, however, that $i^2 = -1$.

Examples Multiply.

5. $(-4i)(3 - 5i) = (-4i) \cdot 3 + (-4i)(-5i)$ Using a distributive law

$\qquad = -12i + 20i^2$

$\qquad = -12i - 20 \qquad i^2 = -1$

$\qquad = -20 - 12i$

6. $(1 + 2i)(1 + 3i) = 1 + 3i + 2i + 6i^2$ Multiplying each term of one number by every term of the other FOIL

$\qquad = 1 + 3i + 2i - 6 \qquad i^2 = -1$

$\qquad = -5 + 5i$ Collecting like terms

DO EXERCISES 9–12.

Examples Simplify.

7. $i^3 = (i^2)i = -1 \cdot i = -i$

8. $i^4 = (i^2)(i^2) = (-1) \cdot (-1) = 1$

9. $(3 - 2i)^2 = 3^2 - 2 \cdot (3)(2i) + (2i)^2$

$\qquad = 9 - 12i + 4i^2$

$\qquad = 9 - 12i - 4$

$\qquad = 5 - 12i$

DO EXERCISES 13–16.

> *Caution!* The property $\sqrt{a}\sqrt{b} = \sqrt{ab}$ does not hold for imaginary numbers. All square roots of negative numbers must be expressed in terms of i before multiplying. For example,
> $$\sqrt{-2} \cdot \sqrt{-5} = i\sqrt{2} \cdot i\sqrt{5} = i^2\sqrt{10} = -\sqrt{10} \text{ is correct.}$$
> But
> $$\sqrt{-2} \cdot \sqrt{-5} = \sqrt{(-2)(-5)} = \sqrt{10} \text{ is wrong!}$$

▪▪ CONJUGATES AND DIVISION

Conjugates of complex numbers are defined as follows.

> The *conjugate* of a complex number $a + bi$ is $a - bi$, and the *conjugate* of $a - bi$ is $a + bi$.

Examples   Find the conjugate.

10. $5 + 7i$ The conjugate is $5 - 7i$.

11. $14 - 3i$ The conjugate is $14 + 3i$.

12. $-3 - 9i$ The conjugate is $-3 + 9i$.

13. $4i$ The conjugate is $-4i$.

DO EXERCISES 17–19.

When we multiply a number by its conjugate, we get a real number.

Examples   Multiply.

14. $(5 + 7i)(5 - 7i) = 5^2 - (7i)^2$
$$= 25 - 49i^2$$
$$= 25 - (-49)$$
$$= 74$$

15. $(2 - 3i)(2 + 3i) = 2^2 - (3i)^2$
$$= 4 - 9i^2$$
$$= 4 - (-9)$$
$$= 13$$

DO EXERCISES 20 AND 21.

We use conjugates in dividing complex numbers.

Example 16   Divide: $\dfrac{-5 + 9i}{1 - 2i}$.

$$\frac{-5 + 9i}{1 - 2i} \cdot \frac{1 + 2i}{1 + 2i} = \frac{(-5 + 9i)(1 + 2i)}{(1 - 2i)(1 + 2i)} \qquad \text{Multiplying by 1}$$

$$= \frac{-23 - i}{5}$$

$$= -\frac{23}{5} - \frac{1}{5}i$$

Note the similarity between this example and rationalizing denominators. The symbol for the number 1 was chosen with reference to the divisor, $1 - 2i$. The symbol for 1 was formed using the conjugate of the divisor for both numerator and denominator.

Example 17   What symbol for 1 would you use to divide?

Division to be done	*Symbol for 1*
$\dfrac{3 + 5i}{4 + 3i}$	$\dfrac{4 - 3i}{4 - 3i}$

DO EXERCISES 22 AND 23.

Find the conjugate.

17. $6 + 3i$

18. $-9 - 5i$

19. $\pi - \dfrac{1}{4}i$

Multiply.

20. $(7 - 2i)(7 + 2i)$

21. $(-3 - i)(-3 + i)$

Divide.

22. $\dfrac{6 + 2i}{1 - 3i}$

23. $\dfrac{2 + 3i}{-1 + 4i}$

24. Determine whether $-i$ is a solution of $x^2 + 1 = 0$.

⠿ SOLUTIONS OF EQUATIONS

The equation $x^2 + 1 = 0$ has no real-number solution, but it has two nonreal solutions.

Example 18   Determine whether i is a solution of the equation $x^2 + 1 = 0$.

We substitute i for x in the equation.

$$
\begin{array}{c|c}
\multicolumn{2}{c}{x^2 + 1 = 0} \\
\hline
i^2 + 1 & 0 \\
-1 + 1 & \\
0 & \\
\end{array}
$$

The number i is a solution.

DO EXERCISE 24.

Any equation consisting of a polynomial in one variable on one side and 0 on the other has complex-number solutions (some may be real). It is not always easy to find the solutions, but they always exist.

Example 19   Determine whether $1 + i$ is a solution of $x^2 - 2x + 2 = 0$.

25. Determine whether $1 - i$ is a solution of $x^2 - 2x + 2 = 0$.

We substitute $1 + i$ for x in the equation.

$$
\begin{array}{c|c}
\multicolumn{2}{c}{x^2 - 2x + 2 = 0} \\
\hline
(1 + i)^2 - 2(1 + i) + 2 & 0 \\
1 + 2i + i^2 - 2 - 2i + 2 & \\
1 + 2i - 1 - 2 - 2i + 2 & \\
(1 - 1 - 2 + 2) + (2 - 2)i & \\
0 + 0i & \\
0 & \\
\end{array}
$$

$1 + i$ is a solution.

Example 20   Determine whether $2i$ is a solution of $x^2 + 3x - 4 = 0$.

$$
\begin{array}{c|c}
\multicolumn{2}{c}{x^2 + 3x - 4 = 0} \\
\hline
(2i)^2 + 3(2i) - 4 & 0 \\
4i^2 + 6i - 4 & \\
-4 + 6i - 4 & \\
-8 + 6i & \\
\end{array}
$$

$2i$ is not a solution.

DO EXERCISE 25.

NAME   CLASS   ANSWERS

EXERCISE SET 6.10

• Express in terms of i.

1. $\sqrt{-15}$ **2.** $\sqrt{-17}$ **3.** $\sqrt{-16}$

4. $\sqrt{-25}$ **5.** $-\sqrt{-12}$ **6.** $-\sqrt{-20}$

• • Add or subtract and simplify.

7. $(3 + 2i) + (5 - i)$ **8.** $(-2 + 3i) + (7 + 8i)$ **9.** $(4 - 3i) + (5 - 2i)$

10. $(-2 - 5i) + (1 - 3i)$ **11.** $(9 - i) + (-2 + 5i)$ **12.** $(6 + 4i) + (2 - 3i)$

13. $(3 - i) - (5 + 2i)$ **14.** $(-2 + 8i) - (7 + 3i)$ **15.** $(4 - 2i) - (5 - 3i)$

16. $(-2 - 3i) - (1 - 5i)$ **17.** $(9 + 5i) - (-2 - i)$ **18.** $(6 - 3i) - (2 + 4i)$

• • • Simplify.

19. $(3 + 2i)(1 + i)$ **20.** $(4 + 3i)(2 + 5i)$ **21.** $(2 + 3i)(6 - 2i)$

22. $(5 + 6i)(2 - i)$ **23.** $(6 - 5i)(3 + 4i)$ **24.** $(5 - 6i)(2 + 5i)$

ANSWERS

1. _____
2. _____
3. _____
4. _____
5. _____
6. _____
7. _____
8. _____
9. _____
10. _____
11. _____
12. _____
13. _____
14. _____
15. _____
16. _____
17. _____
18. _____
19. _____
20. _____
21. _____
22. _____
23. _____
24. _____

ANSWERS

25. _____

26. _____

27. _____

28. _____

29. _____

30. _____

31. _____

32. _____

33. _____

34. _____

35. _____

36. _____

37. _____

38. _____

39. _____

40. _____

41. _____

42. _____

43. _____

44. _____

45. _____

46. _____

25. $(7 - 2i)(2 - 6i)$ **26.** $(-4 + 5i)(3 - 4i)$ **27.** i^6

28. $(-i)^4$ **29.** $(5i)^3$ **30.** $(-3i)^5$

31. $(3 - 2i)^2$ **32.** $(5 - 2i)^2$ **33.** $(2 + 3i)^2$

34. $(4 + 2i)^2$ **35.** $(-2 + 3i)^2$ **36.** $(-5 - 2i)^2$

⠿ Divide.

37. $\dfrac{3 + 2i}{2 + i}$ **38.** $\dfrac{4 + 5i}{5 - i}$ **39.** $\dfrac{5 - 2i}{2 + 5i}$

40. $\dfrac{3 - 2i}{4 + 3i}$ **41.** $\dfrac{8 - 3i}{7i}$ **42.** $\dfrac{3 + 8i}{5i}$

⠿ Determine whether the complex number is a solution of the equation.

43. $1 + 2i$;
$x^2 - 2x + 5 = 0$

44. $1 - 2i$;
$x^2 - 2x + 5 = 0$

45. $1 - i$;
$x^2 + 2x + 2 = 0$

46. $2 + i$;
$x^2 - 4x - 5 = 0$

EXTENSION EXERCISES

SECTION 6.1

1. The distance light travels in 100 years is approximately 5.87×10^{14} miles. How far does light travel in 13 weeks? Write scientific notation for your answer.

2. Compare $8 \cdot 10^{-90}$ and $9 \cdot 10^{-91}$. Which is the larger value? How much larger? Write scientific notation for the difference.

3. Find 8^{-x} when $8^{3x} = 27$.

4. Evaluate: $(4096)^{0.05}(4096)^{0.2}$.

5. What is the unit's digit in the product $(513)^{127}$?

6. Given that $a = 4 - 3^b$ and $c = 2 - 3^{-b}$, express c in terms of a.

Simplify.

7. $[7y(7-8)^{-2} - 8y(8-7)^{-2}]^{-(-2)^2}$

8. $\dfrac{3^{q+3} - 3^2(3^q)}{3(3^{q+4})}$

9. $[3^{-(3s+1)} - 3^{-(3s-1)} + 3^{-3s}]^{-2^2}$

SECTION 6.3

Multiply and simplify.

10. $\sqrt{1.6 \times 10^3}\sqrt{36 \times 10^{-8}}$

11. $\sqrt{968}\sqrt{1014}$

12. $\sqrt[3]{48}\sqrt[3]{63}\sqrt[3]{196}$

In Sections 6.4–6.10 assume that all expressions under radicals represent positive numbers.

SECTION 6.4

Divide and simplify.

13. $\dfrac{7\sqrt{a^2 b}\sqrt{25xy}}{5\sqrt{a^{-4}b^{-1}}\sqrt{49x^{-1}y^{-3}}}$

14. $\dfrac{(\sqrt[3]{81mn^2})^2}{(\sqrt[3]{mn})^2}$

15. $\dfrac{\sqrt{44x^2y^9z}\sqrt{22y^9z^6}}{(\sqrt{11xy^8z^2})^2}$

SECTION 6.5

Add or subtract by collecting like terms.

16. $\sqrt{432} - \sqrt{6125} + \sqrt{845} - \sqrt{4800}$

17. $\sqrt{1250x^3y} - \sqrt{1800xy^3} - \sqrt{162x^3y^3}$

18. $\dfrac{1}{2}\sqrt{36a^5bc^4} - \dfrac{1}{2}\sqrt[3]{64a^4bc^6} + \dfrac{1}{6}\sqrt{144a^3bc^2}$

19. $7x\sqrt{(x+y)^3} - 5xy\sqrt{x+y} - 2y\sqrt{(x+y)^3}$

SECTION 6.6

Multiply and simplify.

20. $\sqrt{9 + 3\sqrt{5}}\sqrt{9 - 3\sqrt{5}}$

21. $(\sqrt{x+2} - \sqrt{x-2})^2$

22. $(\sqrt{3} + \sqrt{5} - \sqrt{6})^2$

23. $\sqrt[3]{y}(1 - \sqrt[3]{y})(1 + \sqrt[3]{y})$

24. $(\sqrt[3]{9} - 2)(\sqrt[3]{9} + 4)$

25. $[\sqrt{3} + \sqrt{2} + \sqrt{1}]^4$

26. $(\sqrt{y+10} - \sqrt{10})(\sqrt{y+10} + \sqrt{10})$

27. $(8\sqrt{a+3} - a\sqrt{3})(2\sqrt{a+3} + 5a\sqrt{3})$

28. $(8\sqrt{3} + 5\sqrt{2})(6\sqrt{7} - 2\sqrt{5})$

29. $(\sqrt{17} + \sqrt{2} - \sqrt{19})(\sqrt{17} + \sqrt{2} + \sqrt{19})$

30. Evaluate: $\sqrt{7 + 4\sqrt{3}} - \sqrt{7 - 4\sqrt{3}}$.

SECTION 6.7

Rationalize the denominator.

31. $\dfrac{\sqrt{5} + \sqrt{10} - \sqrt{6}}{\sqrt{50}}$

32. $\dfrac{3\sqrt{y} + 4\sqrt{yz}}{5\sqrt{y} - 2\sqrt{z}}$

33. $\dfrac{b + \sqrt{b}}{1 + b + \sqrt{b}}$

34. $\dfrac{12}{3 - \sqrt{3 - y}}$

35. $\dfrac{36a^2b}{\sqrt[3]{6a^2b}}$

Rationalize the numerator.

36. $\dfrac{\sqrt{y+18}-\sqrt{y}}{18}$

37. $\dfrac{\sqrt{x+6}-5}{\sqrt{x+6}+5}$

Simplify. (*Hint:* Rationalize the denominator.)

38. $\sqrt{a^2-3}-\dfrac{a^2}{\sqrt{a^2-3}}$

39. $5\sqrt{\dfrac{x}{y}}+4\sqrt{\dfrac{y}{x}}-\dfrac{3}{\sqrt{xy}}$

40. $\dfrac{\dfrac{1}{\sqrt{w}}-\sqrt{w}}{\dfrac{\sqrt{w}+1}{\sqrt{w}}}$

41. $\dfrac{1}{4+\sqrt{3}}+\dfrac{1}{\sqrt{3}}+\dfrac{1}{\sqrt{3}-4}$

SECTION 6.8

Use fractional exponents to write a single radical expression and simplify.

42. $\sqrt[5]{yx^2}\sqrt{xy}$

43. $\sqrt{x^5\sqrt[3]{x^4}}$

44. $\dfrac{\sqrt{(a+b)^3}\sqrt[3]{(a+b)^2}}{\sqrt[4]{a+b}}$

45. $\sqrt[4]{\sqrt[3]{8x^3y^6}}$

Simplify.

46. $(-\sqrt[4]{7}\sqrt[3]{w})^{12}$

47. $\sqrt[12]{p^2+2pq+q^2}$

48. $\dfrac{1}{\sqrt[3]{3}-\sqrt[3]{2}}$

49. $[\sqrt[10]{\sqrt[5]{x^{15}}}]^5[\sqrt[5]{\sqrt[10]{x^{15}}}]^5$

50. $\sqrt[p]{x^{5p}y^{7p+1}z^{p+3}}$

SECTION 6.9

Solve.

51. $\sqrt[3]{\dfrac{z}{4}}-10=2$

52. $\sqrt[4]{z^2+17}=3$

53. $\sqrt{\sqrt{y+49}-\sqrt{y}}=\sqrt{7}$

54. $\sqrt[3]{x^2+x+15}-3=0$

55. $x^2-5x-\sqrt{x^2-5x-2}=4$

56. $\sqrt{8-b}=b\sqrt{8-b}$

57. $\sqrt{x-2}-\sqrt{x+2}+2=0$

58. $6\sqrt{y}+6y^{-1/2}=37$

59. $\sqrt{a^2+30a}=a+\sqrt{5a}$

60. $\sqrt{\sqrt{x+4}}=\sqrt{x}-2$

61. $\dfrac{x-1}{\sqrt{x^2+3x+6}}=\dfrac{1}{4}$

62. $\sqrt{x+1}-\dfrac{2}{\sqrt{x+1}}=1$

63. $\sqrt{11x+\sqrt{6+x}}=6$

64. $\sqrt{2x-2}+\sqrt{7x+4}=\sqrt{13x+10}$

SECTION 6.10

65. A function g is given by $g(x)=\dfrac{x^4-x^2}{x-1}$. Find $g(2i)$.

66. Evaluate $\dfrac{1}{w-w^2}$ when $w=\dfrac{1-i}{10}$.

Express in terms of i.

67. $\dfrac{1}{8}(-24-\sqrt{-1024})$

68. $12\sqrt{-\dfrac{1}{32}}$

69. $7\sqrt{-64}-9\sqrt{-256}$

Simplify.

70. $\dfrac{i^5+i^6+i^7+i^8}{(1-i)^4}$

71. $(1-i)^3(1+i)^3$

72. $\dfrac{5-\sqrt{5}i}{\sqrt{5}i}$

73. $\dfrac{6}{1+\dfrac{3}{i}}$

74. $\left(\dfrac{1}{2}-\dfrac{1}{3}i\right)^2-\left(\dfrac{1}{2}+\dfrac{1}{3}i\right)^2$

75. $\dfrac{i-i^{38}}{1+i}$

76. $\dfrac{2}{1-3i}-\dfrac{3}{4+2i}$

77. $\dfrac{2\sqrt{6}+4\sqrt{5}i}{2\sqrt{6}-4\sqrt{5}i}$

78. $\dfrac{1-4i}{4i(1+4i)^{-1}}$

NAME _____   SCORE _____   ANSWERS

TEST OR REVIEW—CHAPTER 6

Review the material in the chapter. Then test yourself by doing these exercises. If you miss
an item, restudy the section indicated to the left of the item.

1. _____

[6.1, ■ ● ■]   **1.** Simplify: $\dfrac{12^{-3}x^{-4}y}{4^{-2}x^{-3}y^2}$.

2. _____

[6.1, ●●●]   **2.** Multiply and write scientific notation for your answer:
                   $(3.2 \times 10^7)(4.2 \times 10^{-12})$.

3. _____

In Questions 3–7 assume that letters can represent any real number.

[6.2, ● ●]   Simplify.

                   3. $\sqrt{36y^2}$ **4.** $\sqrt{x^2 + 10x + 25}$

4. _____

[6.2, ●●●]   **5.** Simplify: $\sqrt[3]{-8}$.

5. _____

[6.2, ⋮ ⋮]   **6.** Simplify: $\sqrt[10]{(-4)^{10}}$.

6. _____

[6.3, ●●●]   **7.** Multiply and simplify by factoring: $\sqrt[3]{x^4}\sqrt[3]{8x^5}$.

7. _____

[6.3, ⋮ ⋮]   **8.** Simplify and then approximate to the nearest tenth: $\sqrt{148}$.

8. _____

In Questions 9 and 10 assume that all expressions under radical signs represent positive
numbers.

[6.4, ● ●]   **9.** Divide. Then simplify by taking roots if possible: $\sqrt{\dfrac{20x^3y}{4y}}$.

9. _____

[6.4, ●●●]   **10.** Simplify: $(\sqrt[3]{16a^2b})^2$.

10. _____

ANSWERS

[6.5, ●]   **11.** Add. Then simplify by collecting like radical terms: $3\sqrt{128} + 2\sqrt{18} + 2\sqrt{32}$.

11. _____

[6.6, ●]   **12.** Multiply and simplify:

$(\sqrt{20} + 2\sqrt{5})(\sqrt{20} - 3\sqrt{5})$.

12. _____

[6.7, ●●●]   **13.** Rationalize the denominator. Assume that a and b represent positive numbers. $\dfrac{\sqrt{a} + \sqrt{b}}{\sqrt{a} - \sqrt{b}}$.

13. _____

[6.8, ●]   **14.** Rewrite with fractional exponents:

$(\sqrt{5xy^2})^5$.

14. _____

[6.8, ●●]   **15.** Use fractional exponents to write a single radical expression.

$\sqrt[4]{2y}\sqrt{x - 3}$

15. _____

[6.9, ●●]   **16.** Solve: $\sqrt{y - 6} = \sqrt{y + 9} - 3$.

16. _____

[6.10, ●]   **17.** Express in terms of i and simplify: $\sqrt{-18}$.

17. _____

[6.10, ●●]   **18.** Simplify: $(5 + 8i) - (-2 + 3i)$.

18. _____

[6.10, ●●●]   **19.** Determine whether $1 + 2i$ is a solution of $x^2 + 2x + 5 = 0$.

19. _____

[6.10, ●●●●]   **20.** Divide: $\dfrac{-7 + 14i}{6 - 8i}$.

20. _____

QUADRATIC
EQUATIONS
AND FUNCTIONS

7

*Prices of pizzas of various
diameters are related by
equations of second degree,
called quadratic equations.*

READINESS CHECK: SKILLS FOR CHAPTER 7

Solve.

1. $(x - 5)(x + 3) = 0$ 2. $x^2 - 5x - 14 = 0$

3. $4x^2 - 8x = 0$ 4. $x^2 + 10x + 25 = 0$

5. $x^2 - 9 = 0$

6. Graph: $y = 2x + 1$.   7. Simplify: $\sqrt{88}$.

8. Simplify and approximate to the   9. Rationalize the denominator:
   nearest tenth:
   $$\frac{14 - \sqrt{88}}{10}.$$ $$\sqrt{\frac{2}{5}}.$$

10. Express in terms of i: $\sqrt{-20}$.   11. Determine whether $1 + i$ is a
                                                solution of $x^2 + 2x + 1 = 0$.

7.1 QUADRATIC EQUATIONS

▐●▌ EQUATIONS OF THE TYPE $ax^2 = k$

Equations of the second degree are called *quadratic*. The standard form of a quadratic equation is

$$ax^2 + bx + c = 0$$

Here a, b, and c are constants, and $a \neq 0$ (otherwise the equation would not be of second degree). The constants b and c may be 0.

We first consider equations in which $b = 0$, that is, equations $ax^2 + c = 0$ or $ax^2 = -c$, equations in which ax^2 is a constant.

Example 1 Solve: $3x^2 = 6$.

$$3x^2 = 6$$
$$x^2 = 2 \qquad \text{Multiplying by } \tfrac{1}{3}$$

Next, recall that every positive number has two square roots and that they are additive inverses of each other. Then proceed as follows:

$$x^2 = 2$$
$$x = \sqrt{2} \quad \text{or} \quad x = -\sqrt{2} \qquad \text{Taking square roots}$$

We often use the symbol $\pm\sqrt{2}$ to represent both of the numbers, $\sqrt{2}$ and $-\sqrt{2}$.

In this equation we can check both numbers at once.

$$
\begin{array}{c|c}
\multicolumn{2}{c}{3x^2 = 6} \\
\hline
3(\pm\sqrt{2})^2 & 6 \\
3 \cdot 2 & \\
6 & \\
\end{array}
$$

The solutions are $\sqrt{2}$ and $-\sqrt{2}$, or $\pm\sqrt{2}$.

OBJECTIVES

After finishing Section 7.1, you should be able to:

▐●▌ Solve equations of the type $ax^2 = k$.

▐●●▌ Solve equations of the type $ax^2 + bx = 0$.

▐●●●▌ Solve equations of the type $ax^2 + bx + c = 0$.

▐▪▪▌ Translate problems to quadratic equations and solve.

Solve.

1. $5x^2 = 15$

2. $7x^2 = 0$

DO EXERCISES 1 AND 2.

Sometimes we rationalize denominators to simplify answers.

Example 2 Solve: $-5x^2 + 2 = 0$.

$$5x^2 + 2 = 0$$

$$x^2 = \frac{2}{5} \qquad \text{Adding } -2 \text{ and multiplying by } -\tfrac{1}{5}$$

$$x = \sqrt{\frac{2}{5}} \quad \text{or} \quad x = -\sqrt{\frac{2}{5}} \qquad \text{Taking square roots}$$

$$x = \sqrt{\frac{2}{5} \cdot \frac{5}{5}} \quad \text{or} \quad x = -\sqrt{\frac{2}{5} \cdot \frac{5}{5}} \qquad \begin{array}{l}\text{Rationalizing the}\\ \text{denominators}\end{array}$$

$$x = \frac{\sqrt{10}}{5} \quad \text{or} \quad x = -\frac{\sqrt{10}}{5}$$

Check:

$$\begin{array}{c|c} \multicolumn{2}{c}{-5x^2 + 2 = 0} \\ \hline -5\left(\pm\dfrac{\sqrt{10}}{5}\right)^2 + 2 & 0 \\ -5\left(\dfrac{10}{25}\right) + 2 & \\ -2 + 2 & \\ 0 & \end{array}$$

The solutions are $\dfrac{\sqrt{10}}{5}$ and $-\dfrac{\sqrt{10}}{5}$, or $\pm\dfrac{\sqrt{10}}{5}$.

DO EXERCISE 3.

Sometimes we get solutions that are nonreal numbers.

Example 3 Solve: $4x^2 + 9 = 0$.

$$4x^2 + 9 = 0$$

$$x^2 = -\frac{9}{4} \qquad \text{Adding } -9 \text{ and multiplying by } \tfrac{1}{4}$$

$$x = \sqrt{-\frac{9}{4}} \quad \text{or} \quad x = -\sqrt{-\frac{9}{4}} \qquad \text{Taking square roots}$$

$$x = \frac{3}{2}i \quad \text{or} \quad x = -\frac{3}{2}i \qquad \text{Simplifying}$$

Check:

$$\begin{array}{c|c} \multicolumn{2}{c}{4x^2 + 9 = 0} \\ \hline 4(\pm\tfrac{3}{2}i)^2 + 9 & 0 \\ 4(-\tfrac{9}{4}) + 9 & \\ -9 + 9 & \\ 0 & \end{array}$$

The solutions are $\frac{3}{2}i$ and $-\frac{3}{2}i$, or $\pm\frac{3}{2}i$.

DO EXERCISE 4.

3. Solve.

$$-3x^2 + 8 = 0$$

4. Solve.

$$2x^2 + 1 = 0$$

5. Solve.

$4x^2 - 3x = 0$

● ● **EQUATIONS OF THE TYPE** $ax^2 + bx = 0$

When c is 0 but b is not, we can factor and use the principle of zero products.

Example 4   Solve: $3x^2 + 5x = 0$. ←

$$3x^2 + 5x = 0$$

$$x(3x + 5) = 0 \qquad \text{Factoring}$$

$$x = 0 \quad \text{or} \quad 3x + 5 = 0 \qquad \text{Principle of zero products}$$

$$x = 0 \quad \text{or} \qquad 3x = -5$$

$$x = 0 \quad \text{or} \qquad x = -\frac{5}{3}$$

This time the checks cannot be done together.

Check: For 0:   For $-\frac{5}{3}$:

$$\begin{array}{c|c} 3x^2 + 5x = 0 & 0 \\ \hline 3 \cdot (0)^2 + 5(0) & \\ 0 + 0 & \\ 0 & \end{array} \qquad \begin{array}{c|c} 3x^2 + 5x = 0 & 0 \\ \hline 3(-\frac{5}{3})^2 + 5(-\frac{5}{3}) & \\ 3(\frac{25}{9}) + (-\frac{25}{3}) & \\ \frac{25}{3} + (-\frac{25}{3}) & \\ 0 & \end{array}$$

The solutions are 0 and $-\frac{5}{3}$.

DO EXERCISE 5.

> A common error is to multiply on both sides by $\dfrac{1}{3x}$. That procedure is wrong because one of the solutions is lost. You must factor and use the principle of zero products.

6. Solve.

$5x^2 - 8x + 3 = 0$

A quadratic equation of the above type will always have 0 as one solution and a nonzero number as the other solution.

● ● ● **EQUATIONS OF THE TYPE** $ax^2 + bx + c = 0$

When none of the constants a, b, or c is 0, we can try factoring.

Example 5   Solve: $3x^2 + x - 2 = 0$.

$$3x^2 + x - 2 = 0$$

$$(3x - 2)(x + 1) = 0 \qquad \text{Factoring}$$

$$3x - 2 = 0 \quad \text{or} \quad x + 1 = 0$$

$$3x = 2 \quad \text{or} \qquad x = -1$$

$$x = \frac{2}{3} \quad \text{or} \qquad x = -1$$

Check: For -1:   For $\frac{2}{3}$:

$$\begin{array}{c|c} 3x^2 + x - 2 = 0 & 0 \\ \hline 3(-1)^2 + (-1) - 2 & \\ 3 - 1 - 2 & \\ 0 & \end{array} \qquad \begin{array}{c|c} 3x^2 + x - 2 = 0 & 0 \\ \hline 3(\frac{2}{3})^2 + (\frac{2}{3}) - 2 & \\ 3 \cdot \frac{4}{9} + \frac{2}{3} - 2 & \\ \frac{12}{9} + \frac{6}{9} - 2 & \\ \frac{18}{9} - 2 & \\ 0 & \end{array}$$

The solutions are -1 and $\frac{2}{3}$.

DO EXERCISE 6.

Sometimes it helps to find the standard form before factoring.

Example 6 Solve: $(x - 1)(x + 1) = 5(x - 1)$.

$$(x - 1)(x + 1) = 5(x - 1)$$

$$x^2 - 1 = 5x - 5 \qquad \text{Multiplying}$$

$$x^2 - 5x + 4 = 0 \qquad \text{Finding standard form}$$

$$(x - 4)(x - 1) = 0 \qquad \text{Factoring}$$

$$x = 4 \quad \text{or} \quad x = 1 \qquad \text{Principle of zero products}$$

The solutions are 4 and 1 (the check has been left for the reader).

DO EXERCISE 7.

🔳 APPLIED PROBLEMS

Sometimes when we translate a problem to mathematical language, we get a quadratic equation. If so, we proceed as usual to solve the problem. That is, we solve the equation and then check possible answers in the original problem.

Example 7 A rectangular garden is 60 ft by 80 ft. Part of the garden is torn up to install a sidewalk of uniform width around it. The new area of the garden is $\frac{1}{6}$ of the old area. How wide is the sidewalk?

First make a drawing and label it with the known information We don't know how wide the sidewalk is, so we have called the width x.

Remember, the area of a rectangle is $l \times w$ (length times width) Then:

Area of old garden $= 60 \cdot 80$

Area of new garden $= (60 - 2x)(80 - 2x)$

Since the new garden is $\frac{1}{6}$ of the old, we have

$$(60 - 2x)(80 - 2x) = \frac{1}{6} \cdot 60 \cdot 80.$$

We now solve this equation.

$$4800 - 120x - 160x + 4x^2 = 800 \qquad \text{Multiplying binomials on the left}$$

$$4x^2 - 280x + 4000 = 0 \qquad \text{Collecting like terms}$$

$$x^2 - 70x + 1000 = 0 \qquad \text{Multiplying on both sides by } \frac{1}{4}$$

$$(x - 20)(x - 50) = 0 \qquad \text{Factoring}$$

$$x = 20 \quad \text{or} \quad x = 50 \qquad \text{Principle of zero products}$$

7. Solve.

$$6(y - 3) = (y - 3)(y - 2)$$

ANSWER ON PAGE A–25

8. An open box is to be made from a 10-ft-by-20-ft rectangular piece of cardboard by cutting a square from each corner. The area of the bottom of the box is to be 96 sq ft. What is the length of the sides of the squares that are cut from the corners?

9. Three towns A, B, and C are situated as shown. The roads at A form a right angle. The distance from A to B is two miles less than the distance from A to C. The distance from B to C is 10 miles. Find the distance from A to B and the distance from A to C.

We now check in the original problem. We see that 50 is not a solution because when $x = 50$, $60 - 2x = -40$, and the width of the garden cannot be negative.

Let's see whether 20 checks. If the sidewalk is 20 ft wide, the garden itself will have length $80 - 2 \cdot 20$, or 40 ft. The width will be $60 - 2 \cdot 20$, or 20 ft. The new area is thus $20 \cdot 40$, or 800 ft². The old area was $60 \cdot 80$, or 4800 ft². The new area of 800 ft² is $\frac{1}{6}$ of this, so the number 20 checks. The answer is that the sidewalk is 20 ft wide.

DO EXERCISE 8.

Example 8   A wire is stretched from the ground to the top of a building, as shown. The wire is 20 ft long. The height of the building is 4 ft greater than the distance d from the building. Find the distance d and the height of the building.

We first make a drawing and label it. We do not know d or the height of the building, but the height must be $d + 4$.

As we look at the drawing, we see that a right triangle is formed. We can use the Pythagorean theorem, which contains a right triangle formula.* It says that in any right triangle, the square of the longest side (the hypotenuse) is the sum of the squares of the other two sides (the legs).

$$c^2 = a^2 + b^2$$

In this problem then, we have

$$20^2 = d^2 + (d + 4)^2$$

We now solve this equation.

$$
\begin{aligned}
400 &= d^2 + d^2 + 8d + 16 &&\text{Squaring} \\
2d^2 + 8d - 384 &= 0 &&\text{Finding standard form} \\
d^2 + 4d - 192 &= 0 &&\text{Multiplying by } \tfrac{1}{2} \\
(d + 16)(d - 12) &= 0 &&\text{Factoring} \\
d + 16 = 0 \quad \text{or} \quad d - 12 &= 0 &&\text{Principle of zero products} \\
d = -16 \quad \text{or} \quad d &= 12
\end{aligned}
$$

We know that -16 is not an answer because distances are not negative. The number 12 checks (we leave the check to you), so the answer is that $d = 12$ ft, and the height of the building is 16 ft.

DO EXERCISE 9.

* For more detail on this formula, see p. 418.

NAME

CLASS

EXERCISE SET 7.1

■● Solve.

1. $4x^2 = 20$

2. $3x^2 = 21$

3. $25x^2 + 4 = 0$

4. $9x^2 + 16 = 0$

5. $2x^2 - 3 = 0$

6. $3x^2 - 7 = 0$

●● Solve.

7. $14x^2 + 9x = 0$

8. $19x^2 + 8x = 0$

9. $5x^2 + 10x = 0$

10. $3x^2 + 12x = 0$

1. _____

2. _____

3. _____

4. _____

5. _____

6. _____

7. _____

8. _____

9. _____

10. _____

ANSWERS

●●● Solve.

11. $x^2 - 6x + 5 = 0$ **12.** $x^2 - 7x + 6 = 0$ **13.** $6x^2 - x - 2 = 0$

11. _____

12. _____

14. $2x^2 + 13x + 15 = 0$ **15.** $x^2 + 8x + 15 = 0$ **16.** $x^2 + 9x + 14 = 0$

13. _____

14. _____

15. _____

17. $4x(x - 2) - 5x(x - 1) = 2$ **18.** $3x(x + 1) - 7x(x + 2) = 6$

16. _____

17. _____

19. $14(x - 4) - (x + 2) = (x + 2)(x - 4)$ **20.** $11(x - 2) + (x - 5) = (x + 2)(x - 6)$

18. _____

19. _____

20. _____

⠶ Solve.

21. A picture frame measures 14 in. by 20 in.; 160 in² of picture shows. Find the width of the frame.

22. A picture frame measures 12 cm by 20 cm; 84 cm² of picture shows. Find the width of the frame.

21. _____

23. The width of a rectangle is 4 ft less than the length. The area is 12 ft². Find the length and the width.

24. The width of a rectangle is 5 m less than the length. The area is 24 m². Find the length and the width.

22. _____

23. _____

25. The length of a rectangle is twice the width. The area is 288 yd². Find the length and the width.

26. The length of a rectangle is twice the width. The area is 338 cm². Find the length and the width.

24. _____

25. _____

26. _____

27. The hypotenuse of a right triangle is 26 ft long. The length of one leg is 14 ft more than the other. Find the lengths of the legs.

28. The hypotenuse of a right triangle is 25 m long. The length of one leg is 17 m less than the other. Find the lengths of the legs.

27. _____

28. _____

29. The hypotenuse of a right triangle is 5 ft long. One leg is 1 ft shorter than the other. Find the lengths of the legs.

30. The hypotenuse of a right triangle is 13 m long. One leg is 7 m longer than the other. Find the lengths of the legs.

29. _____

30. _____

31. _____

Solve.

32. _____

31. ▦ $25.55x^2 - 1635.2 = 0$

32. $(3x^2 - 7x - 20)(2x - 5) = 0$

7.2 THE QUADRATIC FORMULA

■● SOLVING EQUATIONS WITH THE FORMULA

Some quadratic equations are difficult to solve by factoring. The following formula can be used to find the solutions.

> *The Quadratic Formula:* **The solutions of a quadratic equation $ax^2 + bx + c = 0$ are given by**
> $$x = \frac{-b \pm \sqrt{b^2 - 4ac}}{2a}.$$

To obtain this formula, we solve the equation $ax^2 + bx + c = 0$.

$$4a^2x^2 + 4abx + 4ac = 0 \qquad \text{Multiplying on both sides by } 4a, \text{ to make the first term a square}$$

$$4a^2x^2 + 4abx + (b^2 - b^2) + 4ac = 0 \qquad \text{Adding } b^2 - b^2 \text{ on the left, to complete the square}$$

$$4a^2x^2 + 4abx + b^2 - (b^2 - 4ac) = 0 \qquad \text{Rearranging}$$

$$4a^2x^2 + 4abx + b^2 = b^2 - 4ac \qquad \text{Adding } b^2 - 4ac \text{ on both sides}$$

$$(2ax + b)^2 = b^2 - 4ac \qquad \text{Factoring the trinomial square on the left}$$

We now take square roots:

$$2ax + b = \sqrt{b^2 - 4ac} \quad \text{or} \quad 2ax + b = -\sqrt{b^2 - 4ac}.$$

Solving for x, we get

$$x = \frac{-b + \sqrt{b^2 - 4ac}}{2a} \quad \text{or} \quad x = \frac{-b - \sqrt{b^2 - 4ac}}{2a}.$$

These are the solutions. We abbreviate as follows:

$$x = \frac{-b \pm \sqrt{b^2 - 4ac}}{2a}.$$

Example 1 Solve: $5x^2 - 8x = 3$.

First find standard form and determine a, b, and c.

$$5x^2 - 8x - 3 = 0; \quad a = 5, b = -8, c = -3$$

Then substitute into the quadratic formula.

$$x = \frac{-(-8) \pm \sqrt{(-8)^2 - 4 \cdot 5 \cdot (-3)}}{2 \cdot 5}$$

$$x = \frac{8 \pm \sqrt{64 + 60}}{10} = \frac{8 \pm \sqrt{124}}{10}$$

$$x = \frac{8 \pm \sqrt{4 \cdot 31}}{10} = \frac{8 \pm 2\sqrt{31}}{10}$$

$$x = \frac{2(4 + \sqrt{31})}{2 \cdot 5} = \frac{4 \pm \sqrt{31}}{5}$$

> *Caution!* To avoid a common error in simplifying, remember to *factor numerator and denominator* and then remove a factor of 1.

1. Solve using the quadratic formula.

$$3x^2 + 2x = 7$$

The checking of possible solutions such as these is cumbersome. When using the quadratic formula, no numbers are ever obtained that are not solutions of the original equation unless a mistake has been made. Therefore, it is usually better to check your work, rather than to check by substituting into the original equation.

The solutions are $\dfrac{4 + \sqrt{31}}{5}$ and $\dfrac{4 - \sqrt{31}}{5}$.

DO EXERCISE 1.

Some quadratic equations have solutions that are not real numbers.

Example 2 Solve: $x^2 + x + 1 = 0$.

$$a = 1, b = 1, c = 1$$

$$x = \frac{-b \pm \sqrt{b^2 - 4ac}}{2a}$$

$$x = \frac{-1 \pm \sqrt{1^2 - 4 \cdot 1 \cdot 1}}{2 \cdot 1}$$

$$x = \frac{-1 \pm \sqrt{1 - 4}}{2}$$

$$x = \frac{-1 \pm \sqrt{-3}}{2}$$

$$x = \frac{-1 \pm i\sqrt{3}}{2}$$

The solutions are $\dfrac{-1 + i\sqrt{3}}{2}$ and $\dfrac{-1 - i\sqrt{3}}{2}$.

2. Solve.

$$x^2 - x + 2 = 0$$

DO EXERCISE 2.

Example 3 Solve: $2 + \dfrac{7}{x} = \dfrac{4}{x^2}$.

First find standard form.

$$2x^2 + 7x = 4 \qquad \text{Multiplying by } x^2, \text{ the LCM of the denominators}$$

$$2x^2 + 7x - 4 = 0 \qquad \text{Adding } -4$$

$$a = 2, b = 7, c = -4$$

$$x = \frac{-7 \pm \sqrt{7^2 - 4 \cdot 2 \cdot (-4)}}{2 \cdot 2}$$

$$x = \frac{-7 \pm \sqrt{49 + 32}}{4}$$

$$x = \frac{-7 \pm \sqrt{81}}{4} = \frac{-7 \pm 9}{4}$$

$$x = \frac{-7 + 9}{4} = \frac{1}{2} \quad \text{or} \quad x = \frac{-7 - 9}{4} = -4$$

In this case, since we started with a fractional equation, we *do* need to check.* At least we need to show that neither of the numbers makes a denominator 0. Since neither of them does, the solutions are $\frac{1}{2}$ and -4.

The solutions of a quadratic equation can *always* be found using the quadratic formula. They are not always easy to find by factoring. A general strategy for solving quadratic equations is as follows.

> **To solve a quadratic equation:**
>
> **a) Try factoring.**
>
> **b) If factoring seems difficult, use the quadratic formula; it** *always works!*

DO EXERCISE 3.

◨◧ APPROXIMATING SOLUTIONS

A square root table or a calculator can be used to approximate solutions.

Example 4   Approximate to the nearest tenth the solutions of the equation in Example 1.

From Table 1 in the back of the book, $\sqrt{31} \approx 5.568$.

$$\frac{4 + \sqrt{31}}{5} \approx \frac{4 + 5.568}{5}$$

$$\approx \frac{9.568}{5} \quad \text{Adding}$$

$$\approx 1.913 \quad \text{Dividing}$$

$$\approx 1.9 \quad \text{Rounding to the nearest tenth}$$

$$\frac{4 - \sqrt{31}}{5} \approx \frac{4 - 5.568}{5}$$

$$\approx \frac{-1.568}{5} \quad \text{Subtracting}$$

$$\approx -0.313 \quad \text{Dividing}$$

$$\approx -0.3 \quad \text{Rounding to the nearest tenth}$$

DO EXERCISES 4 AND 5.

3. Solve.

$$3 - \frac{5}{x} = \frac{2}{x^2}$$

4. Approximate to the nearest tenth.

$$\frac{8 \pm \sqrt{47}}{7}$$

5. Approximate the solutions to Margin Exercise 1 on p. 362. Round to the nearest tenth.

---

* The quadratic formula always gives correct results when we start with the standard form. In such cases we need check only to detect errors. In this example, however, we cleared of fractions before obtaining standard form, and this step could possibly introduce numbers that do not check in the original equation.

SOMETHING EXTRA

AN APPLICATION: HANDLING DIMENSION SYMBOLS (PART I)*

In many applications, we multiply, divide, add, or subtract quantities having units, or dimensions, such as ft, km, sec, hr, etc. For example, to find average speed, we divide total distance by total time.

Example   A car travels 150 km in 2 hr. What is its average speed?

$$\text{Speed} = \frac{150 \text{ km}}{2 \text{ hr}} \quad \text{or} \quad 75\frac{\text{km}}{\text{hr}}$$

(The standard abbreviation for km/hr is km/h, but it does not suit our present discussion well.)

The symbol km/hr makes it look as if we are dividing kilometers by hours. It may be argued that we can divide numbers only. Nevertheless, we can treat dimension symbols, such as km, ft, and lb, as if they were numerals or variables, obtaining correct results mechanically. Compare, for example,

$$\frac{150x}{2y} = \frac{150}{2} \cdot \frac{x}{y} = 75\frac{x}{y}$$

with

$$\frac{150 \text{ km}}{2 \text{ hr}} = \frac{150}{2} \frac{\text{km}}{\text{hr}} = 75\frac{\text{km}}{\text{hr}}.$$

Compare

$$3x + 2x = (3 + 2)x = 5x$$

with

$$3 \text{ ft} + 2 \text{ ft} = (3 + 2) \text{ ft} = 5 \text{ ft}.$$

EXERCISES

Add these measures.

1. 45 ft + 23 ft

2. 45 km/hr + 27 km/hr

3. 17 g + 28 g

4. 3.4 lb + 5.2 lb

Find average speeds, given total distance and total time.

5. 45 mi, 6 hr

6. 640 km, 20 hr

7. 9.9 m, 3 sec

8. 76 ft, 4 min

---

* For more on handling dimension symbols, see pp. 440 and 502.

NAME   CLASS

EXERCISE SET 7.2

▪ Solve.

1. $x^2 + 6x + 4 = 0$ **2.** $x^2 - 6x - 4 = 0$ **3.** $3p^2 = -8p - 5$

4. $3u^2 = 18u - 6$ **5.** $x^2 - x + 1 = 0$ **6.** $x^2 + x + 2 = 0$

7. $x^2 + 13 = 4x$ **8.** $x^2 + 13 = 6x$ **9.** $r^2 + 3r = 8$

10. $h^2 + 4 = 6h$ **11.** $1 + \dfrac{2}{x} + \dfrac{5}{x^2} = 0$ **12.** $1 + \dfrac{5}{x^2} = \dfrac{2}{x}$

1. _____

2. _____

3. _____

4. _____

5. _____

6. _____

7. _____

8. _____

9. _____

10. _____

11. _____

12. _____

13. $3x + x(x - 2) = 0$

14. $4x + x(x - 3) = 0$

15. $(2t - 3)^2 + 17t = 15$

16. $2y^2 - (y + 2)(y - 3) = 12$

17. $(x - 2)^2 + (x + 1)^2 = 0$

18. $(x + 3)^2 + (x - 1)^2 = 0$

■ ● ■ Use the square root table, Table 1, in the back of the book to approximate solutions to the nearest tenth.

19. $x^2 + 4x - 7 = 0$

20. $x^2 + 6x + 4 = 0$

21. $2x^2 - 3x - 7 = 0$

22. $3x^2 - 3x - 2 = 0$

Solve.

23. ▦ $2.2x^2 + 0.5x - 1 = 0$

24. ▦ $5.33x^2 - 8.23x - 3.24 = 0$

25. Solve for x in terms of y: $3x^2 + xy + 4y^2 - 9 = 0$.

7.3 FRACTIONAL EQUATIONS AND APPLIED PROBLEMS

● FRACTIONAL EQUATIONS THAT ARE QUADRATIC

In a fractional equation, when we clear of fractions, we sometimes get a quadratic equation.

Example 1　Solve: $\dfrac{14}{x + 2} - \dfrac{1}{x - 4} = 1$.

We multiply by the LCM of the denominators, which is $(x + 2)(x - 4)$.

$$(x + 2)(x - 4) \cdot \left[\frac{14}{x + 2} - \frac{1}{x - 4} \right] = (x + 2)(x - 4) \cdot 1$$

$$(x + 2)(x - 4) \cdot \frac{14}{x + 2} - (x + 2)(x - 4) \cdot \frac{1}{x - 4}$$

$$= (x + 2)(x - 4) \cdot 1 \qquad \text{Using a distributive law}$$

$$14(x - 4) - (x + 2) = (x + 2)(x - 4) \qquad \begin{array}{l}\text{Simplifying by}\\\text{removing factors of 1}\end{array}$$

$$14x - 56 - x - 2 = x^2 - 2x - 8 \qquad \text{Multiplying}$$

$$13x - 58 = x^2 - 2x - 8 \qquad \text{Collecting like terms}$$

$$0 = x^2 - 15x + 50 \qquad \text{Writing standard form}$$

$$0 = (x - 5)(x - 10) \qquad \text{Factoring}$$

$$x = 5 \quad \text{or} \quad x = 10 \qquad \begin{array}{l}\text{Using the principle of}\\\text{zero products}\end{array}$$

Since we started with a fractional equation, we might have introduced some numbers that are not solutions of the original equation when we cleared of fractions. This time, we must check.

Check:　　For 5:　　　　　　　　For 10:

$$\begin{array}{c|c} \dfrac{14}{x + 2} - \dfrac{1}{x - 4} = 1 & \\ \hline \dfrac{14}{5 + 2} - \dfrac{1}{5 - 4} & 1 \\ \dfrac{14}{7} - 1 & \\ 2 - 1 & \\ 1 & 1 \end{array} \qquad \begin{array}{c|c} \dfrac{14}{x + 2} - \dfrac{1}{x - 4} = 1 & \\ \hline \dfrac{14}{10 + 2} - \dfrac{1}{10 - 4} & 1 \\ \dfrac{14}{12} - \dfrac{1}{6} & \\ \dfrac{14}{12} - \dfrac{2}{12} & \\ 1 & 1 \end{array}$$

The solutions are 5 and 10.

> If for a fractional equation, clearing of fractions produces a quadratic equation, we solve that resulting equation as we would solve any quadratic equation. However, it is necessary to check possible solutions by substituting into the original equation.

DO EXERCISE 1.

ANSWER ON PAGE A–25

OBJECTIVES

After finishing Section 7.3, you should be able to:

● Solve fractional equations that are quadratic after clearing of fractions.

●● Solve problems that translate to this kind of equation.

1. Solve.

$$\frac{24}{r - 2} + \frac{24}{r + 2} = 5$$

2. It takes Red Bryck 9 hours longer to build a wall than it takes Mort Arr. If they work together they can build the wall in 20 hours. Working alone, how long would it take Mort to build the wall?

◖◗ APPLIED PROBLEMS

Some problems translate to equations like that in Example 1.

Example 2 Fred Fast and Sam Slow work together and mow a lawn in 4 hours. It would take Sam 6 hours more than Fred to do the job alone. How long would each need to do the job working alone?

Let x represent the time it takes Fred to work alone. Then $x + 6$ is the amount of time it takes Sam to do the work alone. Thus Fred can do $1/x$ of the job in 1 hour and Sam can do $1/(x + 6)$ of the job in 1 hour. They can do $(1/x) + [1/(x + 6)]$ of the job in 1 hour. They can also do $\frac{1}{4}$ of it in 1 hour. Thus we get the equation

$$\frac{1}{x} + \frac{1}{x + 6} = \frac{1}{4}.$$

Now we solve.

$$4x(x + 6) \cdot \left[\frac{1}{x} + \frac{1}{x + 6}\right] = 4x(x + 6) \cdot \frac{1}{4} \quad \text{Multiplying by the LCM}$$

$$4x(x + 6) \cdot \frac{1}{x} + 4x(x + 6) \cdot \frac{1}{x + 6} = 4x(x + 6) \cdot \frac{1}{4}$$

$$4(x + 6) + 4x = x(x + 6)$$
$$4x + 24 + 4x = x^2 + 6x$$
$$8x + 24 = x^2 + 6x$$
$$0 = x^2 - 2x - 24 \quad \text{Standard form}$$
$$0 = (x - 6)(x + 4) \quad \text{Factoring}$$
$$x = 6 \quad \text{or} \quad x = -4 \quad \text{Principle of zero products}$$

Since negative time has no meaning in this problem, -4 is not a solution. Let's see if 6 checks. This is the time for Fred to do the job alone. Then Sam would take 12 hours alone. Thus Fred would do $\frac{4}{6}$ of the job in 4 hours and Sam would do $\frac{4}{12}$ of it. Thus in 4 hours they would do $\frac{4}{6} + \frac{4}{12}$, or $\frac{8}{12} + \frac{4}{12}$ of the job. This is all of it, so the numbers check. The answer is that it would take Fred 6 hours alone and Sam 12 hours alone.

DO EXERCISE 2.

Example 3 A man leaves Homeville and travels the 600 miles to Fartown at a certain speed. He makes the return trip 10 mph slower. His total time for the round trip was 22 hours. How fast did he travel on each part of the trip?

We first make a drawing and label it with the known information.

We let r represent the speed on the first part of the trip. Then $r - 10$ is the speed on the return trip. We can organize these facts in a chart.

	Distance	Speed	Time
Trip out	600	r	t_1
Return trip	600	$r - 10$	t_2

Since we do not know either time, but only the total, we have called the times t_1 and t_2. Since the total was 22 hours, we can write one equation at once:

$$t_1 + t_2 = 22.$$

Looking for a way to substitute into this equation, we recall the definition of speed, $r = d/t$, which we solve for t to get $t = d/r$. Now we go back to the chart. For each line we write $t = d/r$:

$$t_1 = \frac{600}{r} \quad \text{and} \quad t_2 = \frac{600}{r - 10}.$$

Substituting into our first equation, we get

$$\frac{600}{r} + \frac{600}{r - 10} = 22.$$

Next, we solve. The LCM is $r(r - 10)$.

$$r(r - 10) \cdot \left[\frac{600}{r} + \frac{600}{r - 10} \right] = r(r - 10) \cdot 22 \qquad \text{Multiplying by the LCM}$$

$$r(r - 10) \cdot \frac{600}{r} + r(r - 10) \cdot \frac{600}{r - 10} = r(r - 10) \cdot 22$$

$$600(r - 10) + 600r = 22r^2 - 220r$$
$$600r - 6000 + 600r = 22r^2 - 220r$$
$$300r - 3000 + 300r = 11r^2 - 110r \qquad \text{Multiplying by } \tfrac{1}{2}$$
$$0 = 11r^2 - 710r + 3000 \qquad \text{Standard form}$$
$$0 = (11r - 50)(r - 60) \qquad \text{Factoring}$$
$$r = 60 \quad \text{or} \quad r = \frac{50}{11}, \text{ or } 4\frac{6}{11} \qquad \text{Principle of zero products}$$

$4\frac{6}{11}$ is not a solution because it would make the return speed negative, but 60 checks. The answer is that the speed out was 60 mph, and the speed back was 50 mph.

DO EXERCISE 3.

3. A stream flows at 2 mph. A boat travels 24 miles upstream and returns in a total time of 5 hours. What is the speed of the boat in still water? (*Hint:* The speed upstream is $r - 2$ and the speed downstream is $r + 2$, where r is the speed of the boat in still water.)

ANSWER ON PAGE A–25

4. Two ships make the same voyage, a distance of 3000 nautical miles. The faster ship travels 10 knots faster than the slower one (a *knot* is 1 nautical mile per hour). The faster ship makes the voyage in 50 hours less time than the slower one. Find the speeds of the two ships.

Example 4   A car travels 300 miles at a certain speed. If the speed had been 10 mph faster the trip would have been made in 1 hour less. Find the speed.

We make a drawing, labeling it with the known information. We can also organize the information in a table.

Distance	Speed	Time
300	r	t
300	$r + 10$	$t - 1$

From the first line of the table, we obtain

$$r = \frac{300}{t}.$$

From the second line we get

$$r + 10 = \frac{300}{t - 1}.$$

We substitute for r from the first equation into the second, and solve.

$$\frac{300}{t} + 10 = \frac{300}{t - 1}$$

$$t(t - 1)\left[\frac{300}{t} + 10\right] = t(t - 1) \cdot \frac{300}{t - 1} \quad \text{Multiplying by the LCM}$$

$$t(t - 1) \cdot \frac{300}{t} + t(t - 1) \cdot 10 = t(t - 1) \cdot \frac{300}{t - 1}$$

$$300(t - 1) + 10(t^2 - t) = 300t$$

$$10t^2 - 10t - 300 = 0 \quad \text{Standard form}$$

$$t^2 - t - 30 = 0 \quad \text{Multiplying by } \tfrac{1}{10}$$

$$(t - 6)(t + 5) = 0 \quad \text{Factoring}$$

$$t = 6 \quad \text{or} \quad t = -5 \quad \text{Principle of zero products}$$

Since negative time has no meaning in this problem, we try 6 hours. Remembering that $r = d/t$, we get $r = 300/6 = 50$ mph.

To check, we take the speed 10 mph faster, which is 60 mph, and see how long the trip would have taken at that speed.

$$t = \frac{d}{r} = \frac{300}{60} = 5 \text{ hours}$$

This is one hour less than the trip actually took, so we have an answer to the problem. The answer is that the speed was 50 mph.

DO EXERCISE 4.

NAME   CLASS   ANSWERS

EXERCISE SET 7.3

▣ Solve.

1. $\dfrac{1}{x} = \dfrac{x-2}{24}$

2. $\dfrac{x+3}{14} = \dfrac{2}{x}$

3. $\dfrac{1}{2x-1} - \dfrac{1}{2x+1} = \dfrac{1}{4}$

4. $\dfrac{1}{4-x} - \dfrac{1}{2+x} = \dfrac{1}{4}$

5. $\dfrac{50}{x} - \dfrac{50}{x-5} = -\dfrac{1}{2}$

6. $3x + \dfrac{10}{x-1} = \dfrac{16}{x-1}$

7. $\dfrac{x+2}{x} = \dfrac{x-1}{2}$

8. $\dfrac{x}{3} - \dfrac{6}{x} = 1$

9. $x - 6 = \dfrac{1}{x+6}$

10. $x + 7 = \dfrac{1}{x-7}$

ANSWERS

● ● Solve.

11. During the first part of a trip a canoeist travels 80 miles at a certain speed. The canoeist travels 35 miles on the second part of the trip at a speed 5 mph slower. The total time for the trip is 3 hr. What was the speed on each part of the trip?

12. During the first part of a trip a car travels 120 miles at a certain speed. It travels 100 miles on the second part of the trip at a speed 10 mph slower. The total time for the trip is 4 hr. What was the speed on each part of the trip?

11. _____

12. _____

13. A car travels 280 miles at a certain speed. If the speed had been 5 mph faster the trip would have been made in 1 hour less time. Find the speed.

14. A car travels 200 miles at a certain speed. If the speed had been 10 mph faster the trip would have been made in 1 hour less time. Find the speed.

13. _____

14. _____

15. Airplane A travels 2800 km at a certain speed. Plane B travels 2000 km at a speed 50 km/h faster than plane A in 3 hr less time. Find the speed of each plane.

16. Airplane A travels 600 miles at a certain speed. Plane B travels 1000 miles at a speed 50 mph faster, but takes 1 hour longer. Find the speed of each plane.

15. _____

17. Two pipes are connected to the same tank. When working together they can fill the tank in 2 hours. The larger pipe, working alone, can fill the tank in 3 hours less time than the smaller one. How long would the smaller one take, working alone, to fill the tank?

18. Two hoses are connected to a swimming pool. When working together they can fill the pool in 4 hours. The larger hose, working alone, can fill the pool in 6 hours less time than the smaller one. How long would the smaller one take, working alone, to fill the pool?

16. _____

17. _____

18. _____

19. A boat travels 1 km upstream and 1 km back. The time for the roundtrip is 1 hour. The speed of the stream is 2 km/hr. What is the speed of the boat in still water?

20. A boat travels 10 miles upstream and 10 miles back. The time required for the roundtrip is 3 hours The speed of the stream is 5 mph What is the speed of the boat in still water?

19. _____

20. _____

21. ▦ At a factory, smokestack A pollutes the air 2.13 times as fast as smokestack B. Together they yield a certain amount of pollution in 16.3 hr. How long would it take each to pollute this much alone?

22. ▦ At Pizza Perfect, Ron Italio can make 100 large pizza crusts in 1.2 hours less time than Chester Mosserelli. Together they can do the job in 1.8 hours. How long does it take each to do the job alone?

21. _____

22. _____

7.4 SOLUTIONS OF QUADRATIC EQUATIONS

◖●◗ THE DISCRIMINANT

We now consider quadratic equations with real-number coefficients.

From the quadratic formula we know that the solutions x_1 and x_2 of a quadratic equation are given by

$$x_1 = \frac{-b + \sqrt{b^2 - 4ac}}{2a} \quad \text{and} \quad x_2 = \frac{-b - \sqrt{b^2 - 4ac}}{2a}.$$

The expression $b^2 - 4ac$ shows the nature of the solutions. This expression is called the *discriminant*. If it is 0, then it doesn't matter whether we choose the plus or minus sign in the formula; hence there is just one solution. Since there is no negative radicand, the solutions must be real. If the discriminant is positive, there will be two real solutions. If it is negative, we will be taking the square root of a negative number; hence there will be two nonreal solutions, and they will be complex conjugates. We summarize.

Discriminant $b^2 - 4ac$	Nature of solutions
0	Only one solution; it is a real number.
Positive	Two different real-number solutions.
Negative	Two different nonreal number solutions (complex conjugates).

Example 1   Determine the nature of the solutions of $9x^2 - 12x + 4 = 0$.

$$a = 9, \quad b = -12, \quad \text{and} \quad c = 4$$

We compute the discriminant.

$$\begin{aligned} b^2 - 4ac &= (-12)^2 - 4 \cdot 9 \cdot 4 \\ &= 144 - 144 \\ &= 0 \end{aligned}$$

There is just one solution and it is a real number.

Example 2   Determine the nature of the solutions of $x^2 + 5x + 8 = 0$.

$$a = 1, \quad b = 5, \quad \text{and} \quad c = 8$$

We compute the discriminant.

$$\begin{aligned} b^2 - 4ac &= 5^2 - 4 \cdot 1 \cdot 8 \\ &= 25 - 32 \\ &= -7 \end{aligned}$$

Since the discriminant is negative, there are two nonreal solutions.

OBJECTIVES

After finishing Section 7.4, you should be able to:

◖●◗ Determine the nature of the solutions of a quadratic equation with real-number coefficients, without solving it.

◖●●◗ Write a quadratic equation having two numbers specified as solutions.

Determine the nature of the solutions without solving.

1. $x^2 + 5x - 3 = 0$

2. $9x^2 - 6x + 1 = 0$

3. $3x^2 - 2x + 1 = 0$

Find a quadratic equation having the following solutions.

4. -4 and $\dfrac{5}{3}$

5. $5i$ and $-5i$

6. $-2\sqrt{2}$ and $\sqrt{2}$

Example 3 Determine the nature of the solutions of $x^2 + 5x + 6 = 0$.

$$a = 1, \quad b = 5, \quad \text{and} \quad c = 6$$
$$b^2 - 4ac = 5^2 - 4 \cdot 1 \cdot 6 = 1$$

Since the discriminant is positive, there are two solutions and they are real numbers.

DO EXERCISES 1–3.

◼◼ WRITING EQUATIONS FROM SOLUTIONS

We know by the principle of zero products that $(x - 2)(x + 3) = 0$ has solutions 2 and -3. If we know the solutions of an equation, we can write the equation, using this principle in reverse.

Example 4 Find a quadratic equation whose solutions are 3 and $-\frac{2}{5}$.

$$x = 3 \quad \text{or} \qquad x = -\frac{2}{5}$$

$$x - 3 = 0 \quad \text{or} \quad \left(x + \frac{2}{5}\right) = 0 \qquad \text{Getting 0's on one side}$$

$$(x - 3)\left(x + \frac{2}{5}\right) = 0 \qquad \begin{array}{l}\text{Principle of zero products}\\ \text{(multiplying)}\end{array}$$

$$x^2 + \frac{2}{5}x - 3x - 3 \cdot \frac{2}{5} = 0 \qquad \text{Using FOIL}$$

$$x^2 - \frac{13}{5}x - \frac{6}{5} = 0, \quad \text{or}$$

$$5x^2 - 13x - 6 = 0 \qquad \begin{array}{l}\text{Collecting like terms and}\\ \text{multiplying by 5}\end{array}$$

Example 5 Write a quadratic equation whose solutions are $2i$ and $-2i$.

$$x = 2i \quad \text{or} \qquad x = -2i$$
$$x - 2i = 0 \quad \text{or} \quad x + 2i = 0 \qquad \text{Getting 0's on one side}$$
$$(x - 2i)(x + 2i) = 0 \qquad \begin{array}{l}\text{Principle of zero}\\ \text{products}\\ \text{(multiplying)}\end{array}$$
$$x^2 + 2ix - 2ix - (2i)^2 = x^2 - (2i)^2 = 0 \qquad \text{Using FOIL}$$
$$x^2 + 4 = 0$$

Example 6 Write a quadratic equation whose solutions are $\sqrt{3}$ and $-2\sqrt{3}$.

$$x = \sqrt{3} \quad \text{or} \qquad x = -2\sqrt{3}$$
$$x - \sqrt{3} = 0 \quad \text{or} \quad x + 2\sqrt{3} = 0 \qquad \text{Getting 0's on one side}$$
$$(x - \sqrt{3})(x + 2\sqrt{3}) = 0 \qquad \text{Principle of zero products}$$
$$x^2 + 2\sqrt{3}x - \sqrt{3}x - 2(\sqrt{3})^2 = 0 \qquad \text{Using FOIL}$$
$$x^2 + \sqrt{3}x - 6 = 0 \qquad \text{Collecting like terms}$$

DO EXERCISES 4–6.

NAME

CLASS

ANSWERS

EXERCISE SET 7.4

▣ Determine the nature of the solutions of each equation.

1. $x^2 - 6x + 9 = 0$

2. $x^2 + 10x + 25 = 0$

3. $x^2 + 7 = 0$

4. $x^2 + 2 = 0$

5. $x^2 - 2 = 0$

6. $x^2 - 5 = 0$

7. $4x^2 - 12x + 9 = 0$

8. $4x^2 + 8x - 5 = 0$

9. $x^2 - 2x + 4 = 0$

10. $x^2 + 3x + 4 = 0$

11. $9t^2 - 3t = 0$

12. $4m^2 + 7m = 0$

13. $y^2 = \dfrac{1}{2}y + \dfrac{3}{5}$

14. $y^2 + \dfrac{9}{4} = 4y$

15. $4x^2 - 4\sqrt{3}x + 3 = 0$

16. $6y^2 - 2\sqrt{3}y - 1 = 0$

1. _____

2. _____

3. _____

4. _____

5. _____

6. _____

7. _____

8. _____

9. _____

10. _____

11. _____

12. _____

13. _____

14. _____

15. _____

16. _____

● ●   Write a quadratic equation having the given numbers as solutions.

17. $-11,\quad 9$ **18.** $-4,\quad 4$

17. _____

18. _____

19. 7, only solution
[*Hint:* It must be a double solution,
that is, $(x - 7)(x - 7) = 0$.]   **20.** -5, only solution

19. _____

20. _____

21. $-\dfrac{2}{5},\quad \dfrac{6}{5}$ **22.** $-\dfrac{1}{4},\quad -\dfrac{1}{2}$

21. _____

22. _____

23. $\dfrac{c}{2},\quad \dfrac{d}{2}$ **24.** $\dfrac{k}{3},\quad \dfrac{m}{4}$

23. _____

24. _____

25. $\sqrt{2},\quad 3\sqrt{2}$ **26.** $-\sqrt{3},\quad 2\sqrt{3}$

25. _____

26. _____

27. _____

For each equation under the given condition, (a) find k, and (b) find the other solution.

27. $kx^2 - 17x + 33 = 0$;
one solution is 3   **28.** $kx^2 - 2x + k = 0$;
one solution is -3

28. _____

7.5 FORMULAS

■   To solve a formula for a certain letter, we use the principles for solving equations to get that letter alone on one side. When square roots appear, we can usually get rid of the radical signs by squaring both sides.

Example 1   (*A pendulum formula*)   Solve $T = 2\pi\sqrt{\dfrac{l}{g}}$ for l.

$$T = 2\pi\sqrt{\frac{l}{g}}$$

$$T^2 = \left(2\pi\sqrt{\frac{l}{g}}\right)^2 \qquad \text{Principle of powers (squaring)}$$

$$T^2 = 2^2\pi^2\frac{l}{g}$$

$$gT^2 = 4\pi^2 l \qquad \text{Clearing of fractions}$$

$$\frac{gT^2}{4\pi^2} = l \qquad \text{Multiplying by } \frac{1}{4\pi^2}$$

We now have l alone on one side, so the formula is solved for l.

DO EXERCISE 1.

In most formulas the letters represent nonnegative numbers, so we do not need to use absolute value signs when taking square roots.

Example 2   (*A right triangle formula*)   Solve $c^2 = a^2 + b^2$ for a.

$$c^2 - b^2 = a^2 \qquad \text{Adding } -b^2 \text{ to get } a^2 \text{ alone}$$

$$\sqrt{c^2 - b^2} = a \qquad \text{Taking square root}$$

Example 3   (*A motion formula*)   Solve $s = gt + 16t^2$ for t.

This time we use the quadratic formula to get t alone on one side of the equation.

$$16t^2 + gt - s = 0 \qquad \text{Writing the equation in standard form}$$

$$a = 16,\ b = g,\text{ and } c = -s$$

$$t = \frac{-g \pm \sqrt{g^2 - 4\cdot 16\cdot(-s)}}{2\cdot 16} \qquad \text{Using the quadratic formula}$$

Since taking the negative square root would result in a negative answer, we take the positive one.

$$t = \frac{-g + \sqrt{g^2 + 64s}}{32}$$

DO EXERCISES 2 AND 3.

The following list of steps should help you when solving formulas for a given letter. Try to remember that when solving a formula you do the same things you would do to solve any equation. The same principles hold.

OBJECTIVE

After finishing Section 7.5, you should be able to:

■   Solve a formula for a given letter.

1. Solve $A = \sqrt{\dfrac{w_1}{w_2}}$ for w_2.

2. Solve $V = \pi r^2 h$ for r.

3. Solve $Ls^2 + Rs = -\dfrac{1}{C}$ for s.

4. Solve the following formula for b.

$$\frac{b}{\sqrt{a^2 - b^2}} = p$$

To Solve a Formula for a Letter, say b:

1. **Clear of fractions and use the principle of powers, as needed, to eliminate radicals and until b does not appear in any denominator. (In some cases you may clear of fractions first, and in some cases you may use the principle of powers first. But do these until radicals are gone and b is not in any denominator.)**
2. **Collect all terms with b^2 in them. Also collect all terms with b in them.**
3. **If b^2 does not appear, you can finish by using just the addition and multiplication principles. Get all terms containing b on one side of the equation, then factor out b. Dividing on both sides will then get b alone.**
4. **If b^2 appears but b does not appear to be the first power, solve the equation for b^2. Then take the square root on both sides.**
5. **If there are terms containing both b and b^2, put the equation in standard form and use the quadratic formula.**

Example 4   Solve the following formula for a: $q = \dfrac{a}{\sqrt{a^2 + b^2}}$.

In this case, we could either clear of fractions first or use the principle of powers first. Let us clear of fractions. We then have

$$q\sqrt{a^2 + b^2} = a.$$

Now we square both sides and then continue.

$$(q\sqrt{a^2 + b^2})^2 = a^2 \qquad \text{Squaring}$$
$$q^2(a^2 + b^2) = a^2$$
$$q^2a^2 + q^2b^2 = a^2$$
$$q^2b^2 = a^2 - q^2a^2 \qquad \text{Getting all } a^2 \text{ terms together}$$
$$q^2b^2 = a^2(1 - q^2) \qquad \text{Factoring out } a^2$$
$$\frac{q^2b^2}{1 - q^2} = a^2 \qquad \text{Multiplying by } \frac{1}{1 + q^2}$$
$$\sqrt{\frac{q^2b^2}{1 - q^2}} = a \qquad \text{Taking the square root}$$

> *Important!* a appears alone on one side of the equation and *does not* appear on the other side as well.

Now that we have a alone on one side, we have the formula solved for a.

DO EXERCISE 4.

EXERCISE SET 7.5

⬛● Solve each formula for the given letter. Assume that all variables represent non-negative numbers.

1. $A = 6s^2$, for s
    (Area of a cube)

2. $A = 4\pi r^2$, for r
    (Area of a sphere)

3. $F = \dfrac{Gm_1 m_2}{r^2}$, for r

    (Law of gravity)

4. $N = \dfrac{kQ_1 Q_2}{s^2}$, for s

    (Number of phone calls between two cities)

5. $E = mc^2$, for c
    (Energy-mass relationship)

6. $A = \pi r^2$, for r
    (Area of a circle)

7. $a^2 + b^2 = c^2$, for b
    (Pythagorean formula in two dimensions)

8. $a^2 + b^2 + c^2 = d^2$, for c
    (Pythagorean formula in three dimensions)

9. $N = \dfrac{k^2 - 3k}{2}$, for k

    (Number of diagonals of a polygon)

10. $s = v_0 t + \dfrac{gt^2}{2}$, for t

    (A motion formula)

ANSWERS

1. _____

2. _____

3. _____

4. _____

5. _____

6. _____

7. _____

8. _____

9. _____

10. _____

11. $A = 2\pi r^2 + 2\pi rh$, for r
(Area of a cylinder)

12. $A = \pi r^2 + \pi rs$, for r
(Area of a cone)

11. _____

12. _____

13. $T = 2\pi\sqrt{\dfrac{l}{g}}$, for g

(A pendulum formula)

14. $W = \sqrt{\dfrac{1}{LC}}$, for L

(An electricity formula)

13. _____

15. $P_1 - P_2 = \dfrac{32LV}{gD^2}$, for D

16. $N + p = \dfrac{6.2A^2}{pR^2}$, for R

14. _____

15. _____

17. $m = \dfrac{m_0}{\sqrt{1 - \dfrac{v^2}{c^2}}}$, for v

(A relativity formula)

18. Solve the formula given in Exercise 17 for c.

16. _____

17. _____

18. _____

19. _Stopping distance._ The formula $s = -4t^2 + 40t + 2$, giving the stopping distance s of a car in terms of the time after the brakes are applied (see p. 146), can be solved for t. Then it will give the time required to stop in a certain distance. Solve the formula for t. Then use the formula to determine how long it will take to stop in a distance of 50 ft.

19. _____

7.6 EQUATIONS REDUCIBLE TO QUADRATIC

 Certain equations that are not really quadratic can be thought of in such a way that they can be solved as quadratic. For example, consider this fourth-degree equation.

$$x^4 - 9x^2 + 8 = 0$$
$$(x^2)^2 - 9(x^2) + 8 = 0 \quad \text{Thinking of } x^4 \text{ as } (x^2)^2$$
$$u^2 - 9u + 8 = 0 \quad \text{To make this clearer, write } u \text{ instead of } x^2.$$

The equation $u^2 - 9u + 8 = 0$ can be solved by factoring or the quadratic formula. After that, we can find x by remembering that $x^2 = u$. Equations that can be solved like this are said to be *reducible to quadratic*.

Example 1   Solve: $x^4 - 9x^2 + 8 = 0$.

Let $u = x^2$. Then we solve the equation found by substituting u for x^2.

$$u^2 - 9u + 8 = 0$$
$$(u - 8)(u - 1) = 0 \quad \text{Factoring}$$
$$u - 8 = 0 \quad \text{or} \quad u - 1 = 0 \quad \text{Principle of zero products}$$
$$u = 8 \quad \text{or} \quad u = 1$$

Now we substitute x^2 for u and solve these equations.

$$x^2 = 8 \quad \text{or} \quad x^2 = 1$$
$$x = \pm\sqrt{8} \quad \text{or} \quad x = \pm 1$$
$$x = \pm 2\sqrt{2} \quad \text{or} \quad x = \pm 1$$

To check first note that when $x = 2\sqrt{2}$, $x^2 = 8$ and $x^4 = 64$. Also, when $x = -2\sqrt{2}$, $x^2 = 8$ and $x^4 = 64$. Similarly, when $x = 1$, $x^2 = 1$ and $x^4 = 1$, and when $x = -1$, $x^2 = 1$ and $x^4 = 1$. Thus instead of making four checks we need make only two.*

Check:

For $\pm 2\sqrt{2}$:

$$\begin{array}{c|c} x^4 - 9x^2 + 8 = 0 & \\ \hline (\pm 2\sqrt{2})^4 - 9(\pm 2\sqrt{2})^2 + 8 & 0 \\ 64 - 9 \cdot 8 + 8 & \\ 0 & \end{array}$$

For ± 1:

$$\begin{array}{c|c} x^4 - 9x^2 + 8 = 0 & \\ \hline (\pm 1)^4 - 9(\pm 1)^2 + 8 & 0 \\ 1 - 9 + 8 & \\ 0 & \end{array}$$

> A common error is to solve for u and then forget to solve for x. Remember that you must find values for the *original* variable!

The solutions are $1, -1, 2\sqrt{2}$, and $-2\sqrt{2}$.

DO EXERCISE 1.

---

* A check by substituting is necessary when solving equations reducible to quadratic. Reducing the equation to quadratic can sometimes introduce numbers that are not solutions of the original equation.

After finishing Section 7.6, you should be able to:

◼ **Solve equations that are reducible to quadratic.**

1. Solve.

$$x^4 - 10x^2 + 9 = 0$$

2. Solve $x + 3\sqrt{x} - 10 = 0$. Be sure to check.

Example 2 Solve: $x - 3\sqrt{x} - 4 = 0$.

Let $u = \sqrt{x}$. Then we solve the equation found by substituting u for $\sqrt{x}$ (and, of course, u^2 for x).

$$u^2 - 3u - 4 = 0$$
$$(u - 4)(u + 1) = 0$$
$$u = 4 \quad \text{or} \quad u = -1$$

Now we substitute $\sqrt{x}$ for u and solve these equations.

$$\sqrt{x} = 4 \quad \text{or} \quad \sqrt{x} = -1$$

Squaring the first equation we get $x = 16$. The second equation has no real solution since principal square roots are never negative.

The number 16 checks and is the solution.

DO EXERCISE 2.

3. Solve.

$$(x^2 - x)^2 - 14(x^2 - x) + 24 = 0$$

Example 3 Solve: $(x^2 - 1)^2 - (x^2 - 1) - 2 = 0$.

Let $u = x^2 - 1$. Then we solve the equation found by substituting u for $x^2 - 1$.

$$u^2 - u - 2 = 0$$
$$(u - 2)(u + 1) = 0$$
$$u = 2 \quad \text{or} \quad u = -1$$

Now we substitute $x^2 - 1$ for u and solve these equations.

$$x^2 - 1 = 2 \qquad \text{or} \quad x^2 - 1 = -1$$
$$x^2 = 3 \qquad \text{or} \qquad x^2 = 0$$
$$x = \pm\sqrt{3} \quad \text{or} \qquad x = 0$$

The numbers $\sqrt{3}$, $-\sqrt{3}$, and 0 check. They are the solutions.

DO EXERCISE 3.

4. Solve.

$$x^{-2} + x^{-1} - 6 = 0$$

Example 4 Solve: $y^{-2} - y^{-1} - 2 = 0$.

Let $u = y^{-1}$. Then we solve the equation found by substituting u for y^{-1} and u^2 for y^{-2}.

$$u^2 - u - 2 = 0$$
$$(u - 2)(u + 1) = 0$$
$$u = 2 \quad \text{or} \quad u = -1$$

Now we substitute y^{-1} or $\dfrac{1}{y}$ for u and solve these equations:

$$\frac{1}{y} = 2 \quad \text{or} \quad \frac{1}{y} = -1.$$

Solving, we get

$$y = \frac{1}{2} \quad \text{or} \quad y = \frac{1}{(-1)} = -1.$$

The numbers $\dfrac{1}{2}$ and -1 both check. They are the solutions.

DO EXERCISE 4.

EXERCISE SET 7.6

◖●◗ Solve.

1. $x - 10\sqrt{x} + 9 = 0$

2. $2x - 9\sqrt{x} + 4 = 0$

1. _____

2. _____

3. $x^4 - 10x^2 + 25 = 0$

4. $x^4 - 3x^2 + 2 = 0$

3. _____

4. _____

5. $(x^2 - 6x)^2 - 2(x^2 - 6x) - 35 = 0$

6. $(1 + \sqrt{x})^2 + (1 + \sqrt{x}) - 6 = 0$

5. _____

6. _____

7. $(y^2 - 5y)^2 - 2(y^2 - 5y) - 24 = 0$ **8.** $(2t^2 + t)^2 - 4(2t^2 + t) + 3 = 0$

7. _____

8. _____

9. $w^4 - 4w^2 - 2 = 0$ **10.** $t^4 - 5t^2 + 5 = 0$

9. _____

10. _____

11. $x^{-2} - x^{-1} - 6 = 0$ **12.** $4x^{-2} - x^{-1} - 5 = 0$

11. _____

12. _____

13. $2x^{-2} + x^{-1} - 1 = 0$ **14.** $m^{-2} + 9m^{-1} - 10 = 0$

13. _____

14. _____

15. _____

Solve. Check possible solutions by substituting into the original equation.

15. ▦ $6.75x - 35\sqrt{x} - 5.36 = 0$ **16.** ▦ $\pi x^4 - \pi^2 x^2 - \sqrt{99.3} = 0$

16. _____

7.7   QUADRATIC VARIATION

▪ ● ▪ FINDING EQUATIONS OF VARIATION

We have already studied direct and inverse variation in Chapter 5. We now consider variation when the equations involved are of second degree and, in some cases, higher. Consider the equation for the area of a circle, in which A and r are variables and π is a constant:

$$A = \pi r^2.$$

We say that the area varies directly as the square of the radius. When (r_1, A_1) and (r_2, A_2) are solutions of the equation we have $A_1 = \pi r_1^2$ and $A_2 = \pi r_2^2$. Then dividing we get

$$\frac{A_1}{A_2} = \frac{\pi r_1^2}{\pi r_2^2} = \frac{r_1^2}{r_2^2}.$$

A proportion like $\dfrac{A_1}{A_2} = \dfrac{r_1^2}{r_2^2}$ can be helpful in solving problems.

> **y varies directly as the square of x if there is some positive constant k such that $y = kx^2$.**

Example 1   Find an equation of variation where y varies directly as the square of x, and $y = 12$ when $x = 2$.

We write an equation of variation and find k.

$$y = kx^2, \quad \text{so } 12 = k \cdot 2^2 \quad \text{and} \quad 3 = k.$$

Thus $y = 3x^2$.

DO EXERCISE 1.

From the law of gravity, we know that the weight W of an object varies inversely as the square of its distance d from the center of the earth:

$$W = \frac{k}{d^2}.$$

When (d_1, W_1) and (d_2, W_2) are solutions of the equation we have $W_1 = k/d_1^2$ and $W_2 = k/d_2^2$. Then

$$\frac{W_1}{W_2} = \frac{\dfrac{k}{d_1^2}}{\dfrac{k}{d_2^2}} = \frac{k}{d_1^2} \cdot \frac{d_2^2}{k} = \frac{d_2^2}{d_1^2}.$$

A proportion like $\dfrac{W_1}{W_2} = \dfrac{d_2^2}{d_1^2}$ can be helpful in solving problems.

> **y varies inversely as the square of x if there is some positive constant k such that $y = k/x^2$.**

OBJECTIVES

After finishing Section 7.7, you should be able to:

▪ ● ▪ Given a situation of variation, write an equation of variation or a proportion.

▪ ● ● ▪ Solve problems involving variation.

1. Find an equation of variation where y varies directly as the square of x and $y = 175$ when $x = 5$.

2. Find an equation of variation where y varies inversely as the square of x and $y = \frac{1}{4}$ when $x = 6$.

3. Find an equation of variation where y varies jointly as x and z, and $y = 65$ when $x = 10$ and $z = 13$.

4. Find an equation of variation where y varies jointly as x and the square of z and inversely as w, and $y = 80$ when $x = 4$, $z = 10$, and $w = 25$.

Example 2   Find an equation of variation where W varies inversely as the square of d and $W = 3$ when $d = 5$.

$$W = \frac{k}{d^2}, \quad \text{so } 3 = \frac{k}{5^2} \quad \text{and} \quad k = 75$$

Thus $W = \frac{75}{d^2}$.

DO EXERCISE 2.

Consider the equation for the area A of a triangle with height h and base b:

$$A = \frac{1}{2}bh.$$

We say that the area varies *jointly* as the height and the base. From two solutions of the equation we get the proportion

$$\frac{A_1}{A_2} = \frac{\frac{1}{2}b_1 h_1}{\frac{1}{2}b_2 h_2} = \frac{b_1 h_1}{b_2 h_2}.$$

> y **varies jointly as** x **and** z **if there is some positive constant** k **such that** $y = kxz$.

Example 3   Find an equation of variation where y varies jointly as x and z, and $y = 42$ when $x = 2$ and $z = 3$.

$$y = kxz, \quad \text{so } 42 = k \cdot 2 \cdot 3 \quad \text{and} \quad k = 7$$

Thus $y = 7xz$.

DO EXERCISE 3.

The equation

$$y = k \cdot \frac{xz^2}{w}$$

asserts that y varies jointly as x and the square of z, and inversely as w.

Example 4   Find an equation of variation where y varies jointly as x and z and inversely as the square of w, and $y = 105$ when $x = 3$, $z = 20$, and $w = 2$.

$$y = k \cdot \frac{xz}{w^2}, \quad \text{so } 105 = k \cdot \frac{3 \cdot 20}{2^2} \quad \text{and} \quad k = 7.$$

Thus $y = 7 \cdot \frac{xz}{w^2}$.

DO EXERCISE 4.

◦•◦ SOLVING PROBLEMS

Many problem situations can be described with equations of variation.

Example 5   The volume of wood V in a tree varies jointly as the height h and the square of the girth g (girth is distance around). If the volume of a redwood tree is 216 m³ when the height is 30 m and the girth is 1.5 m, what is the height of a tree whose volume is 960 m³ and girth is 2 m?

Method 1.   First find k using the first set of data. Then solve for h using the second set of data.

$$V = khg^2$$
$$216 = k \cdot 30 \cdot 1.5^2$$
$$3.2 = k$$

Then

$$960 = 3.2 \cdot h \cdot 2^2$$
$$75 = h.$$

Method 2.   Let h_1 represent the height of the first tree and h_2 the height of the second tree. Use a proportion to solve for h_2 without first finding the variation constant.

$$\frac{V_1}{V_2} = \frac{h_1 \cdot g_1^2}{h_2 \cdot g_2^2}$$

$$\frac{216}{960} = \frac{30 \cdot 1.5^2}{h_2 \cdot 2^2}$$

$$h_2 = \frac{960 \cdot 67.5}{4 \cdot 216}$$

$$h_2 = 75 \text{ m}$$

Example 6   The intensity I of a TV signal varies inversely as the square of the distance d from the transmitter. If the intensity is 23 watts per square meter (W/m²) at a distance of 2 km, what is the intensity at a distance of 6 km?

We use the proportion.

$$\frac{I_1}{I_2} = \frac{d_2^2}{d_1^2}$$

$$\frac{I_2}{I_1} = \frac{d_1^2}{d_2^2}$$

$$\frac{I_2}{23} = \frac{2^2}{6^2}$$

$$I_2 = \frac{4 \cdot 23}{36}$$

$$I_2 = 2.56 \text{ W/m}^2 \qquad \text{Rounded to the nearest hundredth}$$

DO EXERCISES 5 AND 6.

5. The distance s that an object falls when dropped from some point above the ground varies directly as the square of the time t it falls. If the object falls 19.6 m in 2 seconds, how far will the object fall in 10 seconds?

6. At a fixed temperature, the resistance R of a wire varies directly as the length l and inversely as the square of its diameter d. If the resistance is 0.1 ohm when the diameter is 1 mm and the length is 50 cm, what is the resistance when the length is 2000 cm and the diameter is 2 mm?

SOMETHING EXTRA

AN APPLICATION: DEPRECIATION—DECLINING BALANCE METHOD *

On pp. 94 and 252 we considered depreciation of a copying machine. The cost was $5200 and it was depreciated over a period of 8 years. We now consider another way of computing depreciation, called *declining balance* depreciation. Using the straight-line method, the depreciation was 12.5% each year. Using the declining balance method, we multiply this by some factor, a number from 1 to 2. (If the number is 2, we call the method *double declining balance* depreciation.)

In this example, the depreciation in the first year (let's use 2 as a factor) will be 25% of $5200, or $1300.

Book value is now $5200 − $1300,   or $3900.

Depreciation the second year is 25% of $3900, or $975.

Book value is now $3900 − $975,   or $2925.

Below is part of the depreciation schedule.

Year	Rate of depreciation	Annual depreciation	Total depreciation	Book value, V
0				$5200.00
1	12.5% × 2	$1300.00	$1300.00	3900.00
2	Same	975.00	2275.00	2925.00
3	Same	731.25	3006.25	2193.75
4	Same	548.44	3554.69	1645.31

There is a formula to find the book value, V, using this method:

$$V = C\left(1 - \frac{F}{N}\right)^n,$$

where C is the initial cost, N is the number of years of useful life, n is the number of years after purchase, and F is the multiplier, a number from 1 to 2. Note that you do not need to know the salvage value to compute declining balance depreciation.

EXERCISES

1. Complete the above depreciation schedule and find the salvage value.

2. Compare this schedule with the one for the straight-line method. For how many years would this method be better for a taxpayer?

3. Use the formula to compute V, given $C = \$8700$, $N = 5$ years, and $F = 1.5$, when $n = 4$ years.

---

* For more on depreciation, see pp. 94, 252, and 404.

EXERCISE SET 7.7

▪ Find an equation of variation in which:

1. y varies directly as the square of x and $y = 0.15$ when $x = 0.1$.

2. y varies directly as the square of x and $y = 6$ when $x = 3$.

3. y varies inversely as the square of x and $y = 0.15$ when $x = 0.1$.

4. y varies inversely as the square of x and $y = 6$ when $x = 3$.

5. y varies jointly as x and z and $y = 56$ when $x = 7$ and $z = 8$.

6. y varies directly as x and inversely as z and $y = 4$ when $x = 12$ and $z = 15$.

7. y varies jointly as x and the square of z and $y = 105$ when $x = 14$ and $z = 5$.

8. y varies jointly as x and z and inversely as w and $y = \frac{3}{2}$ when $x = 2$, $z = 3$, and $w = 4$.

9. y varies jointly as x and z and inversely as the product of w and p, and $y = \frac{3}{28}$ when $x = 3$, $z = 10$, $w = 7$, and $p = 8$.

10. y varies jointly as x and z and inversely as the square of w, and $y = \frac{12}{5}$ when $x = 16$, $z = 3$, and $w = 5$.

1. _____

2. _____

3. _____

4. _____

5. _____

6. _____

7. _____

8. _____

9. _____

10. _____

●● Solve.

11. *Stopping distance of a car.* The stopping distance d of a car after the brakes are applied varies directly as the square of the speed r. If a car traveling 60 mph can stop in 200 ft, how many feet will it take the same car to stop when it is traveling 80 mph?

12. *Area of a cube.* The area of a cube varies directly as the square of the length of a side. If a cube has an area 168.54 in^2 when the length of a side is 5.3 in., what will the area be when the length of a side is 10.2 in.?

13. *Weight of an astronaut.* The weight W of an object varies inversely as the square of the distance d from the center of the earth. At sea level (6400 km from the center of the earth) an astronaut weighs 100 kg. Find his weight when he is 200 km above the surface of the earth and the spacecraft is not in motion.

14. *Intensity of light.* The intensity I of light from a light bulb varies inversely as the square of the distance d from the bulb. Suppose I is 90 W/m^2 when the distance is 5 m. Find the intensity at a distance of 10 m.

15. *Earned run average.* A pitcher's earned run average A varies directly as the number R of earned runs allowed and inversely as the number I of innings pitched. In a recent year Tom Seaver had an earned run average of 2.92. He gave up 85 earned runs in 262 innings. How many earned runs would he have given up had he pitched 300 innings with the same average? Round to the nearest whole number.

16. *Volume of a gas.* The volume V of a given mass of a gas varies directly as the temperature T and inversely as the pressure P. If $V = 231$ cm^3 when $T = 42°$ and $P = 20$ kg/cm^2, what is the volume when $T = 30°$ and $P = 15$ kg/cm^2?

7.8 QUADRATIC FUNCTIONS AND THEIR GRAPHS

● GRAPHS OF $f(x) = ax^2$

In Chapter 4 the notion of a *function* was introduced. We graphed equations such as $f(x) = 3x + 2$, or $y = 3x + 2$. Recall that the letter y acts very much like the symbol $f(x)$. In Chapter 4, we considered graphs of linear equations, or linear *functions*, that is, graphs of equations like $y = mx + b$. We now consider equations (or functions) in which the right-hand side is a quadratic polynomial,

$$y = ax^2 + bx + c.$$

Such functions, or equations, are called *quadratic*. To graph such a function, we choose values for x and compute y.*

Example 1   Graph $y = x^2$.

We choose some numbers for x, some positive and some negative, and for each number we compute y.

x	0	1	2	3	-1	-2	-3
y, or x^2	0	1	4	9	1	4	9

We plot these ordered pairs and connect them with a smooth curve.

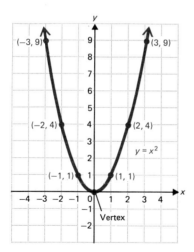

All quadratic functions have graphs similar to the one above.

Such curves are called *parabolas*. They are smooth, cup-shaped curves, with a line of symmetry. In the graph above, the vertical axis is the *line of symmetry*. If the paper were folded on this line, the two halves of the curve would match. The point $(0, 0)$ is known as the *vertex* of the parabola.

* Or $f(x)$. In this section we will sometimes use y and sometimes $f(x)$. For most purposes they are interchangeable.

OBJECTIVES

After finishing Section 7.8, you should be able to:

● Graph functions of the form $y = ax^2$ or $f(x) = ax^2$.

●● Graph functions like $f(x) = a(x - h)^2$.

●●● Graph functions like $f(x) = a(x - h)^2 + k$ and determine the vertex, line of symmetry, and the maximum or minimum function values.

1. Graph $y = \frac{1}{2}x^2$.

Some common errors in drawing graphs of quadratic functions are shown in the figure below.

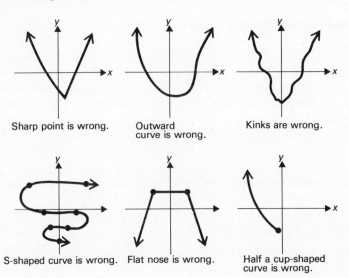

Sharp point is wrong. Outward curve is wrong. Kinks are wrong.

S-shaped curve is wrong. Flat nose is wrong. Half a cup-shaped curve is wrong.

DO EXERCISE 1.

From Margin Exercise 1, you can see that the graph of $y = \frac{1}{2}x^2$ is a flatter parabola than the graph of $y = x^2$. The graph of $y = 2x^2$ is narrower, as shown here, but the vertex and line of symmetry have not changed.

2. Graph $y = 3x^2$.

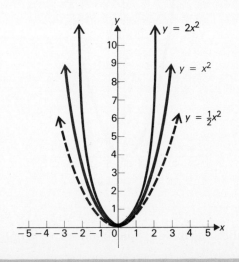

3. Graph $y = -2x^2$.

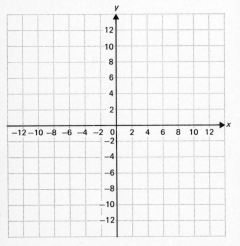

> The graph of $y = ax^2$ is a parabola with the vertical axis as its line of symmetry and its vertex at the origin.
>
> If $|a|$ is greater than 1, the parabola is narrower than $y = x^2$.
>
> If $|a|$ is between 0 and 1, the parabola is flatter than $y = x^2$.
>
> If a is positive, the parabola opens upward; if a is negative, the parabola opens downward.

DO EXERCISES 2 AND 3.

● ● GRAPHS OF $f(x) = a(x - h)^2$

Example 2   Graph $f(x) = 2(x - 3)^2$.

We choose some values of x and compute $f(x)$.

x	0	1	2	3	4	5	6
$f(x)$, or $2(x - 3)^2$	18	8	2	0	2	8	18

Now we plot the points and draw the curve.

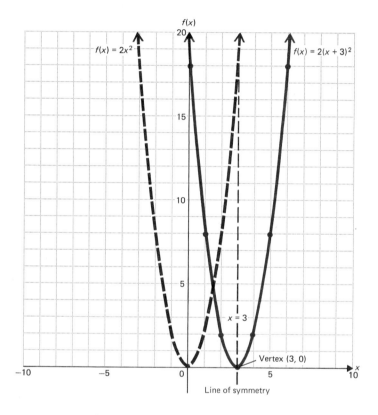

The curve in Example 2 looks just like the graph of $f(x) = 2x^2$, except that it is moved three units to the right. The line of symmetry is now the line $x = 3$ and the vertex is the point $(3, 0)$.

> **The graph of $y = a(x - h)^2$ looks just like the graph of $y = ax^2$, except that it is moved to the right or left. If h is positive, it is moved to the right. If h is negative, it is moved to the left. The vertex is now $(h, 0)$ and the line of symmetry is $x = h$.**

Example 3   Graph $f(x) = 2(x + 3)^2$.

We know that the graph looks like that of $f(x) = 2x^2$, but moved right or left. Think of $x + 3$ as $x - (-3)$. The number we are *subtracting* is negative, so we move *left*. We can draw $f(x) = 2x^2$ lightly, then move it three units to the left. We should compute a few values as a check.

4. a) Graph $y = \frac{1}{2}(x - 4)^2$.

Label the vertex and draw the line of symmetry.

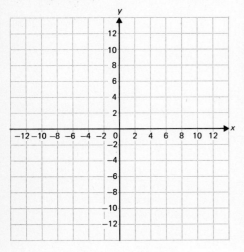

b) Graph $y = -\frac{1}{2}(x - 4)^2$.

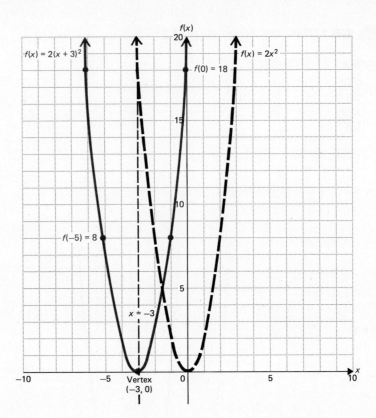

Example 4   Graph $y = -2(x + 2)^2$.

We know that the graph looks like that of $y = 2x^2$, but moved to the left two units, and it will also open downward, because of the negative coefficient, -2.

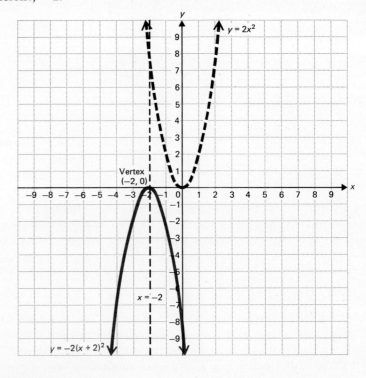

●●● GRAPHS OF $f(x) = a(x - h)^2 + k$

Given a graph of $f(x) - a(x - h)^2$, what happens to it if we add a constant k? Suppose we add 2. This increases each function value $f(x)$ by 2, so the curve is moved up. If we add a negative number the curve is moved down. The vertex of the parabola will be at the point (h, k) and the line of symmetry will be $x = h$.

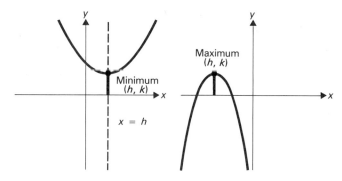

Note that if a parabola opens upward $(a > 0)$ the function value, or y-value, at the vertex is a least, or *minimum* value. That is, it is less than the y-value at any other point. If the parabola opens downward $(a < 0)$ the function value at the vertex will be a greatest, or *maximum* value.

The graph of $f(x) = a(x - h)^2 + k$ looks just like the graph of $f(x) = a(x - h)^2$, except that it is moved up or down. If k is positive the curve is moved up k units. If k is negative the curve is moved down. The vertex is at (h, k) and the line of symmetry is $x = h$. If $a > 0$ then k is the minimum function value. If $a < 0$ then k is the maximum function value.

Example 5   Graph $f(x) = 2(x + 3)^2 - 5$ and find the minimum function value.

We know that the graph looks like that of $f(x) = 2(x + 3)^2$ (this is shown in Example 3) but moved down five units. The vertex is now $(-3, -5)$, and the minimum function value is -5.

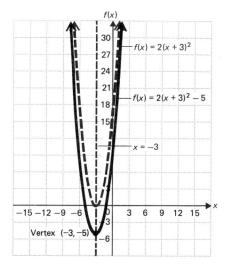

5. a) Graph $y = \frac{1}{2}(x + 2)^2 - 4$.

b) Label the vertex and draw the line of symmetry.

c) Find the minimum function value.

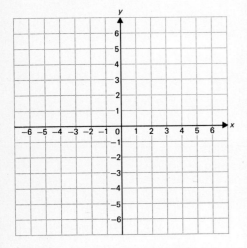

6. a) Graph $f(x) = -2(x - 5)^2 + 3$.

b) Label the vertex and draw the line of symmetry.

c) Find the maximum function value.

Example 6 Graph $f(x) = \frac{1}{2}(x - 3)^2 + 5$ and find the minimum function value.

We know that the graph looks just like that of $f(x) = \frac{1}{2}x^2$, but moved to the right three units and up five units. The vertex is $(3, 5)$ and the line of symmetry is the line $x = 3$. We draw $f(x) = \frac{1}{2}x^2$ and then move the curve over and up. We plot a few points as a check. The minimum function value is 5.

x	$f(x)$
0	$9\frac{1}{2}$
1	7
2	$5\frac{1}{2}$
6	$9\frac{1}{2}$

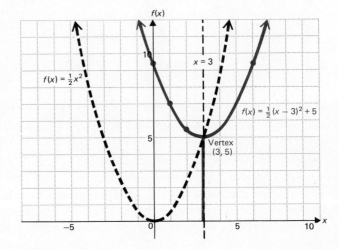

Example 7 Graph $y = -2(x + 3)^2 + 5$ and find the maximum function value.

We know that the graph looks like that of $f(x) = 2x^2$, but moved to the left three units and up five units, except that it now opens downward. The vertex is $(-3, 5)$. This time there is a *greatest* (maximum) function value, 5.

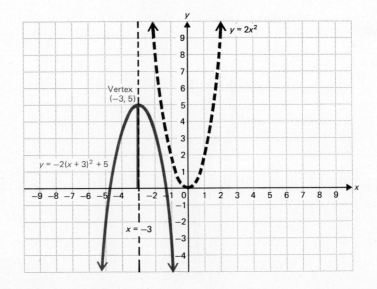

DO EXERCISES 5 AND 6.

NAME

CLASS

EXERCISE SET 7.8

Use graph paper.

■● Graph these functions.

1. $f(x) = 5x^2$

2. $f(x) = 4x^2$

3. $f(x) = \frac{1}{4}x^2$

4. $f(x) = \frac{1}{3}x^2$

5. $f(x) = -\frac{1}{2}x^2$

6. $f(x) = -\frac{1}{4}x^2$

7. $f(x) = -4x^2$

8. $f(x) = -3x^2$

●● For each of the following, graph the function, label the vertex, and draw the line of symmetry.

9. $f(x) = (x - 3)^2$

10. $f(x) = (x - 5)^2$

Graph each of the following.

11. $y = 2(x - 4)^2$

12. $y = -4(x - 6)^2$

13. $f(x) = -2(x + 6)^2$

14. $f(x) = 2(x + 4)^2$

15. $f(x) = 3(x - 1)^2$

16. $f(x) = -4(x - 2)^2$

●●● For each of the following, graph the function, label the vertex, and draw the line of symmetry.

17. $f(x) = (x - 3)^2 + 1$

18. $f(x) = (x + 2)^2 - 3$

For each of the following, graph the function and find the maximum or minimum function value.

19. $y = \frac{1}{2}(x - 1)^2 - 3$

20. $y = \frac{1}{2}(x + 1)^2 + 4$

21. $f(x) = -3(x + 4)^2 + 1$

22. $f(x) = -2(x - 5)^2 - 3$

23. $y = -2(x + 2)^2 - 3$

24. $y = 3(x - 4)^2 + 2$

25. $f(x) = -(x + 1)^2 - 2$

26. $f(x) = -(x + 1)^2 + 1$

7.9 MORE ABOUT GRAPHING QUADRATIC FUNCTIONS

● COMPLETING THE SQUARE

The procedures discussed in Section 7.8 enable us to graph any quadratic function $f(x) = ax^2 + bx + c$. By *completing the square*, we can always rewrite the polynomial $ax^2 + bx + c$ in the form

$$a(x - h)^2 + k.$$

Example 1   Graph: $f(x) = x^2 - 6x + 4$.

$$f(x) = x^2 - 6x + 4$$
$$= (x^2 - 6x) + 4$$

We complete the square inside the parentheses. We take half the x-coefficient:

$$\frac{1}{2} \cdot (-6) = -3.$$

We square it:

$$(-3)^2 = 9.$$

Then we add $9 - 9$ inside the parentheses.

$$f(x) = (x^2 - 6x + 9 - 9) + 4$$
$$= (x^2 - 6x + 9) + (-9 + 4) \qquad \text{Rearranging terms}$$
$$= 1 \cdot (x - 3)^2 - 5$$

The vertex is $(3, -5)$.

The line of symmetry is $x = 3$.

We plot a few points as a check, and draw the curve.

x	$f(x)$
0	4
1	−1
5	−1
6	4

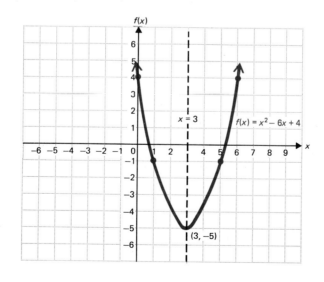

DO EXERCISE 1.

OBJECTIVES

After finishing Section 7.9, you should be able to:

● Given a quadratic function $f(x) = ax^2 + bx + c$, complete the square to put it into the form

$$f(x) = a(x - h)^2 + k,$$

and then graph the function.

●● Find the x-intercepts of a quadratic function.

1. For $f(x) = x^2 - 4x + 7$:
   a) find an equation of the type $f(x) = a(x - h)^2 + k$;
   b) find the vertex and line of symmetry;
   c) graph the function.

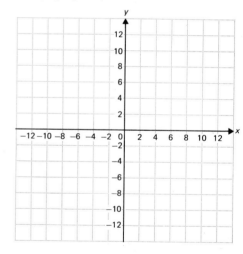

2. For $f(x) = -4x^2 + 12x - 5$:

a) find an equation of the type
$f(x) = a(x - h)^2 + k$;
b) find the vertex and line of symmetry;
c) graph the function.

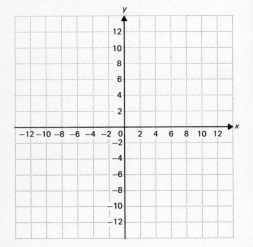

Example 2 Graph: $f(x) = -2x^2 + 10x - 7$.

We first factor the expression $-2x^2 + 10x$. Then we proceed as before:

$$f(x) = -2x^2 + 10x - 7.$$

We "remove" -2 from the first two terms by factoring. This makes the coefficient of x^2 inside the parentheses 1.

$$f(x) = -2(x^2 - 5x) - 7$$

We take half the x-coefficient and square it, to get $\frac{25}{4}$. Then we add $\frac{25}{4} - \frac{25}{4}$ inside the parentheses.

$$f(x) = -2\left(x^2 - 5x + \frac{25}{4} - \frac{25}{4}\right) - 7$$

$$= -2\left(x^2 - 5x + \frac{25}{4}\right) + 2\left(\frac{25}{4}\right) - 7 \quad \text{Multiplying by } -2\text{, using the distributive law, and rearranging terms}$$

$$= -2\left(x - \frac{5}{2}\right)^2 + \frac{11}{2}$$

The vertex is $\left(\frac{5}{2}, \frac{11}{2}\right)$.

The line of symmetry is $x = \frac{5}{2}$.

The coefficient -2 is negative, so the graph opens downward.

We plot a few points as a check and draw the curve.

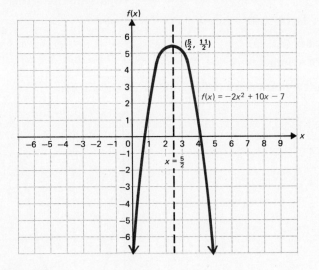

DO EXERCISE 2.

⦿⦿ INTERCEPTS (Optional)

The points at which a graph crosses the x-axis are called its *x-intercepts*. These are of course the points at which $y - 0$.

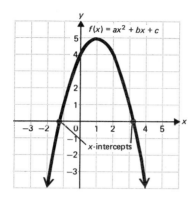

To find the x-intercepts of a quadratic function $f(x) = ax^2 + bx + c$ we solve the equation

$$0 = ax^2 + bx + c.$$

Example 3   Find the x-intercepts of the graph of $f(x) = x^2 - 2x - 2$.

We solve the equation

$$0 = x^2 - 2x - 2.$$

The equation is difficult to factor, so we use the quadratic formula and get $x = 1 \pm \sqrt{3}$. Thus the x-intercepts are $(1 - \sqrt{3}, 0)$ and $(1 + \sqrt{3}, 0)$. For plotting, we approximate, to get $(-0.7, 0)$ and $(2.7, 0)$. We sometimes refer to the x-coordinates as intercepts.

It is useful to have these points when graphing a function.

DO EXERCISE 3.

The discriminant, $b^2 - 4ac$, tells us how many real-number solutions the equation $0 = ax^2 + bx + c$ has, so it also indicates how many intercepts there are. Compare.

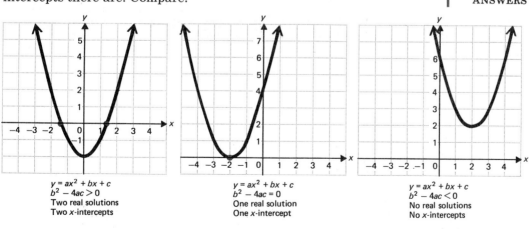

$y = ax^2 + bx + c$
$b^2 - 4ac > 0$
Two real solutions
Two x-intercepts

$y = ax^2 + bx + c$
$b^2 - 4ac = 0$
One real solution
One x-intercept

$y = ax^2 + bx + c$
$b^2 - 4ac < 0$
No real solutions
No x-intercepts

DO EXERCISES 4–6.

3. Find the x-intercepts.
$$f(x) = x^2 - 2x - 5$$

Find the x-intercepts if they exist.

4. $f(x) = x^2 - 2x - 3$

5. $f(x) = x^2 + 8x + 16$

6. $f(x) = -2x^2 - 4x - 3$

ANSWERS ON PAGE A–28

SOMETHING EXTRA

AN APPLICATION: DEPRECIATION—THE SUM OF THE DIGITS METHOD*

On p. 390 we considered a copying machine costing $5200 depreciated over an 8-year period. We shall now calculate the depreciation using a method called the *sum of the digits*. In this method we use a different rate of depreciation each year. Each year's rate is a fraction, calculated as follows.

The numerators are 8, 7, 6, . . . , 1.

We begin with 8, the number of years of useful life, and go down to 1. The denominators are all the same:

$$1 + 2 + 3 + 4 + 5 + 6 + 7 + 8, \quad \text{or} \quad 36.$$

We add the numbers 1, 2, 3, and so on, up to 8, the number of years of useful life. Thus the rates of depreciation are

$$\frac{8}{36}, \frac{7}{36}, \cdots, \frac{1}{36}.$$

In our example, the salvage value is $1300, so the total depreciation is $3900. The depreciation the first year will be $\frac{8}{36}$ of $3900, or $866.67. The second year the depreciation will be $\frac{7}{36}$ of $3900, or $758.33, and so on.

Here is part of the depreciation schedule.

Year	Rate of depreciation	Annual depreciation	Total depreciation	Book value, V
0				$5200.00
1	$\frac{8}{36}$ or 22.2%	$866.67	$ 866.67	4333.33
2	$\frac{7}{36}$ or 19.4%	758.33	1625.00	3575.00
3	$\frac{6}{36}$ or 16.7%	650.00	2275.00	2925.00

EXERCISES

1. Complete the depreciation schedule above.

2. Make a depreciation schedule, using a cost of $8700, a useful life of 5 years, and a salvage value of $1600.

_____

* For more on depreciation, see pp. 94, 252, and 390.

NAME   CLASS

EXERCISE SET 7.9

⬤ Complete the square. Then graph.

1. $f(x) = x^2 - 2x - 3$ **2.** $f(x) = x^2 + 2x - 5$

1. _____

2. _____

3. $f(x) = -x^2 + 4x + 1$ **4.** $f(x) = -x^2 - 4x + 3$

3. _____

4. _____

5. $f(x) = 3x^2 - 24x + 50$ **6.** $f(x) = 4x^2 + 8x - 3$

5. _____

6. _____

7. $f(x) = -2x^2 + 2x + 1$ **8.** $f(x) = -2x^2 - 2x + 3$

7. _____

8. _____

ANSWERS

● ● Find the x-intercepts.

9. $f(x) = x^2 - 4x + 1$ **10.** $f(x) = x^2 + 6x + 10$

9. _____

10. _____

11. $f(x) = -x^2 + 2x + 3$ **12.** $f(x) = -x^2 + 3x + 4$

11. _____

12. _____

13. $f(x) = 2x^2 - 4x + 6$ **14.** $f(x) = 2x^2 + 4x - 1$

13. _____

14. _____

15. $f(x) = 4x^2 + 12x + 9$ **16.** $f(x) = 3x^2 - 6x + 1$

15. _____

16. _____

17. _____

Find the x-intercepts.

18. _____

17. ▦ $f(x) = -0.978x^2 + 2.101x + 3.121$ **18.** ▦ $f(x) = 2.899x^2 - 5.901x + 0.969$

7.10 APPLICATIONS OF QUADRATIC FUNCTIONS

● MAXIMUM AND MINIMUM PROBLEMS (Optional)

For a quadratic function, the value $f(x)$ at the vertex will either be greater than any other $f(x)$ or less than all other $f(x)$. If the graph opens upward, $f(x)$ will be a minimum. If the graph opens downward, $f(x)$ will be a maximum. In certain problems we want to find a maximum or a minimum. If the situation can be translated to a quadratic function, we can solve by finding $f(x)$ at the vertex.

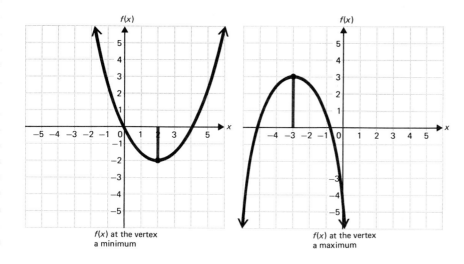

$f(x)$ at the vertex
a minimum

$f(x)$ at the vertex
a maximum

Example 1 What are the dimensions of the largest rectangular pen that can be enclosed with 64 meters of fence?

We make a drawing and label it.
The perimeter must be 64 m, so we have

$$2w + 2l = 64. \qquad (1)$$

We wish to find the maximum area, so we try to find a quadratic function for the area that is an expression in one variable. We know that

$$A = lw. \qquad (2)$$

Solving (1) for l, we get $l = 32 - w$. Substituting in (2), we get a quadratic function:

$$A = (32 - w)w$$
$$A = -w^2 + 32w.$$

Completing the square, we get

$$A = -(w - 16)^2 + 256.$$

The maximum function value (area) is 256. It occurs when $w = 16$. Thus the dimensions are 16 m by 16 m.

DO EXERCISES 1 AND 2.

OBJECTIVES

After finishing Section 7.10, you should be able to:

● Solve maximum and minimum problems.

●● Solve problems by fitting a quadratic function to three data points.

1. What is the maximum product of two numbers whose sum is 30?

2. What are the dimensions of the largest rectangular pen that can be enclosed with 100 meters of fence?

3. Find the quadratic function that fits the data points $(1, 0)$, $(-1, 4)$, and $(2, 1)$.

4. The following table shows the accident records in a city. It has values that a quadratic function will fit.

Age of driver	Number of accidents (in a year)
20	400
40	150
60	400

a) Assuming that a quadratic function will describe the situation, find the number of accidents as a function of age.

b) Use the function to calculate the number of accidents in which a typical 16-year-old is involved.

Every quadratic function is of the type $f(x) = ax^2 + bx + c$. If a is positive, the graph opens upward and the function has a minimum value. If a is negative, the graph opens downward and the function has a maximum value.

▄▄ ▄ FITTING QUADRATIC FUNCTIONS TO DATA

In many problems, a quadratic function can be used to describe the situation. We can find a quadratic function if we know three inputs and their outputs. Each such ordered pair is called a *data point*.

Example 2 Pizza Unlimited has the following prices for pizzas.

Diameter	Price
8″	$3.00
12″	$4.25
16″	$5.75

Is price a quadratic function of diameter? It probably should be, because the price should be proportional to the area, and the area is a quadratic function of the diameter. (The area of a circular region is given by $A = \pi r^2$ or $(\pi/4) \cdot d^2$.) **(a)** Fit a quadratic function to the data points $(8, 3)$, $(12, 4.25)$, and $(16, 5.75)$. **(b)** Use the function to find the price of a 14-inch pizza.

a) We use the three data points to find a, b, and c in $f(x) = ax^2 + bx + c$.

$$3.00 = a(8)^2 + b(8) + c$$
$$4.25 = a(12)^2 + b(12) + c$$
$$5.75 = a(16)^2 + b(16) + c$$

Simplifying, we get the system of equations

$$3.00 = 64a + 8b + c$$
$$4.25 = 144a + 12b + c$$
$$5.75 = 256a + 16b + c.$$

We solve this system, obtaining $(0.0078125, 0.15625, 1.25)$. Thus the function we are looking for is

$$f(x) = 0.0078125x^2 + 0.15625x + 1.25.$$

b) To find the price of a 14-inch pizza, we find $f(14)$.

$$f(14) = 0.0078125(14)^2 + 0.15625(14) + 1.25$$
$$= 4.97, \quad \text{to the nearest cent.}$$

Note that this price function always gives results greater than $1.25, since $f(0) = 1.25$, and diameters are never negative

DO EXERCISES 3 AND 4.

NAME   CLASS   ANSWERS

EXERCISE SET 7.10

■● Solve.

1. A rancher is fencing off a rectangular area with a fixed perimeter of 76 ft. What dimensions would yield the maximum area? What is the maximum area?

2. A carpenter is building a rectangular room with a fixed perimeter of 68 ft. What dimensions would yield the maximum area? What is the maximum area?

1. _____

2. _____

3. What is the maximum product of two numbers whose sum is 22? What numbers yield this product?

4. What is the maximum product of two numbers whose sum is 45? What numbers yield this product?

3. _____

4. _____

5. What is the minimum product of two numbers whose difference is 4? What are the numbers?

6. What is the minimum product of two numbers whose difference is 6? What are the numbers?

5. _____

6. _____

7. What is the minimum product of two numbers whose difference is 5? What are the numbers?

8. What is the minimum product of two numbers whose difference is 7? What are the numbers?

7. _____

8. _____

■ ● ● Find the quadratic function that fits each set of data points.

9. $(1, 4), (-1, -2), (2, 13)$ **10.** $(1, 4), (-1, 6), (-2, 16)$

9. _____

10. _____

11. $(1, 5), (2, 9), (3, 7)$ **12.** $(1, -4), (2, -6), (3, -6)$

11. _____

12. _____

13. *Predicting earnings*. A business earns $38 in the first week, $66 in the second week, and $86 in the third week. The manager graphs the points (1, 38), (2, 66), and (3, 86) and uses a quadratic function to describe the situation.

   a) Find a quadratic function that fits the data.

   b) Using the function, predict the earnings for the fourth week.

13. a) _____

b) _____

14. *Predicting earnings*. A business earns $1000 in its first month, $2000 in the second month, and $8000 in the third month. The manager plots the points (1, 1000), (2, 2000), and (3, 8000) and uses a quadratic function to describe the situation.

   a) Find a quadratic function that fits the data.

   b) Using the function, predict the earnings for the fourth month.

14. a) _____

b) _____

ANSWERS

15. a) _____

b) _____

16. a) _____

b) _____

17. _____

18. _____

15. a) Find a quadratic function that fits the following data.

Travel speed in km/h	Number of daytime accidents (for every 200 million km)
60	200
80	130
100	100

b) Use the function to calculate the number of daytime accidents that occur at 50 km/h.

16. a) Find a quadratic function that fits the following data.

Travel speed in km/h	Number of nighttime accidents (for every 200 million km)
60	400
80	250
100	250

b) Use the function to calculate the number of nighttime accidents that occur at 50 km/h.

▦ Find the maximum or minimum value for each function.

17. $f(x) = 2.31x^2 - 3.105x - 5.98$ 18. $f(x) = -17.7x^2 + 6.29x + 6.08$

EXTENSION EXERCISES

SECTION 7.1

Solve.

1. $(x)(2x^2 + 9x - 56)(3x + 10) = 0$

2. $\left(x - \dfrac{1}{3}\right)\left(x - \dfrac{1}{3}\right) + \left(x - \dfrac{1}{3}\right)\left(x + \dfrac{2}{9}\right) = 0$

3. Boats A and B leave the same point at the same time at right angles. B travels 7 km/h slower than A. After 4 hr they are 68 km apart. Find the speed of each boat.

4. Find three consecutive integers such that the square of the first plus the product of the other two is 67.

SECTION 7.2

Solve.

5. $\dfrac{5}{x} + \dfrac{x}{4} = \dfrac{11}{7}$

6. $2x^2 - x - \sqrt{5} = 0$

7. $\sqrt{3}x^2 + 6x + \sqrt{3} = 0$

8. $ix^2 - x - 1 = 0$

9. $(1 + \sqrt{3})x^2 - (3 + 2\sqrt{3})x + 3 = 0$

10. Solve $3x^2 + xy + 4y^2 - 9 = 0$ for x in terms of y.

11. One solution of $kx^2 + 3x - k = 0$ is -2. Find the other.

SECTION 7.3

Solve.

12. $\dfrac{12}{x^2 - 9} = 1 + \dfrac{3}{x - 3}$

13. $\dfrac{x^2}{x - 2} - \dfrac{x + 4}{2} + \dfrac{2 - 4x}{x - 2} + 1 = 0$

14. $\dfrac{4}{2x + i} - \dfrac{1}{x - i} = \dfrac{2}{x + i}$

15. Find a when the reciprocal of $a - 1$ is $a + 1$.

16. A discount store bought a quantity of beach towels for $250 and sold all but 15 at a profit of $3.50 per towel. With the total amount received the manager could buy 4 more than twice as many as were bought before. Find the cost per towel.

SECTION 7.4

17. Prove:

a) The sum of the solutions of $ax^2 + bx + c = 0$ is $-\dfrac{b}{a}$.

b) The product of the solutions of $ax^2 + bx + c = 0$ is $\dfrac{c}{a}$.

For each equation under the given condition, a) find k, and b) find the other solution.

18. $kx^2 - 17x + 33 = 0$; one solution is 3.

19. $kx^2 - 2x + k = 0$; one solution is -3.

20. Find k if $kx^2 - 4x + (2k - 1) = 0$ and the product of the solutions is 3.

21. Find a quadratic equation for which the sum of the solutions is $\sqrt{3}$ and the product is 8.

22. Write a quadratic equation having the given numbers as solutions:

a) $\dfrac{2 + \sqrt{3}}{2}, \dfrac{2 - \sqrt{3}}{2}$

b) $\dfrac{g}{h}, -\dfrac{h}{g}$

c) $2 - 5i, 2 + 5i$

23. Find h and k, where $3x^2 - hx + 4k = 0$ and the sum of the solutions is -12 and the product of the solutions is 20.

24. One solution of the equation $p(q - r)y^2 + q(r - p)y + r(p - q) = 0$ is 2. Find the other.

25. The sum of the squares of the solutions of $x^2 + 2kx - 5 = 0$ is 26. Find the absolute value of k.

26. The solutions of $x^2 + px + q = 0$ are p and q, where $p \neq 0$ and $q \neq 0$. Find the sum of the solutions.

27. When the solutions of each of the equations $x^2 + kx + 8 = 0$ and $x^2 - kx + 8 = 0$ are listed in increasing order, each solution of the second equation is 6 more than the corresponding solution of the first equation. Find k.

SECTION 7.6

Solve.

28. $\left(\dfrac{y^2 - 1}{y}\right)^2 - 4\left(\dfrac{y^2 - 1}{y}\right) - 12 = 0$

29. $\left(\sqrt{\dfrac{x}{x - 3}}\right)^2 - 24 = 10\sqrt{\dfrac{x}{x - 3}}$

30. $\sqrt{x - 3} - \sqrt[4]{x - 3} = 12$

31. $a^3 - 26a^{3/2} - 27 = 0$

SECTION 7.7

32. Suppose y varies directly as x and x is tripled. What is the effect on y?

33. Suppose y varies inversely as the cube of x and x is multiplied by 0.5. What is the effect on y?

34. *The Gravity Model in Sociology.* It has been determined that the average number of telephone calls in a day N, between two cities, is directly proportional to the populations P_1 and P_2 of the cities and inversely proportional to the square of the distance d between the cities. That is,

$$N = \frac{kP_1P_2}{d^2}.$$

a) The population of Indianapolis is 744,624 and the population of Cincinnati is 452,524, and the distance between the cities is 174 km. The average number of daily phone calls between the two cities is 11,153. Find the value k and write the equation of variation.

b) The average number of daily phone calls between Indianapolis and New York is 4270 and the population of New York is 7,895,563. Find the distance between Indianapolis and New York.

SECTION 7.9

35. Let $y = (x - p)^2 + (x - q)^2$, where p and q are constants. For what value of x is y a minimum?

SECTION 7.10

36. The sum of the base and the height of a triangle is 38 cm. Find the dimensions for which the area is a maximum, and find the maximum area.

37. The perimeter of rectangle $RSTV$ is 44 ft. Find the least possible length of the diagonal RT.

38. A horticulturist has 180 ft of fencing with which to form a rectangular garden. A greenhouse will provide one side of the garden, and the fencing will be used for the other three sides. What is the area of the largest region that can be enclosed that region?

NAME   SCORE   ANSWERS

TEST OR REVIEW—CHAPTER 7

[7.1, ●●●]   **1.** Solve: $4x(x - 2) - 3x(x + 1) = -18$.

1. _____

[7.1, ⫶ ⫶]   **2.** A rectangle is 4.5 inches longer than it is wide. Its area is 34 in². Find its length and width.

2. _____

[7.2, ●]   **3.** Solve: $x^2 + x + 1 = 0$.

3. _____

[7.2, ● ●]   **4.** Solve. Approximate the solutions to the nearest tenth.

$$x^2 + 4x - 2 = 0$$

4. _____

[7.3, ●]   **5.** Solve: $\dfrac{1}{4 - x} + \dfrac{1}{2 + x} = \dfrac{3}{4}$.

5. _____

[7.3, ● ●]   **6.** Two pipes can fill a tank together in $1\frac{1}{2}$ hours. One pipe requires 4 hours longer alone than the other. How long would it take for the faster pipe, working alone?

6. _____

[7.4, ●]   **7.** Determine the nature of the solutions of $x^2 + 5x + 17 = 0$.

7. _____

[7.4, ● ●]   **8.** Write a quadratic equation having solutions $\sqrt{3}$ and $3\sqrt{3}$.

8. _____

[7.5, ●]   **9.** Solve for r: $V = \dfrac{1}{3}\pi(R^2 + r^2)$.

9. _____

[7.6, ■ • ■] **10.** Solve: $x^4 - 5x^2 + 5 = 0$.

10. _____

[7.7, ■ • ■ • ■] **11.** The area of a balloon varies directly as the square of its radius. If the area is 3.4 cm^2 when the radius is 5 cm, what is the area when the radius is 7 cm?

11. _____

[7.8, ■ • • • ■] **12.** a) Graph: $f(x) = 4(x - 3)^2 + 5$.
    b) Label the vertex.
    c) Draw the line of symmetry.
    d) Find the maximum or minimum function value.

12. _____

[7.9, ■ • ■] **13.** Complete the square: $f(x) = 2x^2 + 4x - 6$.

13. _____

[7.9, ■ • • ■] **14.** Find the x-intercepts: $f(x) = x^2 - x - 6$.

14. _____

[7.10, ■ • ■] **15.** What is the minimum product of two numbers having a difference of 8? What are the numbers?

15. _____

[7.10, ■ • • ■] **16.** Find the quadratic function fitting these data points.

$(0, 0)$, $(3, 0)$, $(5, 2)$

16. _____

EQUATIONS OF SECOND DEGREE AND THEIR GRAPHS

8

The motion of objects in the heavens, such as planets, comets, and satellites, can be described using equations of second degree.

Simplify.

1. $\sqrt{169}$ **2.** $\sqrt{48}$

3. Graph: $y = 3x - 1$.

4. Complete the square and then graph.

$y = x^2 - 4x + 8$

5. Solve.   **6.** Solve.

$2x + 3y = 8$ $x^4 - 20x^2 + 64 = 0$
$x - 2y = -3$

OBJECTIVES

After finishing Section 8.1, you should be able to:

⬛• Given the lengths of two sides of a right triangle, find the length of the third side.

⬛•• Use the distance formula to find the distance between two points whose coordinates are known.

8.1 THE DISTANCE FORMULA

⬛• **THE PYTHAGOREAN THEOREM**

A famous theorem about right triangles relates the lengths of the sides.

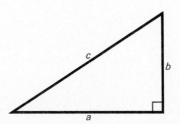

> *The Pythagorean Theorem:*   **In a right triangle, if c is the length of the hypotenuse and a and b are the lengths of the other two sides (the legs), then**
>
> $$a^2 + b^2 = c^2.$$

If we know two sides of a right triangle, we can find the other.

Example 1   Find the length of the hypotenuse of this right triangle.

$$a^2 + b^2 = c^2 \qquad \text{Pythagorean theorem}$$
$$7^2 + 5^2 = c^2 \qquad \text{Substituting}$$
$$49 + 25 = c^2$$
$$74 = c^2$$
$$\sqrt{74} = c \qquad \text{Taking principal square root}$$

The answer is $\sqrt{74}$. We do not consider the negative square root, because lengths are not negative. From Table 1, we can approximate $\sqrt{74}$. We find $c \approx 8.602$.

DO EXERCISE 1.

Example 2 Find the length of the leg of this right triangle.

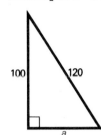

$$a^2 + b^2 = c^2 \quad \text{Pythagorean theorem}$$
$$a^2 + 100^2 = 120^2 \quad \text{Substituting}$$
$$a^2 + 10{,}000 = 14{,}400$$
$$a^2 - 4400$$
$$a = \sqrt{4400}, \quad \text{or} \quad 10\sqrt{44}$$

Approximating, using Table 1, we get

$a \approx 10 \times 6.633, \quad \text{or} \quad 66.33.$

DO EXERCISE 2.

●■● THE DISTANCE FORMULA

We now develop a formula for finding the distance between any two points on a graph when we know their coordinates. First, we consider points on a vertical or horizontal line.

If points are on a vertical line, they have the same first coordinate.

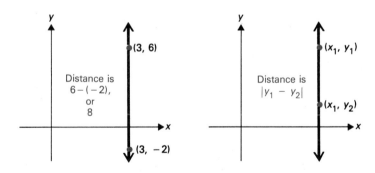

On the left above, we can find the distance between the points by subtracting their second coordinates.

$6 - (-2) = 8$ The distance is 8.

If we subtract the opposite way, we get

$-2 - 6 = -8$ The distance is $|-8|$, or 8.

If we use absolute value, as on the right above, we need not worry about the order. We can subtract either way.

If points are on a horizontal line, we subtract their first coordinates and take the absolute value.

1. Find the length of the hypotenuse of this right triangle. Use Table 1 to find an approximation.

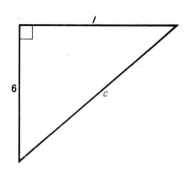

2. Find the length of the leg of this right triangle.

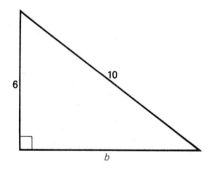

Find the distance between each pair of points.

3. $(7, 12)$ and $(7, -2)$

4. $(6, 2)$ and $(-5, 2)$

Find the distance between each pair of points. Where appropriate, use Table 1 to approximate.

5. $(2, 6)$ and $(-4, -2)$

6. $(-2, 1)$ and $(4, 2)$

Examples   Find the distance between these points.

3. $(-5, 13)$ and $(-5, 2)$

We subtract second coordinates, since the first coordinates are the same. The distance is

$$|13 - 2|, \quad \text{or} \quad 11.$$

4. $(7, -3)$ and $(-5, -3)$

Since the second coordinates are the same, we subtract first coordinates.

$$|-5 - 7| = |-12|, \quad \text{or} \quad 12$$

DO EXERCISES 3 AND 4.

Next we consider two points that are not on either a vertical or horizontal line, such as (x_1, y_1) and (x_2, y_2) in the figure below. By drawing horizontal and vertical lines through these points, we form a right triangle. The vertex of the right angle of this triangle has coordinates (x_2, y_1). The legs have lengths $|x_1 - x_2|$ and $|y_1 - y_2|$. The distance we want is the length of the hypotenuse. By the Pythagorean theorem,

$$d^2 = |x_1 - x_2|^2 + |y_1 - y_2|^2.$$

Since squares of numbers are never negative, we don't really need the absolute value signs. Thus we have

$$d^2 = (x_1 - x_2)^2 + (y_1 - y_2)^2.$$

Taking the principal square root we get the distance formula.

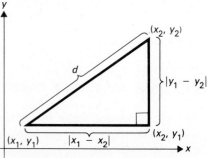

The Distance Formula: **The distance between any two points (x_1, y_1) and (x_2, y_2) is given by**

$$d = \sqrt{(x_1 - x_2)^2 + (y_1 - y_2)^2}.$$

This formula holds even when the two points are on a vertical or horizontal line.

Example 5   Find the distance between $(4, -3)$ and $(-5, 4)$.

We substitute into the distance formula.

$$d = \sqrt{[4 - (-5)]^2 + (-3 - 4)^2} \qquad \text{Substituting}$$
$$d = \sqrt{9^2 + (-7)^2}$$
$$d = \sqrt{130}$$
$$d = \sqrt{10} \cdot \sqrt{13} \approx 3.162 \times 3.606 \approx 11.402$$

> Note that the distance formula has a radical sign in it. Do not make the mistake of writing $d = (x_1 - x_2)^2 + (y_1 - y_2)^2$.

DO EXERCISES 5 AND 6.

NAME

CLASS

ANSWERS

EXERCISE SET 8.1

● In each exercise, the lengths of two sides of a right triangle are given. Find the length of the other side (c is always the length of the hypotenuse). Where appropriate, use Table 1 to find an approximation.

1. $a = 3$, $b = 4$

2. $b = 8$, $c = 10$

3. $a = 16$, $c = 20$

4. $a = 5$, $b = 12$

5. $a = 6$, $b = 8$

6. $b = 20$, $c = 25$

7. $a = 5$, $b = 7$

8. $b = 11$, $c = 12$

9. $a = 1$, $b = 1$

10. $a = 4$, $b = 7$

11. $a = 2$, $b = 2\sqrt{2}$

12. $a = 3$, $b = 3\sqrt{6}$

1. _____

2. _____

3. _____

4. _____

5. _____

6. _____

7. _____

8. _____

9. _____

10. _____

11. _____

12. _____

■ ● ● Find the distance between each pair of points. Where appropriate, use Table 1 to find an approximation.

13. $(9, 5)$ and $(6, 1)$ **14.** $(1, 10)$ and $(7, 2)$ **15.** $(0, -7)$ and $(3, -4)$

13. _____

14. _____

15. _____

16. $(6, 2)$ and $(6, -8)$ **17.** $(2, 2)$ and $(-2, -2)$ **18.** $(5, 21)$ and $(-3, 1)$

16. _____

17. _____

18. _____

19. _____

The converse of the Pythagorean theorem is true. That is, if the sides of a triangle have lengths a, b, and c, and $a^2 + b^2 = c^2$, then the triangle is a right triangle. Determine whether the points are vertices of a right triangle.

20. _____

19. $(9, 6)$, $(-1, 2)$, and $(1, -3)$ **20.** $(-8, -5)$, $(6, 1)$, and $(-4, 5)$

21. Explain why the distance formula holds even if two points are on a vertical or horizontal line.

21. _____

22. _____

22. ▦ Find the distance between these points: $(5.989, 2.001)$ and $(3.712, -7.784)$.

8.2 CONIC SECTIONS: CIRCLES AND ELLIPSES

◖◗ CIRCLES

A circle is defined to be the set of all points in a plane that are at a fixed distance from a point in that plane. Circles are called *conic sections* because they can be formed by cutting a cone by a plane, as shown.

Circle

Let us find an equation for a circle. We call the center (a, b) and we suppose the radius has length r. Suppose (x, y) is any point on the circle. By the distance formula, we have

$$\sqrt{(x - a)^2 + (y - b)^2} = r$$

Squaring both sides gives us an equation of the circle in standard form.

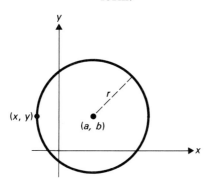

Equation of a circle: **A circle with center (a, b) and radius r has an equation**

$$(x - a)^2 + (y - b)^2 = r^2. \quad \text{(Standard form)}$$

Example 1 Find the center and radius and graph this circle.

$$(x + 2)^2 + (y - 3)^2 = 16$$

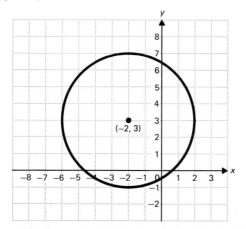

OBJECTIVES

After finishing Section 8.2, you should be able to:

◖◗ Given an equation of a circle in standard form, find its center and radius and graph it.

◖◗◗ Given the center and radius of a circle, write an equation of the circle.

◖◗◗◗ Given an equation of a circle not in standard form, complete the square to find standard form.

◖◗◖◗ Given the standard equation of an ellipse, graph the ellipse.

1. Find the center and radius of the circle $(x - 5)^2 + (y + \frac{1}{2})^2 = 9$. Graph the circle.

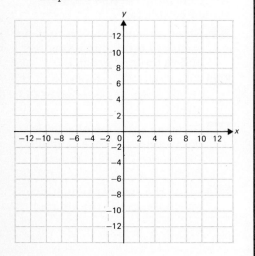

2. Find an equation of a circle with center $(-3, 1)$ and radius 6.

3. Find the center and radius of the circle $x^2 + y^2 = 64$.

4. Show that this is an equation of a circle. Find the center and radius.

$$x^2 + y^2 + 8x - 10y - 8 = 0$$

We rewrite to get the standard form given above.

$$[x - (-2)]^2 + (y - 3)^2 = 4^2$$

Thus the center is $(-2, 3)$ and the radius is 4. The graph is as shown. It can be drawn by locating the center and then using a compass.

DO EXERCISE 1.

◦◦ WRITING EQUATIONS

Example 2   Write an equation of a circle with center $(9, -5)$ and radius $\sqrt{2}$.

We use the standard form $(x - a)^2 + (y - b)^2 = r^2$.

$$(x - 9)^2 + [y - (-5)]^2 = (\sqrt{2})^2 \quad \text{Substituting}$$
$$(x - 9)^2 + (y + 5)^2 = 2 \quad \text{Simplifying}$$

DO EXERCISE 2.

If a circle is centered at the origin, then $(a, b) = (0, 0)$.

> **A circle centered at the origin with radius r has an equation**
> $$x^2 + y^2 = r^2.$$

DO EXERCISE 3.

••• EQUATIONS NOT IN STANDARD FORM

Example 3   Show that this is an equation of a circle. Find the center and radius.

$$x^2 + y^2 + 6x - 4y + 9 = 0$$

We first regroup the terms and then complete the square twice, once with $x^2 + 6x$ and once with $y^2 - 4y$.

$$(x^2 + 6x) + (y^2 - 4y) + 9 = 0$$
$$(x^2 + 6x + 9 - 9) + (y^2 - 4y + 4 - 4) + 9 = 0$$
$$(x^2 + 6x + 9) + (y^2 - 4y + 4) - 9 - 4 + 9 = 0$$
$$(x + 3)^2 + (y - 2)^2 = 4$$
$$[x - (-3)]^2 + (y - 2)^2 = 2^2$$

This shows that the graph of the equation is a circle. The center is $(-3, 2)$ and the radius is 2.

> Remember, from p. 423, that the standard form is
> $$(x - a)^2 + (y - b)^2 = r^2.$$

DO EXERCISE 4.

⊞ ELLIPSES

When a cone is cut at an angle, as shown, the conic section formed is an *ellipse*.

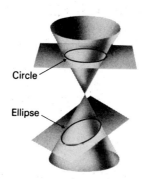

You can draw an ellipse by sticking two tacks in a piece of cardboard. Then tie a string to the tacks, place a pencil as shown, and draw.

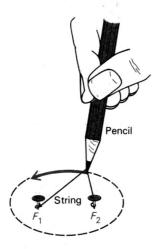

The formal mathematical definition is related to this method of drawing the ellipse. The ellipse is defined to be the set of all points in a plane such that the *sum* of the distances from two fixed points F_1 and F_2 (called *foci*) is constant.

Ellipses have equations as follows.

Equation of an ellipse: **An ellipse with its center at the origin has an equation**

$$\frac{x^2}{a^2} + \frac{y^2}{b^2} = 1, \quad a, b > 0. \qquad \textbf{(Standard form)}$$

We can think of a circle as a special kind of ellipse. A circle is formed when the angle at which the cutting plane cuts the cone is 90°. It is also formed when the foci, F_1 and F_2, are the same point. An ellipse with its foci close together is very nearly a circle.

Planets and comets have orbits around the sun that are ellipses. The sun is located at one focus. (The plural is "foci.")

5. Graph the ellipse

$$\frac{x^2}{9} + \frac{y^2}{4} = 1.$$

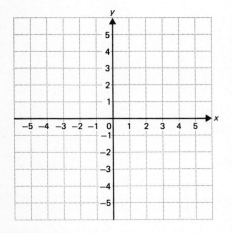

In graphing ellipses, it helps to first find the intercepts. If we replace x by 0, we can find the y-intercepts.

$$\frac{0^2}{a^2} + \frac{y^2}{b^2} = 1$$

$$\frac{y^2}{b^2} = 1$$

$$y^2 = b^2$$

$$y = \pm b$$

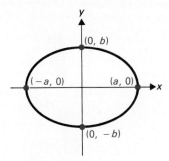

Thus the y-intercepts are $(0, b)$ and $(0, -b)$. Similarly, the x-intercepts are $(a, 0)$ and $(-a, 0)$. Plotting these points and filling in an oval-shaped curve we get a graph of the ellipse. If a more precise graph is desired, we can plot more points.

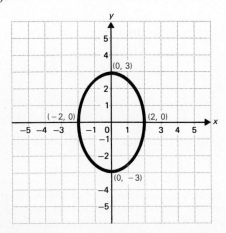

6. Graph the ellipse

$$\frac{x^2}{9} + \frac{y^2}{25} = 1.$$

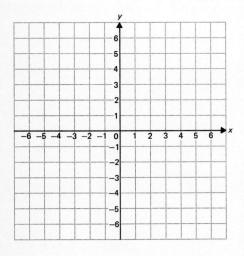

Example 4 Graph the ellipse

$$\frac{x^2}{4} + \frac{y^2}{9} = 1.$$

Note that

$$\frac{x^2}{4} + \frac{y^2}{9} = \frac{x^2}{2^2} + \frac{y^2}{3^2}.$$

Thus the x-intercepts are $(2, 0)$ and $(-2, 0)$, and the y-intercepts are $(0, 3)$ and $(0, -3)$.

DO EXERCISES 5 AND 6.

NAME

CLASS

ANSWERS

EXERCISE SET 8.2

Find the center and radius of each circle. Graph each circle.

1. $(x + 1)^2 + (y + 3)^2 = 4$

2. $(x - 2)^2 + (y + 3)^2 = 1$

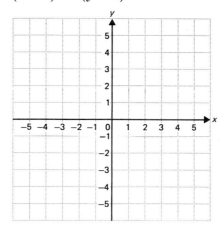

1. _____

2. _____

3. $(x - 3)^2 + y^2 = 2$

4. $x^2 + (y - 1)^2 = 3$

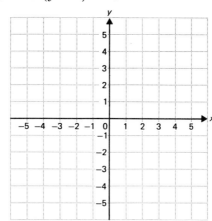

3. _____

4. _____

5. $x^2 + y^2 = 25$

6. $x^2 + y^2 = 9$

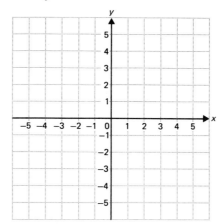

5. _____

6. _____

ANSWERS

Find an equation of the circle having center and radius as given.

7. Center $(0, 0)$, radius 7

8. Center $(0, 0)$, radius 4

9. Center $(-2, 7)$, radius $\sqrt{5}$

10. Center $(5, 6)$, radius $2\sqrt{3}$

Find the center and radius of each circle.

11. $x^2 + y^2 + 8x - 6y - 15 = 0$

12. $x^2 + y^2 + 6x - 4y - 15 = 0$

13. $x^2 + y^2 - 8x + 2y + 13 = 0$

14. $x^2 + y^2 + 6x + 4y + 12 = 0$

15. $x^2 + y^2 - 4x = 0$

16. $x^2 + y^2 + 10y - 75 = 0$

Graph each ellipse.

17. $\dfrac{x^2}{4} + \dfrac{y^2}{1} = 1$

18. $\dfrac{x^2}{1} + \dfrac{y^2}{4} = 1$

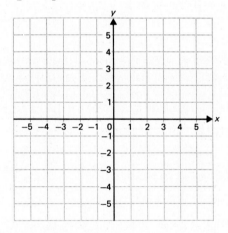

19. $\dfrac{x^2}{9} + \dfrac{y^2}{16} = 1$

20. $\dfrac{x^2}{25} + \dfrac{y^2}{16} = 1$

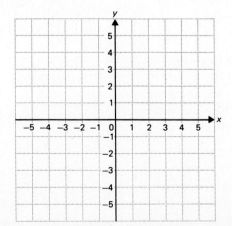

7. _____

8. _____

9. _____

10. _____

11. _____

12. _____

13. _____

14. _____

15. _____

16. _____

17. See graph. _____

18. See graph. _____

19. See graph. _____

20. See graph. _____

8.3   CONIC SECTIONS: PARABOLAS AND HYPERBOLAS

PARABOLAS

When a cone is cut as shown on the left below, the conic section formed is a *parabola*. If both parts of the cone are cut by a plane, as shown on the right, the conic section is called a *hyperbola*. We do not show it here, but graphs of quadratic functions (see Chapter 7) are actually parabolas. General equations of parabolas are quadratic.

Parabola   Hyperbola

Parabolas: **Parabolas have equations as follows.**

$y = ax^2 + bx + c$ **(Line of symmetry parallel to y-axis)**

$x = ay^2 + by + c$ **(Line of symmetry parallel to x-axis)**

Example 1   Graph $y = x^2 - 4x + 8$. Then graph $x = y^2 - 4y + 8$.

a) We complete the square:

$y = (x^2 - 4x + 4) + 4 = (x - 2)^2 + 4.$

The vertex is at $(2, 4)$ and the curve opens upward.

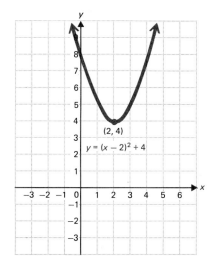

After finishing Section 8.3, you should be able to:

● Graph parabolas.

●● Given an equation of a hyperbola in standard form, graph it.

●●● Given an equation of a hyperbola $xy = c$, graph the hyperbola.

Parabolas have many applications in electricity, mechanics, and optics. The cross section of a radar reflector is a parabola. Cables that support bridges are shaped like parabolas. (Free hanging cables have a different shape, called a "catenary.")

Graph.

1. $y = x^2 + 4x + 7$

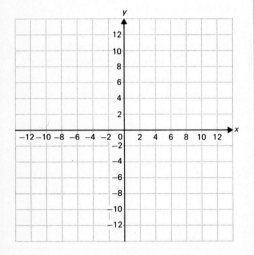

2. $x = y^2 + 4y + 7$

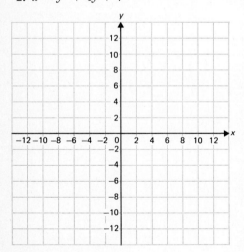

b) To graph $x = y^2 - 4y + 8$ we complete the square on y:

$$x = (y - 2)^2 + 4.$$

Since x and y are merely interchanged, this graph looks just like the first one, but is turned horizontally. The vertex is $(4, 2)$. It helps to plot a few points.

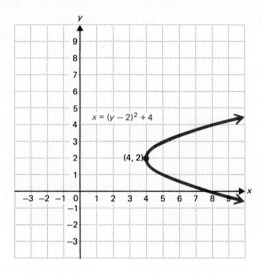

DO EXERCISES 1 AND 2.

●● HYPERBOLAS (STANDARD FORM)

A hyperbola looks somewhat like a pair of parabolas, but the actual shapes are different. A hyperbola has two vertexes, and the line through the vertexes is known as an *axis*. The point halfway between the vertexes is called the *center*. We consider only hyperbolas centered at the origin.

Hyperbolas: **Hyperbolas with their centers at the origin have equations as follows.**

$$\frac{x^2}{a^2} - \frac{y^2}{b^2} = 1 \qquad \text{(Axis horizontal)}$$

$$\frac{y^2}{b^2} - \frac{x^2}{a^2} = 1 \qquad \text{(Axis vertical)}$$

Note carefully that these equations have a 1 on the right.

Axis horizontal   Axis vertical

To graph a hyperbola, it helps to begin by graphing the lines called *asymptotes*.

> *Asymptotes of a hyperbola:*   **For hyperbolas with equations as given above, the asymptotes are the lines**
>
> $$y = \frac{b}{a}x \quad \text{and} \quad y = -\frac{b}{a}x.$$

As a hyperbola gets farther away from the origin, it gets closer and closer to its asymptotes.

> Parabolas do not have asymptotes.

The next thing to do is to find the vertexes. Then it is easy to sketch the curve.

Example 2   Graph: $\dfrac{x^2}{4} - \dfrac{y^2}{9} = 1$.

a) Note that

$$\frac{x^2}{4} - \frac{y^2}{9} = \frac{x^2}{2^2} - \frac{y^2}{3^2},$$

so $a = 2$ and $b = 3$. The asymptotes are thus

$$y = \frac{3}{2}x \quad \text{and} \quad y = -\frac{3}{2}x.$$

We sketch them.

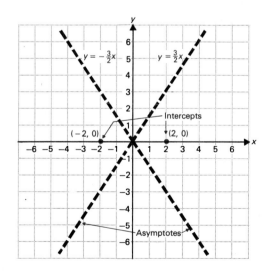

b) Next we find the intercepts. Let $y = 0$ (simply cover up the term containing y), and we see that $x^2/2^2 = 1$, so $x = \pm 2$. The intercepts are $(2, 0)$ and $(-2, 0)$.

> There are intercepts only on one axis. If we cover up the term containing x, we see that $y^2/9 = -1$, and this equation has no real-number solutions.

3. Graph.

$$\frac{x^2}{16} - \frac{y^2}{25} = 1$$

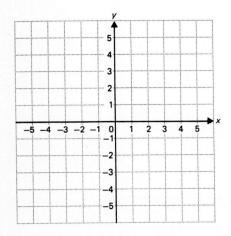

c) We plot the intercepts and sketch the graph. Then through each intercept we draw a smooth curve through each intercept that approaches the asymptotes closely.

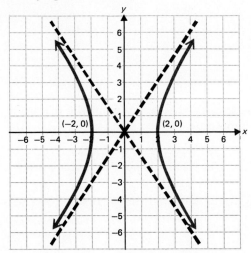

DO EXERCISE 3.

Example 3   Graph: $\dfrac{y^2}{36} - \dfrac{x^2}{4} = 1.$

a) Note that

$$\frac{y^2}{36} - \frac{x^2}{4} = \frac{y^2}{6^2} - \frac{x^2}{2^2} = 1.$$

The intercept distance is this number, in the term without the minus sign. There is y in this term, so the intercepts are on the y-axis.

The slopes of the asymptotes are $\dfrac{6}{2}$ and $-\dfrac{6}{2}$.

b) Using the numbers 6 and 2 we can quickly sketch a rectangle to use as a guide. To find the corners, we go 6 units (up or down) in the y-direction and 2 units (right or left) in the x-direction. The asymptotes go through the corners.

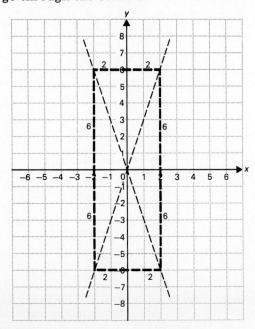

c) Now we draw curves through the intercepts toward the asymptotes as shown.

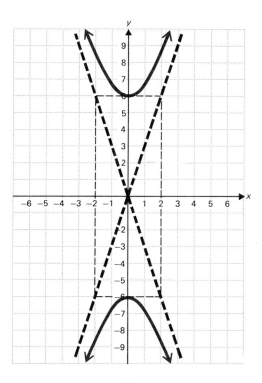

DO EXERCISE 4.

●●●   HYPERBOLAS (NONSTANDARD FORM)

The equations we have already seen for hyperbolas are the standard ones, but there are other hyperbolas. We consider some of them.

> Hyperbolas having the x- and y-axes as asymptotes have equations as follows:
>
> $$xy = c, \quad \text{where } c \text{ is a nonzero constant.}$$

Example 4   Graph: $xy = -8$.

First solve for y:

$$y = -\frac{8}{x}.$$

Next find some solutions, keeping the results in a table. (Note that we cannot use 0 for x.)

x	2	-2	4	-4	1	-1
y	-4	4	-2	2	-8	8

4. Graph.

a) $\dfrac{y^2}{9} - \dfrac{x^2}{49} = 1$

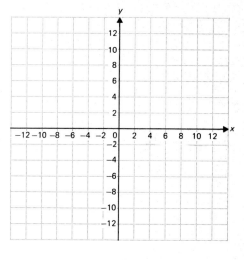

b) $\dfrac{x^2}{49} - \dfrac{y^2}{9} = 1$

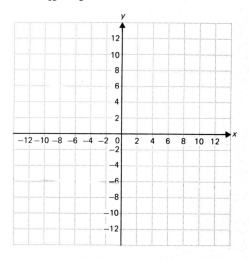

ANSWERS ON PAGE A–33

5. Graph.

$$xy = -12$$

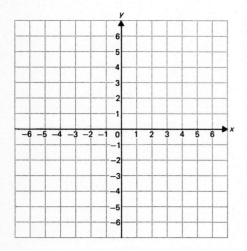

Now plot the points. Connect those points in the second quadrant with a smooth curve. Similarly, connect those in the fourth quadrant. Remember that the x-axis and the y-axis are asymptotes.

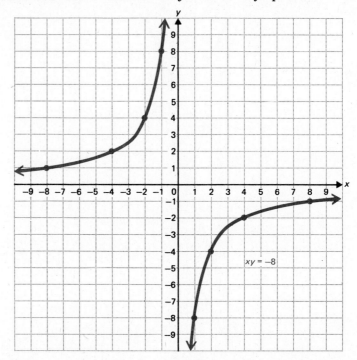

In Example 4, the two parts, or *branches*, of the hyperbola are in the second and fourth quadrants. If the constant c is positive the two branches will be in the first and third quadrants, as in the graph below.

6. Graph.

$$xy = 10$$

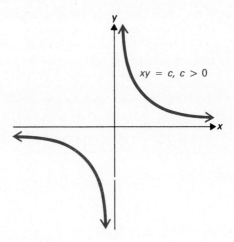

DO EXERCISES 5 AND 6.

Hyperbolas having equations $xy = c$ have the same shape as hyperbolas aligned with the coordinate axes, having equations

$$\frac{x^2}{a^2} - \frac{y^2}{a^2} = 1.$$

The graph is merely rotated 45° from the position of the latter.

NAME

CLASS

EXERCISE SET 8.3

● Graph each of the following.

1. $y = x^2$

2. $x = y^2$

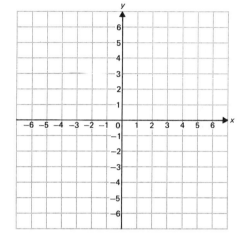

3. $x = y^2 + 4y + 1$

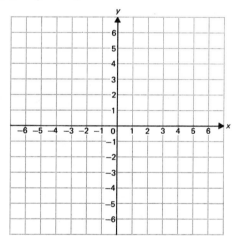

4. $y = x^2 - 2x + 3$

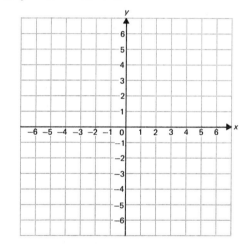

5. $y = -x^2 + 4x - 5$

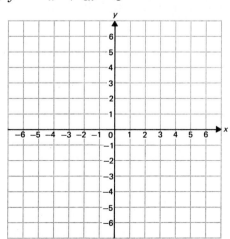

6. $x = 4 - 3y - y^2$

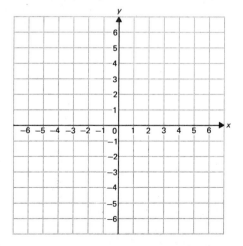

7. $x = -3y^2 - 6y - 1$

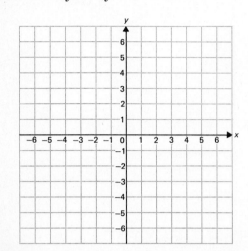

8. $y = -5 - 8x - 2x^2$

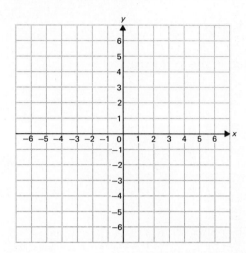

● ● Use graph paper. Graph each of the following.

9. $\dfrac{x^2}{16} - \dfrac{y^2}{16} = 1$

10. $\dfrac{y^2}{9} - \dfrac{x^2}{9} = 1$

11. $\dfrac{y^2}{16} - \dfrac{x^2}{9} = 1$

12. $\dfrac{x^2}{9} - \dfrac{y^2}{4} = 1$

13. $\dfrac{x^2}{25} - \dfrac{y^2}{36} = 1$

14. $\dfrac{y^2}{9} - \dfrac{x^2}{25} = 1$

● ● ● Use graph paper. Graph each of the following.

15. $xy = 6$

16. $xy = -4$

17. $xy = -8$

18. $xy = 3$

---

19. *Horsepower of an engine.* The horsepower of a certain kind of engine is given by the formula

$$H = \frac{D^2 N}{2.5},$$

where N is the number of cylinders and D is the diameter of each piston. Graph this equation, assuming that $N = 6$ (a 6-cylinder engine). Let D run from 2.5 to 8.

8.4 SYSTEMS OF EQUATIONS

● ALGEBRAIC SOLUTIONS

We consider systems of one first-degree and one second-degree equation. For example, the graphs may be a circle and a line. If so, there are three possibilities. (See the figure below.)

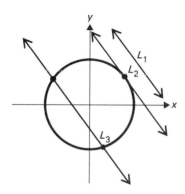

For L_1 there is no point of intersection, hence no solution in real numbers. For L_2 there is one point of intersection, hence one real-number solution. For L_3 there are two points of intersection, hence two real solutions.

These systems can be solved graphically by finding the points of intersection. In solving algebraically we use the substitution method.

Example 1 Solve.

$$x^2 + y^2 = 25 \qquad (1)$$
$$3x - 4y = 0 \qquad (2)$$

First solve the linear equation (2) for x:

$$x = \frac{4}{3}y. \qquad (3)$$

Then substitute $\frac{4}{3}y$ for x in equation (1) and solve for y:

$$\left(\frac{4}{3}y\right)^2 + y^2 = 25$$

$$\frac{16}{9}y^2 + y^2 = 25$$

$$\frac{25}{9}y^2 = 25$$

$$y^2 = 9$$

$$y = \pm 3.$$

Now substitute these numbers into equation (3) and solve for x:

$$x = \frac{4}{3}(3), \quad \text{or } 4; \qquad x = \frac{4}{3}(-3), \quad \text{or } -4.$$

The pairs $(4, 3)$ and $(-4, -3)$ check, so they are solutions.

DO EXERCISES 1 AND 2.

ANSWERS ON PAGE A–35

OBJECTIVES

After finishing Section 8.4, you should be able to:

● Solve systems of one first-degree and one second-degree equation.

●● Solve problems involving such systems of equations.

Solve.

1. $x^2 + y^2 = 25$
   $y - x = -1$

2. $y = x^2 - 2x - 1$
   $y = x + 3$

3. Solve.

$$y + 3x = 1$$
$$x^2 - 2xy = 5$$

Example 2   Solve the system.

$$y + 3 = 2x \qquad (1)$$
$$x^2 + 2xy = -1 \qquad (2)$$

First solve the linear equation (1) for y:

$$y = 2x - 3. \qquad (3)$$

Then substitute $2x - 3$ for y in equation (2) and solve for x:

$$x^2 + 2x(2x - 3) = -1$$
$$5x^2 - 6x + 1 = 0$$
$$(5x - 1)(x - 1) = 0 \qquad \text{Factoring}$$
$$5x - 1 = 0 \qquad x - 1 = 0 \qquad \text{Using the principle of}$$
$$\qquad \qquad \qquad \qquad \qquad \text{zero products}$$
$$x = \frac{1}{5} \quad \text{or} \qquad x = 1.$$

Now substitute these numbers into equation (3) and solve for y:

$$y = 2\left(\frac{1}{5}\right) - 3, \quad \text{or} \quad -\frac{13}{5},$$
$$y = 2(1) - 3, \quad \text{or} \quad -1.$$

The pairs $(\frac{1}{5}, -\frac{13}{5})$ and $(1, -1)$ check, so they are solutions.

DO EXERCISE 3.

●● SOLVING PROBLEMS

Example 3   The sum of two numbers is 14 and the sum of their squares is 106. What are the numbers?

We translate to equations.

The sum of two numbers is 14.
$$\underbrace{\qquad \qquad \qquad \qquad}$$
$$\qquad \quad x + y \qquad \qquad = 14$$

The sum of the squares is 106.
$$\underbrace{\qquad \qquad \qquad \qquad}$$
$$\qquad \quad x^2 + y^2 \qquad \qquad = 106$$

Now we solve the system

$$x + y = 14$$
$$x^2 + y^2 = 106.$$

$$(14 - y)^2 + y^2 = 106 \qquad \text{Solving the first equation for } x$$
$$\qquad \qquad \qquad \qquad \qquad \text{and substituting into the second}$$
$$14^2 - 28y + y^2 + y^2 = 106 \qquad \text{Squaring}$$
$$2y^2 - 28y + 90 = 0 \qquad \text{Standard form}$$
$$y^2 - 14y + 45 = 0 \qquad \text{Multiplying by } \frac{1}{2}$$
$$(y - 5)(y - 9) = 0 \qquad \text{Factoring}$$
$$y = 5 \quad \text{or} \quad y = 9 \qquad \text{Principle of zero products}$$

If $y = 5$, then $x = 9$; if $y = 9$, then $x = 5$.

We get two solutions: (5, 9) and (9, 5). These check, and they give the same answer to the problem. The numbers are 5 and 9.

DO EXERCISE 4.

Example 4   The perimeter of a rectangular field is 204 m and the area is 2565 m². Find the dimensions of the field.

We first translate the conditions of the problem to equations, using w for the width and l for the length.

Perimeter: $2w + 2l = 204$

Area: $lw = 2565$

Now we solve the system

$$2w + 2l = 204$$
$$lw = 2565.$$

$2w + 2\left(\dfrac{2565}{w}\right) = 204$ Solving the second equation for l and substituting into the first

$2w^2 + 2 \times 2565 = 204w$ Multiplying by w

$2w^2 - 204w + 2 \times 2565 = 0$ Standard form

$w^2 - 102w + 2565 = 0$ Multiplying by $\dfrac{1}{2}$

$w = \dfrac{102 \pm \sqrt{102^2 - 4 \cdot (1) \cdot 2565}}{2}$ Quadratic formula

$w = \dfrac{102 \pm 12}{2}$

$w = 57$ or $w = 45$

If $w = 57$, then $l = 45$; if $w = 45$, then $l = 57$.

Since we want w less then l, we have the solution (57, 45).

Now we check in the original problem: The perimeter is $2 \cdot 45 + 2 \cdot 57$, or 204. The area is $45 \cdot 57$, or 2565. The numbers check, so the answer is $l = 57$ m and $w = 45$ m.

DO EXERCISE 5.

4. The difference of two numbers is 4 and the difference of their squares is 72. What are the numbers?

5. The perimeter of a rectangular field is 34 m and the length of a diagonal is 13 m. Find the dimensions of the field.

SOMETHING EXTRA
AN APPLICATION: HANDLING DIMENSION SYMBOLS (PART II)*

We can treat dimension symbols very much like numerals or variables, because we obtain correct results that way.

Example 1 Compare

$$5x \cdot 3x = 15x^2$$

with

$$5 \text{ ft} \cdot 3 \text{ ft} = 15 \text{ ft}^2 \quad \text{(square feet)}.$$

Example 2 Compare

$$5x \cdot 8y = 40x \cdot y$$

with

$$5 \text{ men} \cdot 8 \text{ hours} = 40 \text{ man-hours}$$

(that is, if 5 men work 8 hours, the total amount of labor is 40 man-hours).

Example 3 Compare

$$\frac{300x \cdot 240y}{15w} = 4800 \frac{x \cdot y}{w}$$

with

$$\frac{300 \text{ kW} \cdot 240 \text{ hr}}{15 \text{ da}} = 4800 \frac{\text{kW-hr}}{\text{da}}$$

(that is, if an electrical device uses 300 kilowatts for 240 hours over a period of 15 days, its rate of usage of energy is 4800 kilowatt-hours per day). The standard abbreviation for kilowatt-hours is kWh.

EXERCISES

Perform these calculations.

1. $12 \text{ ft} \cdot 3 \text{ lb}$

2. $3 \text{ km} \cdot 2 \text{ hr}$

3. $\dfrac{3 \text{ in.} \cdot 8 \text{ lb}}{6 \text{ sec}}$

4. $\dfrac{60 \text{ men} \cdot 8 \text{ hr}}{20 \text{ da}}$

---

* For more on handling dimension symbols, see pp. 364 and 502.

NAME	CLASS	ANSWERS

EXERCISE SET 8.4

▮ ● ▮ Solve.

1. $x^2 + y^2 = 25$
   $y - x = 1$

2. $x^2 + y^2 = 100$
   $y - x = 2$

3. $4x^2 + 9y^2 = 36$
   $3y + 2x = 6$

4. $9x^2 + 4y^2 = 36$
   $3x + 2y = 6$

5. $y^2 = x + 3$
   $2y = x + 4$

6. $y = x^2$
   $3x = y + 2$

7. $x^2 - xy + 3y^2 = 27$
$x - y = 2$

8. $2y^2 + xy + x^2 = 7$
$x - 2y = 5$

7. _____

8. _____

●● Solve.

9. The sum of two numbers is 13 and the sum of their squares is 109. What are the numbers?

10. The sum of two numbers is 15 and the difference of their squares is also 15. What are the numbers?

9. _____

10. _____

11. A rectangle has perimeter 28 ft and the length of a diagonal is 10 ft. What are its dimensions?

12. A rectangle has perimeter 6 in. and the length of a diagonal is $\sqrt{5}$ in. What are its dimensions?

11. _____

12. _____

13. _____

13. Find an equation of the circle that passes through the points $(4, 6)$, $(-6, 2)$, and $(1, -3)$.

8.5 SYSTEMS OF SECOND-DEGREE EQUATIONS

◼◦ ALGEBRAIC SOLUTIONS

Two second-degree equations can have common solutions in various ways. If the graphs happen to be a circle and a hyperbola, for example, there are five possibilities, as shown below.

| 4 real solutions | 3 real solutions | 2 real solutions | 1 real solution | 0 real solutions |

To solve systems of two second-degree equations we can use either the substitution method or the addition method.

Example 1 Solve.

$$2x^2 + 5y^2 = 22 \qquad (1)$$
$$3x^2 - y^2 = -1 \qquad (2)$$

Here we use the addition method.

$$2x^2 + 5y^2 = 22$$
$$\underline{15x^2 - 5y^2 = -5} \qquad \text{Multiplying by 5 on both sides of (2)}$$
$$17x^2 = 17 \qquad \text{Adding}$$
$$x^2 = 1$$
$$x = \pm 1$$

If $x = 1$, $x^2 = 1$, and if $x = -1$, $x^2 = 1$, so substituting 1 or -1 for x in equation (2) we have

$$3 \cdot 1 - y^2 = -1$$
$$y^2 = 4$$
$$y = \pm 2.$$

Thus if $x = 1$, $y = 2$ or $y = -2$, and if $x = -1$, $y = 2$ or $y = -2$. The possible solutions are $(1, 2)$, $(1, -2)$, $(-1, 2)$, and $(-1, -2)$.

Check: Since $(2)^2 = 4$, $(-2)^2 = 4$, $(1)^2 = 1$, and $(-1)^2 = 1$, we can check all four pairs at one time.

$2x^2 + 5y^2 = 22$		$3x^2 - y^2 = -1$	
$2(\pm 1)^2 + 5(\pm 2)^2$	22	$3(\pm 1)^2 - (\pm 2)^2$	-1
$2 + 20$		$3 - 4$	
22		-1	

The solutions are $(1, 2)$, $(1, -2)$, $(-1, 2)$, and $(-1, -2)$.

DO EXERCISE 1.

Example 2 Solve the system.

$$x^2 + 4y^2 = 20 \qquad (1)$$
$$xy = 4 \qquad (2)$$

Solve.

1. $2y^2 - 3x^2 = 6$
   $5y^2 + 2x^2 = 53$

2. Solve.

$$x^2 + xy + y^2 = 19$$
$$xy = 6$$

Here we use the substitution method. First we solve equation (2) for y:

$$y = \frac{4}{x}.$$

Then we substitute $\frac{4}{x}$ for y in equation (1) and solve for x:

$$x^2 + 4\left(\frac{4}{x}\right)^2 = 20$$

$$x^2 + \frac{64}{x^2} = 20$$

$$x^4 + 64 = 20x^2 \qquad \text{Multiplying by } x^2$$

$$x^4 - 20x^2 + 64 = 0$$

$$u^2 - 20u + 64 = 0 \qquad \text{Letting } u = x^2$$

$$(u - 16)(u - 4) = 0 \qquad \text{Factoring}$$

$$u = 16 \quad \text{or} \quad u = 4 \qquad \text{Principle of zero products}$$

Then $x = 4$ or $x = -4$ or $x = 2$ or $x = -2$.

Since $y = 4/x$, if $x = 4$, $y = 1$; if $x = -4$, $y = -1$; if $x = 2$, $y = 2$; and if $x = -2$, $y = -2$.

The ordered pairs $(4, 1)$, $(-4, -1)$, $(2, 2)$, $(-2, -2)$ check. They are the solutions.

DO EXERCISE 2.

3. The area of a rectangle is 2 in^2 and the length of a diagonal is $\sqrt{5}$ in. Find the dimensions of the rectangle.

●● SOLVING PROBLEMS

Example 3 The area of a rectangle is 300 yd^2 and the length of a diagonal is 25 yd. Find the dimensions.

First draw a picture.

We use l for the length and w for the width and translate to equations.

From the Pythagorean theorem: $l^2 + w^2 = 25^2$

Area: $lw = 300$

Now we solve the system

$$l^2 + w^2 = 625$$
$$lw = 300.$$

We get $(20, 15)$ and $(-20, -15)$. We check in the original problem: $15^2 + 20^2 = 25^2$ and $15 \cdot 20 = 300$, so $(20, 15)$ is a solution of the problem. Lengths of sides cannot be negative so $(-20, -15)$ is not a solution. The answer is $l = 20$ yd and $w = 15$ yd.

DO EXERCISE 3.

NAME

CLASS

ANSWERS

EXERCISE SET 8.5

● Solve.

1. $x^2 + y^2 = 25$
   $y^2 = x + 5$

2. $y = x^2$
   $x = y^2$

3. $x^2 + y^2 = 9$
   $x^2 - y^2 = 9$

1. _____

2. _____

3. _____

4. $y^2 - 4x^2 = 4$
   $4x^2 + y^2 = 4$

5. $x^2 + y^2 = 5$
   $xy = 2$

6. $x^2 + y^2 = 20$
   $xy = 8$

4. _____

5. _____

6. _____

ANSWERS

7. $xy - y^2 = 2$
    $2xy - 3y^2 = 0$

8. $2xy + 3y^2 = 7$
    $3xy - 2y^2 = 4$

7. _____

8. _____

● ● Solve.

9. The product of two numbers is 156. The sum of their squares is 313. Find the numbers.

10. The product of two numbers is 60. The sum of their squares is 136. Find the numbers.

9. _____

10. _____

11. The area of a rectangle is $\sqrt{3}$ m² and the length of a diagonal is 2 m. Find the dimensions.

12. The area of a rectangle is $\sqrt{2}$ m² and the length of a diagonal is $\sqrt{3}$ m. Find the dimensions.

11. _____

12. _____

EXTENSION EXERCISES

SECTION 8.1

1. Find the distance between the points.
   a) $(-1, 3k)$ and $(6, 2k)$

   b) $(0, \sqrt{7})$ and $(\sqrt{6}, 0)$

   c) $(6m, -7n)$ and $(-2m, n)$

   d) $(\sqrt{d}, -\sqrt{3c})$ and $(\sqrt{d}, \sqrt{3c})$

   e) $\left(\frac{5}{7}, \frac{1}{14}\right)$ and $\left(\frac{1}{7}, \frac{11}{14}\right)$

   f) $(-3\sqrt{3}, 1 - \sqrt{6})$ and $(\sqrt{3}, 1 + \sqrt{6})$

2. Prove: If the endpoints of a segment are (x_1, y_1) and (x_2, y_2), then the coordinates of the midpoint are

$$\left(\frac{x_1 + x_2}{2}, \frac{y_1 + y_2}{2}\right).$$

3. Find the midpoints of the segments having the following endpoints.
   a) $(-12, -7)$ and $(-3, 13)$

   b) $(-a, b)$ and (a, b)

   c) $(2 - \sqrt{3}, 5\sqrt{2})$ and $(2 + \sqrt{3}, 3\sqrt{2})$

   d) $\left(\frac{1}{3}, -\frac{7}{11}\right)$ and $\left(-\frac{2}{3}, \frac{3}{4}\right)$

4. Determine whether the points $(6, 3)$, $(-4, -1)$, and $(-2, -6)$ are vertices of a right triangle.

5. Find the point on the y-axis that is equidistant from $(2, 10)$ and $(6, 2)$.

SECTION 8.2

6. Find an equation of a circle satisfying the given conditions:

   a) Center $(0, 0)$, passing through $\left(\frac{1}{4}, \frac{\sqrt{31}}{4}\right)$,

   b) Center $(-4, 1)$, passing through $(2, -5)$,

   c) Center $(-3, -2)$ and tangent to the y-axis,

   d) The endpoints of a diameter are $(7, 3)$ and $(-1, -3)$.

The equation (standard form) of an ellipse centered at the origin is

$$\frac{x^2}{a^2} + \frac{y^2}{b^2} = 1.$$

The vertices are $(a, 0)$, $(-a, 0)$, $(0, b)$, and $(0, -b)$. The equation (standard form) of an ellipse with center (h, k) is

$$\frac{(x - h)^2}{a^2} + \frac{(y - k)^2}{b^2} = 1.$$

The vertices are $(a + h, k)$, $(-a + h, k)$, $(h, b + k)$, and $(h, -b + k)$.

7. For each ellipse write the equation in standard form, and find the center and vertices. Then graph the ellipse.
   a) $4x^2 + 9y^2 = 36$

   b) $4(x - 1)^2 + 9(y + 2)^2 = 36$

   c) $16x^2 + y^2 + 96x - 8y + 144 = 0$

   d) $4x^2 + 25y^2 - 8x + 50y - 71 = 0$

8. Find the equation of the ellipse with the following vertices.
   a) $(-8, 0)$, $(8, 0)$, $(0, -2)$, $(0, 2)$

   b) $(0, 2)$, $(5, -1)$, $(10, 2)$, $(5, 5)$

9. Find an equation of the ellipse that contains the point $(1, 2\sqrt{3})$ and two of its vertices are $(2, 0)$ and $(-2, 0)$.

SECTION 8.3

10. Find an equation of a parabola satisfying the given conditions:
    a) Line of symmetry parallel to y-axis and passing through the points $(0, 3)$, $(-1, 6)$, $(2, 9)$.
    b) Line of symmetry parallel to x-axis, vertex $(2, 1)$ and passing through the point $(18, -1)$.

11. Classify the graph of each of the following equations as a circle, an ellipse, a parabola, or a hyperbola.
    a) $x^2 + y^2 - 10x + 8y - 40 = 0$
    b) $y + 1 = 2x^2$
    c) $9x^2 - 4y^2 - 36x + 24y - 36 = 0$
    d) $1 - 3y = 2y^2 - x$

    e) $4x^2 + 25y^2 - 8x - 100y + 4 = 0$
    f) $\dfrac{x^2}{7} + \dfrac{y^2}{7} = 1$

SECTION 8.4

12. Find two numbers whose product is 2 and the sum of whose reciprocals is $\frac{33}{8}$.

13. Find an equation of a circle that passes through $(-2, 3)$ and $(-4, 1)$ and whose center is on the line $5x + 8y = -2$.

14. The sum of two numbers is 1, and their product is 1. Find the sum of their cubes. There is a method to solve this problem that is easier than solving a system of one first-degree equation and one second-degree equation. Can you discover it?

SECTION 8.5

15. The square of a certain number exceeds twice the square of another number by $\frac{1}{8}$. The sum of their squares is $\frac{5}{16}$. Find the numbers.

16. Find an equation of a circle that passes through $(4, -20)$, $(10, -2)$ and $(-4, -4)$.

17. Four squares with sides 5 in. long are cut from the corners of a rectangular metal sheet that has an area of 340 in^2. The edges are bent up to form an open box with a volume of 350 in^3. Find the dimensions of the box.

NAME _____   SCORE _____

TEST OR REVIEW—CHAPTER 8

Review the material in the chapter. Then test yourself by doing these exercises. If you miss an item, restudy the section indicated to the left of the item.

[8.1, ▨]   **1.** The legs of a right triangle have lengths 3 and $\sqrt{5}$. Find the length of the hypotenuse.

1. _____

[8.1, ▨]   **2.** Find the distance between $(-6, 2)$ and $(6, 8)$.

[8.2, ▨]   **3.** Write an equation of the circle with center $(2, -7)$ and having a radius of length $3\sqrt{2}$.

[8.2, ▨]   **4.** Find the center and radius of the circle $x^2 + y^2 + 8x - 10y - 8 = 0$.

2. _____

[8.2, ▨]   **5.** Graph: $\dfrac{x^2}{16} + \dfrac{y^2}{4} = 1$.

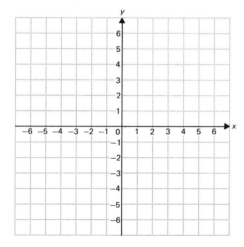

3. _____

4. _____

[8.3, ▨]   **6.** Graph: $y = x^2 - 4x - 1$.

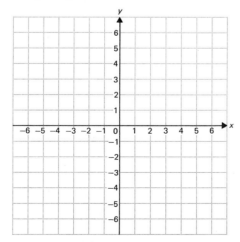

5. _____

6. _____

ANSWERS

[8.3, ●●]   **7.** Graph: $\dfrac{x^2}{9} - \dfrac{y^2}{4} = 1$.

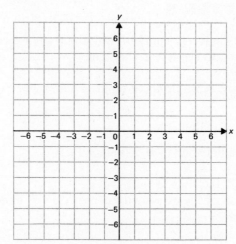

7. _____

[8.3, ●●●]   **8.** Graph: $xy = -5$.

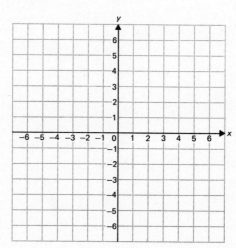

8. _____

9. _____

[8.4, ●]   **9.** Solve.

$$\frac{x^2}{16} + \frac{y^2}{9} = 1$$

$$3x + 4y = 12$$

[8.5, ●]   **10.** Solve.

$$x^2 + y^2 = 16$$

$$\frac{x^2}{16} - \frac{y^2}{9} = 1$$

10. _____

EXPONENTIAL AND LOGARITHMIC FUNCTIONS

9

The speed at which city dwellers walk varies with the size of the city. A formula for finding the walking speed involves a logarithm function.

OBJECTIVES

After finishing Section 9.1, you should be able to:

◖ Graph exponential functions.

◖◗ Graph inverses of exponential functions.

9.1 EXPONENTIAL FUNCTIONS

◖ In Chapter 6 we gave meaning to expressions with rational-number exponents such as

$$5^{1/4}, \quad 3^{-3/4}, \quad 7^{2.34}, \quad 8^{1.73}.$$

For example, $5^{1.73}$, or $5^{173/100}$, means to raise 5 to the 173rd power and then take the 100th root. We shall now give meaning to expressions with irrational exponents, such as

$$5^{\sqrt{2}}, \quad 7^{\pi}, \quad 9^{-\sqrt{3}}.$$

Consider $5^{\sqrt{3}}$. Let us think of rational numbers r close to $\sqrt{3}$ and look at 5^r. As r gets closer to $\sqrt{3}$, 5^r gets closer to some real number.

$\underbrace{r \text{ closes in on } \sqrt{3}.}$ $\underbrace{5^r \text{ closes in on some real number } p.}$

r	5^r
$1 < \sqrt{3} < 2$	$5^1 < p < 5^2$
$1.7 < \sqrt{3} < 1.8$	$5^{1.7} < p < 5^{1.8}$
$1.73 < \sqrt{3} < 1.74$	$5^{1.73} < p < 5^{1.74}$
$1.732 < \sqrt{3} < 1.733$	$5^{1.732} < p < 5^{1.733}$

> The symbol $<$ means "is less than," so $1 < \sqrt{3} < 2$ means that 1 is less than $\sqrt{3}$ and that $\sqrt{3}$ is less than 2.

As r closes in on $\sqrt{3}$, 5^r closes in on some real number p. We define $5^{\sqrt{3}}$ to be the number p.

Any positive irrational exponent can be defined in a similar way. Negative irrational exponents are then defined in the same way as negative integer exponents. Then the expression a^x has meaning for any real number x. The usual laws of exponents still hold, but we will not prove that here. We now define exponential functions.

The function $f(x) = a^x$, where a is a positive constant, is called *the exponential function*, base a.

Example 1 Graph the exponential function $y = 2^x$.

We compute some function values, thinking of y as $f(x)$.

$$f(1) = 2^1 = 2$$
$$f(3) = 2^3 = 8$$

$$f(-2) = 2^{-2} = \frac{1}{2^2} = \frac{1}{4}$$

We can keep a record of values in a table.

x	0	1	2	-1	-2	-3
y, or $f(x)$	1	2	4	$\frac{1}{2}$	$\frac{1}{4}$	$\frac{1}{8}$

Next, we plot these points and connect them with a smooth curve.

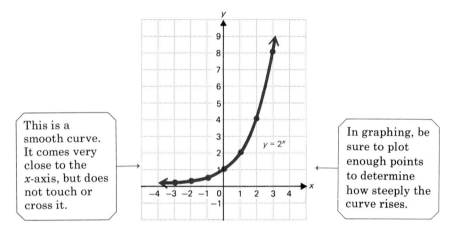

This is a smooth curve. It comes very close to the x-axis, but does not touch or cross it.

In graphing, be sure to plot enough points to determine how steeply the curve rises.

Note that as x increases, the function values increase indefinitely. As x decreases, the function values decrease, becoming very close to 0. The x-axis is an *asymptote*, meaning that the curve comes very close to it.

DO EXERCISE 1.

Example 2 Graph the exponential function $y = (\frac{1}{2})^x$.

We compute some function values, thinking of y as $f(x)$, and keep the results in a table.

$$f(2) = \left(\frac{1}{2}\right)^2 = \frac{1}{4}$$

$$f(3) = \left(\frac{1}{2}\right)^3 = \frac{1}{8}$$

$$f(0) = \left(\frac{1}{2}\right)^0 = 1$$

$$f(-3) = \left(\frac{1}{2}\right)^{-3} = \frac{1}{\left(\frac{1}{2}\right)^3} = \frac{1}{\frac{1}{8}} = 8$$

1. Graph $y = 3^x$.

a) Complete this table of solutions.

x	0	1	2	-1	-2	-3
y						

b) Plot the points from the table and connect them with a smooth curve.

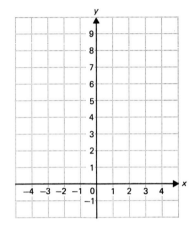

ANSWERS ON PAGE A−38

2. Graph $y = \left(\dfrac{1}{3}\right)^x$.

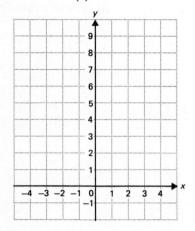

x	-3	-2	-1	0	1	2	3
y, or $f(x)$	8	4	2	1	$\frac{1}{2}$	$\frac{1}{4}$	$\frac{1}{8}$

We plot these points and draw the curve. Note that this graph is a reflection of the one in Example 1.

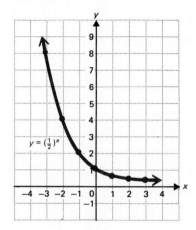

DO EXERCISE 2.

The preceding examples and exercises illustrate exponential functions of various bases. If $a = 1$, the graph is the line $y = 1$. For other positive values of a, the graphs are more interesting.

A When $a > 1$, the function $f(x) = a^x$ increases from left to right. See the colored curves below. The greater the value of a, the steeper the curve.

B When $a < 1$, the function $f(x) = a^x$ decreases from left to right. See the black curve below. As a approaches 1, the curve becomes less steep. All such functions go through the point $(0, 1)$.

Example 3   Graph: $y = 2^{x-2}$.

We construct a table of values and then plot points and connect them. Be sure to note that $x - 2$ is the *exponent*.

x	0	-1	-2	1	2	3	4	5
y	$\frac{1}{4}$	$\frac{1}{8}$	$\frac{1}{16}$	$\frac{1}{2}$	1	2	4	8

ANSWER ON PAGE A–38

The graph looks just like the graph of $y = 2^x$, but is moved to the right.

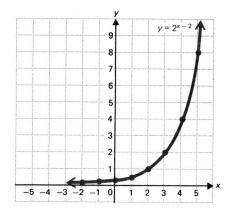

DO EXERCISE 3.

3. Graph $y = 2^{x+2}$.

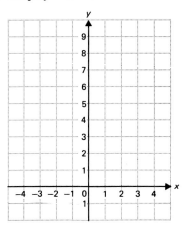

◉◉ INVERSES

If in $y = a^x$ we interchange x and y, we get another kind of function, called the *inverse* of the exponential function.

Example 4   Graph: $x = 2^y$.

To construct a table of values, we can use the one we constructed in Example 1, interchanging x and y.

x	1	2	4	$\frac{1}{2}$	$\frac{1}{4}$	$\frac{1}{8}$
y, or $f(x)$	0	1	2	-1	-2	-3

We plot these points and connect them with a smooth curve.

This curve does not touch or cross the y-axis.

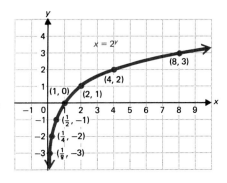

4. Graph $x = 3^y$.

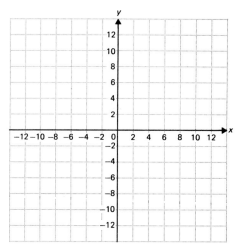

Note that this curve looks just like the graph of $y = 2^x$, except that it is reflected across the line $y = x$, as shown on the right.

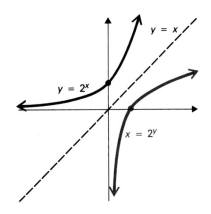

DO EXERCISE 4.

ANSWERS ON PAGE A–38

SOMETHING EXTRA

CALCULATOR CORNER: NUMBER PATTERNS

Do the following calculations and look for patterns. What seems to be happening to your answers as n gets larger and larger?

EXERCISES

1. Complete this table. You will need a calculator with a power key $\boxed{a^b}$. Find decimal notation.

n	$\left(1 + \dfrac{1}{n}\right)^n$
1	
2	
5	
10	
50	
100	
500	

2. Complete this table. Find decimal notation.

n	2^n	2^{-n}
1		
2		
5		
10		
50		
100		

NAME

CLASS

EXERCISE SET 9.1

▪● Graph.

1. $y = 2^x$

2. $y = 3^x$

3. $y = 5^x$

4. $y = 6^x$

5. $y = 2^{x+1}$

6. $y = 2^{x-1}$

7. $y = \left(\dfrac{1}{2}\right)^x$

8. $y = \left(\dfrac{1}{3}\right)^x$

9. $y = \left(\dfrac{1}{4}\right)^x$

10. $y = \left(\dfrac{1}{5}\right)^x$

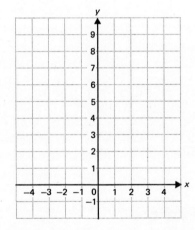

●● Graph.

11. $x = 4^y$

12. $x = 3^y$

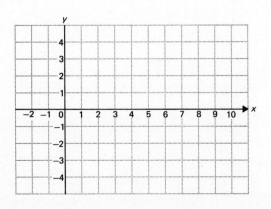

9.2 LOGARITHMIC FUNCTIONS

◘ GRAPHS OF LOGARITHMIC FUNCTIONS

The inverse of an exponential function is called a *logarithmic function*, or *logarithm function*. Given an exponential function $y = a^x$, if we reflect its graph across the line $y = x$, we get the graph of a logarithm function, $x = a^y$. We use the symbol $\log_a (x)$ or $\log_a x$ to name the function values* of a logarithmic function $x = a^y$. Thus we write $y = \log_a x$ rather than $x = a^y$. Almost always, we use a number a that is greater than 1 for a logarithm base.

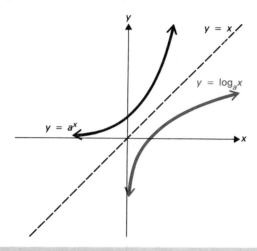

> $y = \log_a x$ **is defined to mean** $x = a^y$, **where** $x > 0$ **and** a **is a positive constant different from 1.**

We read "$\log_a x$" as "the logarithm, base a, of x." Note that $\log_a x$ represents the exponent in the equation $x = a^y$. Thus the logarithm, base a, of a number x is the power to which a is raised to get x.

> Remember: A logarithm is an exponent!

Example 1   Graph: $y = \log_5 x$.

This is equivalent to $5^y = x$. The graph is a reflection of $y = 5^x$ across the line $y = x$. We make a table of values for $y = 5^x$ and then interchange x and y.

For $y = 5^x$

x	0	1	2	-1	-2
y	1	5	25	$\frac{1}{5}$	$\frac{1}{25}$

For $y = \log_5 x$ (or $5^y = x$)

x	1	5	25	$\frac{1}{5}$	$\frac{1}{25}$
y	0	1	2	-1	-2

---

* The parentheses in $\log_a (x)$ are like those in $f(x)$. For logarithm functions we usually omit the parentheses and write $\log_a x$.

After finishing Section 9.2, you should be able to:

◘ Graph logarithm functions.

◘◘ Convert from exponential to logarithmic equations.

◘◘◘ Convert from logarithmic to exponential equations.

⁞⁞ Solve certain logarithmic equations.

1. Graph $y = \log_3 x$.

Convert to logarithmic equations.

2. $6^0 = 1$

3. $10^{-3} = 0.001$

4. $16^{0.25} = 2$

5. $m^T = p$

Convert to exponential equations.

6. $\log_2 32 = 5$

7. $\log_{10} 1000 = 3$

8. $\log_a Q = 7$

9. $\log_t M = x$

We plot the latter set of ordered pairs and connect them with a smooth curve. The graph of $y = 5^x$ has been shown only for reference.

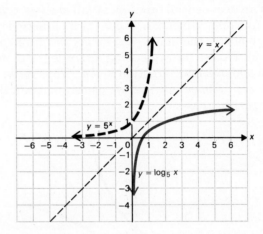

DO EXERCISE 1.

●● FROM EXPONENTIAL TO LOGARITHMIC EQUATIONS

We use the definition on the preceding page to convert from exponential to logarithmic equations.

Examples   Convert to logarithmic equations.

2. $8 = 2^x \rightarrow x = \log_2 8$ The exponent is the logarithm.
The base remains the base.

3. $y^{-1} = 4 \rightarrow -1 = \log_y 4$

4. $a^b = c \rightarrow b = \log_a c$

DO EXERCISES 2–5.

●●● FROM LOGARITHMIC TO EXPONENTIAL EQUATIONS

We also use the definition to convert from logarithmic to exponential equations.

Examples   Convert to exponential equations.

5. $y = \log_3 5 \rightarrow 3^y = 5$ The logarithm is the exponent.
The base does not change.

6. $-2 = \log_a 7 \rightarrow a^{-2} = 7$

7. $a = \log_b d \rightarrow b^a = d$

DO EXERCISES 6–9.

🎛 SOLVING LOGARITHMIC EQUATIONS

Certain equations involving logarithms can be solved by first converting to exponential equations.

Example 8   Solve: $\log_2 x = -3$.

$\log_2 x = -3$

$2^{-3} = x$ Converting to an exponential equation

$\frac{1}{8} = x$ Computing 2^{-3}

This checks, so the solution is $\frac{1}{8}$.

Example 9   Solve: $\log_x 16 = 2$.

$\log_x 16 = 2$

$x^2 = 16$ Converting to an exponential equation

$x = 4$ or $x = -4$ Solving

Check: $\log_4 16 = 2$ because $4^2 = 16$. Thus 4 is a solution.

Since all logarithm bases must be positive, $\log_{-4} 16$ is not defined. Therefore, -4 is not a solution.

DO EXERCISES 10 AND 11.

Solving an equation like $\log_b a = x$ amounts to finding the logarithm, base b, of the number a. You have done this before in graphing logarithmic functions. To think of finding logarithms as solving equations may help in some cases.

Example 10   Find $\log_{10} 1000$.

Let $\log_{10} 1000 = x$. Then,

$10^x = 1000$ Converting to an exponential equation

$x = 3$

Therefore, $\log_{10} 1000 = 3$.

Example 11   Find $\log_{10} 0.01$.

Let $\log_{10} 0.01 = x$. Then,

$10^x = 0.01$ or $10^x = \frac{1}{100}$, or $10^x = 10^{-2}$

$x = -2$

Therefore, $\log_{10} 0.01 = -2$.

Example 12   Find $\log_5 1$.

Let $\log_5 1 = x$. Then,

$5^x = 1$ Converting to an exponential equation

$x = 0$

Therefore, $\log_5 1 = 0$.

DO EXERCISES 12–14.

Solve.

10. $\log_{10} x = 4$

11. $\log_x 81 = 4$

Find the following.

12. $\log_{10} 10{,}000$

13. $\log_{10} 0.0001$

14. $\log_7 1$

NAME   CLASS   ANSWERS

EXERCISE SET 9.2

1. $y = \log_2 x$ **2.** $y = \log_6 x$

          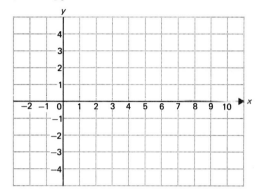

●● Convert to logarithmic equations.

3. $10^3 = 1000$ **4.** $10^2 = 100$ **5.** $5^{-3} = \dfrac{1}{125}$ **6.** $4^{-5} = \dfrac{1}{1024}$

7. $8^{1/3} = 2$ **8.** $16^{1/4} = 2$ **9.** $10^{0.3010} = 2$ **10.** $10^{0.4771} = 3$

●●● Convert to exponential equations.

11. $t = \log_3 8$ **12.** $h = \log_7 10$ **13.** $\log_5 25 = 2$ **14.** $\log_6 6 = 1$

15. $\log_{10} 0.1 = -1$ **16.** $\log_{10} 0.01 = -2$ **17.** $\log_{10} 7 = 0.845$ **18.** $\log_{10} 3 = 0.4771$

1. See graph.

2. See graph.

3. _____

4. _____

5. _____

6. _____

7. _____

8. _____

9. _____

10. _____

11. _____

12. _____

13. _____

14. _____

15. _____

16. _____

17. _____

18. _____

:: Solve.

19. $\log_3 x = 2$ **20.** $\log_4 x = 3$ **21.** $\log_x 16 = 2$ **22.** $\log_x 64 = 3$

23. $\log_2 x = -1$ **24.** $\log_3 x = -2$ **25.** $\log_8 x = \frac{1}{3}$ **26.** $\log_{32} x = \frac{1}{5}$

Find the following.

27. $\log_{10} 100$ **28.** $\log_{10} 100,000$ **29.** $\log_{10} 0.1$ **30.** $\log_{10} 0.001$

31. $\log_{10} 1$ **32.** $\log_{10} 10$ **33.** $\log_5 625$ **34.** $\log_2 64$

35. $\log_5 \frac{1}{25}$ **36.** $\log_2 \frac{1}{16}$ **37.** $\log_3 1$ **38.** $\log_4 4$

9.3   PROPERTIES OF LOGARITHMIC FUNCTIONS

◖•◗ LOGARITHMS OF PRODUCTS

Logarithmic functions are important in many applications and in later mathematics. We now establish some basic properties.

> *Property 1.*   **For any positive numbers M and N,**
>
> $$\log_a M \cdot N = \log_a M + \log_a N.$$
>
> **(The logarithm of a product is the sum of the logarithms of the factors. The number a can be any logarithm base.)**

Example 1   Express $\log_2 (4 \cdot 16)$ as a sum of logarithms.

$$\log_2 (4 \cdot 16) = \log_2 4 + \log_2 16 \qquad \text{By Property 1}$$

Example 2   Express as a single logarithm: $\log_{10} 0.01 + \log_{10} 1000$.

$$\log_{10} 0.01 + \log_{10} 1000 = \log_{10} (0.01 \times 1000) \qquad \text{By Property 1}$$
$$= \log_{10} 10$$

DO EXERCISES 1–4.

A PROOF OF PROPERTY 1 (Optional)

Let $\log_a M = x$ and $\log_a N = y$. Converting to exponential equations, we have $a^x = M$ and $a^y = N$.

Now we multiply the latter two equations, to obtain

$$M \cdot N = a^x \cdot a^y = a^{x+y}.$$

Converting back to a logarithmic equation, we get

$$\log_a M \cdot N = x + y.$$

Remembering what x and y represent, we get

$$\log_a M \cdot N = \log_a M + \log_a N,$$

which was to be shown.

◖•◗ LOGARITHMS OF POWERS

The second basic property is as follows.

> *Property 2.*   **For any positive number M and any real number p,**
>
> $$\log_a M^p = p \cdot \log_a M.$$
>
> **(The logarithm of a power of M is the exponent times the logarithm of M. The number a can be any logarithm base.)**

OBJECTIVES

After finishing Section 9.3, you should be able to:

◖•◗ Express the logarithm of a product as a sum of logarithms, and conversely.

◖••◗ Express a logarithm of a power as a product.

◖•••◗ Express the logarithm of a quotient as a difference of logarithms, and conversely.

◖∷◗ Convert from logarithms of products, quotients, and powers to expressions in terms of individual logarithms, and conversely.

Express as a sum of logarithms.

1. $\log_5 (25 \cdot 5)$

2. $\log_b PQ$

Express as a single logarithm.

3. $\log_3 7 + \log_3 5$

4. $\log_a C + \log_a A + \log_a B + \log_a I + \log_a N$

ANSWERS ON PAGE A–39

Express as a product.

5. $\log_7 4^5$

6. $\log_a \sqrt{5}$

7. Express as a difference of logarithms.

$$\log_b \frac{P}{Q}$$

8. Express as a single logarithm.

$$\log_2 x - \log_2 25$$

Examples Express as a product.

3. $\log_a 9^{-5} = -5\log_a 9$ By Property 2

4. $\log_a \sqrt[4]{5} = \log_a 5^{1/4}$ Writing exponential notation

$\qquad\qquad\; = \frac{1}{4}\log_a 5$ By Property 2

DO EXERCISES 5 AND 6.

A PROOF OF PROPERTY 2 (Optional)

Let $x = \log_a M$. Then convert to an exponential equation, to get $a^x = M$. Raising both sides to the pth power, we obtain

$$(a^x)^p = M^p, \quad \text{or} \quad a^{xp} = M^p.$$

Converting back to a logarithmic equation, we get

$$\log_a M^p = xp.$$

But $x = \log_a M$, so

$$\log_a M^p = (\log_a M)p, \quad \text{or} \quad p \cdot \log_a M,$$

which was to be shown.

▣▣▣ LOGARITHMS OF QUOTIENTS

Here is the third basic property.

> *Property 3.*   **For any positive numbers M and N,**
>
> $$\log_a \frac{M}{N} = \log_a M - \log_a N.$$
>
> **(The logarithm of a quotient is the logarithm of the dividend minus the logarithm of the divisor. The number a can be any logarithm base.)**

Example 5 Express as a difference of logarithms: $\log_t \dfrac{6}{U}$.

$$\log_t \frac{6}{U} = \log_t 6 - \log_t U \qquad \text{By Property 3}$$

Example 6 Express as a single logarithm: $\log_b 17 - \log_b 27$.

$$\log_b 17 - \log_b 27 = \log_b \frac{17}{27} \qquad \text{By Property 3}$$

Example 7 Express as a single logarithm: $\log_{10} 10{,}000 - \log_{10} 100$.

$$\log_{10} 10{,}000 - \log_{10} 100 = \log_{10} \frac{10{,}000}{100} = \log_{10} 100$$

DO EXERCISES 7 AND 8.

▐▌ USING THE PROPERTIES TOGETHER

Examples Express in terms of logarithms of x, y and z.

8. $\log_a \dfrac{x^2 y^3}{z^4} = \log_a (x^2 y^3) - \log_a z^4$ Using Property 3

$\qquad = \log_a x^2 + \log_a y^3 - \log_a z^4$ Using Property 1

$\qquad = 2 \log_a x + 3 \log_a y - 4 \log_a z$ Using Property 2

9. $\log_a \sqrt[4]{\dfrac{xy}{z^3}} = \log_a \left(\dfrac{xy}{z^3}\right)^{1/4}$ Writing exponential notation

$\qquad = \dfrac{1}{4} \cdot \log_a \dfrac{xy}{z^3}$ Using Property 2

$\qquad = \dfrac{1}{4}[\log_a xy - \log_a z^3]$ Using Property 3

$\qquad = \dfrac{1}{4}[\log_a x + \log_a y - 3 \log_a z]$ Using Properties 1 and 2

DO EXERCISE 9.

Examples Express as a single logarithm.

10. $\dfrac{1}{2} \log_a x - 7 \log_a y + \log_a z$

$\qquad = \log_a \sqrt{x} - \log_a y^7 + \log_a z$ Using Property 2

$\qquad = \log_a \dfrac{\sqrt{x}}{y^7} + \log_a z$ Using Property 3

$\qquad = \log_a \dfrac{z\sqrt{x}}{y^7}$ Using Property 1

11. $\log_a \dfrac{b}{\sqrt{x}} + \log_a \sqrt{bx}$

$\qquad = \log_a b - \log_a \sqrt{x} + \log_a \sqrt{bx}$ Using Property 3

$\qquad = \log_a b - \dfrac{1}{2} \log_a x + \dfrac{1}{2} \log_a (bx)$ Using Property 2

$\qquad = \log_a b - \dfrac{1}{2} \log_a x + \dfrac{1}{2}[\log_a b + \log_a x]$ Using Property 1

$\qquad = \log_a b - \dfrac{1}{2} \log_a x + \dfrac{1}{2} \log_a b + \dfrac{1}{2} \log_a x$

$\qquad = \dfrac{3}{2} \log_a b$ Collecting like terms

$\qquad = \log_a b^{3/2}$ Using Property 2

DO EXERCISE 10.

9. Express in terms of logarithms of x, y, and z.

$\log_a \sqrt{\dfrac{z^3}{xy}}$

10. Express as a single logarithm.

$5 \log_a x - \log_a y + \dfrac{1}{4} \log_a z$

Example 11 could also be done as follows:

$$\log \frac{b}{\sqrt{x}} + \log_a \sqrt{bx}$$

$$= \log_a \frac{b}{\sqrt{x}} \sqrt{bx} \qquad \text{Using Property 1}$$

$$= \log_a b \sqrt{b}$$

$$= \log_a b^{3/2}$$

SOMETHING EXTRA

CALCULATOR CORNER: NUMBER PATTERNS

Do the following calculations. Look for patterns. Then figure out what the next lines in the patterns should be. After that, check your answers by doing the actual calculations.

EXERCISES

1. Sums of odd numbers.

1	$1^2 = $ _____
$1 + 3 = $ _____	$2^2 = $ _____
$1 + 3 + 5 = $ _____	$3^2 = $ _____
$1 + 3 + 5 + 7 = $ _____	$4^2 = $ _____
$1 + 3 + 5 + 7 + 9 = $ _____	$5^2 = $ _____
$1 + 3 + 5 + 7 + 9 + 11 = $ _____	$6^2 = $ _____

2. The sum of the first n natural numbers.

1	$\dfrac{1 \cdot 2}{2} = $ _____
$1 + 2 = $ _____	$\dfrac{2 \cdot 3}{2} = $ _____
$1 + 2 + 3 = $ _____	$\dfrac{3 \cdot 4}{2} = $ _____
$1 + 2 + 3 + 4 = $ _____	$\dfrac{4 \cdot 5}{2} = $ _____
$1 + 2 + 3 + 4 + 5 = $ _____	$\dfrac{5 \cdot 6}{2} = $ _____

NAME

CLASS

EXERCISE SET 9.3

⬤ Express as a sum of logarithms.

1. $\log_2 (32 \cdot 8)$ **2.** $\log_3 (27 \cdot 81)$ **3.** $\log_4 (64 \cdot 16)$

4. $\log_5 (25 \cdot 125)$ **5.** $\log_c Bx$ **6.** $\log_t 5Y$

Express as a single logarithm.

7. $\log_a 6 + \log_a 70$ **8.** $\log_b 65 + \log_b 2$

9. $\log_c K + \log_c y$ **10.** $\log_t H + \log_t M$

⬤⬤ Express as a product.

11. $\log_a x^3$ **12.** $\log_b t^5$

⬤⬤⬤ Express as a difference of logarithms.

13. $\log_a \dfrac{67}{5}$ **14.** $\log_t \dfrac{T}{7}$

Express as a single logarithm.

15. $\log_a 15 - \log_a 7$ **16.** $\log_b 42 - \log_b 7$

1. _____

2. _____

3. _____

4. _____

5. _____

6. _____

7. _____

8. _____

9. _____

10. _____

11. _____

12. _____

13. _____

14. _____

15. _____

16. _____

⠶ Express in terms of logarithms of x, y, and z.

17. $\log_a x^2 y^3 z$

18. $\log_a 5xy^4 z^3$

17. _____

18. _____

19. $\log_b \dfrac{xy^2}{z^3}$

20. $\log_c \sqrt[3]{\dfrac{x^4}{y^3 z^2}}$

19. _____

20. _____

Express as a single logarithm and simplify if possible.

21. $\dfrac{2}{3}\log_a x - \dfrac{1}{2}\log_a y$

22. $\dfrac{1}{2}\log_a x + 3\log_a y - 2\log_a x$

21. _____

22. _____

23. $\log_a 2x + 3(\log_a x - \log_a y)$

24. $\log_a x^2 - 2\log_a \sqrt{x}$

23. _____

24. _____

25. $\log_a (x^2 - 4) - \log_a (x - 2)$

26. $\log_a \dfrac{a}{\sqrt{x}} - \log_a \sqrt{ax}$

25. _____

26. _____

9.4   COMMON LOGARITHMS

◖●◗ LOGARITHMS AND COMPUTATION

Base 10 logarithms are known as *common logarithms*. Table 2 at the back of the book contains common logarithms of numbers from 1 to 10, to four decimal places.

Before calculators and computers became so available, common logarithms were used extensively in calculating. Today, computations with logarithms are mainly of historical interest, but logarithm *functions* are of great importance.

The study of logarithms in computation does help you become more familiar with the important properties of logarithm functions.

The following is a short table of powers of 10, or *logarithms* base 10, accurate to four decimal places.

$$
\begin{array}{llll}
1 = 10^{0.0000}, & \text{or} & \log_{10} 1 & = 0.0000 \\
2 \approx 10^{0.3010}, & \text{or} & \log_{10} 2 & \approx 0.3010 \\
3 \approx 10^{0.4771}, & \text{or} & \log_{10} 3 & \approx 0.4771 \\
4 \approx 10^{0.6021}, & \text{or} & \log_{10} 4 & \approx 0.6021 \\
5 \approx 10^{0.6990}, & \text{or} & \log_{10} 5 & \approx 0.6990 \\
6 \approx 10^{0.7782}, & \text{or} & \log_{10} 6 & \approx 0.7782 \\
7 \approx 10^{0.8451}, & \text{or} & \log_{10} 7 & \approx 0.8451 \\
8 \approx 10^{0.9031}, & \text{or} & \log_{10} 8 & \approx 0.9031 \\
9 \approx 10^{0.9542}, & \text{or} & \log_{10} 9 & \approx 0.9542 \\
10 = 10^{1.0000}, & \text{or} & \log_{10} 10 & = 1.0000 \\
11 \approx 10^{1.0414}, & \text{or} & \log_{10} 11 & \approx 1.0414 \\
12 \approx 10^{1.0792}, & \text{or} & \log_{10} 12 & \approx 1.0792 \\
13 \approx 10^{1.1139}, & \text{or} & \log_{10} 13 & \approx 1.1139 \\
14 \approx 10^{1.1461}, & \text{or} & \log_{10} 14 & \approx 1.1461 \\
15 \approx 10^{1.1761}, & \text{or} & \log_{10} 15 & \approx 1.1761 \\
16 \approx 10^{1.2041}, & \text{or} & \log_{10} 16 & \approx 1.2041 \\
\end{array}
$$

To illustrate how logarithms can be used for computation we will use the above table and do some easy calculations. The calculations are very simple, and we are doing them only to illustrate how logarithms work. As you read and work examples, remember that the logarithms in the table are only approximate because they have been rounded. Therefore, our answers will not always be exact.

Example 1   Find 3×4 using the table of exponents.

$3 \times 4 \approx 10^{0.4771} \times 10^{0.6021}$ From the table

$ \approx 10^{1.0792}$ Adding exponents

From the table we see that $10^{1.0792} \approx 12$, so $3 \times 4 \approx 12$.

DO EXERCISE 1.

OBJECTIVES

After finishing Section 9.4, you should be able to:

◖●◗ Use the table on this page to do simple illustrative logarithm calculations.

◖●●◗ Find logarithms, using Table 2 and scientific notation.

◖●●●◗ Find antilogarithms, using Table 2 and scientific notation.

1. Find 2×8 using the table on this page.

ANSWER ON PAGE A–40

Using base 10 logarithms and the table on the preceding page, find the following.

2. 4×2

3. $\dfrac{15}{3}$

4. $\sqrt[3]{8}$

5. 3^2

Note from Example 1 that we find a product by *adding* the logarithms of the factors. Then we find the number having that result as its logarithm. This is called finding the *antilogarithm*. Thus reading the table on p. 552 from right to left gives us antilogarithms. From that table we see that the antilogarithm of a number x is 10^x.

Example 2   Find $\dfrac{14}{2}$ using base 10 logarithms.

$$\log_{10} \frac{14}{2} = \log_{10} 14 - \log_{10} 2 \qquad \text{Using Property 3}$$

$$\approx 1.1461 - 0.3010 \qquad \text{Finding the logs from the table}$$

$$\log_{10} \frac{14}{2} \approx 0.8451 \qquad \text{Subtracting}$$

$$\frac{14}{2} \approx \text{antilog}_{10}\, 0.8451 \approx 7 \qquad \text{Using the table in reverse}$$

Example 3   Find $\sqrt[4]{16}$ using base 10 logarithms.

$$\log_{10} \sqrt[4]{16} = \log_{10} 16^{1/4}$$

$$= \frac{1}{4} \cdot \log_{10} 16 \qquad \text{Using Property 2}$$

$$\approx \frac{1}{4} \cdot 1.2041$$

$$\log_{10} \sqrt[4]{16} \approx 0.3010 \qquad \text{Rounding to four decimal places}$$
$$\sqrt[4]{16} \approx \text{antilog}_{10}\, 0.3010 \approx 2$$

Example 4   Find 2^3 using base 10 logarithms.

$$\log_{10} 2^3 = 3 \cdot \log_{10} 2 \qquad \text{Using Property 2}$$
$$\approx 3 \cdot 0.3010 \approx 0.9030$$
$$\log_{10} 2^3 \approx 0.9030$$
$$2^3 \approx \text{antilog}_{10}\, 0.9030 \approx 8$$

DO EXERCISES 2–5.

●● USING A LOGARITHM TABLE

We will often omit the base, 10, when working with common logarithms. That is,

$\log M$ will be agreed to mean $\log_{10} M$.

Table 2 contains logarithms of numbers from 1 to 10. Part of that table is as follows.

x	0	1	2	3	4	5	6	7	8	9
5.0	0.6990	0.6998	0.7007	0.7016	0.7024	0.7033	0.7042	0.7050	0.7059	0.7067
5.1	0.7076	0.7084	0.7093	0.7101	0.7110	0.7118	0.7126	0.7135	0.7143	0.7152
5.2	0.7160	0.7168	0.7177	0.7185	0.7193	0.7202	0.7210	0.7218	0.7226	0.7235
5.3	0.7243	0.7251	0.7259	0.7267	0.7275	0.7284	0.7292	0.7300	0.7308	0.7316
5.4	0.7324	0.7332	0.7340	0.7348	0.7356	0.7364	0.7372	0.7380	0.7388	0.7396

To illustrate the use of the table, let us find log 5.24. We locate the row headed 5.2, then move across to the column headed 4. We find log 5.24 (the colored entry in the table): log 5.24 = 0.7193.

We can find antilogarithms by reversing this process. For example, antilog 0.7193 = $10^{0.7193} \approx 5.24$. Similarly, antilog 0.7292 $\approx$ 5.36.

DO EXERCISES 6–11.

Using Table 2 and scientific notation we can find logarithms of numbers that are not between 1 and 10. First recall that logarithms are exponents, so

$$\log_a a^k = k \quad \text{for any number } k.$$

Thus,

$$\log_{10} 10^k = k \quad \text{for any number } k.$$

Thus we know that

log 10 = 1   because   10 = 10^1,
log 100 = 2   because   100 = 10^2,
log 1000 = 3   because   1000 = 10^3,
log 0.001 = −3   because   0.001 = 10^{-3}.

Examples

5. log 52.4 = log(5.24×10^1)   Writing scientific notation for 52.4
       = log 5.24 + log 10^1 Using Property 1
       $\approx$ 0.7193 + 1   Using log table

6. log 52,400 = log(5.24×10^4)
        = log 5.24 + log 10^4
        $\approx$ 0.7193 + 4

7. log 0.00524 = log(5.24×10^{-3})
        = log 5.24 + log 10^{-3}
        $\approx$ 0.7193 + (−3)

DO EXERCISES 12 AND 13.

In all of the examples above, 0.7193 is the fractional part of the logarithm. It is called a *mantissa*. The integer part varies and is known as the *characteristic*. Table 2 contains only mantissas. Characteristics must be supplied. The characteristic is the exponent used in writing scientific notation. Note that characteristics may be positive, negative, or zero, but mantissas are never negative.

log 5240 $\approx$ 0.7193 + 3

Mantissa: a   *Characteristic:*
number from 0 to 1   an integer

Use Table 2 to find each logarithm.

6. log 3.14

7. log 9.99

8. log 4.00

Use Table 2 to find each antilogarithm.

9. antilog 0.7589

10. antilog 0.0000

11. antilog 0.5587

Use scientific notation and Table 2 to find each logarithm.

12. log 289

13. log 0.000289

ANSWERS ON PAGE A–40

Find the following using Table 2.

14. log 67800

15. log 609,000,000

16. log 0.0782

17. log 0.000111

The preceding examples illustrate the importance of using the base 10. It allows great economy in the printing of tables. If we know the logarithms of numbers from 1 to 10 we can find the logarithm of any number. For any base other than 10 that would not be the case.

NEGATIVE CHARACTERISTICS

To find the logarithm of a number, we of course add the characteristic and the mantissa. In Example 6, we found that

$$\log 52{,}400 \approx 0.7193 + 4.$$

If we add the characteristic and mantissa, we get

4.7193.

In this case, we can still *see* the characteristic and mantissa. In Example 7, we found that

$$\log 0.00524 \approx 0.7193 + (-3).$$

Again, we can add the characteristic and mantissa to obtain standard notation for the logarithm.

$$\begin{array}{r} -3.0000 \\ 0.7193 \\ \hline -2.2807 \end{array}$$

Caution! When we add, we do *not* get −3.7193.

Thus log 0.00524 ≈ −2.2807.

In this case the mantissa is no longer visible after we do the addition. Neither is the characteristic.

In some applications it does not matter that the characteristic and the mantissa are not preserved. In some cases it does.

If the characteristic and mantissa of a logarithm are to be preserved when the characteristic is negative, they must not be added.

DO EXERCISES 14–17.

●●● ANTILOGARITHMS

To find antilogarithms we reverse the procedure for finding logarithms. Recall that antilog $x = 10^x$ for any number x.

Example 8   Find antilog 2.6085.

$$\begin{aligned} \text{antilog } 2.6085 &= 10^{2.6085} = 10^{(2 + 0.6085)} \\ &= 10^2 \cdot 10^{0.6085} \end{aligned}$$

Now $10^{0.6085} = $ antilog 0.6085 and we can find it from the table. It is 4.06. Thus we have

$$\text{antilog } 2.6085 = 10^2 \times 4.06, \quad \text{or} \quad 406.$$

In Example 8 we, in effect, separated the number 2.6085 into two parts, the characteristic (an integer) and a number between 0 and 1 (the mantissa). We found the antilogarithm of the mantissa from the table and the antilogarithm of the characteristic is 10^2. We then had scientific notation for our answer.

Example 9   Find antilog 3.7118.

From the table we find antilog $0.7118 \approx 5.15$. Thus

$$\text{antilog } 3.7118 \approx 5.15 \times 10^3 \qquad \text{Note that 3 is the characteristic.}$$
$$\approx 5150.$$

Example 10   Find antilog $0.7143 + (-3)$.

The characteristic is -3 and the mantissa is 0.7143. From the table we find that antilog $0.7143 \approx 5.18$. Thus

$$\text{antilog } [0.7143 + (-3)] \approx 5.18 \times 10^{-3}$$
$$\approx 0.00518.$$

DO EXERCISES 18–21.

LOGARITHMS ON A CALCULATOR (Optional) 🖩

Tables of common logarithms contain only mantissas. The reason for this is to save space. When we find a logarithm using a table, we must supply the characteristic ourselves. Scientific calculators that have $\boxed{\text{log}}$ keys contain logarithms, *including* characteristic and mantissa. On a calculator it is not usually important to keep the characteristic and mantissa separate.

Example 11   Use a calculator to find log 475,000.

We enter 475,000 and then press the $\boxed{\text{log}}$ key.

We find that log $475,000 \approx 5.6767$.

The characteristic is 5 and the mantissa is 0.6767.

Example 12   Use a calculator to find log 0.00008952.

We enter 0.00008952 and then press the $\boxed{\text{log}}$ key.

We find that log $0.00008952 \approx -4.0481$.

We do not know the characteristic or the mantissa. If we should need them we can add 5 and then add -5.

Using the calculator we find that $5 + (-4.0481) \approx 0.9519$. Now we add -5 and we have

$$\log 0.00008952 \approx \underset{\underset{\text{mantissa}}{\uparrow}}{0.9519} + \underset{\underset{\text{characteristic}}{\uparrow}}{(-5)}.$$

DO EXERCISES 22 AND 23.

Find the following using Table 2. Try to write only the answer.

18. antilog 4.8069

19. $10^{4.8069}$

20. antilog $0.6284 + (-4)$

21. $10^{[0.2014 + (-2)]}$

🖩 Find the following. Determine the characteristic and mantissa.

22. log 35,720

23. log 0.000008182

Find the following.

24. antilog 4.9285

Calculators do not usually have an ⟨antilog⟩ key. To find an anti-logarithm, you must remember that

$$\text{antilog}_{10} X = 10^X.$$

To find the antilogarithm of a number, remember that the logarithm is an *exponent* and raise 10 to that power.

Example 13   Find antilog 3.2145.

We press the proper keys to find $10^{3.2145}$.
(The key strokes are different on different calculators.)

We find that antilog $3.2145 \approx 1639$.

Example 14   Find antilog $[0.6646 + (-5)]$.

We first use the calculator to do the addition $0.6646 + (-5)$.

We get -4.3354.

Next, we raise 10 to this power (we find $10^{-4.3354}$).

Antilog $(-4.3354) \approx 0.00004620$.

DO EXERCISES 24 AND 25.

25. antilog $[0.4487 + (-6)]$

SOMETHING EXTRA

AN APPLICATION: FORGETTING

Psychologists have developed a formula, as follows. A group of people take a test and make an average score of S. After a time t they take a retest and the average score is $S(t)$. Then

$$S(t) = A - B \log (t + 1),$$

where t is in months and the constants A and B are determined by experiment in various situations.

Students in a zoology class took a final exam, and a retest determined that the constants were as follows:

$$S(t) = 78 - 15 \log (t + 1).$$

The average score when they first took the test (time zero) was

$$S(0) = 78 - 15 \log (0 + 1) = 78.$$

After 4 months ($t = 4$), the average score was

$$
\begin{aligned}
S(4) &= 78 - 15 \log (4 + 1) \\
&= 78 - 15 \log 5 = 78 - 15 \times 0.6990 \\
&= 78 - 10.49 = 67.51.
\end{aligned}
$$

EXERCISES

In a business math class, students were tested at the end of the course and then again after 6 months. The forgetting formula was determined to be

$$S(t) = 62 - 18 \log (t + 1).$$

1. What was the average score on the final exam?

2. What was the average score after 6 months?

NAME   CLASS   ANSWERS

EXERCISE SET 9.4

▮• Find each of the following using the table of exponents. Show your work.

1. 7×2 **2.** 3×5 **3.** $\dfrac{12}{3}$ **4.** $\dfrac{16}{8}$

▮•▮ Find these logarithms using Table 2.

5. log 2.46   **6.** log 7.65   **7.** log 5.31   **8.** log 8.57

9. log 1.07   **10.** log 4.60   **11.** log 347   **12.** log 8720

13. log 52.5   **14.** log 20.6   **15.** log 3870   **16.** log 62,400

17. log 0.64   **18.** log 0.00216   **19.** log 0.0000404   **20.** log 0.00347

1. _____

2. _____

3. _____

4. _____

5. _____

6. _____

7. _____

8. _____

9. _____

10. _____

11. _____

12. _____

13. _____

14. _____

15. _____

16. _____

17. _____

18. _____

19. _____

20. _____

●●● Find these antilogarithms using Table 2.

21. antilog 3.3674   **22.** antilog 4.9222   **23.** antilog 1.2553   **24.** antilog 2.6294

21. _____

22. _____

23. _____

24. _____

25. antilog 9.7875 − 10   **26.** antilog 8.9881 − 10   **27.** antilog 7.5391 − 10

25. _____

26. _____

27. _____

28. antilog 7.7774 − 10   **29.** $10^{1.4014}$ **30.** $10^{2.5391}$

28. _____

29. _____

30. _____

31. $10^{7.9881-10}$ **32.** $10^{8.5391-10}$ **33.** $10^{6.7875-10}$ **34.** $10^{4.6294-10}$

31. _____

32. _____

33. _____

34. _____

9.5 CALCULATIONS WITH LOGARITHMS (Optional)

● Before calculators and computers were readily available, logarithms were a very important tool for doing computations. Today, they are not important for that purpose. Their importance is in working certain kinds of problems and in mathematical theory. Therefore, this section is mainly of historical interest. Doing calculations with logarithms can also serve to fix the ideas and properties of logarithms in one's mind.

In calculating with logarithms, it is necessary to preserve both mantissas and characteristics. When negative characteristics occur, we write the logarithm as a difference, in which 10 or a multiple of 10 is subtracted.

Example 1   Write $0.7196 + (-3)$ as a positive number minus a multiple of 10.

We add 7 and subtract 7, and we get

$7.7196 - 10$.   Note that the mantissa is still visible.

Suppose we wish to divide a logarithm by a number, say 5. We then use 50 for our multiple of 10, as in Example 2.

Example 2   Write $0.1425 + (-6)$ as a positive number minus 50.

We add 44 and subtract 44, and we get

$44.1425 - 50$.

We could now divide the number in Example 2 by 5 and still have a multiple of 10 subtracted.

Example 3   Divide $44.1425 - 50$ by 5. Find the characteristic and the mantissa of the result.

```
      8.8285   - 10
  5│44.1425   - 50
    40
    ‾‾
     4 1
     4 0
     ‾‾
      14
      10
      ‾‾
      42
      40
      ‾‾
      25          Mantissa 0.8285; characteristic (−2).
```

DO EXERCISES 1–3.

The kinds of calculations in which logarithms can be used are multiplication, division, raising to powers, and taking roots.

Example 4   Find: $\dfrac{0.0578 \times 32.7}{8460}$.

We write a *plan* for the use of logarithms, and then look up all of the mantissas at one time.

Let

$$N = \frac{0.0578 \times 32.7}{8460}.$$

OBJECTIVE

After finishing Section 9.5, you should be able to:

● **Do certain kinds of calculations using logarithms.**

1. Write $0.1589 + (-6)$ as a positive number minus 10.

2. Write $0.1589 + (-6)$ as a positive number minus 30.

3. Look at your answer to Exercise 2. Multiply that answer by $\frac{1}{3}$.

Use logarithms to calculate.

4. $\dfrac{78.6 \times 0.00642}{9.45}$

5. $\sqrt[4]{0.325} \times (4.23)^2$

Then, using Properties 1 and 3, we get

$$\log N = \log 0.0578 + \log 32.7 - \log 8460.$$

This gives us the plan. We use a straight line to indicate addition and a wavy line to indicate subtraction.

Completion of plan:

$$
\begin{aligned}
\log 0.0578 &\approx 8.7619 - 10 \\
\log 32.7 &\approx \underline{1.5145} \\
\log \text{numerator} &\approx 10.2764 - 10 \qquad \text{Sum of the logs above} \\
\log 8460 &\approx 3.9274 \\
\log N &\approx 6.3490 - 10 \qquad \text{Subtracting logs} \\
N &\approx 0.0002234 \qquad \text{Taking the antilog}
\end{aligned}
$$

This answer was found from the table. Sometimes we had to estimate the fourth digit.

Example 5 Use logarithms to find $(92.8)^3 \times \sqrt[5]{0.986}$.

Let $N = (92.8)^3 \times \sqrt[5]{0.986}$.

Then

$$
\begin{aligned}
\log N &= \log (92.8)^3 + \log \sqrt[5]{0.986} \qquad \text{Using Property 1} \\
&= 3 \log 92.8 + \frac{1}{5} \log (0.986). \qquad \text{Using Property 2}
\end{aligned}
$$

Powers and roots are involved. In such cases it is easier to first find the following logs:

$$\log 92.8 \approx 1.9675,$$

$$\log 0.986 \approx 9.9939 - 10.$$

Consider $9.9939 - 10$. We are going to multiply this number by $\frac{1}{5}$, so we rename this number as follows:

$$\log 0.986 \approx 49.9939 - 50.$$

Completion of plan and work:

$$
\begin{aligned}
\log 92.8 &\approx 1.9675 \\
3 \log 92.8 &\approx 5.9025 \qquad \text{Multiplying by 3} \\
\log 0.986 &\approx 49.9939 - 50 \\
\\
\frac{1}{5} \log 0.986 &\approx 9.9988 - 10 \qquad \text{Multiplying by } \frac{1}{5} \\
\log N &\approx 15.9013 - 10 \qquad \text{Adding logs} \\
N &\approx 796,700 \qquad \text{Taking the antilog (We estimated the fourth digit.)}
\end{aligned}
$$

DO EXERCISES 4 AND 5.

NAME   CLASS   ANSWERS

EXERCISE SET 9.5

▪ Use logarithms to calculate.

1. 3.14×60.4 **2.** 541×0.0152

1. _____

2. _____

3. $286 \div 1.05$ **4.** $12.8 \div 81.6$

3. _____

4. _____

5. $\sqrt{76.9}$ **6.** $\sqrt{678}$

5. _____

6. _____

ANSWERS

7. $\sqrt[3]{56.9}$

8. $\sqrt[4]{2600}$

7. _____

8. _____

9. $(1.36)^{4.2}$

10. $(0.727)^{3.6}$

9. _____

10. _____

11. $\sqrt[3]{\dfrac{3.24 \times (3.16)^2}{78.4 \times 24.6}}$

12. $\dfrac{70.7 \times (10.6)^2}{18.6 \times \sqrt{276}}$

11. _____

12. _____

9.6 INTERPOLATION

● LOGARITHMS

In Section 9.4 we used Table 2 when three digits occur, as in 4.58 or 0.0458. We can also use the table when four digits occur, by a process called *interpolation*. The simplest kind of interpolation is called *linear interpolation*. What we say here applies to tables of many different functions.

Let us consider how a table of values for any function is made. We select numbers of the domain x_1, x_2, x_3, and so on. Then we compute or somehow determine the corresponding function values (outputs), $f(x_1)$, $f(x_2)$, $f(x_3)$, and so on. Then we tabulate the results. We might also graph the results.

x	x_1	x_2	x_3	x_4	$\cdots$
$f(x)$	$f(x_1)$	$f(x_2)$	$f(x_3)$	$f(x_4)$	$\cdots$

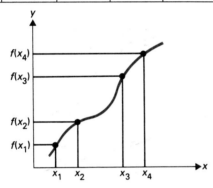

Suppose we want to find an output $f(x)$ for an x not in the table. If x is halfway between x_1 and x_2, then we can take the number halfway between $f(x_1)$ and $f(x_2)$ as an approximation to $f(x)$. If x is one-fifth of the way between x_2 and x_3, we can take the number that is one-fifth of the way between $f(x_2)$ and $f(x_3)$ as an approximation to $f(x)$. What we do is divide the length from x_2 to x_3 in a certain ratio, and then divide the length from $f(x_2)$ to $f(x_3)$ in the same ratio. This is *linear interpolation*.

We can show this geometrically. The length from x_1 to x_2 is divided in a certain ratio by x. The length from $f(x_1)$ to $f(x_2)$ is divided in the same ratio by y. The number y approximates $f(x)$ with the noted error.

Note the slanted line in the figure. The approximation y comes from this line. This explains the use of the term *linear interpolation*.

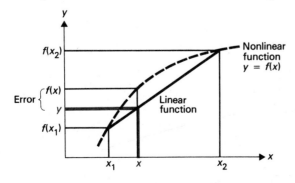

1. Find log 4562.

Example 1   Find: log 34870.

a) Find the characteristic. Since $34870 = 3.487 \times 10^4$, the characteristic is 4.

b) Find the mantissa. From Table 2 we have:

The tabular difference (the difference between consecutive values in the table) is 0.0012. Now 3.487 is $\frac{7}{10}$ of the way from 3.480 to 3.490. So we take 0.7 of 0.0012, which is 0.00084, and round it to 0.0008. We add this to 0.5416. The mantissa is 0.5424.

c) Add the characteristic and mantissa:

  log 34870 ≈ 4.5424.

With practice you will take 0.7 of 12, forgetting the zeros, but adding in the same way.

DO EXERCISE 1.

2. Find log 0.02387.

Example 2   Find: log 0.009543.

a) Find the characteristic. Since $0.009543 = 9.543 \times 10^{-3}$, the characteristic is −3, or 7 − 10.

b) Find the mantissa. From Table 2 we have:

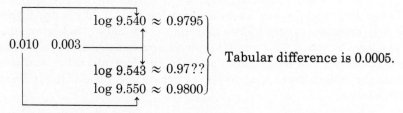

Now 9.543 is $\frac{3}{10}$ of the way from 9.540 to 9.550, so we take 0.3 of 0.0005, which is 0.00015, and round it to 0.0002. We add this to 0.9795. The mantissa that results is 0.9797.

c) Add the characteristic and the mantissa:

  log 0.009543 ≈ 7.9797 − 10.

DO EXERCISE 2.

⚫⚫⚫ ANTILOGARITHMS

We interpolate when finding antilogarithms, using the table in reverse.

Example 3   Find: antilog 4.9164.

a) The characteristic is 4. The mantissa is 0.9164.

b) Find the antilog of the mantissa, 0.9164. From Table 2 we have:

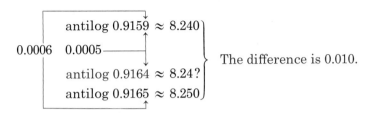

The difference between 0.9159 and 0.9165 is 0.0006. Thus 0.9164 is $\frac{0.0005}{0.0006}$, or $\frac{5}{6}$, of the way between 0.9159 and 0.9165. Then antilog 0.9164 is $\frac{5}{6}$ of the way between 8.240 and 8.250, so we take $\frac{5}{6}(0.010)$, which is 0.00833..., and round it to 0.008. Thus the antilog of the mantissa is 8.248.

So antilog $4.9164 \approx 8.248 \times 10^4 = 82{,}480$.

DO EXERCISE 3.

Example 4   Find: antilog $(7.4122 - 10)$.

a) The characteristic is -3. The mantissa is 0.4122.

b) Find the antilog of the mantissa, 0.4122. From Table 2 we have:

The difference between 0.4116 and 0.4133 is 0.0017. Thus 0.4122 is $\frac{0.0006}{0.0017}$, or $\frac{6}{17}$, of the way between 0.4116 and 0.4133. Then antilog 0.4122 is $\frac{6}{17}$ of the way between 2.580 and 2.590, so we take $\frac{6}{17}(0.010)$, which is 0.0035 to four places. We round it to 0.004. Thus the antilog of the mantissa is 2.584.

So antilog $(7.4122 - 10) \approx 2.584 \times 10^{-3} = 0.002584$.

DO EXERCISE 4.

3. Find antilog 3.7749.

4. Find antilog $(6.4557 - 10)$.

SOMETHING EXTRA
AN APPLICATION: WALKING SPEED

It has been found that the speed at which people walk in a city is a function of the size of the city. Here is a function, experimentally determined:

$V = 1.98 \log P + 0.05.$

V is the velocity in ft/sec, and P is the population of the city *in thousands*.

Example   The population of Lincoln, Nebraska, is 175,000. The average walking speed there should be

$$V = 1.98 \log 175 + 0.05$$
$$\approx 1.98 \times 2.2430 + 0.05$$
$$\approx 4.49 \text{ ft/sec.}$$

EXERCISES

Find walking speeds in the following cities.

1. Pierre, South Dakota: population 10,300

2. Key West, Florida: population 29,600

3. Rome, New York: population 50,400

4. Lansing, Michigan: population 134,000

5. Albuquerque, New Mexico: population 290,000

6. Seattle, Washington: population 490,000

NAME _____

CLASS _____

EXERCISE SET 9.6

■ ● Find the following logarithms using interpolation.

1. log 41.63

2. log 472.1

3. log 2.944

4. log 21.76

5. log 650.2

6. log 37.37

7. log 0.1425

8. log 0.09045

9. log 0.004257

10. log 4518

11. log 0.1776

12. log 0.08356

13. log 600.6

14. log 800.1

1. _____

2. _____

3. _____

4. _____

5. _____

6. _____

7. _____

8. _____

9. _____

10. _____

11. _____

12. _____

13. _____

14. _____

● ● Find the following antilogarithms using interpolation.

15. antilog 1.6350   **16.** antilog 2.3512   **17.** antilog 0.6478

15. _____

16. _____

17. _____

18. antilog 1.1624   **19.** antilog 0.0342   **20.** antilog 4.8453

18. _____

19. _____

20. _____

21. antilog 9.8564 − 10   **22.** antilog 8.9659 − 10   **23.** antilog 7.4128 − 10

21. _____

22. _____

23. _____

24. antilog 9.7278 − 10   **25.** antilog 8.2010 − 10   **26.** antilog 7.8630 − 10

24. _____

25. _____

26. _____

27. _____

Use Table 2 and interpolation to find the following.

28. _____

27. $\sqrt{23.37}$ **28.** $\sqrt[3]{12.93}$

9.7 EXPONENTIAL AND LOGARITHMIC EQUATIONS

◼◦ EXPONENTIAL EQUATIONS

Equations with variables in exponents, such as $3^{2x-1} = 4$, are called *exponential equations*. We solve such equations by taking the logarithm on both sides and then using Property 2.

Example 1   Solve: $2^{3x-5} = 16$.

$$\log 2^{3x-5} = \log 16 \qquad \text{Taking log on both sides}$$

$$(3x - 5)\log 2 = \log 16 \qquad \text{Using Property 2}$$

$$3x - 5 = \frac{\log 16}{\log 2}$$

$$x = \frac{\dfrac{\log 16}{\log 2} + 5}{3} \qquad \text{Solving for } x$$

$$x \approx \frac{\dfrac{1.2041}{0.3010} + 5}{3} \qquad \text{Finding logs}$$

$$x \approx 3.0001 \qquad \text{Calculating}$$

The answer is approximate because the logarithms are approximate.

Example 2   Solve: $2^{3x} \cdot 2^{-5} = 16$.

$$2^{3x-5} = 16 \qquad \text{Multiplying (adding exponents) on the left}$$

$$\log 2^{3x-5} = \log 16 \qquad \text{Taking log on both sides}$$

$$(3x - 5)\log 2 = \log 16 \qquad \text{Using Property 2}$$

$$3x - 5 = \frac{\log 16}{\log 2}$$

$$x = \frac{\dfrac{\log 16}{\log 2} + 5}{3} \approx \frac{\dfrac{1.2041}{0.3010} + 5}{3} \qquad \begin{array}{l}\text{Solving for } x \text{ and} \\ \text{finding logarithms}\end{array}$$

$$x \approx 3.0001 \qquad \text{Calculating}$$

DO EXERCISE 1.

◼◦◦ LOGARITHMIC EQUATIONS

Equations containing logarithmic expressions are called *logarithmic equations*. To solve them, we obtain a single logarithmic expression on one side and then take the antilogarithm.

Example 3   Solve: $\log x + \log (x - 3) = 1$.

$$\log x(x - 3) = 1 \qquad \begin{array}{l}\text{Using Property 1 to obtain} \\ \text{a single logarithm}\end{array}$$

$$x(x - 3) = 10^1 \qquad \text{Taking the antilog on both sides}$$

$$x^2 - 3x - 10 = 0$$

$$(x + 2)(x - 5) = 0 \qquad \text{Factoring}$$

$$x = -2 \quad \text{or} \quad x = 5 \qquad \text{Principle of zero products}$$

OBJECTIVES

After finishing Section 9.7, you should be able to:

◼◦ Solve exponential equations.

◼◦◦ Solve logarithmic equations.

◼◦◦◦ Solve certain applied problems involving exponential and logarithmic equations.

1. Solve $4^{2x-3} = 64$.

ANSWER ON PAGE A–40

2. Solve $\log x + \log (x + 3) = 1$.

3. Suppose \$5000 principal is invested at 8% interest compounded and yields \$8569. How many years was it invested?

4. The earthquake in Anchorage, Alaska, on March 27, 1964, had an intensity $10^{8.4}$ times I_0. What was its magnitude on the Richter scale?

Check:

$$\frac{\log x + \log (x - 3) = 1}{\log (-2) + \log (-2 - 3) \mid 1}$$

$$\frac{\log x + \log (x - 3) = 1}{\log 5 + \log (5 - 3) \mid 1}$$
$$\log 5 + \log 2$$
$$\log 10$$
$$1$$

The number -2 is not a solution because negative numbers do not have logarithms.

The solution is 5.

DO EXERCISE 2.

●●● APPLICATIONS

There are many applications of exponential and logarithmic equations and functions.

Example 4 (*Compound interest*). The amount A that principal P will be worth after t years at interest rate r, compounded annually, is given by the formula

$$A = P(1 + r)^t.$$

Suppose \$4000 principal is invested at 12% interest compounded annually and yields \$10,706. How many years was it invested?

Using the formula $A = P(1 + r)^t$, we have

$$10,706 = 4000(1 + 0.12)^t \text{ or } 10,706 = 4000(1.12)^t.$$

We solve for t.

$$\log 10,706 = \log 4000(1.12)^t = \log 4000 + t \log 1.12$$

$$\frac{\log 10,706 - \log 4000}{\log 1.12} = t$$

$$\frac{4.0296 - 3.6021}{0.0492} \approx t$$

$$8.69 \approx t.$$

The money was invested for about $8\frac{2}{3}$ years.

DO EXERCISE 3.

Example 5 (*Earthquake magnitude*). The magnitude R (on the Richter scale) of an earthquake of intensity I is defined as follows:

$$R = \log \frac{I}{I_0},$$

where I_0 is a minimum intensity used for comparison.

An earthquake has an intensity $10^{8.6}$ times I_0. What is its magnitude on the Richter scale?

We substitute into the formula:

$$R = \log \frac{10^{8.6} \cdot I_0}{I_0} = \log 10^{8.6}$$

$$R = 8.6.$$

DO EXERCISE 4.

EXERCISE SET 9.7

● Solve.

1. $2^x = 9$ **2.** $2^x = 30$ **3.** $2^x = 10$ **4.** $2^x = 33$

5. $5^{4x-7} = 125$ **6.** $4^{3x+5} = 16$ **7.** $3^{x^2} \cdot 3^{4x} = \dfrac{1}{27}$

8. $3^{5x} \cdot 9^{x^2} = 27$ **9.** $4^x = 7$ **10.** $8^x = 10$

1. _____

2. _____

3. _____

4. _____

5. _____

6. _____

7. _____

8. _____

9. _____

10. _____

ANSWERS

●● Solve.

11. $\log x + \log (x - 9) = 1$

12. $\log x + \log (x + 9) = 1$

11. _____

13. $\log x - \log (x + 3) = -1$

14. $\log (x + 9) - \log x = 1$

12. _____

13. _____

15. $\log_2 (x + 1) + \log_2 (x - 1) = 3$

16. $\log_4 (x + 3) - \log_4 (x - 5) = 2$

14. _____

15. _____

●●● Solve.

17. (*Doubling time*). How many years will it take an investment of $1000 to double itself when interest is compounded annually at 10%?

18. (*Tripling time*). How many years will it take an investment of $1000 to triple itself when interest is compounded annually at 10%?

16. _____

17. _____

19. The Los Angeles earthquake of 1971 had an intensity $10^{6.7}$ times I_0. What was its magnitude on the Richter scale?

20. The San Francisco earthquake of 1906 had an intensity $10^{8.25}$ times I_0. What was its magnitude on the Richter scale?

18. _____

19. _____

20. _____

EXTENSION EXERCISES

SECTION 9.1

Graph both equations using the same set of axes.

1. $y = \left(\dfrac{1}{4}\right)^x$, $x = \left(\dfrac{1}{4}\right)^y$

2. $y = 3^{-(x-1)}$, $x = 3^{-(y-1)}$

Graph.

3. $y = 2^{2x-1}$

4. $y = \left(\dfrac{1}{2}\right)^x - 1$

5. $y = 3^x + 3^{-x}$

6. $y = 2^{-(x-1)^2}$

7. $y = |2^{x^2} - 1|$

8. $y = |2^x - 2|$

SECTION 9.2

Graph both equations using the same set of axes.

9. $y = \left(\dfrac{3}{2}\right)^x$, $y = \log_{3/2} x$

Graph.

10. $y = \log_2 (x - 1)$

11. $y = \log_3 |x + 1|$

Solve.

12. $|\log_3 x| = 3$

13. $\log_{125} x = \dfrac{2}{3}$

14. $\log_\pi \pi^4 = x$

15. $\log_{\sqrt{5}} x = -3$

16. $\log_b b^{2x^2} = x$

17. $\log_4 (3x - 2) = 2$

18. $\log_8 (2x + 1) = -1$

19. $\log_x \sqrt[5]{36} = \dfrac{1}{10}$

20. $\log_{10} (x^2 + 21x) = 2$

Simplify.

21. $\log_{27} 9$

22. $\log_{1/4} \dfrac{1}{64}$

23. $\log_3 3^{x^2}$

24. $\log_{81} 3 \cdot \log_3 81$

25. $\log_{10} (\log_4 (\log_3 81))$

26. $8^{\log_8 (3x-8)}$

27. $\log_2 (\log_2 (\log_4 256))$

28. $\log_{\sqrt{3}} \dfrac{1}{81}$

29. $\log_{1/5} 25$

SECTION 9.3

True or false.

30. $\dfrac{\log_a M}{c} = \log_a M^{1/c}$

31. $\log_N (M \cdot N)^x = x \log_N M + x$

32. $\log_a \dfrac{1}{8} x = 8 \log_a x$

33. $\log_a (9x^5) = 45 \log_a x$

Express as a single logarithm and simplify if possible.

34. $\log_a (x^8 - y^8) - \log_a (x^2 + y^2)$

35. $\log_a (x + y) + \log_a (x^2 - xy + y^2)$

Express as a sum or difference of logarithms.

36. $\log_a \sqrt{1 - s^2}$

37. $\log_a \dfrac{c - d}{\sqrt{c^2 - d^2}}$

Given that $\log_a 2 = 0.301$, $\log_a 6 = 0.778$, and $\log_a 7 = 0.845$, find:

38. $\log_a \dfrac{2}{7}$

39. $\log_a \sqrt{24}$

40. $\log_a 21$

41. If $\log_a x = 2$, $\log_a y = 3$, and $\log_a z = 4$, what is $\log_a \dfrac{\sqrt[3]{x^2 z}}{\sqrt[3]{y^2 z^{-2}}}$?

SECTION 9.7
Solve.

42. $8^x = 16^{3x + 9}$

43. $27^x = 81^{2x - 3}$

44. $\log_2 x + \log_2 (x - 2) = 3$

45. $\log_4 (x + 6) - \log_4 x = 2$

46. $\log_6 (\log_2 x) = 0$

47. $\log_x (\log_3 27) = 3$

48. $\log_5 \sqrt{x^2 - 9} = 1$

49. $x \log \dfrac{1}{8} = \log 8$

50. $\log (\log x) = 5$

51. $2^{x^2 + 4x} = \dfrac{1}{8}$

52. $3^{\log_3 (8x - 4)} = 5$

53. $\log x^2 = (\log x)^2$

54. $10^{\log 14} = 3x - 19$

55. $\log_5 |x| = 4$

56. $\log x^{\log x} = 25$

57. $\log \sqrt{2x} = \sqrt{\log 2x}$

58. $\log_a a^{x^2 + 4x} = 21$

59. $(\log_a x)^{-1} = \log_a x$

60. $x^{\log_{10} x} = \dfrac{x^{-4}}{1000}$

61. $3^{2x} - 8 \cdot 3^x + 15 = 0$

62. $(81^{x-2})(27^{x+1}) = 9^{2x-3}$

63. $25^{x+2} = 3120 + 25^x$

64. $3^{2x} - 3^{2x - 1} = 18$

65. True or false: If $r^a = s^d$ and $s^b = r^c$, then $ab = cd$.

66. Given that $2^y = 16^{x-3}$ and $3^{y+2} = 27^x$, find the value of $x + y$.

67. If $x = (\log_{125} 5)^{\log_5 125}$, find the value of $\log_3 x$.

68. Solve this system of equations: $w^{2u} = v^u$, $5^u = 5 \cdot 25^w$, $u + v + w = 131$.

69. ▦ Suppose an amount P_0 is invested in a savings account at interest rate k, compounded continuously. The balance P, after t years, is given by

$$P = P_0 e^{kt}, \text{ where } e \approx 2.718282.$$

You will need a calculator with $\boxed{e^x}$ and $\boxed{\ln x}$ keys, where $\log_e x = \ln x$.

Suppose $1000 is invested at 14%.
a) What is the balance after 1 year? 2 years? 10 years?
b) After what amount of time will the $1000 double itself?
c) Suppose P_0 dollars are invested at interest rate k and that amount grows to $2P_0$ dollars in time T. Find a formula relating T and k.

NAME   SCORE   ANSWERS

TEST OR REVIEW—CHAPTER 9

[9.1, ▪●▪]   **1.** Graph: $y = 2^{x+3}$.

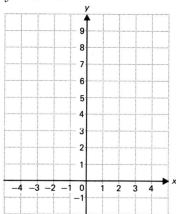

1. See graph.

[9.1, ●●]   **2.** Graph: $x = 2^y$.

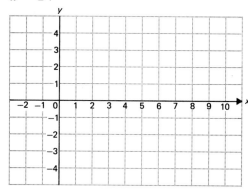

2. See graph.

3. See graph.

[9.2, ▪●▪]   **3.** Graph: $y = \log_3 x$.

4.

[9.2, ●●]   **4.** Convert to a logarithmic equation: $4^{-3} = x$.

[9.2, ●●●]   **5.** Convert to an exponential equation: $\log_4 16 = 2$.

5.

ANSWERS

[9.3, ▓▓] 6. Express in terms of logarithms of a, b, and c: $\log \dfrac{a^3 b^{1/2}}{c^2}$.

6. _____

[9.3, ▓▓] 7. Express as a single logarithm: $\dfrac{1}{3} \log_a x - 3 \log_a y + 2 \log_a z$.

7. _____

[9.5, ▓] 8. Use logarithms to compute:

$$\sqrt[3]{\dfrac{4.76 \times 1.82^2}{72.1^4}}.$$

8. _____

[9.4, ▓] 9. Use Table 2 to find log 0.0123.

9. _____

[9.4, ▓] 10. Use Table 2 to find antilog 5.6484.

10. _____

[9.6, ▓] 11. Use Table 2 to find antilog 0.2614 + (−7).

11. _____

[9.6, ▓] 12. Use Table 2 to find log 0.01234.

12. _____

[9.7, ▓] 13. Solve: $7^x = 1.2$.

13. _____

[9.7, ▓] 14. Solve: $\log (x^2 - 1) - \log (x - 1) = 1$.

14. _____

INEQUALITIES
AND SETS

10

The better of two modes of payment for craftsmen, such as masons, can sometimes be determined using inequalities.

Simplify.   Solve.

1. $|-16|$ **2.** $-|-4|$ **3.** $3x = 9$ **4.** $-3x + 5 = 14$

Use graph paper. Graph these equations.

5. $5y - 4 = 2x$ **6.** $3x + y = 8$

Solve. Use the principle of zero products.

7. $(x + 4)(x - 2) = 0$ **8.** $y(y - 9) = 0$

10.1 THE ADDITION PRINCIPLE FOR SOLVING INEQUALITIES

■∙ ORDER AND INEQUALITIES

The order of the real numbers is often pictured on a number line.

If a number occurs to the left of another, we say that the first number *is less than* the second, or that the second *is greater than* the first. We use the symbols $<$ and $>$ to mean *is less than* and *is greater than*.

Examples   Insert the proper symbol, $<$, $>$, or $=$.

1. 1   5   Since 1 occurs to the left of 5, $1 < 5$.

2. 5   1   Since $1 < 5$, we also know that $5 > 1$.

3. -3 7   Since -3 occurs to the left of 7, $-3 < 7$.

4. 7 -3 Since $-3 < 7$, we also know that $7 > -3$.

5. -9 -2 Since -9 occurs to the left of -2, $-9 < -2$.

6. -2 -9 Since $-9 < -2$, we also know that $-2 > -9$.

7. $5 - 3$ $1 + 1$ Since $5 - 3$ and $1 + 1$ are the same number, $5 - 3 = 1 + 1$.

DO EXERCISES 1–3.

The symbol $\leqslant$ means *is less than or equal to*, and the symbol $\geqslant$ means *is greater than or equal to*.

Examples   Which of the following are true?

8. $-1 \leqslant 7$ True, because -1 is less than 7 ($-1 < 7$ is true).

9. $3 \leqslant 3$ True, because $3 = 3$ is true.

10. $-5 \geqslant 7$ False, because $-5 > 7$ and $-5 = 7$ are not true.

DO EXERCISES 4–6.

An *inequality* is any sentence having one of the verbs $<$, $>$, $\leqslant$, or $\geqslant$. As for equations, a *solution* of an inequality is any number making it a true sentence.

OBJECTIVES

After finishing Section 10.1, you should be able to:

■∙ Determine whether a number is a solution of an inequality.

■∙∙ Solve simple inequalities in one variable, using the addition principle, and graph them on a number line.

■∙∙∙ Solve and graph compound inequalities using the addition principle.

Insert the proper symbol, $<$, $>$, or $=$, to make a true sentence.

1. 1   8

2. 5 -2

3. -3 -8

Which of the following are true?

4. $-5 \leqslant -5$

5. $8 \geqslant -1$

6. $-5 \leqslant -10$

ANSWERS ON PAGE A-42

Examples   Determine whether the given number is a solution of the inequality.

11. $x + 3 < 6$,   5

We substitute and get $5 + 3 < 6$, or $8 < 6$, a false sentence. Therefore 5 is not a solution.

12. $2x - 3 > -3$,   1

We substitute and get $2(1) - 3 > -3$, or $-1 > -3$, a true sentence. Therefore 1 is a solution.

13. $4x - 1 \leqslant 3x + 2$,   3

We substitute and get $4(3) - 1 \leqslant 3(3) + 2$, or $11 \leqslant 11$, a true sentence. Therefore 3 is a solution.

DO EXERCISES 7–9.

●● SOLVING INEQUALITIES

There is an addition principle for inequalities, similar to the one for solving equations.

> *The Addition Principle for Inequalities.* **If any number is added on both sides of a true inequality, another true inequality is obtained.**

To solve inequalities using the addition principle, we do almost exactly what we would do to solve it if it were an equation.

Example 14   Solve $x + 3 > 6$. Then graph.

$$x + 3 + (-3) > 6 + (-3) \qquad \text{Using the addition principle;}$$
$$\text{adding } -3$$
$$x > 3$$

Any number greater than 3 makes the last sentence true. Hence there are many solutions.

> To *solve* an inequality or equation means to find all of its solutions. That is, to find all of the numbers that make it true. When you have found all of the solutions of an equation or inequality, then you have *solved* it, *no matter how* you found the solutions.

We graph the solutions as follows.

Note that there is an open circle on the graph at 3. This indicates that the number 3 is not a solution.

> Remember that the graph of any equation or inequality is a picture of its solution set.

DO EXERCISE 10.

Determine whether the given number is a solution of the inequality.

7. $3 - x < 2$,   4

8. $3x + 2 > -1$, -2

9. $3x + 2 \leqslant 4x - 3$,   5

10. Solve $x + 4 < 7$. Then graph.

11. Solve $3x - 1 \geqslant 2x - 3$. Then graph.

CHECKING

In Example 14 there was an unlimited number of solutions. We cannot check them by substituting into the original inequality, as we do for equations. There are too many of them. We do not really need to check, however. Let us see why. Consider the first and last inequalities in Example 14:

$$x + 3 > 6, \qquad x > 3.$$

Any number that makes the first one true must make the last one true. We know this by the addition principle. Now the question is, will any number that makes the last one true also be a solution of the first one? Let us use the addition principle again.

$$x > 3$$
$$x + \boxed{3} > 3 + \boxed{3} \qquad \text{Using the addition principle; adding 3}$$
$$x + 3 > 6$$

Now we know that any number that makes $x > 3$ true will also make $x + 3 > 6$ true. Whenever we use the addition principle, the first and last inequalities will have the same solutions.

Example 15   Solve $5x - 2 \leqslant 4x - 1$. Then graph.

$$5x - 2 + (\boxed{-4x}) \leqslant 4x - 1 + (\boxed{-4x}) \qquad \text{Adding } -4x$$
$$x - 2 \leqslant -1$$
$$x - 2 + 2 \leqslant -1 + 2 \qquad \text{Adding 2}$$
$$x \leqslant 1$$

12. Graph $-1 < x < 4$.

The graph consists of all points to the left of 1, and also the point for 1. The solid dot at 1 indicates that the number 1 is one of the solutions.

DO EXERCISE 11.

●●● COMPOUND INEQUALITIES

A compound inequality is made up of two or more inequalities. For example, the sentence $-2 < x < 3$ is an abbreviation for $-2 < x$ *and* $x < 3$. In order for a number to be a solution, it must satisfy both of the simple inequalities.

Example 16   Graph: $-2 < x < 1$.

The inequality as stated means $-2 < x$ *and* $x < 1$. The graph consists of all points that are to the right of -2 *and also* to the left of 1.

DO EXERCISE 12.

Example 17 Graph: $-2 \leqslant x < 3$.

The inequality as stated means $-2 \leqslant x$ *and* $x < 3$. The graph consists of all points between -2 and 3, and also the point -2.

DO EXERCISE 13.

Example 18 Solve: $-3 < x + 5 < 7$.

Method 1. We first express this with the word *and*. Then we solve each inequality separately.

$$-3 < x + 5 \qquad and \qquad x + 5 < 7$$
$$-3 + (\boxed{-5}) < x + 5 + (\boxed{-5}) \quad and \quad x + 5 + (\boxed{-5}) < 7 + (\boxed{-5})$$
$$-8 < x \qquad and \qquad x < 2$$
$$-8 < x < 2 \qquad \text{Translating back}$$

Method 2. Note that in Method 1 we did the same thing to each inequality. We can abbreviate as follows.

$$-3 + (\boxed{-5}) < x + 5 + (\boxed{-5}) < 7 + (\boxed{-5}) \qquad \text{Adding} -5$$
$$-8 < x < 2$$

DO EXERCISE 14.

Consider the sentence $x < -3$ *or* $x \geqslant 3$. Any sentence with *or* like this will be true when one or both parts are true. In this case both parts cannot be true at the same time. The solution set consists of those numbers that are less than -3, together with 3 and the numbers greater than 3.

An inequality with the word *or* in it *cannot* be shortened as in Example 18.

Example 19 Graph: $x < -3$ *or* $x \geqslant 3$.

The graph consists of all points to the left of -3 and also all points to the right of 3. The point for 3 is also included.

DO EXERCISES 15 AND 16.

Example 20 Solve: $x - 5 < -2$ or $x - 5 > 2$.

We solve each inequality separately but keep writing the word "or."

$$x - 5 + \boxed{5} < -2 + \boxed{5} \quad or \quad x - 5 + \boxed{5} > 2 + \boxed{5} \qquad \text{Adding} 5$$
$$x < 3 \qquad or \qquad x > 7$$

—We leave *or* in the answer.

DO EXERCISE 17.

13. Graph $-1 < x \leqslant 2$.

14. Consider $-2 < x + 4 < 7$.

a) Solve using Method 1.
b) Solve using Method 2.

Graph.

15. $x < -3$ *or* $x \geqslant 2$

16. $x \leqslant -2$ *or* $x \geqslant 3$

17. Solve.

$x - 4 < -3$ *or* $x - 4 > 3$

SOMETHING EXTRA

AN APPLICATION: HANDLING DIMENSION SYMBOLS (PART III)*
CHANGING UNITS

We can treat dimension symbols much like numerals or variables, because we obtain correct results that way. We can change units using this idea.

Example 1   Convert 3 feet to inches.

Method 1.   We will substitute.

We have 3 ft. We know that 1 ft = 12 in., so we will substitute 12 in. for ft:

$$3 \text{ ft} = 3 \cdot 12 \text{ in.} = 36 \text{ in.}$$

Method 2.   We will multiply by 1.

$$3 \text{ ft} = 3 \text{ ft} \cdot \boxed{\frac{12 \text{ in.}}{1 \text{ ft}}} \qquad \text{Multiplying by 1 to eliminate } ft$$

$$= \frac{3 \cdot 12}{1} \cdot \boxed{\frac{\text{ft}}{\text{ft}}} \cdot \text{in.}$$

$$= 36 \text{ in.}$$

Example 2   Convert 60 mi/hr to ft/sec.

$$60\frac{\text{mi}}{\text{hr}} = 60\frac{\text{mi}}{\text{hr}} \cdot \boxed{\frac{5280 \text{ ft}}{1 \text{ mi}}} \qquad \text{Multiplying by 1 to eliminate } mi$$

$$= 60\frac{\text{mi}}{\text{hr}} \cdot \frac{5280 \text{ ft}}{1 \text{ mi}} \cdot \boxed{\frac{1 \text{ hr}}{60 \text{ min}}} \cdot \boxed{\frac{1 \text{ min}}{60 \text{ sec}}} \qquad \begin{array}{l}\text{Multiplying by 1 to} \\ \text{change hr to sec}\end{array}$$

$$= \frac{60 \cdot 5280}{60 \cdot 60} \boxed{\frac{\text{mi}}{\text{mi}}} \cdot \boxed{\frac{\text{hr}}{\text{hr}}} \cdot \boxed{\frac{\text{min}}{\text{min}}} \cdot \frac{\text{ft}}{\text{sec}}$$

$$= 88\frac{\text{ft}}{\text{sec}}.$$

EXERCISES

Make these unit changes.

1. 3.2 lb to oz

2. 6.2 km to m

3. 35 mi/hr to ft per min

4. \$375 per day, to dollars per minute

---

* For more on dimension symbols, see pp. 364 and 440.

NAME

CLASS

EXERCISE SET 10.1

●● Solve. Then graph.

1. $x - 7 > 5$

1. _____

2. $x + 3 < 2$

2. _____

3. $x - 3 \geqslant -1$

3. _____

4. $x + 2 \leqslant 3$

4. _____

Solve. You need not graph.

5. $x - \dfrac{3}{4} \geqslant \dfrac{11}{8}$

6. $\dfrac{3}{8} \leqslant x - \dfrac{5}{8}$

5. _____

6. _____

7. $3x - 2 \leqslant 2x + 1$

8. $4x + 4 \geqslant 3x - 1$

7. _____

8. _____

ANSWERS

 Graph.

9. $-2 \leqslant x < 5$

9. See graph.

10. $3 > x \geqslant 0$

10. See graph.

11. $x > 5$ *or* $x \leqslant 1$

11. See graph.

12. $x \leqslant -2$ *or* $x > 2$

12. See graph.

Solve. You need not graph.

13.

13. $2 < x + 3 \leqslant 9$

14. $-6 \leqslant x + 1 < 9$

14.

15.

15. $x + 7 \leqslant -2$ *or* $x + 7 > 2$

16. $x + 9 < -4$ *or* $x + 9 \geqslant 4$

16.

17.

🖩 Solve.

17. $x + 6.922 \leqslant -1.998$ *or* $x + 7.0123 > 2.212$

18.

18. $-5.888 \leqslant x + 1.003 < 9.354$

10.2 SETS

■·■ NOTATION FOR SETS

To name a set, we can write the names of the members separated by commas. Then we enclose the names within braces. The names can be written in any order.

Example 1   Write set notation for the set of whole numbers less than 5.

{0, 1, 2, 3, 4}

DO EXERCISE 1.

For some infinite (unending) sets we can write names for some of the members and then write dots to indicate the rest of the members.

Example 2   Write set notation for the set of even whole numbers.

{0, 2, 4, 6, 8, . . .}

DO EXERCISE 2.

Letters are also used to name sets. The set of natural numbers, {1, 2, 3, . . .}, is often named N and the set of integers, {. . . , -3, -2, -1, 0, 1, 2, 3, . . .}, is often named Z. The symbol ∈ is used to denote membership in a set.

Examples   Determine whether the following are true or false.

3. $5 \in N$

This is true, because 5 is a member of the set of natural numbers.

4. $\frac{1}{2} \in \{0, 3, 7, 9\}$

This is false, because $\frac{1}{2}$ is not a member of the set named.

DO EXERCISES 3 AND 4.

For some sets, it is necessary to use a different kind of notation. Consider, for example, the set of all real numbers less than $\sqrt{2}$. We again use braces, but instead of trying to list members, we state a defining property, as follows.

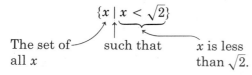

$\{x \mid x < \sqrt{2}\}$

The set of   such that x is less
all x than $\sqrt{2}$.

Example 5   Write set notation for the set of all real numbers greater than $-\pi$.

$\{x \mid x > -\pi\}$

DO EXERCISE 5.

OBJECTIVES

After finishing Section 10.2, you should be able to:

■·■ Write and read set notation.

■·■·■ Write set notation for solution sets of equations and inequalities.

■■■ Find all the subsets of a small set.

■■·■■ Find intersections and unions of sets, and graph certain inequalities as intersections or unions.

1. Write set notation for the set of whole numbers less than 8.

2. Write set notation for the set of odd whole numbers, beginning with 1.

Are the following true or false?

3. $5 \in \{-3, 2, 4, 5, 7\}$

4. $3 \in \{0, 2, 4, 6, 8, . . .\}$

5. Write set notation for the set of all real numbers greater than or equal to $\sqrt{2}$.

Write set notation for the solution sets.

6. $(x - 2)(x + 3) = 0$

7. $4x + 4 > 3x - 1$

8. Is the following true or false?

$\{0, 1, 7\} \subset \{0, 1, 8, 13\}$

◐ ◐ **SOLUTION SETS**

Given an equation or inequality, the set of numbers that make it true is called its *solution set*.

Example 6   Find the solution set of the equation $(x - 1)(x + 2) = 0$.

By the principle of zero products we know that the numbers 1 and -2 make the equation true. Thus the solution set is

$\{1, -2\}$.

Example 7   Find the solution set of the inequality $3x - 2 \leqslant 2x + 1$.

We solve the inequality.

$$3x - 2 \leqslant 2x + 1$$
$$3x + (\boxed{-2x}) - 2 \leqslant 2x + (\boxed{-2x}) + 1 \qquad \text{Adding } -2x$$
$$x - 2 + \boxed{2} \leqslant 1 + \boxed{2} \qquad \text{Adding 2}$$
$$x \leqslant 3$$

All numbers less than or equal to 3 are solutions. Thus the solution set is

$\{x \mid x \leqslant 3\}$.

DO EXERCISES 6 AND 7.

◐ ◐ ◐ **SUBSETS AND THE EMPTY SET**

When all the members of one set are also members of another, we say that the first set is a *subset* of the second. We use the symbol $\subset$ to indicate that one set is a subset of another. For example,

$A \subset B$

means that A is a subset of B.

Examples   Determine whether the following are true or false.

8. $\{1, 2, 3, 4\} \subset \{0, 1, 2, 3, 4, 5\}$

This is true, because every member of the first set is also a member of the second.

9. $\{x \mid x \leqslant 3\} \subset \{x \mid x < 3\}$

This is false, because there is one member of the first set that is not a member of the second. It is the number 3.

10. $\{y \mid y > -6\} \subset \{y \mid y > -10\}$

This is true, because any number greater than -6 is also greater than -10.

DO EXERCISE 8.

The set without any members is called the *empty set*, and is named ∅. The empty set is a subset of every set.

Example 11   List all the subsets of {1, 2, 3}.

{1, 2, 3}   Every set is a subset of itself.

{1, 2}

{1, 3}   There are three subsets with two members each.

{2, 3}

{1}

{2}   There are three subsets with one member each.

{3}

∅   The empty set is a subset of any set.

DO EXERCISE 9.

▓▓ INTERSECTIONS AND UNIONS

The *intersection* of two sets A and B is the set of all members that are common to A and B. We denote the intersection of sets A and B as

$A \cap B$.

The intersection of two sets is often pictured as follows.

Example 12   Find: $\{1, 2, 3, 4, 5\} \cap \{-2, -1, 0, 1, 2, 3\}$.

The numbers 1, 2, and 3 are common to the two sets, so the intersection is {1, 2, 3}.

DO EXERCISE 10.

The *union* of two sets is formed by putting them together. We denote the union of sets A and B as

$A \cup B$.

The union of two sets is often pictured as follows.

Example 13   Find: $\{2, 3, 4\} \cup \{3, 5, 7\}$.

The numbers in either or both sets are 2, 3, 4, 5, and 7, so the union is {2, 3, 4, 5, 7}.

DO EXERCISE 11.

9. List all the subsets of $\{a, b\}$.

10. Find $\{0, 3, 5, 7\} \cap \{0, 1, 3, 11\}$.

11. Find $\{0, 1, 3, 4\} \cup \{0, 1, 7, 9\}$.

12. Graph $-2 < x < 3$.

In many situations, the word *and* corresponds to intersection and the word *or* to union.

Example 14   Graph: $-3 < x < 1$.

The inequality as stated means $-3 < x$ *and* $x < 1$. We graph $-3 < x$ and also graph $x < 1$.

Then we take the intersection of the graphs.

The last graph is the intersection of the two preceding graphs.

DO EXERCISE 12.

Example 15   Graph: $x < -3$ *or* $x \geqslant 2$.

13. Graph $x < -2$ *or* $x > 4$.

We graph $x < -3$. We also graph $x \geqslant 2$. Then we take the union of the two graphs.

A compound inequality using the word *or*, as in this example, *cannot* be shrunk to look like the one in Example 14. Remember, when the word *or* occurs, you must keep that word.

DO EXERCISE 13.

NAME   CLASS   ANSWERS

EXERCISE SET 10.2

■ ●   Write set notation for the following.

1. The even whole numbers less than 10

2. The odd whole numbers less than 10

3. The odd whole numbers greater than 10

4. The even whole numbers greater than 10

5. The real numbers greater than $\sqrt{3}$

6. The real numbers less than $-\sqrt{5}$

■ ● ●   Find the solution sets.

7. $(x + 5)(x - 5) = 0$

8. $(x + 4)(x + 7) = 0$

9. $x^2 - x - 6 = 0$

10. $x^2 + x - 12 = 0$

11. $5x + 3 > 4x - 2$

12. $7x - 7 < 6x + 5$

■ ● ● ●   Which of the following are true and which are false?

13. $\{1, 5, 8\} \subset \{1, 2, 5, 7, 8\}$

14. $\{2, 4\} \subset \{0, 2, 3, 4, 7\}$

15. $\{1, 2, 3, 4, 5\} \subset \{2, 3, 4\}$

16. $\{8, 9, 13, 17\} \subset \{4, 5, 17\}$

17. $\{3, 8, 13\} \subset \{6, 8, 25, 38\}$

18. $\{9, 14, 27\} \subset \{9, 27, 38, 50\}$

19. $\emptyset \subset \{1, 2\}$

20. $\emptyset \subset \emptyset$

21. $\{x \mid x > 3\} \subset \{x \mid x > 2\}$

22. $\{x \mid x > 3\} \subset \{x \mid x > 4\}$

23. $\{x \mid x < 5\} \subset \{x \mid x \leqslant 5\}$

24. $\{x \mid x < 5\} \subset \{x \mid x < 5\}$

ANSWERS

1. _____
2. _____
3. _____
4. _____
5. _____
6. _____
7. _____
8. _____
9. _____
10. _____
11. _____
12. _____
13. _____
14. _____
15. _____
16. _____
17. _____
18. _____
19. _____
20. _____
21. _____
22. _____
23. _____
24. _____

25. _____

26. _____

27. _____

28. _____

29. _____

30. _____

31. _____

32. _____

33. _____

34. _____

35. _____

36. _____

37. See graph. _____

38. See graph. _____

39. See graph. _____

40. See graph. _____

41. _____

List all the subsets.

25. $\{-4, 7\}$ **26.** $\{a, b, c\}$

■■ Find these unions and intersections.

27. $\{5, 6, 7, 8\} \cap \{4, 6, 8, 10\}$ **28.** $\{9, 10, 27\} \cap \{8, 10, 38\}$

29. $\{4, 5, 6, 7, 8\} \cup \{1, 4, 6, 11\}$ **30.** $\{8, 9, 27\} \cup \{2, 8, 27\}$

31. $\{2, 4, 6, 8\} \cap \{1, 3, 5\}$ **32.** $\emptyset \cap \emptyset$

33. $\{1, 2, 3, 4\} \cap \{1, 2, 3, 4\}$ **34.** $\{8, 9, 10\} \cap \emptyset$

35. $\{4, 8, 11\} \cup \emptyset$ **36.** $\emptyset \cup \emptyset$

Graph.

37. $-5 < x \leqslant 2$

```
◄─┼─┼─┼─┼─┼─┼─┼─┼─┼─┼─┼─┼─►
 -8 -7 -6 -5 -4 -3 -2 -1  0  1  2  3  4
```

38. $3 \leqslant x < 7$

```
◄─┼─┼─┼─┼─┼─┼─┼─┼─┼─┼─┼─┼─►
 -2 -1  0  1  2  3  4  5  6  7  8  9  10
```

39. $x < -3 \quad or \quad x \geqslant 5$

```
◄─┼─┼─┼─┼─┼─┼─┼─┼─┼─┼─┼─┼─►
 -6 -5 -4 -3 -2 -1  0  1  2  3  4  5  6
```

40. $x \leqslant 5 \quad or \quad x > 10$

```
◄─┼─┼─┼─┼─┼─┼─┼─┼─┼─┼─┼─┼─►
  0  1  2  3  4  5  6  7  8  9 10 11 12
```

---

41. If a set has n members, how many subsets does it have?

10.3 THE MULTIPLICATION PRINCIPLE FOR SOLVING INEQUALITIES

� THE MULTIPLICATION PRINCIPLE

Consider the true inequality

$4 < 9.$

If we multiply both numbers by 2 we get another true inequality:

$8 < 18.$

If we multiply both numbers by -3 we get a false inequality:

$-12 < -27.$

However, if we now reverse the inequality symbol we get a true inequality:

$-12 > -27.$

> *The Multiplication Principle for Inequalities.* **If we multiply on both sides of a true inequality by a positive number, we get another true inequality. If we multiply by a negative number and reverse the inequality symbol, we get another true inequality.**

When we solve an inequality using the multiplication principle, we do not need to check, provided that we do not multiply by an expression containing a variable.

Examples   Solve.

1. $3y < \dfrac{3}{4}$

$\dfrac{1}{3} \cdot 3y < \dfrac{1}{3} \cdot \dfrac{3}{4}$ Multiplying by $\dfrac{1}{3}$

$y < \dfrac{1}{4}$

Any number less than $\frac{1}{4}$ is a solution. The solution set is $\{y \mid y < \frac{1}{4}\}$.

2. $-4x < \dfrac{4}{5}$

$-\dfrac{1}{4} \cdot (-4x) > -\dfrac{1}{4} \cdot \dfrac{4}{5}$ Multiplying by $-\dfrac{1}{4}$ and reversing the inequality sign

$x > -\dfrac{1}{5}$

Any number greater than $-\frac{1}{5}$ is a solution. The solution set is $\{x \mid x > -\frac{1}{5}\}$.

DO EXERCISES 1 AND 2.

OBJECTIVES

After finishing Section 10.3, you should be able to:

◆ Solve inequalities using the multiplication principle.

◆◆ Solve inequalities using the addition and multiplication principles together.

◆◆◆ Solve problems involving inequalities.

Solve.

1. $5y \le \dfrac{3}{2}$

2. $-2y > \dfrac{5}{6}$

3. Solve $-\dfrac{1}{3}x \leqslant -4$.

Solve.

4. $6 - 5y \geqslant 7$

5. $3x + 5x < 4$

6. $17 - 5y \leqslant 8y - 5$

Example 3   Solve: $-5x \geqslant -80$.

$$-\frac{1}{5} \cdot (-5x) \leqslant -\frac{1}{5} \cdot (-80) \qquad \text{Multiplying by } -\frac{1}{5} \text{ and}$$
$$\text{reversing } \geqslant$$

Remember, always reverse the inequality sign when multiplying by a negative number.

$$x \leqslant 16$$

Any number less than 16 is a solution, and 16 is also a solution. The solution set is

$$\{x \mid x \leqslant 16\}.$$

DO EXERCISE 3.

◖◗ USING THE PRINCIPLES TOGETHER

We use the addition and multiplication principles together in solving inequalities in much the same way as for equations.

Example 4   Solve: $16 - 7y \geqslant 10y - 4$.

$$-16 + 16 - 7y \geqslant -16 + 10y - 4 \qquad \text{Adding } -16$$
$$-7y \geqslant 10y - 20$$
$$-10y - 7y \geqslant -10y + 10y - 20 \qquad \text{Adding } -10y$$
$$-17y \geqslant -20$$
$$-\frac{1}{17} \cdot (-17y) \leqslant -\frac{1}{17} \cdot (-20) \qquad \text{Multiplying by } -\frac{1}{17} \text{ and}$$
$$\text{reversing the inequality}$$
$$\text{sign}$$
$$y \leqslant \frac{20}{17}$$

The solution set is $\{y \mid y \leqslant \frac{20}{17}\}$.

In Example 4, we could have begun by adding $7y$, as follows.

$$16 - 7y + 7y \geqslant 10y - 4 + 7y \qquad \text{Adding } 7y$$
$$16 \geqslant 17y - 4$$
$$16 + 4 \geqslant 17y - 4 + 4 \qquad \text{Adding } 4$$
$$20 \geqslant 17y$$
$$\frac{20}{17} \geqslant y \qquad \text{Multiplying by } \frac{1}{17}$$

The solution set can be named $\{y \mid \frac{20}{17} \geqslant y\}$, or $\{y \mid y \leqslant \frac{20}{17}\}$.

DO EXERCISES 4–6.

Examples   Solve.

5. $-1 < 2x + 3 \leqslant 11$

$-1 +$ (-3) $< 2x + 3 +$ (-3) $\leqslant 11 +$ (-3) Adding -3

$\qquad -4 < 2x \leqslant 8$

$\dfrac{1}{2} \cdot (-4) < \dfrac{1}{2} \cdot 2x \leqslant \dfrac{1}{2} \cdot 8$ Multiplying by $\dfrac{1}{2}$

$\qquad -2 < x \leqslant 4$

The solution set is $\{x \mid -2 < x \leqslant 4\}$.

6. $3 \leqslant 5 - 2x < 7$

$3 - 5 \leqslant -2x < 7 - 5$ Adding -5

$\qquad -2 \leqslant -2x < 2$

$-\dfrac{1}{2} \cdot (-2) \geqslant -\dfrac{1}{2} \cdot (-2x) > -\dfrac{1}{2} \cdot 2$ Multiplying by $-\dfrac{1}{2}$ and reversing the inequality signs

$\qquad 1 \geqslant x > -1$

The solution set is $\{x \mid 1 \geqslant x > -1\}$.

DO EXERCISES 7 AND 8.

▫▫▫   SOLVING PROBLEMS

Certain problems translate quite naturally to inequalities, rather than equations.

Example 7   In a business course, there will be three tests. You must get a total score of 270 for an A. You get 91 and 86 on the first two tests. What scores on the last test will give you an A?

We translate. For an A,

Total score $\geqslant 270$

$91 + 86 + x \geqslant 270$.

We have used x for your score on the last test. Solving the inequality, we get

$x \geqslant 93$.

Thus a score of 93 or higher will give you an A.

DO EXERCISE 9.

Example 8   On your new job, you can be paid one of two ways.

Plan A: A salary of $600 per month, plus a commission of 4% of gross sales.

Plan B: A salary of $800 per month, plus a commission of 6% of gross sales over $10,000.

Solve.

7. $7 \leqslant 4x - 5 < 11$

8. $9 < 5 - 4x \leqslant 13$

9. In a psychology course, there will be five tests. To get a B, 400 points are needed. You get scores of 91, 86, 73, and 79 on the first four tests. What scores on the last test will give you at least a B?

10. A painter can be paid in two ways.

Plan A: $500 plus $4 per hour.

Plan B: Straight $9 per hour.

Suppose the job takes *n* hours. For what values of *n* is Plan A better for the painter?

For what gross sales is Plan A better than Plan B, assuming that gross sales are always more than $10,000?

To translate, we will write an inequality that states that the income from Plan A is greater than that from Plan B.

Income from A > Income from B

$600 + 4\% \cdot x > 800 + (x - 10{,}000) \cdot 6\%.$

We have used x to represent the gross sales for the month. Solving this inequality, we get

$x < 20{,}000.$

Thus for gross sales under $20,000, Plan A is better.

Example 9   A mason can be paid in two ways.

Plan A: $300 plus $3 per hour.

Plan B: Straight $8.50 per hour.

Suppose that the job takes *n* hours. For what values of *n* is Plan B better for the mason?

We translate:

Income from B > Income from A

$8.50n > 300 + 3 \cdot n$

Solving this inequality, we get

$n > 54.55.$

Thus if the job takes over 54.55 hours (54 hr, 33 min), Plan B is better.

DO EXERCISE 10.

SOMETHING EXTRA
CALCULATOR CORNER: NUMBER PATTERNS

Do the following calculations. Look for patterns. Then figure out what the next lines in the pattern should be. After that, check your answers by doing the actual calculations.

EXERCISE

Sums of cubes

Add.	Take the square root of the answer at the left.
1 $1 + 2^3 =$ _____ $1 + 2^3 + 3^3 =$ _____ $1 + 2^3 + 3^3 + 4^3 =$ _____ $1 + 2^3 + 3^3 + 4^3 + 5^3 =$ _____	

NAME   CLASS   ANSWERS

EXERCISE SET 10.3

⬤ Solve.

1. $3x \leqslant 18$ **2.** $4x > 24$ **3.** $-3y > \dfrac{3}{8}$ **4.** $-8y \leqslant -\dfrac{32}{5}$

⬤ ⬤ Solve.

5. $\dfrac{3}{4} + 7x < \dfrac{5}{8}$ **6.** $6.2 + 4.2x \geqslant 11.1$ **7.** $2x - 7 \geqslant 5x - 5$

8. $4 - 5x > 2x + 3$ **9.** $8 < 3x + 2 \leqslant 14$ **10.** $-7 < 5x - 2 < 12$

1. _____

2. _____

3. _____

4. _____

5. _____

6. _____

7. _____

8. _____

9. _____

10. _____

●●● Solve.

11. A car rents for $13.95 per day, plus 10¢ per mile. You are on a daily budget of $76. What mileages will allow you to stay within the budget?

12. You are taking a history course. There will be 4 tests. You have scores of 89, 92, and 95 on the first three. You must make a total of 360 to get an A. What scores on the last test will give you an A?

11. _____

13. A person is going to invest $25,000, part at 7% and part at 8%. What is the most that can be invested at 7% in order to make at least $1800 interest per year?

14. A person is going to invest $20,000, part at 6% and part at 8%. What is the most that can be invested at 6% in order to make at least $1500 interest per year?

12. _____

13. _____

15. On your new job, you can be paid in one of two ways.

    Plan A: A salary of $500 per month, plus a commission of 4% of gross sales.

    Plan B: A salary of $750 per month plus a commission of 5% of gross sales over $8000.

    For what gross sales is Plan B better than Plan A, assuming that gross sales are always more than $8000?

16. A mason can be paid in two ways.

    Plan A: $500 plus $3 per hour.

    Plan B: Straight $8.00 per hour.

    Suppose the job takes *n* hours. For what values of *n* is Plan A better for the mason than Plan B?

14. _____

15. _____

16. _____

10.4 ABSOLUTE VALUE

• DISTANCE

The number line above shows that the distance between -3 and 2 is 5. Another way to find the distance between two numbers on a number line is to subtract and take the absolute value, as follows.

$$|-3 - 2| = |-5| = 5, \quad \text{or} \quad |2 - (-3)| = |5| = 5$$

Note that the order in which we subtract does not matter since we are taking the absolute value afterward.

> **For any numbers a and b, the distance between them is $|a - b|$.**

Example 1　Find the distance between -8 and -92.

$$|-8 - (-92)| = |84| = 84$$

Example 2　Find the distance between x and 0.

$$|x - 0| = |x|$$

DO EXERCISES 1–3.

•• EQUATIONS AND INEQUALITIES WITH ABSOLUTE VALUE

Example 3　Solve $|x| = 4$. Then graph, using a number line.

Note that $|x| = |x - 0|$, so that $|x - 0|$ is the distance from x to 0. The solutions of the equation are those numbers x whose distance from 0 is 4. Those numbers are 4 and -4. The solution set is $\{4, -4\}$. The graph consists of just two points, as shown.

$|x| = 4$

DO EXERCISES 4 AND 5.

> Remember, to *solve* an equation or inequality means to find all of its solutions (its solution set). There may be many ways of doing it, including guessing. To *graph* an equation or inequality means to make a picture of its solution set. For inequalities in one variable we make that picture on a number line.

OBJECTIVES

After finishing Section 10.4, you should be able to:

• Find the distance between numbers using absolute value.

•• Solve equations and inequalities with absolute value.

Find the distance between each pair of numbers.

1. $-6, \quad -35$

2. $19, \quad 14$

3. $-3, \quad 17$

Solve. Then graph, using a number line.

4. $|x| = 6$

5. $|x| = \dfrac{1}{2}$

Solve. Then graph, using a number line.

6. $|x| < 6$

7. $|x| \leq \dfrac{1}{2}$

Solve. Then graph.

8. $|x| \geq 6$

9. $|x| > \dfrac{1}{2}$

Example 4   Solve $|x| < 4$. Then graph.

Again, note that $|x| = |x - 0|$, so $|x - 0|$ is the distance from x to 0. Thus the solutions of $|x - 0| < 4$ are those numbers x whose distance from 0 is less than 4. The solutions are those numbers x such that $-4 < x < 4$. The solution set is $\{x \mid -4 < x < 4\}$. The graph is as follows.

DO EXERCISES 6 AND 7.

Example 5   Solve $|x| \geq 4$. Then graph.

Since $|x| = |x - 0|$, the solutions of $|x - 0| \geq 4$ are those numbers x whose distance from 0 is greater than or equal to 4, in other words, those numbers x such that $x \leq -4$ *or* $x \geq 4$. The solution set is $\{x \mid x \leq -4 \text{ or } x \geq 4\}$. The graph is as follows.

DO EXERCISES 8 AND 9.

Examples 3 through 5 illustrate three cases of solving inequalities with absolute value. The expression inside of the absolute value signs can be something besides a single letter. We may wish to solve $|5x - 2| < 4$, for example. The following statement gives the general principles.

For any positive number b, and any expression X:

1. **The solutions of $|X| = b$ are those numbers that satisfy**

   $X = b$ **or** $X = -b.$

For example, the solutions of $|5x - 1| = 8$ are those numbers x for which

   $5x - 1 = 8$ or $5x - 1 = -8.$

2. **The solutions of $|X| < b$ are those numbers that satisfy**

   $-b < X < b.$

For example, the solutions of $|6x + 7| < 5$ are those numbers x for which

   $-5 < 6x + 7 < 5.$

3. **The solutions of $|X| > b$ are those numbers that satisfy**

   $X < -b$ **or** $X > b.$

For example, the solutions of $|2x - 9| > 4$ are those numbers x for which

   $2x - 9 < -4$ or $2x - 9 > 4.$

The above still hold if $<$ is replaced by $\leq$ and $>$ is replaced by $\geq$.

Example 6   Solve $|5x - 4| = 11$. Then graph.

We use Property 1. In this case, b is 11 and X is $5x - 4$.

$$|X| = b$$
$$|5x - 4| = 11 \qquad \text{Replacing } X \text{ by } 5x - 4 \text{ and } b \text{ by } 11$$
$$5x - 4 = 11 \quad \text{or} \quad 5x - 4 = -11 \qquad \text{Property 1}$$
$$5x = 15 \quad \text{or} \qquad 5x = -7 \qquad \text{Adding 4}$$
$$x = 3 \quad \text{or} \qquad x = -\frac{7}{5} \qquad \text{Multiplying by } \frac{1}{5}$$

The solution set is $\{3, -\frac{7}{5}\}$.

The graph is as follows.

$$|5x - 4| = 11$$

DO EXERCISE 10.

Example 7   Solve $|3x - 2| < 4$. Then graph.

We use Property 2. In this case, b is 4 and X is $3x - 2$.

$$|X| < b$$
$$|3x - 2| < 4 \qquad \text{Replacing } X \text{ by } 3x - 2 \text{ and } b \text{ by } 4$$
$$-4 < 3x - 2 < 4 \qquad \text{Property 2}$$
$$-2 < 3x < 6 \qquad \text{Adding 2}$$
$$-\frac{2}{3} < x < 2 \qquad \text{Multiplying by } \frac{1}{3}$$

The graph is as follows

$$|3x - 2| < 4$$

DO EXERCISE 11.

Example 8   Solve $|4x + 2| \geqslant 6$. Then graph.

We use Property 3. In this case, b is 6 and X is $4x + 2$.

$$|X| \geqslant b$$
$$|4x + 2| \geqslant 6 \qquad \text{Replacing } X \text{ by } 4x + 2 \text{ and } b \text{ by } 6$$
$$4x + 2 \leqslant -6 \quad \text{or} \quad 4x + 2 \geqslant 6 \qquad \text{Property 3}$$
$$4x \leqslant -8 \quad \text{or} \qquad 4x \geqslant 4 \qquad \text{Adding } -2$$
$$x \leqslant -2 \quad \text{or} \qquad x \geqslant 1 \qquad \text{Multiplying by } \frac{1}{4}$$

Solve. Then graph.

10. $|3x + 2| = 8$

```
◄──┼───┼───┼───┼───┼───┼───►
  -6  -4  -2   0   2   4   6
```

Solve. Then graph.

11. $|2x + 3| < 5$

```
◄──┼───┼───┼───┼───┼───┼───►
  -6  -4  -2   0   2   4   6
```

ANSWERS ON PAGE A–43

Solve. Then graph.

12. $|3x + 2| \geqslant 5$

The graph is as follows.

$$|4x + 2| \geq 6$$

DO EXERCISE 12.

> In Example 8, we have the symbol $\geqslant$. Whenever $>$ or $\geqslant$ occurs, we must use the word *or* in solving. A common error is to try to construct a compound inequality without the word *or*.

SOMETHING EXTRA

CALCULATOR CORNER: APPROXIMATING π

Each of the following is an approximation to the number π. As the patterns are extended by including more terms, the approximation becomes better.

EXERCISES

Approximate π with each of the following. Add more terms and watch your approximation become better. Compare your answer with the following: $\pi \approx 3.1415926535$.

1. $\pi = 3 + 4\left[\dfrac{1}{2 \cdot 3 \cdot 4} - \dfrac{1}{4 \cdot 5 \cdot 6} + \dfrac{1}{6 \cdot 7 \cdot 8} - \dfrac{1}{8 \cdot 9 \cdot 10} + \cdots\right]$

2. $\pi = 16\left[\dfrac{1}{8} + \dfrac{1}{1 \cdot 3 \cdot 5} + \dfrac{1}{5 \cdot 7 \cdot 9} + \dfrac{1}{9 \cdot 11 \cdot 13} + \cdots\right]$

3. $\pi = 6\left[\dfrac{1}{2} + \dfrac{1}{2 \cdot 3 \cdot 2^3} + \dfrac{1 \cdot 3}{2 \cdot 4 \cdot 5 \cdot 2^5} + \dfrac{1 \cdot 3 \cdot 5}{2 \cdot 4 \cdot 6 \cdot 7 \cdot 2^7}\right.$

$\left. + \dfrac{1 \cdot 3 \cdot 5 \cdot 7}{2 \cdot 4 \cdot 6 \cdot 8 \cdot 9 \cdot 2^9} + \cdots\right]$

| NAME | CLASS | ANSWERS |

EXERCISE SET 10.4

 Find the distance between each pair of numbers.

1. -8, -42 **2.** -9, -36 **3.** 26, 15   **4.** 54, 18

5. -9, 24   **6.** -18, 37   **7.** -2, 0   **8.** 15, 0

Solve. Then graph.

9. $|x| = 3$

10. $|x| = 5$

11. $|x| < 3$

12. $|x| \leqslant 5$

13. $|x| \geqslant 2$

14. $|y| > 8$

ANSWERS

1. _____

2. _____

3. _____

4. _____

5. _____

6. _____

7. _____

8. _____

For 9–14, also
see graphs.

9. _____

10. _____

11. _____

12. _____

13. _____

14. _____

ANSWERS

For 15–22, also see graphs.

15. _____

16. _____

17. _____

18. _____

19. _____

20. _____

21. _____

22. _____

23. _____

24. _____

Solve. Then graph.

15. $|x - 3| < 12$

16. $|x - 2| < 6$

17. $|2x + 3| \leq 4$

18. $|5x + 2| \leq 3$

19. $|2y - 7| > 5$

20. $|3y - 4| > 8$

21. $|4x - 9| \geq 14$

22. $|9y - 2| \geq 17$

Solve.

23. $\left| \dfrac{2x - 1}{3} \right| \leq 1$

24. $\left| \dfrac{3x - 2}{5} \right| \geq 1$

10.5 INEQUALITIES IN TWO VARIABLES

●● SOLUTIONS

The solutions of inequalities in two variables are ordered pairs, as is the case with equations in two variables.

Example 1   Determine whether $(-3, 2)$ is a solution of $5x - 4y > 13$.

We replace x by -3 and y by 2, using alphabetical order.

$$
\begin{array}{c|c}
5x - 4y > 13 & \\
\hline
5\,(-3) - 4 \cdot 2 & 13 \\
-15 - 8 & \\
-23 & \\
\end{array}
$$

Since $-23 > 13$ is false, $(-3, 2)$ is not a solution.

DO EXERCISE 1.

●● GRAPHS

Inequalities in two variables are graphed on a plane.

Example 2   Graph: $y < x$.

For comparison, we first graph the line $y = x$. We draw it dashed. Every solution of $y = x$ is an ordered pair having the same first and second coordinates.

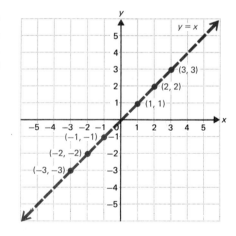

Now look at a vertical line and some ordered pairs on it. The one shown goes through $(2, 2)$.

For any point here, the y-value is greater than the x-value.

For any point here, the y-value is less than the x-value.

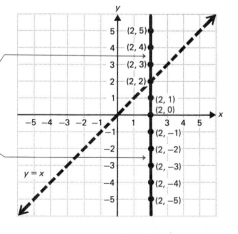

ANSWER ON PAGE A–44

OBJECTIVES

After finishing Section 10.5, you should be able to:

● Determine whether an ordered pair is a solution of an inequality in two variables.

●● Graph, on a plane, linear inequalities in two variables.

1. Determine whether $(4, -3)$ is a solution of $-2x + 3y \leqslant 6$.

2. Graph $y > x + 1$. (*Hint:* First graph $y = x + 1$.)

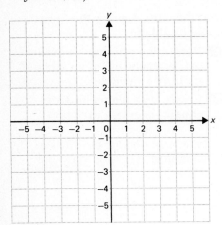

For all points below $y = x$, the second coordinate is less than the first, that is, $y < x$. For points above the line, $y > x$. The same thing happens for any vertical line. Thus all points below the line $y = x$ are solutions. We shade the half-plane below $y = x$. This is the graph of the inequality $y < x$. The points on the line $y = x$ are not included. That is why we drew the line dashed.

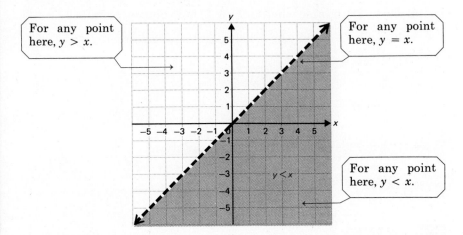

For any point here, $y > x$.

For any point here, $y = x$.

For any point here, $y < x$.

$y < x$

DO EXERCISE 2.

RECOGNIZING AND GRAPHING LINEAR INEQUALITIES

In Chapter 2 we learned to recognize linear equations. If the inequality sign in a sentence is replaced by an equal sign and the result is a linear equation, then we say that the inequality is a *linear inequality*. The following are linear inequalities:

$$y < 3x + 2, \qquad 5y - 4x - 2 \geqslant 0, \qquad 6x + 3 \leqslant 8y - 2.$$

The following are not linear inequalities:

$$y < x^2 + 2, \qquad xy^3 \geqslant 5 - 2y, \qquad \frac{1}{y} \leqslant x + 2y.$$

To graph a linear inequality, we first replace the inequality sign by an equals sign. Then we graph the resulting line. If the inequality symbol is $<$ or $>$, we draw the line dashed. If the inequality symbol is $\leqslant$ or $\geqslant$, we draw the line solid. The graph consists of a half-plane as in Example 2 and, if the line is solid, the line as well.

Example 3   Graph: $6x - 2y < 12$.

a) We first graph the line $6x - 2y = 12$. Since the inequality symbol is $<$, we draw it dashed. The line is most easily graphed using intercepts.

b) We determine which half-plane to shade by trying some point off the line. The point $(0, 0)$ is easy to check unless the line goes through the origin.

$$\frac{6x - 2y < 12}{\begin{array}{c|c} 6 \cdot 0 - 2 \cdot 0 & 12 \\ 0 & \end{array}}$$

Any point not on the line can be used as a check. The origin is an *easy* one to use.

Since $0 < 12$, the point is in the graph. We shade that half-plane. The graph consists of the shaded half-plane, but not the dashed line.

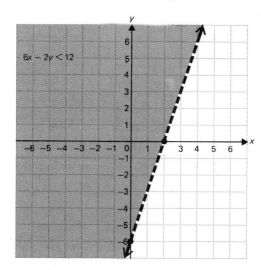

3. Graph $4x + 3y \geqslant 12$.

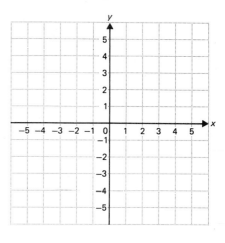

Example 4   Graph: $8x + 3y \geqslant 24$.

a) We graph the line $8x + 3y = 24$. Since the inequality symbol is $\geqslant$, we draw the line solid.

b) The point $(0, 0)$ is not in the graph, because $8 \cdot 0 + 3 \cdot 0 = 0$, which is less than 24. We shade the other half-plane. The graph consists of the shaded half-plane and the solid line as well.

4. Graph $6x - 3y < 18$.

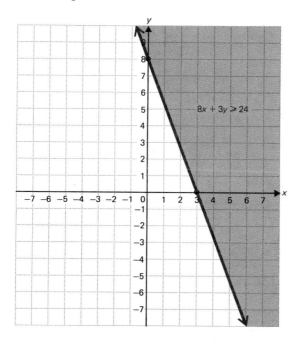

DO EXERCISES 3 AND 4.

ANSWERS ON PAGE A–44

5. Graph $x \leqslant 4$.

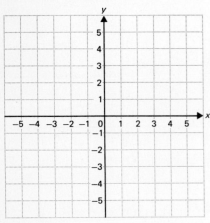

Example 5 Graph: $y > 3$.

a) We first graph $y = 3$, using a dashed line.

b) Checking any point above this line, we see that it satisfies the inequality.

c) We shade the upper half-plane.

Example 6 Graph: $-2 < x \leqslant 5$.

a) We think of this as a system of inequalities:

$-2 < x$

$\quad x \leqslant 5$.

6. Graph $-3 \leqslant y < 4$.

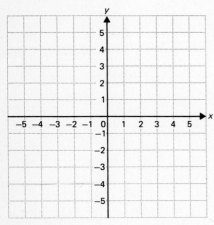

b) We graph the equation $-2 = x$, and see that the graph of the first inequality is the half-plane to the right of the line $-2 = x$ (see the graph on the left below).

c) We graph the second inequality, starting with the line $x = 5$, and find that its graph is the line and also the half-plane to the left of it (see the graph on the right below).

 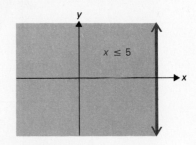

d) We shade the intersection of these graphs.

DO EXERCISES 5 AND 6.

NAME   CLASS   ANSWERS

EXERCISE SET 10.5

[•] Determine whether the ordered pair is a solution of the inequality.

1. $5y + 4x < 1$, $(-2, 1)$

2. $5y + 4x < 1$, $(2, -1)$

3. $y \geqslant 3x - 4$, $(2, 2)$

4. $y \geqslant 3x - 4$, $(2, 3)$

1. _____

2. _____

3. _____

4. _____

[• •] Graph the following on a plane.

5. $y > 2x$

6. $y < 3x$

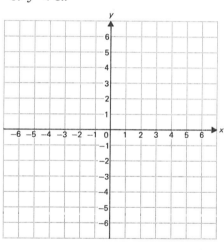

7. $y \leqslant x + 1$

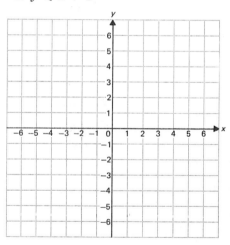

8. $y \geqslant x + 3$

9. $x + y < 4$

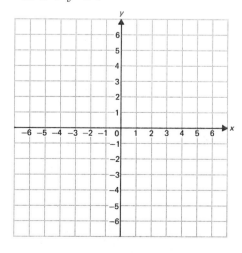

10. $x + y \geqslant 5$

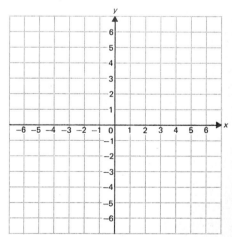

11. $3x + 4y \geqslant 12$

12. $2x + 3y \leqslant 6$

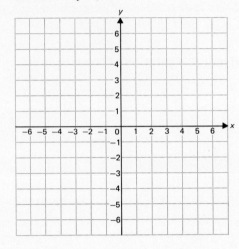

13. $7x + 2y \leqslant 21$

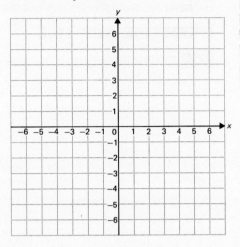

14. $5x + 4y > 20$

15. $x < -4$

16. $y \geqslant 2$

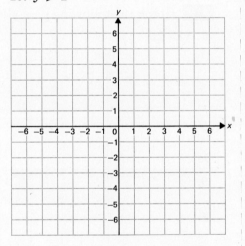

17. $-4 < y < -1$

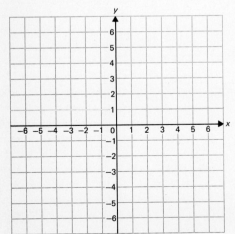

18. $-2 \leqslant x < 3$

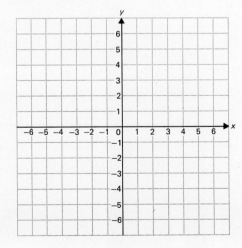

EXTENSION EXERCISES

SECTION 10.1

Solve.

1. $-\dfrac{1}{3} < x - \dfrac{1}{6} \leqslant 6$

2. $0.9x > 1.8 - 0.1x$

3. $3x - \dfrac{1}{8} \leqslant \dfrac{3}{8} + 2x$

4. $2x + 3 \leqslant x - 6$ or $3x - 2 \leqslant 4x + 5$

5. $x + 4 < 2x - 6 \leqslant x + 12$

6. $2x - 3 < \dfrac{13}{4}x + 10 - 1.25x$

7. $7 - 3x \geqslant 5.7x + 8 - 8.7x$

8. $x + \dfrac{1}{10} \leqslant -\dfrac{1}{10}$ or $x + \dfrac{1}{10} \geqslant \dfrac{1}{10}$

SECTION 10.2

Use the following sets when answering questions 9–32.

$A = \{x | -4 < x \leqslant 2\}$

$B = \emptyset$

$C = \{\ldots -4, -2, 0, 2, 4, \ldots\}$

$D = \{0, 1, 2, 3, 4, 5, 6, 7, 8, 9\}$

$E = \{\ldots -5, -3, -1, 1, 3, 5, \ldots\}$

$F = \{x | x \leqslant 3\}$

$G = \{0, 1\}$

$H = \{x | x < -10 \text{ or } x \geqslant 5\}$

$J = \{\{0\}, \{1\}, \{0, 1\}\}$

$K = \{x | x \leqslant -9 \text{ or } x \geqslant 9\}$

$L = \left\{x | x > -\dfrac{10}{3}\right\}$

$M = \{0\}$

$N = \{1, 2, 3, \ldots\}$

$P = \{1\}$

$S = \{1, 2, 3\}$

$T = \{x | -1 \leqslant x < 1\}$

$W = \{0, 1, 2, 3, \ldots\}$

$Z = \{\ldots -3, -2, -1, 0, 1, 2, 3, \ldots\}$

Determine whether the following are true or false. (The symbol $\notin$ means "is not a member of.")

9. $W \subset D$

10. $B \subset J$

11. $1 \in J$

12. $\{1\} \in J$

13. $-\dfrac{2}{3} \notin L$

14. $A \subset F$

15. $-\dfrac{11}{2} \in E$

16. $S \subset N \subset W \subset Z$

17. $2^7 \in C$

18. $G \subset T$

19. $1 \subset G$

20. $L \subset A$

Find these unions and intersections.

21. $C \cup E$

22. $E \cap C$

23. $M \cap G \cap P$

24. $M \cap B$

25. $T \cup P$

26. $A \cap C$

27. $A \cap F$

28. $K \cup H$

29. $K \cap H$

30. $H \cap Z$

31. $M \cup B$

32. $F \cap L$

SECTION 10.3

Solve.

33. $\dfrac{5m - 3}{4} \leqslant m + 1$

34. $\dfrac{3}{7}x - \dfrac{3}{14} < \dfrac{73}{70} - \dfrac{4}{35}x$

35. $x - 10 < 5x + 6 \leqslant x + 10$

36. $\dfrac{2}{3} \leqslant -\dfrac{4}{5}(u - 2) < 1$

37. $2 - (3 - 4a) < a - (1 - 3a)$

38. $2[5(3 - y) - 2(y - 2)] > y + 4$

39. $-\dfrac{2}{15} \leqslant \dfrac{2}{3}x - \dfrac{2}{5} \leqslant \dfrac{2}{15}$

40. $2x - \dfrac{3}{4} < -\dfrac{1}{10}$ or $2x - \dfrac{3}{4} > \dfrac{1}{10}$

41. $3x < 4 - 5x < 5 + 3x$

42. $(x + 6)(x - 4) > (x + 1)(x - 3)$

True or false. Let a, b, and c represent real numbers.

43. If $b > c$, then $b \nleqslant c$.

44. If $-b < -a$, then $a < b$.

45. If $c \neq a$, then $a < c$.

46. If $a < c$ and $c < b$, then $b \ngeqslant a$.

47. If $a < c$ and $b < c$, then $a < b$.

48. If $-a < c$ and $-c > b$, then $a < b$.

SECTION 10.4
Solve.

49. $|3x| = 7$

50. $\left|4x - \dfrac{1}{2}\right| \leqslant \dfrac{1}{12}$

51. $|3x - 4| > -2$

52. $|x - 6| \leqslant -8$

53. $\left|\dfrac{5}{9} + 3x\right| < \dfrac{1}{6}$

54. $1 - \left|\dfrac{1}{4}x + 8\right| > \dfrac{3}{4}$

55. $|x + 5| > x$

56. $2 \leqslant |x - 1| \leqslant 5$

57. $|7x - 2| = x + 4$

58. $|x + 1| \leqslant |x - 3|$

SECTION 10.5
Graph.

59. $4x - 6 > 2x + 4$

60. $y \leqslant |x|$

61. $y > x^2 - 3$

62. $y \leqslant \dfrac{1}{4}x^2 - 1$

63. Solve $2x^2 + x < 6$ by graphing.

Graph each system of inequalities.

(To graph the solution set of a system of inequalities, graph each inequality separately, and find their intersection.)

64. $x + y \leqslant 3$
    $x - y \leqslant 1$

65. $3x + 4y \geqslant 12$
    $5x + 6y \leqslant 30$

66. $2x + y \leqslant 5$
    $x + 2y \leqslant 4$
    $x \geqslant 0$
    $y \geqslant 0$

67. $y - x \leqslant 2$
    $x - y \leqslant 2$
    $2 \leqslant y \leqslant 3$

NAME

SCORE

TEST OR REVIEW—CHAPTER 10

[10.1,] **1.** Solve and graph: $6 < 3x + 2 \leqslant 12$.

[10.3,]

$$\longleftarrow \mid \; \mid \; \mid \; \mid \; \mid \; \mid \; \mid \; \mid \; \mid \; \mid \; \mid \; \mid \; \mid \; \mid \; \mid \; \mid \; \mid \; \mid \; \mid \longrightarrow$$
$$\begin{array}{cccccccccccccccccc} -9 & -8 & -7 & -6 & -5 & -4 & -3 & -2 & -1 & 0 & 1 & 2 & 3 & 4 & 5 & 6 & 7 & 8 & 9 \end{array}$$

1. _____

[10.2,] Find the solution set and write set notation for it.

2. $(x + 2)(x - 3) = 0$

2. _____

3. $4x + 3 > 3x - 2$

3. _____

4. $-16x < 48$

4. _____

[10.2,] **5.** Which of the following are true?

a) $\{2, 3\} \subset \{2, 3, 4\}$
b) $\emptyset \subset \{1, 2, 3\}$
c) $\{x \mid x > 3\} \subset \{r \mid r > 0\}$

5. _____

[10.2,] **6.** Find the intersection: $\{1, 3, 5, 7, 9\} \cap \{3, 5, 11, 13\}$.

6. _____

7. Find the union: $\{1, 3, 5, 7, 9\} \cup \{3, 5, 11, 13\}$.

7. _____

ANSWERS

[10.1,]

8. Graph: $2 < x \leqslant 5$.

9. Graph: $x \leqslant 2 \quad or \quad x > 5$.

8. See graph.

10. Also see graph.

9. See graph.

[10.4, • •]

Solve. Then graph.

10. $|3x - 2| < 7$

11. $|2x - 3| \geqslant 6$

11. Also see graph.

[10.5, • •]

12. Graph: $2x + 3y < 12$

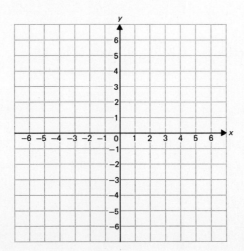

12. See graph.

NAME   SCORE   ANSWERS

FINAL EXAMINATION

This is a practice test. Be sure to review for several days before you take it. You can use the chapter tests as a guide.

1. _____

When you finish this test, check your answers in the back of the book. If you miss an item, go back and review. The bracketed numbers in the answers give you the appropriate page numbers.

2. _____

Chapter 1

1. Simplify: $|-18|$.

2. Evaluate $\dfrac{ab - ac}{bc}$ when $a = -2$, $b = 3$, and $c = -4$.

3. _____

3. Simplify:

$2y - [3 - 4(5 - 2y) - 3y]$.

4. Simplify:

$(-9x^2y^3) \cdot (5x^4y^{-7})$.

4. _____

Chapter 2

5. Solve:

$9(x - 1) - 3(x - 2) = 1$.

6. Find three consecutive even integers whose sum is 198.

5. _____

7. Solve $A = P + Prt$, for r.

8. Find an equation of the line containing the point $(-1, 4)$ and perpendicular to the line whose equation is $x - y = 6$.

6. _____

9. Graph: $3x - y = 6$.

7. _____

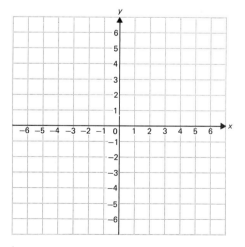

8. _____

9. See graph.

Chapter 3

10. Evaluate:

$$\begin{vmatrix} -5 & -7 \\ 4 & 6 \end{vmatrix}.$$

11. Solve:

$$5x + 3y = 2$$
$$3x + 5y = -2.$$

10. _____

12. Solve:

$$x + y - z = 0$$
$$3x + y + z = 6$$
$$x - y + 2z = 5.$$

11. _____

12. _____

13. The difference between two numbers is 6. Three times the smaller minus twice the larger is 7. Find the numbers.

13. _____

14. A bartender has two kinds of beer. Beer A is 2% alcohol and beer B is 6% alcohol. He wants to mix the two beers to get an 80-liter mixture that is 3.2% alcohol. How many liters of each should he use?

14. _____

Chapter 4

15. Add:

$$-3x^2 + 4x^3 - 5x - 1 \quad \text{and}$$
$$9x^3 - 4x^2 + 7 - x.$$

16. Subtract:

$$(5a^2 - 3ab - 7b^2) -$$
$$(2a^2 + 5ab + 8b^2).$$

15. _____

16. _____

Multiply.

17. $(2a - 1)(3a + 5)$

18. $(3x^2 - 5y)^2$

Factor.

19. $4x^2 - 12x + 9$

20. $27a^3 - 8$

21. $a^2 + 2ab + b^2 - 25$

22. Solve: $x^2 - 2x = 48$.

23. Solve for x: $ax - bx = c^2$.

24. For the function described by $f(x) = 3x^2 - 4x$, find $f(-2)$.

Chapter 5

25. Solve:
$$\frac{6}{x} + \frac{6}{x + 2} = \frac{5}{2}.$$

26. Solve $I = \dfrac{E}{R + r}$, for R.

ANSWERS

17. _____

18. _____

19. _____

20. _____

21. _____

22. _____

23. _____

24. _____

25. _____

26. _____

Perform the following operations and simplify.

27. $\dfrac{1}{x-2} - \dfrac{4}{x^2-4} + \dfrac{3}{x+2}$

28. $\dfrac{x^2-6x+8}{3x+9} \cdot \dfrac{x+3}{x^2-4}$

27. _____

29. $\dfrac{3x+3y}{5x-5y} \div \dfrac{3x^2+3y^2}{5x^2-5y^2}$

30. $\dfrac{x - \dfrac{a^2}{x}}{1 + \dfrac{a}{x}}$

28. _____

29. _____

31. Use synthetic division to find the quotient and remainder:

$(7x^4 - 5x^3 + x^2 - 4) \div (x - 2)$.

30. _____

32. An airplane can fly 190 miles with the wind in the same time it takes to fly 160 miles against the wind. The speed of the wind is 30 mph. How fast can the plane fly in still air?

31. _____

33. A can do a certain job in 21 min. B can do the same job in 14 min. How long would it take to do the job if the two worked together?

32. _____

33. _____

Chapter 6

34. Multiply. Write scientific notation for the answer.

$(8.9 \times 10^{-17})(7.6 \times 10^4)$

34. _____

35. Multiply and simplify:

$\sqrt{8x} \cdot \sqrt{8x^3y}.$

35. _____

36. Divide and simplify:

$\dfrac{\sqrt[3]{15x}}{\sqrt[3]{3y^2}}.$

36. _____

37. Rationalize the denominator:

$\dfrac{1 - \sqrt{x}}{1 + \sqrt{x}}.$

38. Write a single radical expression:

$\dfrac{\sqrt[3]{(x + 1)^5}}{\sqrt{(x + 1)^3}}.$

37. _____

38. _____

39. Solve:

$\sqrt{x - 5} = 5 - \sqrt{x}.$

40. Multiply these complex numbers:

$(3 + 2i)(4 - 7i).$

39. _____

40. _____

Chapter 7

Solve.

41. $7x^2 + 14 = 0$

42. $x^2 + 4x = 3$

41. _____

42. _____

43. Write a quadratic equation whose solutions are $5\sqrt{2}$ and $-5\sqrt{2}$.

44. Solve:

$$x^4 - 29x^2 + 100 = 0.$$

45. Graph $f(x) = -2(x - 3)^2 + 1$.

a) Label the vertex.
b) Draw the line of symmetry.
c) Find the maximum or minimum value.

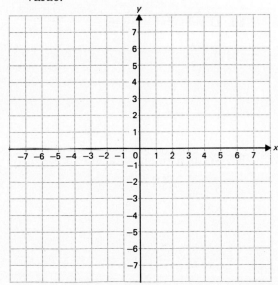

43. _____

44. _____

46. The centripetal force F of an object moving in a circle varies directly as the square of the velocity v and inversely as the radius r of the circle. If $F = 8$ when $v = 1$ and $r = 10$, what is F when $v = 2$ and $r = 16$?

45. See graph. _____

47. A farmer wants to fence in a rectangular area next to a river. (Note that no fence will be needed along the river.) What is the area of the largest region that can be fenced in with 100 ft of fencing?

46. _____

River

w w

$100 - 2w$

47. _____

Chapter 8

48. Solve:

$$x^2 + y^2 = 8$$
$$x^2 - y^2 = 2.$$

49. The perimeter of a rectangle is 34 ft. The length of a diagonal is 13 ft. Find its dimensions.

48. _____

50. Find the center and radius of the circle:

$$x^2 + y^2 - 4x + 6y - 23 = 0.$$

51. Graph:

$$\frac{x^2}{25} + \frac{y^2}{4} = 1.$$

49. _____

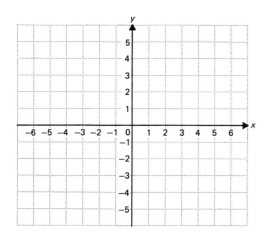

50. _____

51. See graph.

Chapter 9

52. Graph: $y = \log_2 x.$

53. Express as a single logarithm:

$$\frac{2}{3} \log_a x - \frac{1}{2} \log_a y + 5 \log_a z.$$

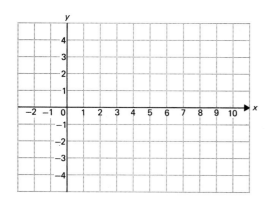

52. See graph.

53. _____

54. Write an exponential equation with the same meaning as $\log_a c = 5$.

55. Solve:
$$\log (x^2 - 1) - \log (x + 1) = 2.$$

54. _____

Find each of the following, using Table 2.

56. $\log 14.1$

57. antilog $7.7808 - 10$

55. _____

58. Solve:
$$5^x = 8.$$

59. Solve:
$$\log (x^2 - 25) - \log (x + 5) = 3.$$

56. _____

57. _____

Chapter 10

60. Find the intersection: $\{-2, -1, 0, 1, 2\} \cap \{0, 2, 4, 6\}$.

58. _____

61. Solve. Then graph. $|2x - 1| \leqslant 5$.

62. Solve. Then graph. $|2y + 3| > 7$.

59. _____

63. Graph: $2x - 3y < -6$.

60. _____

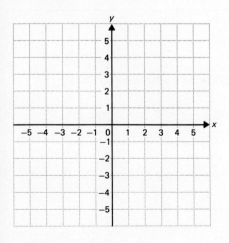

61. _____

62. _____

63. See graph. _____

APPENDIXES AND TABLES

APPENDIX 1: GEOMETRIC FORMULAS

Plane Geometry:

Rectangle

Area: $A = lw$

Perimeter: $P = 2l + 2w$

Square

Area: $A = s^2$

Perimeter: $P = 4s$

Triangle

Area: $A = \frac{1}{2}bh$

Sum of Angle Measures:

$A + B + C = 180°$

Right Triangle

Pythagorean Theorem:

$a^2 + b^2 = c^2$

Parallelogram

Area: $A = bh$

Trapezoid

Area: $A = \frac{1}{2}h(a + b)$

Circle

Area: $A = \pi r^2$

Circumference:

$C = \pi D = 2\pi r$

($\frac{22}{7}$ and 3.14 are different approximations for π)

Solid Geometry:

Rectangular Solid

Volume: $V = lwh$

Cube

Volume: $V = s^3$

Right Circular Cylinder

Volume: $V = \pi r^2 h$

Lateral Surface Area: $L = 2\pi rh$

Total Surface Area:

$S = 2\pi rh + 2\pi r^2$

Right Circular Cone

Volume: $V = \frac{1}{3}\pi r^2 h$

Lateral Surface Area: $L = \pi rs$

Total Surface Area: $S = \pi r^2 + \pi rs$

Slant Height: $s = \sqrt{r^2 + h^2}$

Sphere

Volume: $V = \frac{4}{3}\pi r^3$

Surface Area: $S = 4\pi r^2$

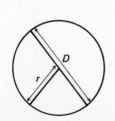

APPENDIX 2: MORE ABOUT FUNCTIONS

REVIEW OF FUNCTIONS

Before studying this appendix you may wish to review the ideas of functions presented earlier in the book. They are to be found on the following pages.

213–215
393–398
401–402
453–455
459

▣ FINDING FUNCTION VALUES

When a function is given by a formula or equation, function values are found by substituting into the formula. That is true even if the inputs are not specific numbers. We may use such things as $2m$ or $a + h$ for inputs, but we still proceed by substituting.

Example 1   The function g given by $g(x) = x^2 - 1$ takes an input x and squares it. Then it subtracts 1 from the result. Find the following function values.

a) $g(-3) = (-3)^2 - 1$ Substituting the input, -3

$\qquad = 9 - 1$ Squaring -3

$\qquad = 8$ Simplifying

b) $g(3m) = (3m)^2 - 1$ Substituting the input, $3m$

$\qquad = 9m^2 - 1$

c) $g(a + h) = (a + h)^2 - 1$ Substituting the input $a + h$

$\qquad = a^2 + 2ah + h^2 - 1$ Squaring $a + h$

d) $g(a) + g(a + h)$

$\qquad = [a^2 - 1] + [(a + h)^2 - 1]$ Substituting the inputs a, and $a + h$

$\qquad = [a^2 - 1] + [a^2 + 2ah + h^2 - 1]$ Squaring $a + h$

$\qquad = a^2 - 1 + a^2 + 2ah + h^2 - 1$ Removing parentheses

$\qquad = 2a^2 + 2ah + h^2 - 2$ Collecting like terms

DO EXERCISES 1–3.

1. For the function f given by $f(x) = 5x - 3$, find the following.

   a) $f(6)$

   b) $f(-4)$

   c) $f(h)$

   d) $f(a + h)$

2. For the function g given by $g(x) = x^2 + 5$, find the following.

   a) $g(2.1)$

   b) $g(-7)$

   c) $g(2y)$

   d) $g(h)$

   e) $g(a + h)$

3. For the function F given by $F(x) = 3x^2 + 1$, find the following.

   a) $F(-2)$

   b) $F(h)$

   c) $F(a + h)$

   d) $F(a + h) - F(a)$

● ● INVERSES

On page 455 we considered the inverse of

$$y = 2^x.$$

The inverse can be obtained by interchanging x and y, to obtain

$$x = 2^y.$$

If for any equation with two variables, such as x and y, we interchange the variables, we obtain an equation of the inverse relation. That inverse may or may not be a function. If the inverse of a function f is also a function, it can be named f^{-1} (read "f-inverse").

The -1 in f^{-1} is not an exponent.

Example 2   Given $f(x) = x + 1$, find a formula for $f^{-1}(x)$.

a) Let us think of this as $y = x + 1$.

b) To find the inverse we interchange x and y: $x = y + 1$.

c) Now we solve for y: $y = x - 1$.

d) Thus $f^{-1}(x) = x - 1$.

Note in Example 2 that f maps any x onto $x + 1$ (this function adds 1 to each number of the domain). Its inverse, f^{-1}, maps any number x onto $x - 1$ (this inverse function subtracts 1 from each member of its domain). Thus the function and its inverse do opposite things.

Example 3   Let $g(x) = 2x + 3$. Find a formula for $g^{-1}(x)$.

a) Let us think of this as $y = 2x + 3$.

b) To find the inverse we interchange x and y: $x = 2y + 3$.

c) Now we solve for y: $y = (x - 3)/2$.

d) Thus $g^{-1}(x) = (x - 3)/2$.

Note in Example 3 that g maps any x onto $2x + 3$ (this function doubles each input and then adds 3). Its inverse, g^{-1}, maps any input onto $(x - 3)/2$ (this inverse function subtracts 3 from each input and then divides it by 2). Thus the function and its inverse do opposite things.

To find a formula for the inverse of a function, if the inverse is also a function,

a) write y instead of $f(x)$;
b) interchange x and y;
c) solve for y;
d) write $f^{-1}(x)$ instead of y.

Example 4   Consider $f(x) = x^2$. Let us think of this as $y = x^2$.

To find the inverse we interchange x and y: $x = y^2$. Now we solve for y: $y = \pm\sqrt{x}$. Note that for each positive x we now get two y's. For example, if x is 4, y can be either 2 or -2, as shown in the graph. Note also that the inverse fails the vertical line test.

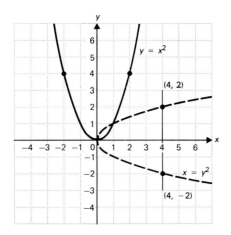

If we restrict the domain of $f(x) = x^2$ to nonnegative numbers, then its inverse is a function, $f^{-1}(x) = \sqrt{x}$.

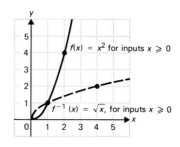

DO EXERCISES 4 AND 5.

Let us consider inverses of functions in terms of function machines. Suppose that the function f programmed into a machine has an inverse that is also a function. Suppose then that the function machine has a reverse switch. When the switch is thrown, the machine is then programmed to do the inverse mapping f^{-1}. Inputs then enter at the opposite end and the entire process is reversed.

4. Let $g(x) = 3x - 2$. Find a formula for $g^{-1}(x)$.

5. Let $f(x) = x^2 - 1$, where the domain of f is the set of all real numbers greater than -1. Find a formula for $f^{-1}(x)$.

●●● TRANSLATIONS

In Section 7.8 we considered graphs of quadratic functions. We compared graphs of

$$f(x) = ax^2 \quad \text{and} \quad f(x) = a(x - h)^2,$$

where h is some constant. We noted that replacing x by $x - h$ moved the graph to the right or left. If h is positive, as in $x - 2$ or $x - 5$ (a positive number being *subtracted* from x), then the graph is moved to the right (in the positive direction). We compare, for example, the graphs of $f(x) = x^2$ and $f(x) = (x - 3)^2$.

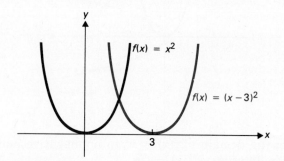

If the constant h is negative, as in $x + 3$ or $x + 7$ [a negative number being *subtracted* from x, as in $x - (-3)$], then the graph is moved in the negative direction.

A similar thing happens with the graph of any function.

> **If in $y = f(x)$ we replace x by $x - h$, where h is a constant, the graph will be moved according to h. If h is positive, the graph is moved in the positive direction, a distance h. If h is negative, the graph is moved in the negative direction.**

Example 5   Graph $f(x) = 2^{x-1}$.

We know what the graph of $f(x) = 2^x$ looks like. (See page 453.)

The graph of $f(x) = 2^{x-1}$ will look just like it, except that it will be moved one unit to the right.

What we have said about replacing x by $x - h$ also holds for replacing y by $y - k$, or $f(x)$ by $f(x) - k$, where k is some constant.

> If in $y = f(x)$ we replace y by $y - k$, where k is a constant, the graph will be moved according to k. If k is positive, the graph is moved in the positive direction (up) a distance k. If k is negative, the graph is moved in the negative direction.

Example 6   Graph $y = x^2 + 3$.

$$y = x^2 \mid 3$$

$$y - 3 = x^2 \qquad \text{Subtracting 3 to get an expression } y - k$$

We know how to graph $y = x^2$. To graph $y - 3 = x^2$, we move that graph up (in the positive y-direction) a distance of 3.

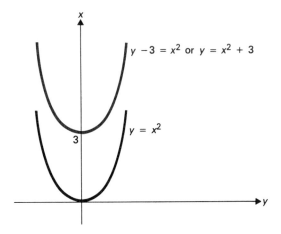

For functions we are more inclined to write $y = x^2 + 3$ than $y - 3 = x^2$. If you wish, you can remember the rule stated accordingly, as follows.

> If in $y = f(x)$, we replace $f(x)$ by $f(x) + k$, where k is some constant, the graph will be moved according to k. If k is positive, the graph is moved upward a distance k. If k is negative, the graph is moved downward.

DO EXERCISES 6 AND 7.

6. Graph $f(x) = 2^x$ and

   a) $f(x) = 2^{x-3}$

   b) $f(x) = 2^{x+1}$

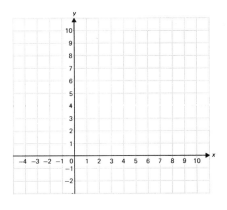

7. Graph $y = x^2$ and

   a) $y = x^2 + 5$

   b) $y = x^2 - 3$

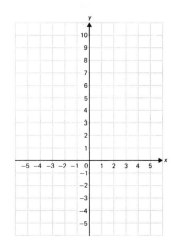

ANSWERS ON PAGE A−48

NAME   CLASS

EXERCISE SET (APPENDIX 2)

● Find the following function values.

1. $f(x) = 5x^2 + 4x$.
   a) $f(0)$;   b) $f(-1)$;
   c) $f(3)$;   d) $f(t)$;

   e) $f(t-1)$;   f) $\dfrac{f(a+h)-f(a)}{h}$.

2. $g(x) = 3x^2 - 2x + 1$.
   a) $g(0)$;   b) $g(-1)$;
   c) $g(3)$;   d) $g(t)$;

   e) $g(a+h)$;   f) $\dfrac{g(a+h)-g(a)}{h}$.

1. _____

2. _____

3. $f(x) = 2|x| + 3x$.
   a) $f(1)$;   b) $f(-2)$;
   c) $f(-4)$;   d) $f(2y)$;

   e) $f(a+h)$;   f) $\dfrac{f(a+h)-f(a)}{h}$.

4. $g(x) = x^3 - 2x$.
   a) $g(1)$;   b) $g(-2)$;
   c) $g(-4)$;   d) $g(3y)$;

   e) $g(2+h)$;   f) $\dfrac{g(2+h)-g(2)}{h}$.

3. _____

4. _____

5. ▦ $f(x) = 4.3x^2 - 1.4x$.
   a) $f(1.034)$;   b) $f(-3.441)$;
   c) $f(27.35)$;   d) $f(-16.31)$.

6. ▦ $g(x) = \sqrt{2.2|x| + 3.5}$.
   a) $g(17.3)$;   b) $g(-64.2)$;
   c) $g(0.095)$;   d) $g(-6.33)$.

5. _____

6. _____

7. $f(x) = \dfrac{x^2 - x - 2}{2x^2 - 5x - 3}$.

   a) $f(0)$;   b) $f(4)$;
   c) $f(-1)$;   d) $f(3)$.

8. $s(x) = \sqrt{\dfrac{3x - 4}{2x + 5}}$.

   a) $s(10)$;   b) $s(2)$;
   c) $s(1)$;   d) $s(-1)$.

7. _____

8. _____

●● Find inverses of the following.

9. $f(x) = 3x + 2$

10. $g(x) = 5x - 3$

11. $g(x) = \dfrac{2x + 1}{3}$

12. $f(x) = \dfrac{3x + 2}{5}$

13. $g(x) = 2x^2 + 3$
(Restrict x to being greater than 3.)

14. $f(x) = 3x^2 - 2$
(Restrict x to being greater than -2.)

15. $f(x) = 10^x$

16. $g(x) = \log_{10} x$

●●● Graph the following.

17. a) $f(x) = 3^x$
b) $f(x) = 3^{x+2}$
c) $f(x) = 3^{x-1}$

18. a) $g(x) = 2x^2$
b) $g(x) = 2x^2 + 3$
c) $g(x) = 2x^2 - 4$

19. a) $y = x^2$
b) $y = (x - 3)^2$
c) $y = (x + 4)^2$

20. a) $y = 2x + 5$
b) $y = 2(x + 4) + 5$
c) $y = 2(x - 3) + 5$

21. a) $f(x) = \log_{10} x$
b) $f(x) = \log_{10}(x - 5)$
c) $f(x) = \log_{10}(x + 4)$

22. a) $g(x) = \log_2 x$
b) $g(x) = \log_2(x - 4)$
c) $g(x) = \log_2(x + 6)$

23. a) $y = 2x^2$
b) $y = 2(x - 4)^2 + 5$
c) $y = 2(x + 3)^2 - 4$

24. a) $y = \log_{10} x$
b) $y = \log_{10}(x + 2) + 4$
c) $y = \log_{10}(x - 5) - 7$

For each function find $\dfrac{f(x + h) - f(x)}{h}$.

25. $f(x) = \dfrac{1}{x}$

26. $f(x) = \dfrac{1}{x^2}$

27. $f(x) = \sqrt{x}$ (Rationalize the numerator.)

TABLE 1

POWERS, ROOTS, AND RECIPROCALS

n	n^2	n^3	$\sqrt{n}$	$\sqrt[3]{n}$	$\dfrac{1}{n}$	n	n^2	n^3	$\sqrt{n}$	$\sqrt[3]{n}$	$\dfrac{1}{n}$
1	1	1	1.000	1.000	1.0000	51	2,601	132,651	7.141	3.708	.0196
2	4	8	1.414	1.260	.5000	52	2,704	140,608	7.211	3.733	.0192
3	9	27	1.732	1.442	.3333	53	2,809	148,877	7.280	3.756	.0189
4	16	64	2.000	1.587	.2500	54	2,916	157,464	7.348	3.780	.0185
5	25	125	2.236	1.710	.2000	55	3,025	166,375	7.416	3.803	.0182
6	36	216	2.449	1.817	.1667	56	3,136	175,616	7.483	3.826	.0179
7	49	343	2.646	1.913	.1429	57	3,249	185,193	7.550	3.849	.0175
8	64	512	2.828	2.000	.1250	58	3,364	195,112	7.616	3.871	.0172
9	81	729	3.000	2.080	.1111	59	3,481	205,379	7.681	3.893	.0169
10	100	1,000	3.162	2.154	.1000	60	3,600	216,000	7.746	3.915	.0167
11	121	1,331	3.317	2.224	.0909	61	3,721	226,981	7.810	3.936	.0164
12	144	1,728	3.464	2.289	.0833	62	3,844	238,328	7.874	3.958	.0161
13	169	2,197	3.606	2.351	.0769	63	3,969	250,047	7.937	3.979	.0159
14	196	2,744	3.742	2.410	.0714	64	4,096	262,144	8.000	4.000	.0156
15	225	3,375	3.873	2.466	.0667	65	4,225	274,625	8.062	4.021	.0154
16	256	4,096	4.000	2.520	.0625	66	4,356	287,496	8.124	4.041	.0152
17	289	4,913	4.123	2.571	.0588	67	4,489	300,763	8.185	4.062	.0149
18	324	5,832	4.243	2.621	.0556	68	4,624	314,432	8.246	4.082	.0147
19	361	6,859	4.359	2.668	.0526	69	4,761	328,509	8.307	4.102	.0145
20	400	8,000	4.472	2.714	.0500	70	4,900	343,000	8.367	4.121	.0143
21	441	9,261	4.583	2.759	.0476	71	5,041	357,911	8.426	4.141	.0141
22	484	10,648	4.690	2.802	.0455	72	5,184	373,248	8.485	4.160	.0139
23	529	12,167	4.796	2.844	.0435	73	5,329	389,017	8.544	4.179	.0137
24	576	13,824	4.899	2.884	.0417	74	5,476	405,224	8.602	4.198	.0135
25	625	15,625	5.000	2.924	.0400	75	5,625	421,875	8.660	4.217	.0133
26	676	17,576	5.099	2.962	.0385	76	5,776	438,976	8.718	4.236	.0132
27	729	19,683	5.196	3.000	.0370	77	5,929	456,533	8.775	4.254	.0130
28	784	21,952	5.292	3.037	.0357	78	6,084	474,552	8.832	4.273	.0128
29	841	24,389	5.385	3.072	.0345	79	6,241	493,039	8.888	4.291	.0127
30	900	27,000	5.477	3.107	.0333	80	6,400	512,000	8.944	4.309	.0125
31	961	29,791	5.568	3.141	.0323	81	6,561	531,441	9.000	4.327	.0123
32	1,024	32,768	5.657	3.175	.0312	82	6,724	551,368	9.055	4.344	.0122
33	1,089	35,937	5.745	3.208	.0303	83	6,889	571,787	9.110	4.362	.0120
34	1,156	39,304	5.831	3.240	.0294	84	7,056	592,704	9.165	4.380	.0119
35	1,225	42,875	5.916	3.271	.0286	85	7,225	614,125	9.220	4.397	.0118
36	1,296	46,656	6.000	3.302	.0278	86	7,396	636,056	9.274	4.414	.0116
37	1,369	50,653	6.083	3.332	.0270	87	7,569	658,503	9.327	4.431	.0115
38	1,444	54,872	6.164	3.362	.0263	88	7,744	681,472	9.381	4.448	.0114
39	1,521	59,319	6.245	3.391	.0256	89	7,921	704,969	9.434	4.465	.0112
40	1,600	64,000	6.325	3.420	.0250	90	8,100	729,000	9.487	4.481	.0111
41	1,681	68,921	6.403	3.448	.0244	91	8,281	753,571	9.539	4.498	.0110
42	1,764	74,088	6.481	3.476	.0238	92	8,464	778,688	9.592	4.514	.0109
43	1,849	79,507	6.557	3.503	.0233	93	8,649	804,357	9.644	4.531	.0108
44	1,936	85,184	6.633	3.530	.0227	94	8,836	830,584	9.695	4.547	.0106
45	2,025	91,125	6.708	3.557	.0222	95	9,025	857,375	9.747	4.563	.0105
46	2,116	97,336	6.782	3.583	.0217	96	9,216	884,736	9.798	4.579	.0104
47	2,209	103,823	6.856	3.609	.0213	97	9,409	912,673	9.849	4.595	.0103
48	2,304	110,592	6.928	3.634	.0208	98	9,604	941,192	9.899	4.610	.0102
49	2,401	117,649	7.000	3.659	.0204	99	9,801	970,299	9.950	4.626	.0101
50	2,500	125,000	7.071	3.684	.0200	100	10,000	1,000,000	10.000	4.642	.0100

TABLE 2

COMMON LOGARITHMS

x	0	1	2	3	4	5	6	7	8	9
1.0	.0000	.0043	.0086	.0128	.0170	.0212	.0253	.0294	.0334	.0374
1.1	.0414	.0453	.0492	.0531	.0569	.0607	.0645	.0682	.0719	.0755
1.2	.0792	.0828	.0864	.0899	.0934	.0969	.1004	.1038	.1072	.1106
1.3	.1139	.1173	.1206	.1239	.1271	.1303	.1335	.1367	.1399	.1430
1.4	.1461	.1492	.1523	.1553	.1584	.1614	.1644	.1673	.1703	.1732
1.5	.1761	.1790	.1818	.1847	.1875	.1903	.1931	.1959	.1987	.2014
1.6	.2041	.2068	.2095	.2122	.2148	.2175	.2201	.2227	.2253	.2279
1.7	.2304	.2330	.2355	.2380	.2405	.2430	.2455	.2480	.2504	.2529
1.8	.2553	.2577	.2601	.2625	.2648	.2672	.2695	.2718	.2742	.2765
1.9	.2788	.2810	.2833	.2856	.2878	.2900	.2923	.2945	.2967	.2989
2.0	.3010	.3032	.3054	.3075	.3096	.3118	.3139	.3160	.3181	.3201
2.1	.3222	.3243	.3263	.3284	.3304	.3324	.3345	.3365	.3385	.3404
2.2	.3424	.3444	.3464	.3483	.3502	.3522	.3541	.3560	.3579	.3598
2.3	.3617	.3636	.3655	.3674	.3692	.3711	.3729	.3747	.3766	.3784
2.4	.3802	.3820	.3838	.3856	.3874	.3892	.3909	.3927	.3945	.3962
2.5	.3979	.3997	.4014	.4031	.4048	.4065	.4082	.4099	.4116	.4133
2.6	.4150	.4166	.4183	.4200	.4216	.4232	.4249	.4265	.4281	.4298
2.7	.4314	.4330	.4346	.4362	.4378	.4393	.4409	.4425	.4440	.4456
2.8	.4472	.4487	.4502	.4518	.4533	.4548	.4564	.4579	.4594	.4609
2.9	.4624	.4639	.4654	.4669	.4683	.4698	.4713	.4728	.4742	.4757
3.0	.4771	.4786	.4800	.4814	.4829	.4843	.4857	.4871	.4886	.4900
3.1	.4914	.4928	.4942	.4955	.4969	.4983	.4997	.5011	.5024	.5038
3.2	.5051	.5065	.5079	.5092	.5105	.5119	.5132	.5145	.5159	.5172
3.3	.5185	.5198	.5211	.5224	.5237	.5250	.5263	.5276	.5289	.5302
3.4	.5315	.5328	.5340	.5353	.5366	.5378	.5391	.5403	.5416	.5428
3.5	.5441	.5453	.5465	.5478	.5490	.5502	.5514	.5527	.5539	.5551
3.6	.5563	.5575	.5587	.5599	.5611	.5623	.5635	.5647	.5658	.5670
3.7	.5682	.5694	.5705	.5717	.5729	.5740	.5752	.5763	.5775	.5786
3.8	.5798	.5809	.5821	.5832	.5843	.5855	.5866	.5877	.5888	.5899
3.9	.5911	.5922	.5933	.5944	.5955	.5966	.5977	.5988	.5999	.6010
4.0	.6021	.6031	.6042	.6053	.6064	.6075	.6085	.6096	.6107	.6117
4.1	.6128	.6138	.6149	.6160	.6170	.6180	.6191	.6201	.6212	.6222
4.2	.6232	.6243	.6253	.6263	.6274	.6284	.6294	.6304	.6314	.6325
4.3	.6335	.6345	.6355	.6365	.6375	.6385	.6395	.6405	.6415	.6425
4.4	.6435	6444	.6454	.6464	.6474	.6484	.6493	.6503	.6513	.6522
4.5	.6532	.6542	.6551	.6561	.6571	.6580	.6590	.6599	.6609	.6618
4.6	.6628	.6637	.6646	.6656	.6665	.6675	.6684	.6693	.6702	.6712
4.7	.6721	.6730	.6739	.6749	.6758	.6767	.6776	.6785	.6794	.6803
4.8	.6812	.6821	.6830	.6839	.6848	.6857	.6866	.6875	.6884	.6893
4.9	.6902	.6911	.6920	.6928	.6937	.6946	.6955	.6964	.6972	.6981
5.0	.6990	.6998	.7007	.7016	.7024	.7033	.7042	.7050	.7059	.7067
5.1	.7076	.7084	.7093	.7101	.7110	.7118	.7126	.7135	.7143	.7152
5.2	.7160	.7168	.7177	.7185	.7193	.7202	.7210	.7218	.7226	.7235
5.3	.7243	.7251	.7259	.7267	.7275	.7284	.7292	.7300	.7308	.7316
5.4	.7324	.7332	.7340	.7348	.7356	.7364	.7372	.7380	.7388	.7396
x	0	1	2	3	4	5	6	7	8	9

TABLE 2 *(cont.)*

x	0	1	2	3	4	5	6	7	8	9
5.5	.7404	.7412	.7419	.7427	.7435	.7443	.7451	.7459	.7466	.7474
5.6	.7482	.7490	.7497	.7505	.7513	.7520	.7528	.7536	.7543	.7551
5.7	.7559	.7566	.7574	.7582	.7589	.7597	.7604	.7612	.7619	.7627
5.8	.7634	.7642	.7649	.7657	.7664	.7672	.7679	.7686	.7694	.7701
5.9	.7709	.7716	.7723	.7731	.7738	.7745	.7752	.7760	.7767	.7774
6.0	.7782	.7789	.7796	.7803	.7810	.7818	.7825	.7832	.7839	.7846
6.1	.7853	.7860	.7868	.7875	.7882	.7889	.7896	.7903	.7910	.7917
6.2	.7924	.7931	.7938	.7945	.7952	.7959	.7966	.7973	.7980	.7987
6.3	.7993	.8000	.8007	.8014	.8021	.8028	.8035	.8041	.8048	.8055
6.4	.8062	.8069	.8075	.8082	.8089	.8096	.8102	.8109	.8116	.8122
6.5	.8129	.8136	.8142	.8149	.8156	.8162	.8169	.8176	.8182	.8189
6.6	.8195	.8202	.8209	.8215	.8222	.8228	.8235	.8241	.8248	.8254
6.7	.8261	.8267	.8274	.8280	.8287	.8293	.8299	.8306	.8312	.8319
6.8	.8325	.8331	.8338	.8344	.8351	.8357	.8363	.8370	.8376	.8382
6.9	.8388	.8395	.8401	.8407	.8414	.8420	.8426	.8432	.8439	.8445
7.0	.8451	.8457	.8463	.8470	.8476	.8482	.8488	.8494	.8500	.8506
7.1	.8513	.8519	.8525	.8531	.8537	.8543	.8549	.8555	.8561	.8567
7.2	.8573	.8579	.8585	.8591	.8597	.8603	.8609	.8615	.8621	.8627
7.3	.8633	.8639	.8645	.8651	.8657	.8663	.8669	.8675	.8681	.8686
7.4	.8692	.8698	.8704	.8710	.8716	.8722	.8727	.8733	.8739	.8745
7.5	.8751	.8756	.8762	.8768	.8774	.8779	.8785	.8791	.8797	.8802
7.6	.8808	.8814	.8820	.8825	.8831	.8837	.8842	.8848	.8854	.8859
7.7	.8865	.8871	.8876	.8882	.8887	.8893	.8899	.8904	.8910	.8915
7.8	.8921	.8927	.8932	.8938	.8943	.8949	.8954	.8960	.8965	.8971
7.9	.8976	.8982	.8987	.8993	.8998	.9004	.9009	.9015	.9020	.9025
8.0	.9031	.9036	.9042	.9047	.9053	.9058	.9063	.9069	.9074	.9079
8.1	.9085	.9090	.9096	.9101	.9106	.9112	.9117	.9122	.9128	.9133
8.2	.9138	.9143	.9149	.9154	.9159	.9165	.9170	.9175	.9180	.9186
8.3	.9191	.9196	.9201	.9206	.9212	.9217	.9222	.9227	.9232	.9238
8.4	.9243	.9248	.9253	.9258	.9263	.9269	.9274	.9279	.9284	.9289
8.5	.9294	.9299	.9304	.9309	.9315	.9320	.9325	.9330	.9335	.9340
8.6	.9345	.9350	.9555	.9360	.9365	.9370	.9375	.9380	.9385	.9390
8.7	.9395	.9400	.9405	.9410	.9415	.9420	.9425	.9430	.9435	.9440
8.8	.9445	.9450	.9455	.9460	.9465	.9469	.9474	.9479	.9484	.9489
8.9	.9494	.9499	.9504	.9509	.9513	.9518	.9523	.9528	.9533	.9538
9.0	.9542	.9547	.9552	.9557	.9562	.9566	.9571	.9576	.9581	.9586
9.1	.9590	.9595	.9600	.9605	.9609	.9614	.9619	.9624	.9628	.9633
9.2	.9638	.9643	.9647	.9652	.9657	.9661	.9666	.9671	.9675	.9680
9.3	.9685	.9689	.9694	.9699	.9703	.9708	.9713	.9717	.9722	.9727
9.4	.9731	.9736	.9741	.9745	.9750	.9754	.9759	.9763	.9768	.9773
9.5	.9777	.9782	.9786	.9791	.9795	.9800	.9805	.9809	.9814	.9818
9.6	.9823	.9827	.9832	.9836	.9841	.9845	.9850	.9854	.9859	.9863
9.7	.9868	.9872	.9877	.9881	.9886	.9890	.9894	.9899	.9903	.9908
9.8	.9912	.9917	.9921	.9926	.9930	.9934	.9939	.9943	.9948	.9952
9.9	.9956	.9961	.9965	.9969	.9974	.9978	.9983	.9987	.9991	.9996
x	0	1	2	3	4	5	6	7	8	9

MARGIN EXERCISES, SECTION 1.1, pp. 2–5

1. 1, 17   **2.** 1, 0, 17   **3.** 1, -5, 0, 17   **4.** 0.875   **5.** $0.63\overline{63}$ **6.** $1.\overline{285714}$ **7.** $\dfrac{59}{10}$ **8.** $\dfrac{125}{1000}$, or $\dfrac{1}{8}$ **9.** 0.035, $\dfrac{35}{1000}$, or $\dfrac{7}{200}$

10. 0.384, $\dfrac{384}{1000}$ **11.** 0.375, 37.5%   **12.** 0.85, 85%   **13.** 0.35%   **14.** 2.35   **15.** $-5 < -4$ **16.** $-\dfrac{1}{4} < \dfrac{1}{2}$ **17.** $87 > 67$

18. $-9.8 < -4.5$ **19.** 2   **20.** $\dfrac{3}{2}$ **21.** 0   **22.** $\dfrac{1}{4}$

SOMETHING EXTRA—REPEATING DECIMALS, p. 6

1. $\dfrac{5}{9}$ **3.** $\dfrac{1}{6}$ **5.** $\dfrac{4}{33}$ **7.** $\dfrac{113}{900}$

EXERCISE SET 1.1, pp. 7–8

1. 2, 14   **3.** All are rational.   **5.** 0.375   **7.** 1.666...   **9.** $\dfrac{27}{10}$ **11.** $\dfrac{145}{1000}$ **13.** 80%   **15.** 37.5%   **17.** 240%   **19.** 3.75%

21. 41.2%   **23.** 0.1%   **25.** 125%   **27.** 0.567, $\dfrac{567}{1000}$ **29.** 0.0685, $\dfrac{685}{10,000}$, or $\dfrac{137}{2000}$ **31.** 1.2, $\dfrac{12}{10}$, or $\dfrac{6}{5}$ **33.** 0.005, $\dfrac{5}{1000}$, or $\dfrac{1}{200}$

35. $55 > 40$ **37.** $2.3 < 3.2$ **39.** $-10 < -4$ **41.** $-2.3 > -3.2$ **43.** 16   **45.** 32   **47.** 0

MARGIN EXERCISES, SECTION 1.2, pp. 9–12

1. 2   **2.** 4   **3.** -4 **4.** 0   **5.** -18 **6.** -17 **7.** -29 **8.** -18.6 **9.** $-\dfrac{29}{5}$ **10.** $-\dfrac{7}{10}$ **11.** $-\dfrac{3}{2}$ **12.** 0   **13.** 0

14. 0   **15.** -3 **16.** 12   **17.** -3.3 **18.** $\dfrac{7}{3}$ **19.** $-\dfrac{1}{4}$ **20.** $-\dfrac{1}{8}$ **21.** $\dfrac{1}{10}$ **22.** -9 **23.** 23   **24.** -5 **25.** 0

26. (a) -11; (b) 17; (c) 0; (d) $-x$; (e) x **27.** -5 **28.** -3 **29.** 3   **30.** 17   **31.** -16 **32.** -29.6 **33.** 1   **34.** $-\dfrac{9}{10}$

35. $\dfrac{3}{2}$ **36.** (a) -6; (b) -40; (c) 6   **37.** 53   **38.** 92   **39.** Associative law of addition   **40.** Commutative law of addition

41. Associative law of addition

EXERCISE SET 1.2, pp. 13–14

1. -28 **3.** -16 **5.** 5   **7.** 4   **9.** -7 **11.** -24 **13.** 1.2   **15.** -8.86 **17.** $\dfrac{1}{7}$ **19.** $-\dfrac{4}{3}$ **21.** $\dfrac{1}{10}$ **23.** -7 **25.** 2.8

27. -10 **29.** 0   **31.** -2 **33.** -12 **35.** 5   **37.** 15   **39.** -11.6 **41.** -29.25 **43.** $-\dfrac{7}{2}$ **45.** $-\dfrac{5}{12}$

47. Commutative law of addition   **49.** Associative law of addition   **51.** Commutative and associative laws of addition

MARGIN EXERCISES, SECTION 1.3, pp. 15–18

1. 10, 5, 0, -5, -10, -15, -20, -25, -30 **2.** -24 **3.** -28.35 **4.** -8 **5.** -10, -5, 0, 5, 10, 15, 20, 25   **6.** 72   **7.** $\dfrac{8}{15}$

8. 42.77   **9.** $\dfrac{21}{28}$ **10.** $\dfrac{-8}{-6}$ **11.** $\dfrac{9}{-2}$ **12.** $\dfrac{-14}{5}$ **13.** $\dfrac{17}{25}$ **14.** 2   **15.** -25 **16.** -3 **17.** 0.2   **18.** Not possible

19. Yes; answer is 0   **20.** Not possible   **21.** Not possible   **22.** $\dfrac{8}{3}$ **23.** $-\dfrac{5}{4}$ **24.** $\dfrac{1}{18}$ **25.** $-\dfrac{1}{4.3}$, or $-\dfrac{10}{43}$ **26.** $\dfrac{1}{0.5}$, or 2

$\dfrac{4}{5}$	$-\dfrac{4}{5}$	$\dfrac{5}{4}$
$-\dfrac{3}{4}$	$\dfrac{3}{4}$	$-\dfrac{4}{3}$
0.25	-0.25	4
8	-8	$\dfrac{1}{8}$
-5	5	$-\dfrac{1}{5}$
0	0	Does not exist

27. (table above)

28. $-\dfrac{12}{5}$ **29.** $\dfrac{5}{8}$ **30.** $-\dfrac{3}{10}$ **31.** $-\dfrac{6}{7}$ **32.** $\dfrac{36}{7}$

EXERCISE SET 1.3, pp. 19–20

1. -21 **3.** -8 **5.** 16 **7.** -126 **9.** 34.2 **11.** 26.46 **13.** 2 **15.** 60 **17.** $-\dfrac{12}{35}$ **19.** 1 **21.** $-\dfrac{8}{27}$ **23.** -2

25. -7 **27.** 7 **29.** 0.3 **31.** Not possible **33.** 0 **35.** Not possible **37.** $\dfrac{4}{3}$ **39.** $-\dfrac{8}{7}$ **41.** $\dfrac{1}{15}$, or $0.06\overline{6}$

43. $\dfrac{1}{4.4}$, or $0.227\overline{27}$ **45.** $-\dfrac{6}{77}$ **47.** 25 **49.** -6 **51.** 5 **53.** $46{,}871{,}451$ **55.** $\dfrac{3}{10}$

MARGIN EXERCISES, SECTION 1.4, pp.

1. 30 **2.** -50 **3.** (a) 50; (b) 50 **4.** (a) 5841; (b) 5841 **5.** $5x + 45$ **6.** $8y - 80$ **7.** $ax + ay - az$ **8.** $40x - 60y + 5z$

9. $9(x + y)$ **10.** $a(c - y)$ **11.** $6(x - 2)$ **12.** $5(7x - 5y + 3w + 1)$ **13.** $b(s + t - w)$ **14.** $20x$ **15.** $-7x$ **16.** $6x$

17. $-6x$ **18.** $23.4x + 3.9y$ **19.** $0.92P$

SOMETHING EXTRA—CALCULATOR CORNER: COMPUTING INTEREST, p. 24

1. $\$5.47$ **2.** $\$323.54$

EXERCISE SET 1.4, pp. 25–26

1. -8 **3.** -4 **5.** -18 **7.** $\$127.20$ **9.** $3a + 6$ **11.** $4x - 4y$ **13.** $-10a - 15b$ **15.** $2ab - 2ac + 2ad$

17. $2\pi rh + 2\pi r$ **19.** $\dfrac{1}{2}ha + \dfrac{1}{2}hb$ **21.** $8(x + y)$ **23.** $9(p - q)$ **25.** $7(x - 3)$ **27.** $x(y + z)$ **29.** $2(x - y + z)$

31. $3(x + 2y - 3z)$ **33.** $a(b + c - d)$ **35.** $\pi r(r + s)$ **37.** $9a$ **39.** $-3b$ **41.** $15y$ **43.** $11a$ **45.** $-8t$ **47.** $10x$

49. $8x - 8y$ **51.** $2c + 10d$ **53.** $22x + 18y$ **55.** $2x - 33y$

MARGIN EXERCISES, SECTION 1.5, pp. 27–30

1. -24 **2.** 0 **3.** 10 **4.** $-9x$ **5.** $24t$ **6.** $-7 + y$ or $y - 7$ **7.** $-x + y$ or $y - x$ **8.** $-9x - 6y - 11$

9. $-23x + 7y + 2$ **10.** $3x + 2y + 1$ **11.** $2x + 5z - 24$ **12.** $-3x + 2y$, or $2y - 3x$ **13.** $-\dfrac{1}{4}t - 41w + 5d + 23$

14. $-5x, -7y, 67t, -\dfrac{4}{5}$ **15.** $3x - 8$ **16.** $4y + 1$ **17.** $-2x - 9y - 6$ **18.** $-x - 2y$ **19.** $23x - 10y$ **20.** -25

21. 6 **22.** $23x + 52$ **23.** $12a + 12$

EXERCISE SET 1.5, pp. 31–32

1. $4b$ **3.** $-a - 2$ **5.** $-b + 3$ **7.** $-t + y$ or $y - t$ **9.** $-a - b - c$ **11.** $-8x + 6y - 13$ **13.** $2c - 5d + 3e - 4f$

15. $4a, -5b, 6$ **17.** $2x, -3y, -2z$ **19.** $3a + 5$ **21.** $m + 1$ **23.** $5d - 12$ **25.** $-7x + 14$ **27.** $-9x + 21$

29. $44a - 22$ **31.** -190 **33.** $-12y - 145$ **35.** $0.01583x + 0.00001y$

SOMETHING EXTRA—AN APPLICATION: SUCCESSIVE DISCOUNTS, p. 30

1. \$400, \$340, \$323   **2.** \$405, \$364.50, \$291.60   **3.** \$55.10, \$52.35, \$39.26   **4.** \$189, \$170.10, \$153.09

MARGIN EXERCISES, SECTION 1.6, pp. 33–34

1. 8^4 **2.** m^3 **3.** $(4y)^5$ **4.** $3 \cdot 3 \cdot 3 \cdot 3$, or 81   **5.** $y \cdot y$ **6.** $5x \cdot 5x \cdot 5x \cdot 5x$ **7.** $25y^2$ **8.** $-8x^3$ **9.** 8   **10.** -31

11. 1   **12.** 1   **13.** 1   **14.** $1/10^4$ **15.** $1/(-4)^3$ **16.** 4^{-3} **17.** $(-5)^{-4}$

EXERCISE SET 1.6, pp. 35–36

1. 4^5 **3.** 5^6 **5.** m^4 **7.** $(3a)^4$ **9.** $5^2 c^3 d^4$ **11.** $2 \cdot 2 \cdot 2 \cdot 2 \cdot 2$, or 32   **13.** $(-3) \cdot (-3) \cdot (-3) \cdot (-3)$, or 81

15. $x \cdot x \cdot x \cdot x$ **17.** $4b \cdot 4b \cdot 4b$ **19.** $ab \cdot ab \cdot ab \cdot ab$ **21.** 5   **23.** $\dfrac{7}{8}$ **25.** 1   **27.** 1   **29.** $\dfrac{1}{6^3}$ **31.** $\dfrac{1}{9^5}$ **33.** $-\dfrac{1}{11}$

35. 3^{-4} **37.** 10^{-3} **39.** $(-16)^{-2}$ **41.** $1 + 2 + 3 + 4 + 5 + 6 + 7 + 8 \times 9 = 100$

MARGIN EXERCISES, SECTION 1.7, pp. 37–40

1. 8^4 **2.** y^5 **3.** $-18x^{11}$ **4.** $-75x^{-14}$ **5.** $-10x^{-12}y^2$ **6.** $60y$ **7.** 4^3 **8.** 5^6 **9.** 10^{-6} **10.** $-2y^{10}x^{-4}$, or $-\dfrac{2y^{10}}{x^4}$

11. $\dfrac{3}{2}a^3 b^2$ **12.** 3^{42} **13.** 8^{-14} **14.** $8x^3 y^3$ **15.** $4^2 x^{-4} y^{14}$, $16x^{-4}y^{14}$, or $\dfrac{16y^{14}}{x^4}$ **16.** $(-2)^5 x^{20} y^{10}$, or $-32x^{20}y^{10}$

17. $10^3 x^{-12} y^{21} z^{-6}$, $1000 x^{-12} y^{21} z^{-6}$, or $\dfrac{1000y^{21}}{x^{12}z^6}$ **18.** $x^9 y^{12}$ **19.** $\dfrac{9}{4}x^4 y^{-16}$, or $\dfrac{9x^4}{4y^{16}}$

SOMETHING EXTRA—CALCULATOR CORNER: DIVIDING AND RECIPROCALS, p. 40

1. 0.96, 0.22222222, 0.95999999   **2.** 2.6, 0.00307692, 2.5999974   **3.** 76.146010, 0.08488964, 76.146007

4. 0.00309734, 0.08849557, 0.00309734

EXERCISE SET 1.7, pp. 41–42

1. 5^9 **3.** 8^{-4} **5.** 8^{-6} **7.** b^{-3} **9.** a^3 **11.** $72x^5$ **13.** $-28m^5 n^5$ **15.** $-14x^{-11}$, or $\dfrac{-14}{x^{11}}$ **17.** 6^5 **19.** 4^5 **21.** 10^{-9}

23. 9^2 **25.** a^5 **27.** 1   **29.** $-\dfrac{4}{3}x^9 y^{-2}$, or $\dfrac{-4x^9}{3y^2}$ **31.** $\dfrac{3}{2}x^3 y^{-2}$, or $\dfrac{3x^3}{2y^2}$ **33.** 4^6 **35.** 8^{-12} **37.** 6^{12} **39.** $27x^6 y^6$

41. $\dfrac{1}{4}x^{-6} y^8$, or $\dfrac{y^8}{4x^6}$ **43.** $\dfrac{1}{36}a^4 b^{-6} c^{-2}$, or $\dfrac{a^4}{36b^6 c^2}$ **45.** $\dfrac{4^{-9}}{3^{12}}$, or $\dfrac{1}{4^9 \cdot 3^{12}}$ **47.** $\dfrac{8}{27}x^9 y^3$

EXTENSION EXERCISES, pp. 43–44

1. $0.2\overline{142857}$ **3.** $0.42\overline{51}$ **5.** $0.\overline{213456789}$ **7.** The denominator, when factored into a product of primes, will contain a

factor different from, or in addition to, 2 or 5.   **9.** $\dfrac{1}{8}\%$, 0.3%, 0.009, 1%, 1.1%, $\dfrac{9}{100}$, $\dfrac{1}{11}$, $\dfrac{99}{1000}$, 0.11, $\dfrac{1}{8}$, $\dfrac{2}{7}$, 0.286   **11.** 2   **13.** 19

15. -43 **17.** 0   **19.** -19.45 **21.** 0.18   **23.** $3(3 + 10 - 8)$ **25.** $(20 \div 2)(5 - 2) + 14$ **27.** $-11\dfrac{5}{8}$, or $-\dfrac{93}{8}$

29. $12x + 24y - 48z$ **31.** $-0.99703x + 0.99699y$ **33.** $\dfrac{5}{3}$ **35.** $\dfrac{328}{3}$ **37.** False   **39.** 8^{120} or 2^{360} **41.** $a^{-14}b^{-27}$

TEST OR REVIEW—CHAPTER 1, pp. 45–46

1. -2.6 **2.** $\dfrac{326}{100}$, or $\dfrac{163}{50}$ **3.** 3.6%   **4.** 18.75%   **5.** 0.0571   **6.** 0.006   **7.** $-4.5 > -8.7$ **8.** $\dfrac{7}{8}$ **9.** 13.4   **10.** 0   **11.** -2

12. -13.1 **13.** $-\dfrac{17}{4}$ **14.** -8 **15.** 13   **16.** 0   **17.** -1 **18.** -29.7 **19.** $\dfrac{25}{4}$ **20.** -33.62 **21.** $\dfrac{3}{4}$ **22.** -264

23. -5 **24.** 15   **25.** $-\dfrac{7}{5}$, or -1.4 **26.** $-\dfrac{28}{15}$ **27.** $\dfrac{22}{7}$ **28.** -1.5 **29.** $-6a + 8b$ **30.** $3\pi rs + 3\pi r$ **31.** $a(b - c + 2d)$

32. $h(2a + 1)$ **33.** $10y - 5x$ **34.** $21a + 14$ **35.** $9x - 7y + 22$ **36.** $-7x + 14$ **37.** $-11y + 30$ **38.** $10x - 21$

39. $-a + 68$ **40.** $-6a^9b^{-5}$, or $\dfrac{-6a^9}{b^5}$ **41.** $-\dfrac{3}{2}x^{-4}y^2$, or $\dfrac{-3y^2}{2x^4}$ **42.** $\dfrac{1}{81}a^{12}b^{-8}c^{-4}$, or $\dfrac{a^{12}}{81b^8c^4}$

43. $-\dfrac{125}{27}x^{-15}y^{24}$, or $\dfrac{-125\,y^{24}}{27\,x^{15}}$

CHAPTER 2

READINESS CHECK, p. 48

1. $[1.2, \blacksquare] -12$ **2.** $[1.2, \blacksquare] -4.6$ **3.** $[1.2, \blacksquare] 7$ **4.** $[1.2, \blacksquare] 0$ **5.** $[1.2, \blacksquare] -8$ **6.** $[1.2, \blacksquare] 13$

7. $[1.2, \blacksquare] -20$ **8.** $[1.3, \blacksquare] -10$ **9.** $[1.3, \blacksquare] 46$ **10.** $[1.3, \blacksquare] \dfrac{1}{5}$ **11.** $[1.3, \blacksquare] -\dfrac{5}{4}$

12. $[1.4, \blacksquare] 5(x - 2y + 3)$ **13.** $[1.4, \blacksquare] 6x - 9y + 12$ **14.** $[1.4, \blacksquare] -x$ **15.** $[1.4, \blacksquare] -5$

MARGIN EXERCISES, SECTION 2.1, pp. 48–52

1. (a), (f)   **2.** (b), (c)   **3.** (d), (e)   **4.** $89 + 2$ and $3 + 4$ represent the same number.

5. $6 - 7$ and $4 - 5$ represent the same number.   **6.** $5, 8, 0, 2$ **7.** 3 **8.** -7 **9.** $-\dfrac{4}{4}$, or -1 **10.** 38 **11.** 140.3

12. $\dfrac{10}{8}$, or $\dfrac{5}{4}$ **13.** -49 **14.** $\dfrac{6}{28}$, or $\dfrac{3}{14}$ **15.** -3 **16.** $\dfrac{4}{3}$ **17.** $4 - 5y = 2$ **18.** $63x - 91 = 30$ **19.** 3 **20.** $-\dfrac{29}{5}$ **21.** $\dfrac{17}{2}$

EXERCISE SET 2.1, pp. 53–54

1. 9 **3.** -4 **5.** 23 **7.** -39 **9.** 1135 **11.** 4 **13.** -22 **15.** $\dfrac{1}{6}$ **17.** 32 **19.** -6 **21.** 18 **23.** 11 **25.** -12

27. 8 **29.** 2 **31.** 21 **33.** -12 **35.** 2 **37.** -1 **39.** $\dfrac{18}{5}$ **41.** 0 **43.** 1 **45.** 0.007

MARGIN EXERCISES, SECTION 2.2, pp. 55–56

1. -2 **2.** $-\dfrac{19}{8}$ **3.** $19, -5$ **4.** $0, \dfrac{17}{3}$ **5.** $-\dfrac{2}{9}, \dfrac{1}{2}$ **6.** No solution   **7.** All real numbers are solutions.

EXERCISE SET 2.2, pp. 57–58

1. 2 **3.** 2 **5.** 7 **7.** 5 **9.** $-\dfrac{3}{2}$ **11.** 5 **13.** $-2, 5$ **15.** $8, 9$ **17.** $\dfrac{3}{2}, \dfrac{2}{3}$ **19.** $0, 8$ **21.** $-4, 3$ **23.** $0, 1, -2$

25. 8.414

MARGIN EXERCISES, SECTION 2.3, pp. 59–62

1. 8 ft; 24 ft   **2.** $\dfrac{1}{2}$ **3.** 14% **4.** $80°, 40°, 60°$ **5.** $\$1.92$ **6.** $\$124$ **7.** $17; 19$ **8.** $l = 13$ m; $w = 6\dfrac{1}{2}$ m

EXERCISE SET 2.3, pp. 63–66

1. 8 in.; 4 in.   **3.** $1\dfrac{3}{5}$ m; $2\dfrac{2}{5}$ m   **5.** $\$14.75$ **7.** $\$59.50$ **9.** 4 **11.** 5 **13.** $\$45$ **15.** $\$650$ **17.** $32°, 96°, 52°$

19. $l = 31$ m; $w = 17$ m   **21.** $12, 13, 14$ **23.** $\$8000$ **25.** $\$1644$

MARGIN EXERCISES, SECTION 2.4, pp. 67–68

1. $b = \dfrac{2A}{h}$ **2.** $\dfrac{5}{3}P - 10 = c$ **3.** $m = \dfrac{H - 2r}{3}$ **4.** $Q = \dfrac{T}{1 + iy}$

EXERCISE SET 2.4, pp. 69–70

1. $l = \dfrac{A}{w}$ **3.** $I = \dfrac{W}{E}$ **5.** $m = \dfrac{F}{a}$ **7.** $t = \dfrac{I}{Pr}$ **9.** $m = \dfrac{E}{c^2}$ **11.** $l = \dfrac{P - 2w}{2}$ **13.** $a^2 = c^2 - b^2$ **15.** $r^2 = \dfrac{A}{\pi}$

17. $h = \dfrac{2}{11}W + 40$ **19.** $r^3 = \dfrac{3V}{4\pi}$ **21.** $h = \dfrac{2A}{a+b}$ **23.** $m = \dfrac{Fr}{v^2}$ **25.** $\dfrac{4}{5}$ yr

MARGIN EXERCISES, SECTION 2.5, pp. 71–74

1–9.

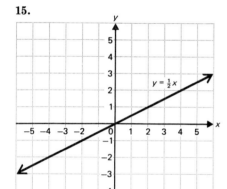

10. Both negative   **11.** First positive, second negative   **12.** No

13. No   **14.** Yes

15.

16.

17.

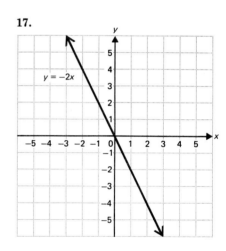

SOMETHING EXTRA—AN APPLICATION: FIELD GOAL COMPLETIONS, p. 74

1. 58.64%   **2.** 66.56%   **3.** 70.52%   **4.** 74.48%

EXERCISE SET 2.5, pp. 75–76

1.

3.

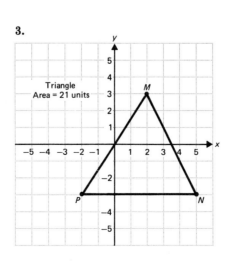

5. Yes   **7.** No

9.

11.

13.

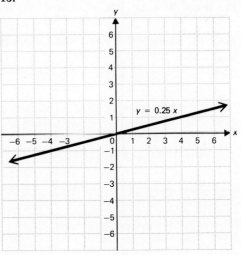

MARGIN EXERCISES, SECTION 2.6, pp. 77–82

1.

2.

3.

4.

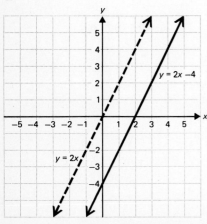

5. The point $(0, 7)$

6. The point $(0, -17)$

7. Slope is $\frac{3}{4}$; y-intercept is $(0, -2)$

8. Slope is $\frac{3}{2}$; y-intercept is $(0, 1)$

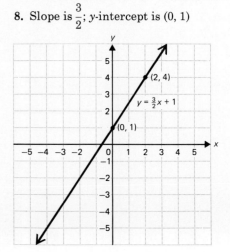

9. Slope is $-\dfrac{3}{5}$; y-intercept is $(0, 5)$

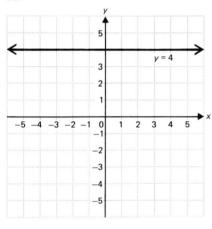

10. Slope is $-\dfrac{5}{3}$; y-intercept is $(0, -4)$

11.

12.

13.

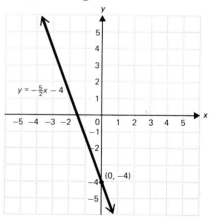

14. Linear, equivalent to $y = 2x + 3$ **15.** Linear, equivalent to $y = -5$ **16.** Not linear

EXERCISE SET 2.6, pp. 83–84

1. Slope is 4; y-intercept is $(0, 5)$ **3.** Slope is -2; y-intercept is $(0, -6)$ **5.** Slope is $-\dfrac{6}{16}$; y-intercept is $(0, -0.2)$

7. Slope is 15.3; y-intercept is $(0, 4.04)$ **9.** Slope is $\dfrac{5}{2}$; y-intercept is (0.1) **11.** Slope is $-\dfrac{5}{2}$; y-intercept is $(0, -4)$

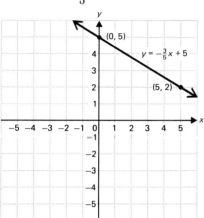

13. Slope is 2; y-intercept is $(0, 5)$.

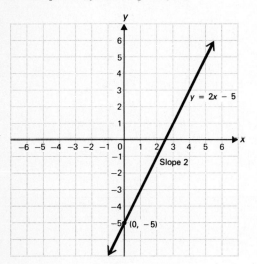

15. Slope is $\frac{1}{3}$; y-intercept is $(0, 6)$.

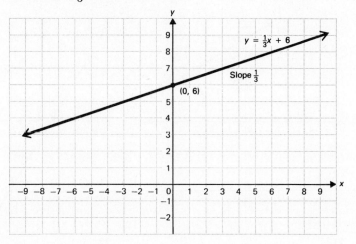

17. Slope is $-0.25x$; y-intercept is $(0, 2)$.

19.

21.

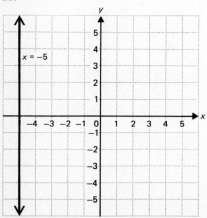

23. Linear   **25.** Linear   **27.** Not linear   **29.** Not linear

MARGIN EXERCISES, SECTION 2.7, pp. 85–88

1.

2.

3.

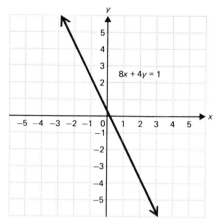

4. $\dfrac{7}{5}$ **5.** $m = \dfrac{\text{rise}}{\text{run}} = \dfrac{3}{4}$ **6.** $-\dfrac{13}{8}$ **7.** $\dfrac{1}{3}$ **8.** $-\dfrac{8.9}{12.5}$, or -0.712 **9.** 0 **10.** Does not exist

11. Zero slope: (a), (e); no slope: (b), (c)

EXERCISE SET 2.7, pp. 89–90

1.

3.

5.

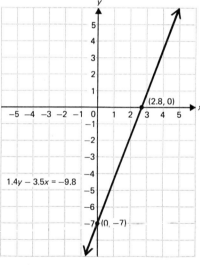

7. -2 **9.** 5 **11.** $-\dfrac{37}{111}$ **13.** Does not exist   **17.** Slope does not exist.   **19.** Zero slope   **21.** Zero slope

23. Slope does not exist.   **27.** Slope does not exist.   **29.** Zero slope

MARGIN EXERCISES, SECTION 2.8, pp. 91–94

1. $y + 2 = 3(x - 1),\ y + 2 = 3x - 3$, or $y = 3x - 5$ **2.** $y - \dfrac{3}{4} = -\dfrac{4}{3}(x + 5),\ y - \dfrac{3}{4} = -\dfrac{4}{3}x - \dfrac{20}{3}$, or $y = -\dfrac{4}{3}x - \dfrac{71}{12}$

3. $y + 3 = -\dfrac{5}{3}x + \dfrac{20}{3},\ y - 2 = -\dfrac{5}{3}x + \dfrac{5}{3}$, or $y = -\dfrac{5}{3}x + \dfrac{11}{3}$ **4.** $y + 5 = -17x - 51,\ y - 12 = -17x - 68$, or $y = -17x - 56$

5. $y + 1 = \dfrac{8}{7}x - \dfrac{16}{7}$, or $y = \dfrac{8}{7}x - \dfrac{23}{7}$ **6.** $y - 4 = -2x + 10$, or $y = -2x + 14$

SOMETHING EXTRA—DEPRECIATION: STRAIGHT LINE METHOD, p. 94

$3250, $2762.50, $2275, $1787.50, $1300

EXERCISE SET 2.8, pp. 95–96

1. $y - 2 = 4x - 12$, or $y = 4x - 10$ **3.** $y - 7 = -2x + 8$, or $y = -2x + 15$ **5.** $y + 4 = 3x + 6$, or $y = 3x + 2$

7. $y = -2x + 16$ **9.** $y + 7 = 0$, or $y = -7$ **11.** $y = \frac{1}{2}x + \frac{7}{2}$ **13.** $y = x$ **15.** $y = \frac{5}{2}x + 5$ **17.** $y = \frac{3}{2}x$ **19.** $y = \frac{2}{5}x$

21. $y = -\frac{1}{2}x + \frac{17}{2}$ **23.** $y = \frac{5}{7}x - \frac{17}{7}$ **25.** $y = \frac{1}{2}x + 4$ **27.** $y = \frac{4}{3}x - 6$ **29.** $y = -\frac{7}{3}x + \frac{22}{3}$

31. The side that contains $(3, 4)$ and $(6, 9)$ has slope $\frac{5}{3}$. The side that contains $(-2, 7)$ and $(3, 4)$ has slope $-\frac{3}{5}$. Since $\frac{5}{3} \cdot \left(-\frac{3}{5}\right) = -1$, these two sides are perpendicular.

MARGIN EXERCISES, SECTION 2.9, pp. 97–98

1. (a) $R = -0.01t + 10.43$; (b) 9.81, 9.73; (c) 2063

EXERCISE SET 2.9, pp. 99–100

1. (a) $E = 0.15t + 72$; (b) 76.5, 77.25   **3.** (a) $H = \frac{3}{20}W + \frac{905}{20}$; (b) $64\frac{3}{4}$ in.   **5.** (a) $R = -0.0075t + 3.85$; (b) 3.475, 3.445;

(c) 2003   **7.** (a) $C = 0.25m + 1.25$; (b) \$3.00

EXTENSION EXERCISES, pp. 101–102

1. $\frac{17}{6}$ **3.** $\frac{155}{8}$ **5.** 5   **7.** $\frac{1}{9}$ **9.** 0   **11.** $-\frac{4}{7}$, 3   **13.** $-\frac{10}{31}$ **15.** 25% increase   **17.** \$100   **19.** $a = \frac{2(s - v, t)}{t^2}$

21. $V_1 = \frac{P_2 V_2 T_1}{P_1 T_2}$ **23.** $(-1, -2), (13, 10)$, or $(-19, -2)$ **25.** a, d **27.** $m = -8$, y-intercept $= -\frac{7}{2}$

29. $m = -\frac{3}{7}$, y-intercept $= -\frac{6}{77}$ **31.** (a) $-\frac{5c}{4b}$; (b) No slope; (c) $\frac{a + d}{f}$ **33.** $-\frac{2}{3}, \frac{4}{3}, -\frac{8}{3}$ **35.** (1) and (4)   **37.** $k = 7$

39. $y = -0.5x - 0.6$

TEST OR REVIEW—CHAPTER 2, pp. 103–104

1. 22   **2.** $\frac{2}{3}$ **3.** $\frac{11}{5}$ **4.** -2 **5.** $-1, 3$ **6.** $W = \frac{11}{2}(h - 40)$

7. $y = -\frac{2}{3}x + 2$; slope is $-\frac{2}{3}$; y-intercept is $(0, 2)$ **8.**

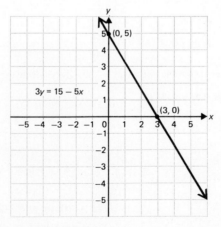

9. $-\frac{9}{7}$ **10.** $y = 2x + 1$ **11.** $y = -\frac{3}{2}x$ **12.** $y = -\frac{5}{7}x + \frac{8}{7}$ **13.** $y = \frac{1}{3}x + \frac{1}{3}$ **14.** 10   **15.** \$12.50

CHAPTER 3

READINESS CHECK, p. 106

1. [1.2, ⬤] -12 **2.** [1.2, ⬤] -4.6 **3.** [1.2, ⬤] 7 **4.** [1.2, ⬤] 0 **5.** [1.2, ⬤] 13 **6.** [1.2, ⬤] -22.9

7. [1.3, ⬤] -47 **8.** [1.3, ⬤] 6 **9.** [1.3, ⬤] $\dfrac{3}{2}$ **10.** [1.3, ⬤] $-\dfrac{4}{5}$ **11.** [1.3, ⬤] $\dfrac{1}{6}$ **12.** [1.3, ⬤] $-\dfrac{1}{5}$

13. [1.4, ⬤] $3(x - 14y + 4)$ **14.** [1.4, ⬤] $(3x^2 + 4y)(3x^2 - 4y)$ **15.** [1.4, ⬤] $8x + 8y - 8z$

16. [1.4, ⬤] $3x^2 + 6xy$ **17.** [1.4, ⬤] $x^2 - \dfrac{5}{3}y$ **18.** [1.4, ⬤] $-x + 3y$ **19.** [1.4, ⬤] $-y - 5$ **20.** [2.1, ⬤] -3

21. [2.1, ⬤] -35

MARGIN EXERCISES, SECTION 3.1, pp. 106–107

1. Yes   **2.** No   **3.**

4.

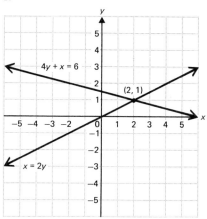

SOMETHING EXTRA—AN APPLICATION: ANNUAL PERCENTAGE RATE, p. 108

1. 19.78%   **2.** 21.24%

EXERCISE SET 3.1, pp. 109–110

1. Yes   **3.** No   **5.** Yes   **7.** No

9.

11.

13.

15.

17.

19.

21.

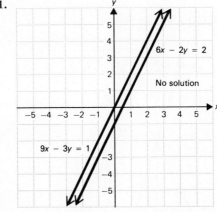

MARGIN EXERCISES, SECTION 3.2, pp. 111–114

1. $(2, 4)$ **2.** $(5, 2)$ **3.** $(-3, 2)$ **4.** $(13, 16)$ **5.** $(1, 4)$ **6.** $\left(\dfrac{1}{3}, \dfrac{1}{2}\right)$ **7.** $(2, 3)$ **8.** $(5, 3)$ **9.** $132x + 12y = 10$
$42x - 6y = 12$

10. $9x + 10y = 5$
$9x - 4y = 3$

EXERCISE SET 3.2, pp. 115–116

1. $(-4, 3)$ **3.** $(-3, -15)$ **5.** $(2, -2)$ **7.** $(-2, 1)$ **9.** $(1, 2)$ **11.** $(3, 0)$ **13.** $(-1, 2)$ **15.** $\left(\dfrac{128}{31}, -\dfrac{17}{31}\right)$ **17.** $(6, 2)$

19. $\left(\dfrac{140}{13}, -\dfrac{50}{13}\right)$ **21.** $(4, 6)$ **23.** $(23.118878, -12.039964)$

MARGIN EXERCISES, SECTION 3.3, pp. 117–120

1. 35 and 140   **2.** 22 white, 8 red   **3.** 30 L of 5%, 70 L of 15%   **4.** $1800 at 7%, $1900 at 9%   **5.** $l = 32.5$ ft, $w = 12.5$ ft

EXERCISE SET 3.3, pp. 121–124

1. 5 and -47 **3.** 24 and 8   **5.** $l = 160$ m, $w = 154$ m   **7.** 8 white, 22 yellow

9. 150 lb of soybean meal, 200 lb of corn meal   **11.** $4100 at 7%, $4700 at 8%   **13.** $30,000 at 5%, $40,000 at 6%

15. Ann 40, son 20   **17.** John 28, Sue 20   **19.** $l = 31$ in., $w = 12$ in.   **21.** 32 at $8.50, 13 at $9.75

23. $825 at 6%, $325 at 5%   **25.** $4\dfrac{4}{7}$ liters

MARGIN EXERCISES, SECTION 3.4, pp. 125–128

1. (a) No; (b) yes   **2.** $(2, 1, -1)$ **3.** $\left(2, -2, \dfrac{1}{2}\right)$ **4.** $(20, 30, 50)$

EXERCISE SET 3.4, pp. 129–130

1. Yes   **3.** $(1, 2, 3)$ **5.** $(-1, 5, -2)$ **7.** $(3, 1, 2)$ **9.** $(-3, -4, 2)$ **11.** $(2, 4, 1)$ **13.** $(-3, 0, 4)$ **15.** $(0, 2, -1, 1)$

MARGIN EXERCISES, SECTION 3.5, pp. 131–132

1. $2, -3, 4$ **2.** $64°, 32°, 84°$ **3.** A, 112; B, 90; C, 85

EXERCISE SET 3.5, pp. 133–136

1. $17, 9, 79$ **3.** $4, 2, -1$ **5.** $34°, 104°, 42°$ **7.** $25°, 50°, 105°$ **9.** Thurs., \$21; Fri., \$18; Sat., \$27   **11.** $82, 76, 97$

13. A, 1500; B, 1900; C, 2300   **15.** A, 900 gph; B, 1300 gph; C, 1500 gph

MARGIN EXERCISES, SECTION 3.6, pp. 137–140

1. $3 - 8$, or -5 **2.** -7 **3.** -6 **4.** 93 **5.** $(3, -1)$ **6.** $(1, 3, -2)$

EXERCISE SET 3.6, pp. 141–142

1. 3 **3.** 36 **5.** -10 **7.** -3 **9.** 5 **11.** $(2, 0)$ **13.** $\left(-\dfrac{25}{2}, -\dfrac{11}{2}\right)$ **15.** $\left(\dfrac{3}{2}, \dfrac{13}{14}, \dfrac{33}{14}\right)$ **17.** $(2, -1, 4)$ **19.** $(1, 2, 3)$

21. An equation of the line through (x_1, y_1) and (x_2, y_2) is

$$y - y_1 = \frac{y_2 - y_1}{x_2 - x_1}(x - x_1),$$

which is equivalent to $yx_2 - yx_1 - y_1x_2 + y_1x_1 = y_2x - y_2x_1 - y_1x + y_1x_1$ and $y_2x_1 + y_1x - y_2x + yx_2 - yx_1 - y_1x_2 = 0$;

$$\begin{vmatrix} x & y & 1 \\ x_1 & y_1 & 1 \\ x_2 & y_2 & 1 \end{vmatrix} = 0$$

is equivalent to $x(y_1 - y_2) - x_1(y - y_2) + x_2(y - y_1) = 0$ and $xy_1 - xy_2 - x_1y + x_1y_2 + x_2y - x_2y_1 = 0$

MARGIN EXERCISES, SECTION 3.7, pp. 143–145

1. (a) $P = 16x - 80,000$; (b) 5000 units   **2.** $(\$12.96, 352)$

SOMETHING EXTRA—AN APPLICATION: STOPPING DISTANCE, p. 146

1. $s = -5t^2 + 50t + 4$ **2.** 124 ft

EXERCISE SET 3.7, pp. 147–148

1. (a) $P = 20x - 600,000$; (b) 30,000 units   **3.** (a) $P = 50x - 120,000$; (b) 2400 units   **5.** (a) $P = 80x - 10,000$; (b) 125 units

7. $(\$10, 1400)$ **9.** $(\$22, 474)$ **11.** $(\$50, 6250)$ **13.** (a) $C = 40x + 22,500$; (b) $R = 85x$; (c) $P = 45x - 22,500$;

(d) profit of \$112,500; (e) 500 units

MARGIN EXERCISES, SECTION 3.8, pp. 149–150

1. (a) Inconsistent; (b) consistent   **2.** (a) Inconsistent; (b) consistent   **3.** (a) Dependent; (b) independent

4. (a) Dependent; (b) independent

EXERCISE SET 3.8, pp. 151–152

1. Inconsistent   **3.** Consistent   **5.** Consistent   **7.** Inconsistent   **9.** Consistent   **11.** Dependent   **13.** Independent

15. Dependent   **17.** Independent   **19.** Dependent

EXTENSION EXERCISES, pp. 153–154

1. $\left(\dfrac{1}{2}, -\dfrac{1}{2}\right)$ **3.** $\left(-\dfrac{4}{3}, -\dfrac{19}{3}\right)$ **5.** $\left(\dfrac{1000}{11}, -\dfrac{1000}{11}\right)$ **7.** $\left(-\dfrac{32}{17}, \dfrac{38}{17}\right)$ **9.** $a = 5, b = 2$ **11.** Mike: 32; Barbara: 14

13. $136\frac{3°}{4}$ **15.** $\frac{7}{19}$ **17.** $\left(\frac{1}{4}, -\frac{1}{2}, -\frac{1}{4}\right)$ **19.** $(0, -1, 10)$ **21.** 20   **23.** Student, 2050; adult, 845; children, 210   **25.** 12

27. 10

TEST OR REVIEW—CHAPTER 3, pp. 155–156

1.

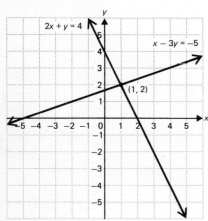

2. $(-1, 2)$ **3.** $l = 235\,\text{m}; w = 120\,\text{m}$ **4.** $(-3, 0, 3)$ **5.** 72, 85, 91   **6.** -2

7. $P = 35x - 90{,}000; 2572$ **8.** Inconsistent   **9.** Independent

CHAPTER 4

READINESS CHECK, p. 158

1. [1.2,] 8.3   **2.** [1.2, ●●●] $\frac{7}{5}$ **3.** [1.4, ●●●] $5(2x + 3y - 1)$ **4.** [1.4, ●●●] $3y - 6$ **5.** [1.4, ●●] $5x$

6. [1.5, ●●] $3y - 2$ **7.** [1.7, ●●] $15y^6$ **8.** [1.7, ●●] x^{12} **9.** [2.2, ●●] 3, -5 **10.** [1.4, ●] -2

MARGIN EXERCISES, SECTION 4.1, pp. 158–161

1. 14¢   **2.** 24   **3.** 105   **4.** $9xy + (-5x^3) + (-y) + 4$ **5.** $-92x^5 + (-8x^4) + y^2 + 5$ **6.** $8pq, -7pqr^2, -34xyz^6$

7. $5, -4, -2, 1, -1, -5$ **8.** $1, 0, 1, 7, 3; 7$ **9.** $3y - 4x + 4xy^2$ **10.** $8xy^3 + 2x^3y - 9x^2y$

11. $-6x^7 + 3x^5 - 7x^4 + 6x^3 - 5x^2 + x$ **12.** $9 + 4y^2 - 2y^3 - 5y^4 - 7y^5$ **13.** $7xy^4 - 2xy^3 - 8xy^2 + 3xy$

14. $5xy^2 + 4x^2yz + 5x^3yz^2$ **15.** None of these   **16.** Monomial   **17.** Trinomial   **18.** Binomial

EXERCISE SET 4.1, pp. 163–164

1. $8x^2 + (-2x) + (-5); 5$ **3.** $18xy + (-8x^3) + (-y) + 50; 20$ **5.** $5x^3, 7x^2, -3x, -9; 5, 7, -3, -9$

7. $-3xyz, 7x^2y^2, -5xy^2z, 4xyz^2; -3, 7, -5, 4$ **9.** $2, 5, 7, 0; 7$ **11.** $5, 6, 2, 1, 0; 6$ **13.** $2x^2$ **15.** $3x + y$ **17.** $a + 6$

19. $-6a^2b - 2b^2$ **21.** $9x^2 + 2xy + 15y^2$ **23.** $-x^2y + 4y + 9xy^2$ **25.** $x^3 - 3x^2 + x + 1$

27. $5a^5 - a^3 + 6a^2 + a - 9$ **29.** $-7 - y + 3y^2 + 2y^3 + y^4$ **31.** $-x + x^2 - 9x^3 + 18x^4 + 5x^5$

33. $-xy^3 + x^2y^2 + x^3y + 1$ **35.** $3xy^3 + x^2y^2 - 9x^3y + 2x^4$ **37.** $0.81125x^2y^2 - 0.07111x^3y$

MARGIN EXERCISES, SECTION 4.2, pp. 165–166

1. $-4x^3 + 2x^2 - 4x + 2$ **2.** $10y^5 - 4y^2 + 5$ **3.** $5p^2q^4 - 8p^2q^2 + 5$ **4.** $-4x^3 + 5x^2 - \frac{1}{4}x + 10$

5. $-8xy^2 + 4x^3y^2 + 9x + \frac{1}{5}$ **6.** $9y^5 + 8y^4 - \frac{1}{2}y^3 + y^2 - y + 1$ **7.** $3x^2 + 5$ **8.** $14y^3 - 2y + 4$ **9.** $p^2 - 6p - 2$

10. $3y^5 - 3y^4 + 5y^3 - 2y^2 - 3$ **11.** $9p^4q - 10p^3q^2 + 4p^2q^3 + 9q^4$ **12.** $y^3 - y^2 + \frac{4}{3}y + 0.1$

EXERCISE SET 4.2, pp. 167–168

1. $5x^2 + 2y^2 + 5$ **3.** $6a + b + c$ **5.** $-4a^2 - b^2 + 3c^2$ **7.** $-2x^2 + x - xy - 1$ **9.** $5x^2y - 4xy^2 + 5xy$

11. $9r^2 + 9r - 9$ **13.** $-5x^3 + 7x^2 - 3x + 6$ **15.** $13x - 6$ **17.** $-4x^2 - 3x + 13$ **19.** $2a - 4b + 3c$ **21.** $-2x^2 + 6x$

23. $-4a^2 + 8ab - 5b^2$ **25.** $27.94x^2 + 9.105y^2 + 8.15x$

MARGIN EXERCISES, SECTION 4.3, pp. 169–170

1. $-18y^3$ **2.** $24x^8y^3$ **3.** $-90x^4y^9z^{12}$ **4.** $-6y^2 - 18y$ **5.** $8xy^3 - 10xy$ **6.** $5x^3 + 15x^2 - 4x - 12$ **7.** $6y^2 + y - 12$

8. $p^4 + p^3 - 12p^2 - 5p + 15$ **9.** $2x^4 - 8x^3 + 4x^2 - 21x + 20$ **10.** $8x^5 + 12x^4 - 20x^3 + 4x^2 - 15x + 6$

EXERCISE SET 4.3, pp. 171 172

1. $10y^3$ **3.** $-20x^3y$ **5.** $-10x^5y^6$ **7.** $6x - 2x^2$ **9.** $3a^2b + 3ab^2$ **11.** $15c^3d^2 - 25c^2d^3$ **13.** $6x^2 + x - 12$

15. $s^2 - 9t^2$ **17.** $x^2 - 2xy + y^2$ **19.** $2y^2 + 9xy - 56x^2$ **21.** $a^4 - 5a^2b^2 + 6b^4$ **23.** $x^3 - 64$ **25.** $x^3 + y^3$

27. $a^4 + 5a^3 - 2a^2 - 9a + 5$ **29.** $4a^3b^2 + 4a^3b - 10a^2b^2 - 2a^2b + 3ab^3 + 7ab^2 - 6b^3$ **31.** $x^2 + 4xh$

MARGIN EXERCISES, SECTION 4.4, pp. 173–175

1. $y^2 + 6y - 40$ **2.** $2p^2 + 7pq - 15q^2$ **3.** $x^3y^3 + x^2y^3 + 2x^2y^2 + 2xy^2$ **4.** $a^2 - 2ab + b^2$ **5.** $x^2 + 16x + 64$

6. $9x^2 - 42x + 49$ **7.** $16x^2 - 40xy + 25y^2$ **8.** $4y^4 + 24x^2y^3 + 36x^4y^2$ **9.** $x^2 - 64$ **10.** $16y^2 - 49$ **11.** $49x^4y^2 - 4y^2$

12. $4x^2 + 12x + 9 - 25y^2$ **13.** $25t^2 - 4x^6y^4$

SOMETHING EXTRA—CALCULATOR CORNER, p. 176

1. (a) 79081.1635; (b) -25.2870 **2.** (a) 30872.7104; (b) 63.5744   **3.** (a) 50586.456; (b) 60.104

4. (a) -435562.1833; (b) 14.1690

EXERCISE SET 4.4, pp. 177–178

1. $a^2 + 5a + 6$ **3.** $y^2 + y - 6$ **5.** $b^2 - \dfrac{5}{6}b + \dfrac{1}{6}$ **7.** $2x^2 + 13x + 18$ **9.** $4x^2 - 4xy - 3y^2$ **11.** $x^2 + 6x + 9$

13. $4a^2 + \dfrac{4}{3}a + \dfrac{1}{9}$ **15.** $x^2 - 4xy + 4y^2$ **17.** $4x^4 - 12x^2y^2 + 9y^4$ **19.** $a^4b^4 + 2a^2b^2 + 1$ **21.** $c^2 - 4$ **23.** $4a^2 - 1$

25. $9m^2 - 4n^2$ **27.** $x^4 - y^2z^2$

EXERCISE SET 4.4A, pp. 179–180

1. $8x^2 - 6xy + y^2$ **3.** $4c^2 + 28cd + 49d^2$ **5.** $9c^2 - 4d^2$ **7.** $2x^4 + 3x^2y - 2y^2$ **9.** $2m^4 + 5m^2n^2 - 3n^4$

11. $9x^2 - 24xy + 16y^2$ **13.** $a^4 - b^2$ **15.** $\dfrac{1}{25} - x^2$ **17.** $8x^4 + 2x^2y^2 - 15y^4$ **19.** $\dfrac{1}{4}x^2 - 6x + 36$ **21.** $x^4 - 1$

23. $a^4 - b^4$ **25.** $a^2 + 2ab + b^2 - 1$ **27.** $4x^2 + 12xy + 9y^2 - 16$ **29.** \$1404.93

MARGIN EXERCISES, SECTION 4.5, pp. 181–182

1. $3(x^2 - 2)$ **2.** $4x^3(x^2 - 2)$ **3.** $3y^2(3y^2 - 5y + 1)$ **4.** $3x^2y(2 - 7xy + y^2)$ **5.** $(p + q)(2x + y + 2)$

6. $(y + 3)(2y - 11)$ **7.** $(x - 4)(x + 5)$ **8.** $(y - 2)(5y + 2)$ **9.** $(y + 2)(y - 2)$ **10.** $(7x^2 + 5y^5)(7x^2 - 5y^5)$

11. $(5xy + 2a)(5xy - 2a)$ **12.** $(3x + 4y^3)(3x - 4y^3)$ **13.** $5(2x + y)(2x - y)$ **14.** $y^2(9x^2 + 4)(3x + 2)(3x - 2)$

EXERCISE SET 4.5, pp. 183–184

1. $2a(2a + 1)$ **3.** $3(y^2 - y - 3)$ **5.** $3x^2(2 - x^2)$ **7.** $2a(2b - 3c + 6d)$ **9.** $5(2a^4 + 3a^2 - 5a - 6)$

11. $(a + c)(b - 2)$ **13.** $(x - 2)(2x + 13)$ **15.** $2a^2(x - y)$ **17.** $(a + b)(c + d)$ **19.** $(b^2 + 2)(b - 1)$

21. $(y - 1)(y - 8)$ **23.** $(x + 4)(x - 4)$ **25.** $6(x + y)(x - y)$ **27.** $4x(y^2 + z^2)(y + z)(y - z)$

29. $a^2(3a + 5b^2)(3a - 5b^2)$ **31.** $\left(\dfrac{1}{5} + x\right)\left(\dfrac{1}{5} - x\right)$ **33.** $(0.2x + 0.3y)(0.2x - 0.3y)$ **35.** 5

MARGIN EXERCISES, SECTION 4.6, pp. 185–188

1. a, b, d   **2.** $(x + 7)^2$ **3.** $(3y - 5)^2$ **4.** $(4x + 9y)^2$ **5.** $(4x^2 - 5y^3)^2$ **6.** $-2(2a - 3b)^2$ **7.** $(x + 7)(x - 2)$

8. $(y - 2)(y + 1)$ **9.** $(3x + 2)(x + 1)$ **10.** $(2x + 3)(2x - 1)$ **11.** $2(4y - 1)(3y - 5)$ **12.** $4(2x - 1)(2x + 3)$

13. $(pq + 4)(pq + 3)$ **14.** $(2x^2y^3 + 5)(x^2y^3 - 4)$ **15.** $(3x + 8)(x - 7)$ **16.** $(4x + 1)(x + 9)$

EXERCISE SET 4.6, pp. 189–190

1. $(y - 3)^2$ **3.** $(x + 7)^2$ **5.** $y(y - 9)^2$ **7.** $3(2a + 3)^2$ **9.** $2(x - 10)^2$ **11.** $(0.5x + 0.3)^2$ **13.** $(x + 5)(x + 4)$

15. $2(y - 4)^2$ **17.** $(x - 9)(x + 3)$ **19.** $y(8 - y)(4 + y)$ **21.** $(t + 5)(t + 3)$ **23.** $(x^2 + 16)(x^2 - 5)$

25. $(3x + 2)(x - 6)$ **27.** $x(3x - 5)(2x + 3)$ **29.** $(3a - 4)(a - 2)$ **31.** $(5y + 2)(7y + 4)$ **33.** $2(5t - 3)(t + 1)$

35. $4(2x + 1)(x - 4)$ **37.** 263.76 cm²

MARGIN EXERCISES, SECTION 4.7, pp. 191–192

1. $(x + 1 + p)(x + 1 - p)$ **2.** $(y - 4 + 3m)(y - 4 - 3m)$ **3.** $(x + 4 - 10)(x + 4 + 10)$, or $(x - 6)(x + 14)$

4. $[8 + (x + 4)][8 - (x + 4)]$, or $(x + 12)(-x + 4)$ **5.** $x^2 + 14x + 49$ **6.** $y^2 - 10by + 25b^2$ **7.** $x^2 - \frac{2}{5}x + \frac{1}{25}$

8. $x^2 + 4.2x + 4.41$ **9.** $(x + 20)(x + 6)$ **10.** $2(x - 3)(x - 21)$ **11.** $(x + 6.2)(x + 2.2)$

EXERCISE SET 4.7, pp. 193–194

1. $(a + b + 3)(a + b - 3)$ **3.** $(r - 1 + 2s)(r - 1 - 2s)$ **5.** $2(m + n + 5b)(m + n - 5b)$ **7.** $[3 + (a + b)][3 - (a + b)]$

9. $x^2 + 16x + 64$ **11.** $x^2 - 4.2x + 4.41$ **13.** $x^2 + \frac{2}{3}bx + \frac{1}{9}b^2$ **15.** $(x + 17)(x + 7)$ **17.** $(x - 5)(x - 21)$

19. $2(x + 19)(x + 9)$ **21.** $2(x - 12a)(x - 4a)$ **23.** $(x + 2.3)(x + 0.3)$ **25.** $\left(x + \frac{4}{10}\right)\left(x + \frac{2}{10}\right)$, or $\left(x + \frac{2}{5}\right)\left(x + \frac{1}{5}\right)$

27. $(x + 5.766)(x - 1.284)$

MARGIN EXERCISES, SECTION 4.8, pp. 195–196

1.

n	n^3
1	1
−1	−1
2	8
−2	−8
3	27
−3	−27
4	64
−4	−64
5	125
−5	−125
6	216
−6	−216
7	343
−7	−343
8	512
−8	−512
9	729
−9	−729
10	1000
−10	−1000

2. $x^3 - 8$ **3.** $a^3 + 125$ **4.** $(x - 2)(x^2 + 2x + 4)$ **5.** $(4 - y)(16 + 4y + y^2)$

6. $(3x + y)(9x^2 - 3xy + y^2)$ **7.** $(2y + z)(4y^2 - 2yz + z^2)$ **8.** $(2xy)(2x^2 + 3y^2)(4x^4 - 6x^2y^2 + 9y^4)$

9. $(3x + 2y)(9x^2 - 6xy + 4y^2)(3x - 2y)(9x^2 + 6xy + 4y^2)$ **10.** $(x - 0.3)(x^2 + 0.3x + 0.09)$

EXERCISE SET 4.8, pp. 197–198

1. $(x + 2)(x^2 - 2x + 4)$ **3.** $2(y - 4)(y^2 + 4y + 16)$ **5.** $3(2a + 1)(4a^2 - 2a + 1)$ **7.** $(2 - 3b)(4 + 6b + 9b^2)$

9. $(2x + 3)(4x^2 - 6x + 9)$ **11.** $\left(a + \dfrac{1}{2}\right)\left(a^2 - \dfrac{1}{2}a + \dfrac{1}{4}\right)$ **13.** $r(s + 4)(s^2 - 4s + 16)$ **15.** $5(x - 2z)(x^2 + 2xz + 4z^2)$

17. $(x + 0.1)(x^2 - 0.1x + 0.01)$ **19.** $8(2x^2 - t^2)(4x^4 + 2x^2t^2 + t^4)$ **21.** $x^2 - 16$

MARGIN EXERCISES, SECTION 4.9, pp. 199–200

1. $3y(y + 2x)(y - 2x)$ **2.** $7(a - 1)(a^2 + a + 1)$ **3.** $(2x + 3y)(4x^2 - 6xy + 9y^2)(2x - 3y)(4x^2 + 6xy + 9y^2)$ **4.** $(x - 2)(3 - bx)$

5. $5(y^4 + 4x^6)$ **6.** $3(x - 2)(2x + 3)$ **7.** $(a - b)^2(a + b)$ **8.** $3(x + 3a)^2$ **9.** $2(x - 5 + 3b)(x - 5 - 3b)$

EXERCISE SET 4.9, pp. 201–202

1. $(x + 12)(x - 12)$ **3.** $(2x + 3)(x + 4)$ **5.** $3(x^2 + 2)(x^2 - 2)$ **7.** $(a + 5)^2$ **9.** $2(x - 11)(x + 6)$

11. $(3x + 5y)(3x - 5y)$ **13.** $(2c - d)^2$ **15.** $(x^2 + 2)(2x - 7)$ **17.** $2(5x - 4y)(25x^2 + 20xy + 16y^2)$

19. $(m^3 + 10)(m^3 - 2)$ **21.** $(c - b)(a + d)$ **23.** $(m + 1)(m^2 - m + 1)(m - 1)(m^2 + m + 1)$

25. $(x + y + 3)(x - y + 3)$ **27.** $(6y - 5)(6y + 7)$ **29.** $(a^4 + b^4)(a^2 + b^2)(a + b)(a - b)$ **31.** $ab(a + 4b)(a - 4b)$

33. $2(x + 2)(x - 2)(x + 3)$ **35.** $2(2x + 3y)(4x^2 - 6xy + 9y^2)$

MARGIN EXERCISES, SECTION 4.10, pp. 203–204

1. $4, 2$ **2.** $\dfrac{1}{2}, -3$ **3.** $0, 2$ **4.** $8, -6$ **5.** $l = 8$ cm; $w = 3$ cm

EXERCISE SET 4.10, pp. 205–208

1. $-7, 4$ **3.** 4 **5.** $-5, -4$ **7.** $0, -8$ **9.** $-6, 6$ **11.** $-5, -9$ **13.** $7, 4$ **15.** $\dfrac{1}{2}, 7$ **17.** $7, -2$ **19.** $-\dfrac{3}{2}, \dfrac{7}{2}$

21. $l = 12$ in.; $w = 7$ in.   **23.** 9 and 11 **25.** $-13, 12$ **27.** $l = 12$ cm; $w = 8$ cm   **29.** 2 **31.** $h = 7$ cm; $b = 16$ cm

33. $-10, -8,$ and -6; $6, 8,$ and 10 **35.** $l = 28$ cm; $w = 14$ cm

MARGIN EXERCISES, SECTION 4.11, pp. 209–210

1. $x = \dfrac{4 - a^2}{c}$ **2.** $y = \dfrac{-6}{5ab}$ **3.** $x = -2$ **4.** $x = \dfrac{5 + ab}{a - b}$ **5.** $M = \dfrac{Fd^2}{km}$ **6.** $t = \dfrac{5L}{r - w}$ **7.** $v^2 = \dfrac{2M}{m} - V^2$

EXERCISE SET 4.11, pp. 211–212

1. $y = \dfrac{15 - b}{3 + a}$ **3.** $z = a$ **5.** $x = -\dfrac{a}{9}$ **7.** $x = 11$ **9.** $x = b + a$ **11.** $r^2 = \dfrac{V}{\pi h}$ **13.** $t^2 = \dfrac{2s}{g}$ **15.** $h = \dfrac{A - \pi r^2}{2\pi r}$

17. $r = \dfrac{A - P}{Pt}$

MARGIN EXERCISES, SECTION 4.12, pp. 213–215

1. (a) Yes; (b) yes; (c) no, because a person can have more than one brother.

2. Domain: the set of items; Range: the set of prices.   **3.** 17 **4.** 15 **5.** 2 **6.** Yes   **7.** No   **8.** No   **9.** Yes

SOMETHING EXTRA—POLYNOMIALS AND MULTIPLICATION, p. 216

1. 529 **2.** 676 **3.** 9409 **4.** $10,404$ **5.** 621 **6.** 9996 **7.** $10,201$ **8.** 729

EXERCISE SET 4.12, pp. 217–218

1. Yes   **3.** 2 **5.** -7 **7.** 3 **9.** 12 **11.** 6 **13.** 4 **15.** $4\sqrt{3}$ **17.** 36π **19.** Yes   **21.** Yes   **23.** Yes

EXTENSION EXERCISES, pp. 219–220

1. $\dfrac{11}{4}$ **3.** 0.000000798 **5.** $-3a^2 - 55ab - 40b^2$ **7.** $0.657u^2 + 5.003v^2$ **9.** $-\dfrac{4}{3}x^6y^{11}$ **11.** $-\dfrac{9}{4}r^{22}s^{14}$ **13.** $\dfrac{8}{21}x^8y^5$

15. $-16s^8t + 24s^6t^5 - 8s^4t^8 + 8s^3t^{11}$ **17.** $16x^4 - 32x^3 + 16x^2$ **19.** $-a^4 - 2a^3b + 25a^2 + 2ab^3 - 25b^2 + b^4$

21. $r^8 - 2r^4s^4 + s^8$ **23.** $a^2 + 2ac - b^2 - 2bd + c^2 - d^2$ **25.** $10y^2 - 38xy + 24x^2 - 212y + 306x + 930$

27. $16x^4 + 4x^2y^2 + y^4$ **29.** $x^{4a} - y^{4b}$ **31.** $a^3 - b^3 + 3b^2 - 3b + 1$ **33.** $\dfrac{1}{81}x^{12} - \dfrac{8}{81}x^6y^4 + \dfrac{16}{81}y^8$ **35.** $x^{a^2-b^2}$

37. $(2x - 3)(10x + 13)$ **39.** $4(2a - 1)(9a + 40)$ **41.** $7\left(x - \dfrac{1}{2}\right)\left(x^2 + \dfrac{1}{2}x + \dfrac{1}{4}\right)$ **43.** $(c - 2d)^2(c^2 - cd + d^2)^2$

45. $(x+1)(x-1)(x+7)(x-7)$ **47.** $(s-3t)(s+3t)(s^2+3st+9t^2)(s^2-3st+9t^2)$ **49.** $(3x^{2s}+4y^t)(9x^{4s}-12x^{2s}y^t+16y^{2t})$

51. $(2x+y-r+3s)(2x+y+r-3s)$ **53.** $(c^2d^2+a^8)(cd-a^4)(cd+a^4)$ **55.** $c(c^w+1)^2$ **57.** $y(y^4+1)(y^2+1)(y+1)(y-1)$

59. $3(a + b - c - d)(a + b + c + d)$ **61.** $4(3a^2 + 4)$ **63.** 0

65. Length = 40 inches, width = 30 inches, depth = 20 inches   **67.** 54 cm^2 **69.** $x = \dfrac{b^2 - 3b - c}{2b - 3}$

TEST OR REVIEW—CHAPTER 4, pp. 221–222

1. (a) $5x^5y^4 - 2x^4y - 4x^2y + 3xy^3$; (b) 9   **2.** $3xy + 3xy^2$ **3.** $3x^2y^4 - 2x^2y^2 + 2x^2y - 3xy^2 - 5x + 4$

4. $2y^2 + 5y + y^3$ **5.** $-8x^5 - 12x^4 + 24x^3 - 19x + 15$ **6.** $9x^2 - 24xy + 16y^2$ **7.** $9x^2 - 25$ **8.** $3a^2b(2 - 7ab + b^2)$

9. $(y - 4)(y + 5)$ **10.** $5(2a - b)(2a + b)$ **11.** $(3x - 5)^2$ **12.** $2(4x - 1)(3x - 5)$ **13.** $(y + 4 + 10t)(y + 4 - 10t)$

14. $x^2 - 5x + \dfrac{25}{4}$ **15.** $(x + 4.2)(x + 0.2)$ **16.** $2ab(2a^2 + 3b^2)(4a^4 - 6a^2b^2 + 9b^4)$ **17.** $\dfrac{1}{2}, -3$ **18.** $l = 8 \text{ cm}; w = 5 \text{ cm}$

19. $r = \dfrac{mv^2}{F}$ **20.** 15

CHAPTER 5

READINESS CHECK, p. 224

1. [1.3, ▣] $\dfrac{1}{3}$ **2.** [1.3, ▣] $\dfrac{2}{5}$ **3.** [1.3, ▣] $\dfrac{5}{3}$ **4.** [1.3, ▣▣] $\dfrac{3}{2}$ **5.** [1.3, ▣▣] $-\dfrac{4}{5}$ **6.** [1.3, ▣▣] $\dfrac{1}{7}$ **7.** [1.3, ▣] $\dfrac{16}{9}$

8. [1.3, ▣] $-\dfrac{14}{15}$ **9.** [1.3, ▣▣] $-\dfrac{21}{8}$ **10** [1.2, ▣] $\dfrac{19}{15}$ **11.** [1.2, ▣] $\dfrac{34}{60}$, or $\dfrac{17}{30}$ **12.** [1.2, ▣▣▣] $\dfrac{7}{60}$

13. [1.2, ▣] $-\dfrac{1}{9}$ **14.** [1.2, ▣▣▣] $-\dfrac{16}{15}$ **15.** [2.2, ▣] 3   **16.** [4.2, ▣▣▣] $-5x^4 + x^3 + x^2 + 5$

17. [4.3, ▣▣▣] $a^5 + 3a^4 - 3a^3 - 9a^2 + 2a + 6$ **18.** [4.5, ▣, ▣▣▣] $3y^2(4y - 1)(y - 1)$

19. [4.5, ▣▣▣] $(x + 5)(x - 5)$ **20.** [4.6, ▣] $(x - 7)^2$

MARGIN EXERCISES, SECTION 5.1, pp. 224–228

1. $\dfrac{x^2 - 4}{5x + 20}$ **2.** $\dfrac{x^2 + 2xy + y^2}{x^2 - 9}$ **3.** $\dfrac{3x^2 + 2xy}{5x^2 + 4xy}$ **4.** $\dfrac{6x^3 + 4x^2 - 3xy - 2y}{9x^2 + 18x + 8}$ **5.** $\dfrac{5 - 2a}{b - a}$ **6.** $7x$ **7.** $2a + 3$ **8.** $\dfrac{3x + 2}{2x + 4}$

9. $\dfrac{2y + 4}{y - 1}$ **10.** $\dfrac{3x^2 - 6xy + 3y^2}{x + y}$ **11.** $a - b$ **12.** $\dfrac{x - 5}{x + 3}$ **13.** $\dfrac{1}{x + 7}$ **14.** $y^3 - 9$ **15.** $\dfrac{x + 5}{2x - 10}$ **16.** $\dfrac{2a^2b + 2ab^2}{a - b}$

EXERCISE SET 5.1, pp. 229–232

1. $\dfrac{3x(x + 1)}{3x(x + 3)}$ **3.** $\dfrac{(t - 3)(t + 3)}{(t + 2)(t + 3)}$ **5.** $\dfrac{3y}{5}$ **7.** $a - 3$ **9.** $\dfrac{y - 3}{y + 3}$ **11.** $\dfrac{t + 4}{t - 4}$ **13.** $\dfrac{x^2 - 16}{x^2 + 3x}$ **15.** $\dfrac{y + 4}{2}$ **17.** $\dfrac{2x^2 + 13x + 15}{7x}$

19. $c - 2$ **21.** $\dfrac{1}{x + y}$ **23.** 3   **25.** $\dfrac{y^2 - y - 6}{y}$ **27.** $\dfrac{2a + 1}{a + 2}$ **29.** $\dfrac{x^2 + 6x + 8}{3x - 15}$ **31.** $\dfrac{y^3 + 3y}{y^2 + y - 6}$ **33.** $\dfrac{x^2 + 4x + 16}{x^2 + 8x + 16}$

35. $\dfrac{2s}{r + 2s}$ **37.** $\dfrac{21{,}934.2x^2}{y^2 - 182{,}499.84}$

MARGIN EXERCISES, SECTION 5.2, pp. 233–234

1. 90   **2.** 72   **3.** $\dfrac{47}{60}$ **4.** $\dfrac{97}{72}$ **5.** $5a^3b^2$ **6.** $(y + 3)(y + 4)(y + 4)$ **7.** $2x^2(x - 3)(x + 3)(x + 2)$

8. $2(a + b)(a - b)$, or $2(a + b)(b - a)$

EXERCISE SET 5.2, pp. 235–236

1. 36   **3.** 144   **5.** 72   **7.** 45   **9.** $\dfrac{11}{10}$ **11.** $\dfrac{43}{36}$ **13.** $\dfrac{79}{60}$ **15.** $12x^2y$ **17.** $3(y-3)(y+3)$ **19.** $30a^3b^2$

21. $5(y-3)(y-3)$ **23.** $(x+2)(x-2)$, or $(x+2)(2-x)$ **25.** $(2r+3)(r-4)(3r-1)(r+4)$

MARGIN EXERCISES, SECTION 5.3, pp. 237–240

1. $\dfrac{12+y}{y}$ **2.** $3x+1$ **3.** $\dfrac{a-b}{b+2}$ **4.** $\dfrac{y+12}{x^2+y^2}$ **5.** $\dfrac{1-b^2}{3}$ **6.** $\dfrac{2x^2+11}{x-5}$ **7.** $\dfrac{3+7x}{4y}$ **8.** $\dfrac{11x^2}{2x-y}$ **9.** $\dfrac{9x^2+28y}{21x}$

10. $\dfrac{3}{x+y}$ **11.** $\dfrac{34y^2+21y-1}{3y(y-1)(y+1)}$ **12.** $\dfrac{a+12}{a^2+3a}$ **13.** $\dfrac{3y^2+12y+3}{(y-4)(y-3)(y+5)}$ **14.** $\dfrac{2}{x-1}$

SOMETHING EXTRA—CALCULATOR CORNER: NUMBER PATTERNS, p. 240

1. 9, 1089, 110889, 11108889; 1111088889   **2.** 36, 4356, 443556, 44435556; 4444355556   **3.** 111, 222, 333, 444; 555

4. 111, 1221, 12321, 123321; 1233321

EXERCISE SET 5.3, pp. 241–244

1. 2   **3.** $\dfrac{3y+5}{y-2}$ **5.** $a+b$ **7.** $\dfrac{11}{x}$ **9.** $\dfrac{1}{x+5}$ **11.** $\dfrac{2y^2+22}{y^2-y-20}$ **13.** $\dfrac{x+y}{x-y}$ **15.** $\dfrac{3x-4}{x^2-3x+2}$ **17.** $\dfrac{8x+1}{x^2-1}$

19. $\dfrac{2x-14}{15(x+5)}$ **21.** $\dfrac{-a^2+7ab-b^2}{a^2-b^2}$ **23.** $\dfrac{y}{y^2-5y+6}$ **25.** $\dfrac{3y-10}{y^2-y-20}$ **27.** $\dfrac{3y^2-3y-29}{(y+8)(y-3)(y-4)}$

29. $\dfrac{2x^2-13x+7}{(x+3)(x-1)(x-3)}$ **31.** 0   **33.** $\dfrac{3}{x+2}$ **35.** $\dfrac{-3x^2-3x-4}{x^2-1}$

MARGIN EXERCISES, SECTION 5.4, pp. 245–246

1. $\dfrac{7(2y+1)}{2(7y-1)}$ **2.** $\dfrac{x}{x+1}$ **3.** $\dfrac{b+a}{b-a}$ **4.** $\dfrac{a^2b^2}{b^2+ab+a^2}$

EXERCISE SET 5.4, pp. 247–248

1. $\dfrac{1+4x}{1-3x}$ **3.** $\dfrac{x^2-1}{x^2+1}$ **5.** $\dfrac{3y+4x}{4y-3x}$ **7.** $\dfrac{x+y}{x}$ **9.** $\dfrac{a^2(b-3)}{b^2(a-1)}$ **11.** $\dfrac{1}{a-b}$ **13.** $\dfrac{-1}{x(x+h)}$ **15.** $\dfrac{y-3}{y+5}$

MARGIN EXERCISES, SECTION 5.5, pp. 249–251

1. $\dfrac{x^2}{2}+8x+3$ **2.** $5y^3-2y^2+6y$ **3.** $\dfrac{x^2}{2}+5x+8$ **4.** $x+5$ **5.** $3y^3-2y^2+6y-4$

6. $y^2-8y-24$, R -66; or $y^2-8y-24+\dfrac{-66}{y-3}$ **7.** $x^2+10x+10$, R 5; or $x^2+10x+10+\dfrac{5}{x-1}$

8. $y-11$, R $3y-27$; or $y-11+\dfrac{3y-27}{y^2-3}$

SOMETHING EXTRA—DEPRECIATION: STRAIGHT LINE METHOD (PART II), p. 252

Year	Rate of depreciation	Annual depreciation	Total depreciation	Book value, V
0				$9400
1	$\frac{1}{10}$ or 10%	$720	$720	8680
2	Same	Same	1440	7960
3	Same	Same	2160	7240
4	Same	Same	2880	6520
5	Same	Same	3600	5800
6	Same	Same	4320	5080
7	Same	Same	5040	4360
8	Same	Same	5760	3640
9	Same	Same	6480	2920
10	Same	Same	7200	2200

EXERCISE SET 5.5, pp. 253–254

1. $6x^4 - 3x^2 + 8$ **3.** $y^3 - 2y^2 + 3y$ **5.** $x + 7$ **7.** $a - 12$, R 32, or $a - 12 + \dfrac{32}{a + 4}$ **9.** $x + 2$, R 4; or $x + 2 + \dfrac{4}{x + 5}$

11. $2y^2 - y + 2$, R 6; or $2y^2 - y + 2 + \dfrac{6}{2y + 4}$ **13.** $2y^2 + 2y - 1$, R 8; or $2y^2 + 2y - 1 + \dfrac{8}{5y - 2}$

15. $2x^2 - x - 9$, R $(3x + 12)$; or $2x^2 - x - 9 + \dfrac{3x + 12}{x^2 + 2}$

MARGIN EXERCISES, SECTION 5.6, pp. 255–256

1. $2x^2 + 2x + 14$, R 34 **2.** $x^2 - 4x + 13$, R -30 **3.** $y^2 - y + 1$

EXERCISE SET 5.6, pp. 257–258

1. $x^2 - x + 1$, R -4 **3.** $a + 7$, R -47 **5.** $x^2 - 5x - 23$, R -43 **7.** $3x^2 - 2x + 2$, R -3 **9.** $y^2 + 2y + 1$, R 12

11. $3x^3 + 9x^2 + 2x + 6$ **13.** $x^2 + 3x + 9$ **15.** $y^3 + y^2 + y + 1$

17. $3.41x^3 + 8.2181x^2 - 4.444379x - 10.710953$, R -39.283396

MARGIN EXERCISES, SECTION 5.7, pp. 259–262

1. $\dfrac{2}{3}$ **2.** No solution **3.** 3 **4.** -3 **5.** 4, -3 **6.** 7 **7.** -13

EXERCISE SET 5.7, pp. 263–266

1. $\dfrac{51}{2}$ **3.** -2 **5.** 144 **7.** $-5, -1$ **9.** 2 **11.** $\dfrac{17}{4}$ **13.** 11 **15.** No solution **17.** 2 **19.** $\dfrac{3}{5}$ **21.** 5 **23.** -145

25. $-\dfrac{10}{3}$ **27.** -3 **29.** $-6, 5$ **31.** No solution

MARGIN EXERCISES, SECTION 5.8, pp. 267–270

1. $2\dfrac{2}{5}$ hours **2.** Pipe A, 32 hr; Pipe B, 96 hr **3.** 280 miles **4.** 35.5 mph

EXERCISE SET 5.8, pp. 271–274

1. $3\dfrac{3}{14}$ hours **3.** $8\dfrac{4}{7}$ hours **5.** $3\dfrac{9}{52}$ hours **7.** A, 10 da; B, 40 da **9.** 7 mph **11.** 375 km **13.** Train A, 46 mph;

train B, 58 mph **15.** 9 km/h **17.** $21\dfrac{9}{11}$ min after 4:00

MARGIN EXERCISES, SECTION 5.9, p. 275–276

1. $T = \dfrac{PV}{k}$ **2.** $t = \dfrac{uv}{v + u}$ **3.** $T = \dfrac{A + Vnt}{Vn}$

EXERCISE SET 5.9, pp. 277–278

1. $d_1 = \dfrac{d_2 W_1}{W_2}$ **3.** $t = \dfrac{2s}{v_1 + v_2}$ **5.** $r_2 = \dfrac{Rr_1}{r_1 - R}$ **7.** $s = \dfrac{Rg}{g - R}$ **9.** $p = \dfrac{qf}{q - f}$ **11.** $r = \dfrac{nE - IR}{In}$

13. $H = Sm(t_1 - t_2)$ **15.** $e = \dfrac{rE}{R + r}$

MARGIN EXERCISES, SECTION 5.10, pp. 279–282

1. $\dfrac{2}{5}$; $y = \dfrac{2}{5}x$ **2.** 0.7; $y = 0.7x$ **3.** 50 volts **4.** 176,250 tons **5.** 0.6; $y = \dfrac{0.6}{x}$ **6.** $7\dfrac{1}{2}$ hr

EXERCISE SET 5.10, pp. 283–284

1. 8; $y = 8x$ **3.** 3.6; $y = 3.6x$ **5.** $\dfrac{8}{5}$; $y = \dfrac{8}{5}x$ **7.** 6 amperes **9.** 125,000 **11.** 60; $y = \dfrac{60}{x}$ **13.** 36; $y = \dfrac{36}{x}$

15. $0.32; y = \dfrac{0.32}{x}$ **17.** $\dfrac{2}{9}$ ampere   **19.** $685\dfrac{5}{7}$ kg

EXTENSION EXERCISES, pp. 285–286

1. $\dfrac{x-3}{(x+1)(x+3)}$ **3.** $\dfrac{m-t}{m+t+1}$ **5.** $\dfrac{x^2+xy+y^2+x+y}{x-y}$ **7.** $\dfrac{-2x}{x-1}$

9. $x^4(x-1)(x+1)(x^2+x+1)(x^2-x+1)(x^2+1)$ **11.** $\dfrac{2y^2+3-7x^3y}{x^2y^2}$ **13.** $\dfrac{5y+23}{5-2y}$ **15.** $\dfrac{-4a-4}{8a+5}$ **17.** $\dfrac{x^4}{81}$

19. $\dfrac{x}{x^3-1}$ **21.** $\dfrac{1}{a^2-ab+b^2}$ **23.** $2y^3+6y^2+17y+45$, R $116y-48$ **25.** a^2+ab

27. $a^6-a^5b+a^4b^2-a^3b^3+a^2b^4-ab^5+b^6$ **29.** All reals except 1 and -1 **31.** Boat: 14 km/h; stream: 10 km/h

33. 3.75 km/h   **35.** 48 km/h   **37.** $t=\dfrac{2}{3}$ hr   **39.** 96 fl oz at $2.99

41. (a) Inversely; (b) Neither; (c) Directly; (d) Directly; (e) Directly

TEST OR REVIEW—CHAPTER 5, pp. 287–288

1. $\dfrac{2x+10}{x-2}$ **2.** $\dfrac{y+4}{2}$ **3.** $(x+3)(x-2)(x+5)$ **4.** $x+y$ **5.** $\dfrac{3x}{(x-1)(x+1)}$ **6.** $\dfrac{x+1}{x}$ **7.** $x-7$, R -46

8. $6y^3+3y^2-6y-16$, R 38   **9.** 9   **10.** $2\dfrac{2}{5}$ hours   **11.** 11 km/h   **12.** $t=\dfrac{mr}{2-m}$ **13.** $495   **14.** $2\dfrac{5}{8}$ hours

CHAPTER 6

READINESS CHECK, p. 290

1. [1.1, ▨] 8   **2.** [1.7, ▨] $12x^5y^{-2}$ **3.** [1.7, ▨] $8x^{-1}y^{13}$ **4.** [1.7, ▨] $256x^8y^{-12}$ **5.** [4.5, ▨] $(2x-7)(2x+7)$

6. [4.5, ▨], [4.6, ▨] $2(x-4)(x-9)$ **7.** [4.6, ▨] $(3y-2)^2$ **8.** [4.4, ▨] $4x^2-9$ **9.** [4.4, ▨] $2a^2-5ab-3b^2$

10. [2.2, ▨] $0, \dfrac{5}{3}$ **11.** [2.2, ▨] 8

MARGIN EXERCISES, SECTION 6.1, pp. 290–293

1. $-9ab^2$ **2.** 1   **3.** $-2y^{-9}$, or $\dfrac{-2}{y^9}$ **4.** $-28y^{-3}z^2$, or $\dfrac{-28z^2}{y^3}$ **5.** $180a^{-5}$, or $\dfrac{180}{a^5}$ **6.** $-20a^{-1}b^3$, or $\dfrac{-20b^3}{a}$

7. $-4x^{-8}y^{-2}$, or $\dfrac{-4}{x^8y^2}$ **8.** $\dfrac{2}{3}a^{-2}b^{-5}c^3$, or $\dfrac{2c^3}{3a^2b^5}$ **9.** y^{-24}, or $\dfrac{1}{y^{24}}$ **10.** $512a^6b^{-12}$, or $\dfrac{512a^6}{b^{12}}$ **11.** $\dfrac{a^4}{b^4}$ **12.** $\dfrac{x^8y^{-12}}{81a^4}$, or $\dfrac{x^8}{81a^4y^{12}}$

13. $\dfrac{y^8}{x^{-12}}$, or y^8x^{12} **14.** 4.6×10^{11} **15.** 9.3×10^7 **16.** 1.235×10^{-8} **17.** 1.7×10^{-24} **18.** 789,300,000,000

19. 0.0000567   **20.** 7.462×10^{-13} **21.** 5.6×10^{-15} **22.** 2×10^3 **23.** 5.5×10^2 **24.** 3×10^{-4}

SOMETHING EXTRA—AN APPLICATION: COMPOUND INTEREST, p. 294

1. $1169.86   **2.** $1172.89   **3.** $1083.28   **4.** $1083.29

EXERCISE SET 6.1, pp. 295–296

1. 1   **3.** x^2+y **5.** $-15r^{-4}s^2$, or $\dfrac{-15s^2}{r^4}$ **7.** $-8m^{-6}n^9$, or $\dfrac{-8n^9}{m^6}$ **9.** $\dfrac{x^6}{5}$ **11.** $-12a^{-2}b^2$ **13.** $a^{-2}b^{-1}$, or $\dfrac{1}{a^2b}$

15. $-31,104y^6z^{-8}$, or $\dfrac{-31,104y^6}{z^8}$ **17.** $\dfrac{x^{12}y^6}{8}$ **19.** $27a^{-6}b^3$, or $\dfrac{27b^3}{a^6c^6}$ **21.** 4.7×10^{10} **23.** 8.6… 1.6×10^{-8}

27. 7×10^{-11} **29.** 673,000,000   **31.** 0.0000000008923   **33.** 9.66×10^{-5} **35.** 1.3338×10^{-11} **37.** 2.5 … 10^3

39. 5×10^{-4} **41.** Answers may vary: 4.2×10^2

MARGIN EXERCISES, SECTION 6.2, pp. 297–299

1. $3, -3$ **2.** $6, -6$ **3.** $11, -11$ **4.** 1 **5.** 6 **6.** $\dfrac{9}{10}$ **7.** 0.08 **8.** (a) 4; (b) -4; (c) Does not exist

9. (a) 7; (b) -7; (c) Does not exist   **10.** (a) 12; (b) -12; (c) Does not exist   **11.** $28 + x$ **12.** $\dfrac{y}{y+3}$ **13.** $|y|$ **14.** 24

15. $|5y|$ or $5|y|$ **16.** $|4y|$ or $4|y|$ **17.** $|x+7|$ **18.** $|2(x-2)|$ or $2|x-2|$ **19.** $|7(y+5)|$ or $7|y+5|$ **20.** $|x-3|$ **21.** -4

22. $3y$ **23.** $2(x+2)$ **24.** $-\dfrac{7}{4}$ **25.** $-2x$ **26.** $3x+2$ **27.** (a) 3; (b) -3; (c) Does not exist   **28.** $2|x-2|$ **29.** $|x|$

30. $|x+3|$

EXERCISE SET 6.2, pp. 301–302

1. $4, -4$ **3.** $12, -12$ **5.** $20, -20$ **7.** $-\dfrac{7}{6}$ **9.** 14 **11.** $-\dfrac{4}{9}$ **13.** 0.3 **15.** -0.07 **17.** p^2+4 **19.** $\dfrac{x}{y+4}$

21. $|4x|$ or $4|x|$ **23.** $|7c|$ or $7|c|$ **25.** $|a+1|$ **27.** $|x-2|$ **29.** $|2x+7|$ **31.** 3 **33.** $-4x$ **35.** -6 **37.** $0.7(x+1)$ **39.** 5

41. -1 **43.** $-\dfrac{2}{3}$ **45.** $|x|$ **47.** $|5a|$ or $5|a|$ **49.** 6 **51.** $|a+b|$ **53.** y **55.** $x-2$

57. (a) 12.5; (b) 15; (c) 17.5; (d) 20

MARGIN EXERCISES, SECTION 6.3, pp. 303–306

1. $\sqrt{133}$ **2.** $\sqrt{x^2-4y^2}$ **3.** $\sqrt[4]{2821}$ **4.** $\sqrt[3]{8x^5+40x}$ **5.** $4\sqrt{2}$ **6.** $2\sqrt[3]{10}$ **7.** $10\sqrt{3}$ **8.** $6y$ **9.** $(x+2)\sqrt{3}$

10. $2bc\sqrt{3ab}$ **11.** $2\sqrt[3]{2}$ **12.** $3xy^2\sqrt[3]{3xy^2}$ **13.** $(a+b)\sqrt[3]{a+b}$ **14.** $3\sqrt{2}$ **15.** $6y\sqrt{7}$ **16.** $3x\sqrt[3]{4y}$ **17.** $7\sqrt{3ab}$

18. $2(y+5)\sqrt[3]{(y+5)^2}$ **19.** 12.6 **20.** 18.5 **21.** -3.5 **22.** 9.1 **23.** Rational   **24.** Rational   **25.** Rational

26. Rational   **27.** Irrational   **28.** Irrational   **29.** Rational

SOMETHING EXTRA—AN APPLICATION: WIND CHILL, p. 306

1. $0°F$ **2.** $-29°F$ **3.** $-10°F$ **4.** $-22°F$ **5.** $-64°F$ **6.** $-94°F$

EXERCISE SET 6.3, pp. 307–308

1. $\sqrt{6}$ **3.** $\sqrt[3]{10}$ **5.** $\sqrt[4]{72}$ **7.** $\sqrt{30ab}$ **9.** $\sqrt[5]{18t^3}$ **11.** $\sqrt{x^2-a^2}$ **13.** $2\sqrt{2}$ **15.** $2\sqrt{6}$ **17.** $2\sqrt{10}$ **19.** $6x^2\sqrt{5}$

21. $3x^2\sqrt[3]{2x^2}$ **23.** $3\sqrt{10}$ **25.** $3\sqrt[3]{2}$ **27.** $30\sqrt{3}$ **29.** $5bc^2\sqrt{2b}$ **31.** $2y^3\sqrt[3]{2}$ **33.** $(b+3)^2$ **35.** $4a^3b\sqrt{6ab}$ **37.** 13.4

39. 7.5 **41.** Rational   **43.** Rational   **45.** Rational   **47.** Irrational   **49.** (a) 20 mph; (b) 37.4 mph; (c) 42.4 mph

MARGIN EXERCISES, SECTION 6.4, pp. 309–310

1. $\dfrac{5}{6}$ **2.** $\dfrac{x}{10}$ **3.** $\dfrac{3x\sqrt[3]{2x^2}}{5}$ **4.** 5 **5.** $56\sqrt{xy}$ **6.** $5a$ **7.** $\dfrac{20}{7}$ **8.** $\dfrac{2ab}{3}$ **9.** 25 **10.** $6y\sqrt{6y}$

EXERCISE SET 6.4, pp. 311–312

1. $\dfrac{4}{5}$ **3.** $\dfrac{4}{3}$ **5.** $\dfrac{7}{y}$ **7.** $\dfrac{5y\sqrt{y}}{x^2}$ **9.** $\dfrac{2x\sqrt[3]{x^2}}{3y}$ **11.** $\sqrt{7}$ **13.** 3 **15.** $y\sqrt{5y}$ **17.** $2\sqrt[3]{a^2b}$ **19.** $3\sqrt{xy}$ **21.** $\sqrt{x^2+xy+y^2}$

23. $6a\sqrt{6a}$ **25.** $4b\sqrt[3]{4b}$ **27.** $54a^3b\sqrt{2b}$ **29.** $2c\sqrt[3]{18cd^2}$ **31.** (a) 1.62 sec; (b) 1.99 sec; (c) 2.20 sec

MARGIN EXERCISES, SECTION 6.5, p. 313

1. $13\sqrt{2}$ **2.** $10\sqrt[4]{5x}-\sqrt{7}$ **3.** $19\sqrt{5}$ **4.** $(3y+4)\sqrt[3]{y^2}+2y^2$ **5.** $2\sqrt{x-1}$

SOMETHING EXTRA—CALCULATOR CORNER: CALCULATING SQUARE ROOTS, p. 314

1. 7.54983 **2.** 9.38083 **3.** 10.90871 **4.** 18.00000

EXERCISE SET 6.5, pp. 315–316

1. $8\sqrt{3}$ **3.** $3\sqrt[3]{5}$ **5.** $13\sqrt[3]{y}$ **7.** $7\sqrt{2}$ **9.** $6\sqrt[3]{3}$ **11.** $21\sqrt{3}$ **13.** $38\sqrt{5}$ **15.** $122\sqrt{2}$ **17.** $9\sqrt[3]{2}$ **19.** $29\sqrt{2}$

21. $(1+6a)\sqrt{5a}$ **23.** $(2-x)\sqrt[3]{3x}$ **25.** $3\sqrt{2y-2}$ **27.** $(x+3)\sqrt{x-1}$

MARGIN EXERCISES, SECTION 6.6, p. 317

1. $5\sqrt{6} + 3\sqrt{14}$ **2.** $a\sqrt[3]{3} - \sqrt[3]{2a^2}$ **3.** $-4 - 9\sqrt{6}$ **4.** $3\sqrt{ab} - 4\sqrt{3a} + 6\sqrt{3b} - 24$ **5.** -3 **6.** $20 - 4y\sqrt{5} + y^2$

7. $58 + 12\sqrt{6}$

SOMETHING EXTRA—CALCULATOR CORNER: MONTHLY MORTGAGE PAYMENTS, p. 318

1. $428.58   **2.** $633.50   **3.** No   **4.** Yes

EXERCISE SET 6.6, pp. 319–320

1. $2\sqrt{6} - 18$ **3.** $\sqrt{6} - \sqrt{10}$ **5.** $2\sqrt{15} - 6\sqrt{3}$ **7.** -6 **9.** $3a\sqrt[3]{2}$ **11.** 1 **13.** -12 **15.** $a - b$ **17.** $1 + \sqrt{5}$

19. $7 + 3\sqrt{3}$ **21.** -6 **23.** $a + \sqrt{3a} + \sqrt{2a} + \sqrt{6}$ **25.** $2\sqrt[3]{9} - 3\sqrt[3]{6} - 2\sqrt[3]{4}$ **27.** $7 + 4\sqrt{3}$ **29.** (a) 19.48528; (b) 19.485281

MARGIN EXERCISES, SECTION 6.7, pp. 321–324

1. $\dfrac{\sqrt{6}}{3}$ **2.** $\dfrac{\sqrt{70}}{7}$ **3.** $\dfrac{\sqrt[3]{4}}{2}$ **4.** $\dfrac{2\sqrt{3ab}}{3b}$ **5.** $\dfrac{2x^2\sqrt{3xy}}{3y^2}$ **6.** $\dfrac{\sqrt[3]{28}}{2}$ **7.** $\dfrac{x\sqrt[3]{12x^2y^2}}{2y}$ **8.** $\dfrac{11}{\sqrt{66}}$ **9.** $\dfrac{7}{\sqrt[3]{196}}$ **10.** $\dfrac{\sqrt{3} + y}{\sqrt{3} + y}$

11. $\dfrac{\sqrt{2} - \sqrt{3}}{\sqrt{2} - \sqrt{3}}$ **12.** $-5(1 + \sqrt{2})$ **13.** $\dfrac{5 + \sqrt{7}}{5 + \sqrt{7}}$ **14.** $\dfrac{4}{3\sqrt{2} - \sqrt{10} - 3\sqrt{6} + \sqrt{30}}$

EXERCISE SET 6.7, pp. 325–328

1. $\dfrac{\sqrt{30}}{5}$ **3.** $\dfrac{\sqrt{70}}{7}$ **5.** $\dfrac{2\sqrt{15}}{5}$ **7.** $\dfrac{2\sqrt[3]{6}}{3}$ **9.** $\dfrac{\sqrt[3]{75ac^2}}{5c}$ **11.** $\dfrac{y\sqrt[3]{9yx^2}}{3x^2}$ **13.** $\dfrac{\sqrt[3]{x^2y^2}}{xy}$ **15.** $\dfrac{7}{\sqrt{21x}}$ **17.** $\dfrac{2}{\sqrt{6}}$ **19.** $\dfrac{52}{3\sqrt{91}}$

21. $\dfrac{7}{\sqrt[3]{98}}$ **23.** $\dfrac{7x}{\sqrt{21xy}}$ **25.** $\dfrac{5y^2}{x\sqrt[3]{150x^2y^2}}$ **27.** $\dfrac{ab}{3\sqrt{ab}}$ **29.** $\dfrac{5(8 + \sqrt{6})}{58}$ **31.** $-2\sqrt{7}(\sqrt{5} + \sqrt{3})$

33. $-\dfrac{\sqrt{15} + 20 - 6\sqrt{2} - 8\sqrt{30}}{77}$ **35.** $\dfrac{x - 2\sqrt{xy} + y}{x - y}$ **37.** $\dfrac{3\sqrt{6} + 4}{2}$ **39.** $\dfrac{-11}{4(\sqrt{3} - 5)}$ **41.** $\dfrac{-22}{\sqrt{6} + 5\sqrt{2} + 5\sqrt{3} + 25}$

43. $\dfrac{x - y}{x + 2\sqrt{xy} + y}$ **45.** $\dfrac{7}{43\sqrt{2} + 66}$ **47.** 6.708

MARGIN EXERCISES, SECTION 6.8, pp. 329–332

1. $\sqrt[4]{y}$ **2.** $\sqrt{3a}$ **3.** $\sqrt[4]{16}$, or 2   **4.** $\sqrt[3]{125}$, or 5   **5.** $\sqrt[5]{a^3b^2c}$ **6.** $19^{1/3}$ **7.** $(abc)^{1/2}$ **8.** $\left(\dfrac{x^2y}{16}\right)^{1/5}$

9. $\sqrt{x^3}$, or $(\sqrt{x})^3$, which simplifies to $x\sqrt{x}$ **10.** $\sqrt[3]{8^2}$ coa or $(\sqrt[3]{8})^2$, which simplifies to 4

11. $\sqrt{4^5}$, or $(\sqrt{4})^5$, which simplifies to 32   **12.** $(7abc)^{4/3}$ **13.** $6^{7/5}$ **14.** $\dfrac{1}{2}$ **15.** $\dfrac{1}{(3xy)^{7/8}}$ **16.** $7^{14/15}$ **17.** $5^{2/6}$, or $5^{1/3}$

18. $9^{2/5}$ **19.** $\sqrt{a}$ **20.** x **21.** $\sqrt{2}$ **22.** ab^2 **23.** xy^3 **24.** $\sqrt[4]{xy^2}$ **25.** $\sqrt[4]{63}$ **26.** $\sqrt[4]{u - b}$ **27.** $\sqrt[6]{x^4y^3z^5}$ **28.** $\sqrt[4]{ab}$

EXERCISE SET 6.8, pp. 333–334

1. $\sqrt[4]{x}$ **3.** 2   **5.** $\sqrt[5]{a^2b^2}$ **7.** 8   **9.** $20^{1/3}$ **11.** $(xy^2z)^{1/5}$ **13.** $(3mn)^{3/2}$ **15.** $(8x^2y)^{5/7}$ **17.** $\dfrac{1}{x^{1/3}}$ **19.** $x^{2/3}$ **21.** $5^{7/8}$

23. $7^{1/4}$ **25.** $8.3^{7/20}$ **27.** $10^{3/20}$ **29.** $\sqrt[3]{a^2}$ **31.** $2y^2$ **33.** $2\sqrt[4]{2}$ **35.** $\sqrt[3]{2x}$ **37.** $2c^2d^3$ **39.** $\dfrac{m^2n^4}{2}$ **41.** $\sqrt[4]{r^2s}$ **43.** $3ab^3$

45. $\sqrt[6]{200}$ **47.** $\sqrt[6]{4x^5}$ **49.** $\sqrt[6]{x^5 - 4x^4 + 4x^3}$ **51.** $\sqrt[6]{a + b}$ **53.** $\sqrt[12]{a^8b^9}$ **55.** $\sqrt[4]{st^4} = t\sqrt[4]{s}$

MARGIN EXERCISES, SECTION 6.9, pp. 335–337

1. 100   **2.** No solution   **3.** 4   **4.** 17   **5.** 9   **6.** 5

SOMETHING EXTRA—AN APPLICATION: ROAD PAVEMENT SIGNS, p. 338

1. 27.4 m   **2.** 21.9 m   **3.** 45.9 m   **4.** 79.4 m

EXERCISE SET 6.9, pp. 339–340

1. 168   **3.** $\dfrac{80}{3}$ **5.** -27 **7.** No solution   **9.** -6 **11.** 5   **13.** 9   **15.** 7   **17.** $\dfrac{80}{9}$ **19.** 6, 2   **21.** $-\dfrac{8}{9}$

MARGIN EXERCISES, SECTION 6.10, pp. 341-344

1. $\sqrt{7}i$ 2. $5i$ 3. $-\sqrt{11}i$ 4. $-6i$ 5. $15 - 3i$ 6. $-12 + 6i$ 7. $3 - 5i$ 8. $12 - 7i$ 9. $-9 - 12i$

10. $-35 - 25i$ 11. $-14 + 8i$ 12. $11 + 10i$ 13. i 14. $-8i$ 15. $5 + 12i$ 16. $-21 - 20i$ 17. $6 - 3i$

18. $-9 + 5i$ 19. $\pi + \frac{1}{4}i$ 20. 53 21. 10 22. $2i$ 23. $\frac{10}{17} - \frac{11}{17}i$ 24. Yes   25. Yes

EXERCISE SET 6.10, pp. 345-346

1. $\sqrt{15}i$ 3. $4i$ 5. $-2\sqrt{3}i$ 7. $8 + i$ 9. $9 - 5i$ 11. $7 + 4i$ 13. $-2 - 3i$ 15. $-1 + i$ 17. $11 + 6i$ 19. $1 + 5i$

21. $18 + 14i$ 23. $38 + 9i$ 25. $2 - 46i$ 27. -1 29. $-125i$ 31. $5 - 12i$ 33. $-5 + 12i$ 35. $-5 - 12i$

37. $\frac{8}{5} + \frac{1}{5}i$ 39. $-i$ 41. $-\frac{21}{49} - \frac{56}{49}i$, or $-\frac{3}{7} - \frac{8}{7}i$ 43. Yes   45. No

EXTENSION EXERCISES, pp. 347-348

1. 1.4675×10^{12} 3. $\frac{1}{3}$ 5. 7 7. y^{-4} 9. $\frac{3^{12s+4}}{625}$ 11. $572\sqrt{3}$ 13. a^3bxy^2 15. $2yz\sqrt{2z}$ 17. $(25x - 30y - 9xy)\sqrt{2xy}$

19. $(7x^2 - 2y^2)\sqrt{x + y}$ 21. $2x - 2\sqrt{x^2 - 4}$ 23. $\sqrt[3]{y} - y$ 25. $12 + 6\sqrt{3}$ 27. $16(a + 3) + 38a\sqrt{3(a + 3)} - 15a^2$

29. $2\sqrt{34}$ 31. $\frac{\sqrt{10} + 2\sqrt{5} - 2\sqrt{3}}{10}$ 33. $\frac{b^2 + \sqrt{b}}{b^2 + b + 1}$ 35. $6a\sqrt[3]{36ab^2}$ 37. $\frac{x - 19}{x + 31 + 10\sqrt{x + 6}}$

39. $\left(\frac{5x + 4y^{-3}}{xy}\right)\sqrt{xy}$ 41. $\frac{7\sqrt{3}}{39}$ 43. $x^3\sqrt[6]{x}$ 45. $\sqrt[4]{2xy^2}$ 47. $\sqrt[6]{p + q}$ 49. x^3 51. 6912 53. 0 55. $-1, 6$ 57. 2

59. $0, \frac{125}{4}$ 61. 2 63. 3 65. $-4 - 8i$ 67. $-3 - 4i$ 69. $-88i$ 71. 8 73. $\frac{3}{5} + \frac{9}{5}i$ 75. 1 77. $-\frac{7}{13} + \frac{2\sqrt{30}}{13}i$

TEST OR REVIEW—CHAPTER 6, pp. 349-350

1. $\frac{x^1y^{-1}}{108}$, or $\frac{1}{108xy}$ 2. 1.344×10^{-4} 3. $|6y|$ or $6|y|$ 4. $|x + 5|$ 5. -2 6. 4 7. $2x^3$ 8. 12.2 9. $x\sqrt{5x}$

10. $4a\sqrt[3]{4ab^2}$ 11. $38\sqrt{2}$ 12. -20 13. $\frac{a + 2\sqrt{ab} + b}{a - b}$ 14. $(5xy^2)^{5/2}$ 15. $\sqrt[4]{2x^2y - 12xy + 18y}$ 16. 7 17. $3\sqrt{2}i$

18. $7 + 5i$ 19. No   20. $-\frac{77}{50} + \frac{14}{50}i$

CHAPTER 7

READINESS CHECK, p. 352

1. [2.2, ⬛] $5, -3$ 2. [4.10, ⬛] $7, -2$ 3. [4.10, ⬛] $0, 2$ 4. [4.10, ⬛] -5 5. [4.5, ⬛] $3, -3$

6. [2.6, ⬛]

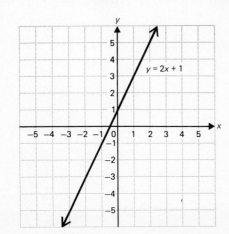

$y = 2x + 1$

7. [6.3, ⬛] $2\sqrt{22}$ 8. [6.3, ⬛] 0.5 9. [6.7, ⬛] $\frac{\sqrt{10}}{5}$

10. [6.10, ⬛] $2\sqrt{5}i$ 11. [6.10, ⬛] No

MARGIN EXERCISES, SECTION 7.1, pp. 352–356

1. $\sqrt{3}, -\sqrt{3}$; or $\pm\sqrt{3}$ **2.** 0   **3.** $\dfrac{2\sqrt{6}}{3}, -\dfrac{2\sqrt{6}}{3}$; or $\pm\dfrac{2\sqrt{6}}{3}$ **4.** $\dfrac{i\sqrt{2}}{2}, -\dfrac{i\sqrt{2}}{2}$; or $\pm\dfrac{i\sqrt{2}}{2}$ **5.** $0, \dfrac{3}{4}$ **6.** $\dfrac{3}{5}, 1$ **7.** 3, 8   **8.** 2 ft

9. A to B is 6 mi. A to C is 8 mi.

EXERCISE SET 7.1, pp. 357–360

1. $\pm\sqrt{5}$ **3.** $\pm\dfrac{2}{5}i$ **5.** $\pm\dfrac{\sqrt{6}}{2}$ **7.** $0, -\dfrac{9}{14}$ **9.** $0, -2$ **11.** 5, 1   **13.** $\dfrac{2}{3}, -\dfrac{1}{2}$ **15.** $-5, -3$ **17.** $-2, -1$ **19.** 10, 5

21. 2 in.   **23.** Length is 6 ft; width is 2 ft   **25.** Length is 24 yd; width is 12 yd   **27.** 10 ft, 24 ft   **29.** 4 ft, 3 ft   **31.** ± 8

MARGIN EXERCISES, SECTION 7.2, pp. 361–363

1. $\dfrac{-1+\sqrt{22}}{3}, \dfrac{-1-\sqrt{22}}{3}$; or $\dfrac{-1\pm\sqrt{22}}{3}$ **2.** $\dfrac{1+i\sqrt{7}}{2}, \dfrac{1-i\sqrt{7}}{2}$; or $\dfrac{1\pm i\sqrt{7}}{2}$ **3.** $2, -\dfrac{1}{3}$ **4.** 2.1, 0.2   **5.** 1.2, -1.9

SOMETHING EXTRA—AN APPLICATION: HANDLING DIMENSION SYMBOLS (PART I), p. 364

1. 68 ft   **2.** 72 km/hr   **3.** 45 g   **4.** 8.6 lb   **5.** 7.5 mi/hr   **6.** 32 km/hr   **7.** 3.3 m/sec   **8.** 19 ft/min

EXERCISE SET 7.2, pp. 365–366

1. $-3 \pm \sqrt{5}$ **3.** $-1, -\dfrac{5}{3}$ **5.** $\dfrac{1 \pm i\sqrt{3}}{2}$ **7.** $2 \pm 3i$ **9.** $\dfrac{-3 \pm \sqrt{41}}{2}$ **11.** $-1 \pm 2i$ **13.** $0, -1$ **15.** $\dfrac{3}{4}, -2$

17. $\dfrac{1 \pm 3i}{2}$ **19.** 1.3, -5.3 **21.** 2.8, -1.3 **23.** 0.5700731, -0.7973458 **25.** $\dfrac{-y \pm \sqrt{-47y^2 + 108}}{6}$

MARGIN EXERCISES, SECTION 7.3, pp. 367–370

1. $-\dfrac{2}{5}, 10$ **2.** 36 hr   **3.** 10 mph   **4.** 20 knots, 30 knots

EXERCISE SET 7.3, pp. 371–374

1. 6, -4 **3.** $\pm\dfrac{3}{2}$ **5.** 25, -20 **7.** 4, -1 **9.** $\pm\sqrt{37}$ **11.** First part, 40 mph; second part, 35 mph   **13.** 35 mph

15. A, 350 km/h; B, 400 km/h   **17.** 6 hours   **19.** 3.24 km/h   **21.** A, 24.0 hours; B, 51.0 hours

MARGIN EXERCISES, SECTION 7.4, pp. 375–376

1. Two real   **2.** One real   **3.** Two nonreal   **4.** $3x^2 + 7x - 20 = 0$ **5.** $x^2 + 25 = 0$ **6.** $x^2 + \sqrt{2}x - 4 = 0$

EXERCISE SET 7.4, pp. 377–378

1. One real   **3.** Two nonreal   **5.** Two real   **7.** One real   **9.** Two nonreal   **11.** Two real   **13.** Two real

15. One real   **17.** $x^2 + 2x - 99 = 0$ **19.** $x^2 - 14x + 49 = 0$ **21.** $25x^2 - 20x - 12 = 0$

23. $4x^2 - 2(c + d)x + cd = 0$ **25.** $x^2 - 4\sqrt{2}x + 6 = 0$ **27.** (a) $k = 2$; (b) $\dfrac{11}{2}$

MARGIN EXERCISES, SECTION 7.5, p. 379–380

1. $w_2 = \dfrac{w_1}{A^2}$ **2.** $r = \sqrt{\dfrac{V}{\pi h}}$ **3.** $s = \dfrac{-R + \sqrt{R^2 - \dfrac{4L}{C}}}{2L}$ **4.** $b = \sqrt{\dfrac{p^2 a^2}{1 + p^2}}$

EXERCISE SET 7.5, pp. 381–382

1. $s = \sqrt{\dfrac{A}{6}}$ **3.** $r = \sqrt{\dfrac{Gm_1 m_2}{F}}$ **5.** $c = \sqrt{\dfrac{E}{m}}$ **7.** $b = \sqrt{c^2 - a^2}$ **9.** $k = \dfrac{3 + \sqrt{9 + 8N}}{2}$ **11.** $r = \dfrac{-\pi h + \sqrt{\pi^2 h^2 + 2\pi A}}{2\pi}$

13. $g = \dfrac{4\pi^2 l}{T^2}$ **15.** $D = \sqrt{\dfrac{32LV}{g(P_1 - P_2)}}$ **17.** $v = \dfrac{c\sqrt{m^2 - m_0^2}}{m}$ **19.** $t = \dfrac{10 \pm \sqrt{102 - s}}{2}$, 1.4 sec

MARGIN EXERCISES, SECTION 7.6, pp. 383–384

1. $\pm 3, \pm 1$ **2.** 4 **3.** $4, 2, -1, -3$ **4.** $\dfrac{1}{2}, -\dfrac{1}{3}$

EXERCISE SET 7.6, pp. 385–386

1. $81, 1$ **3.** $\pm\sqrt{5}$ **5.** $7, -1, 5, 1$ **7.** $4, 1, 6, -1$ **9.** $\pm\sqrt{2+\sqrt{6}}, \pm\sqrt{2-\sqrt{6}}$ **11.** $\dfrac{1}{3}, -\dfrac{1}{2}$ **13.** $2, -1$ **15.** 28.5

MARGIN EXERCISES, SECTION 7.7, pp. 387–389

1. $y = 7x^2$ **2.** $y = \dfrac{9}{x^2}$ **3.** $y = \dfrac{1}{2}xz$ **4.** $y = \dfrac{5xz^2}{w}$ **5.** 490 m   **6.** 1 ohm

SOMETHING EXTRA—DEPRECIATION: DECLINING BALANCE METHOD, p. 390

1. $520.59   **2.** 3 **3.** $2088.87

EXERCISE SET 7.7, pp. 391–392

1. $y = 15x^2$ **3.** $y = \dfrac{0.0015}{x^2}$ **5.** $y = xz$ **7.** $y = \dfrac{3}{10}xz^2$ **9.** $y = \dfrac{xz}{5wp}$ **11.** $355\dfrac{5}{9}$ ft   **13.** 94.03 kg   **15.** 97

MARGIN EXERCISES, SECTION 7.8, pp. 393–398

1.

2.

3.

4. (a)

(b)

5.

6.

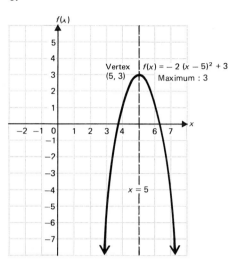

Vertex (5, 3)

$f(x) = -2(x-5)^2 + 3$

Maximum : 3

$x = 5$

EXERCISE SET 7.8, pp. 399–400

1.

$f(x) = 5x^2$

3.

$f(x) = \frac{1}{4}x^2$

5.

$f(x) = -\frac{1}{2}x^2$

7.

$f(x) = -4x^2$

9.

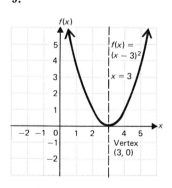

$f(x) = (x-3)^2$

$x = 3$

Vertex (3, 0)

11.

$f(x) = 2(x-4)^2$

13.

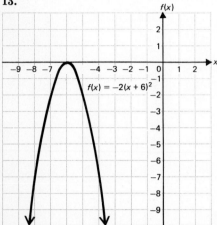

$f(x) = -2(x + 6)^2$

15.

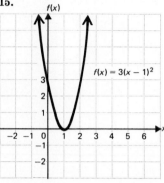

$f(x) = 3(x - 1)^2$

17.

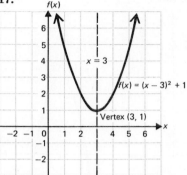

$x = 3$

$f(x) = (x - 3)^2 + 1$

Vertex (3, 1)

19.

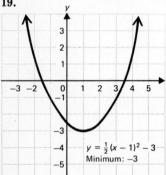

$y = \frac{1}{2}(x - 1)^2 - 3$
Minimum: -3

21.

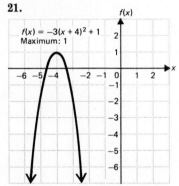

$f(x) = -3(x + 4)^2 + 1$
Maximum: 1

23.

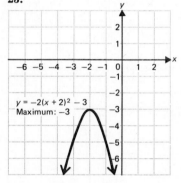

$y = -2(x + 2)^2 - 3$
Maximum: -3

25.

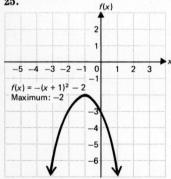

$f(x) = -(x + 1)^2 - 2$
Maximum: -2

MARGIN EXERCISES, SECTION 7.9, pp. 401–403

1.

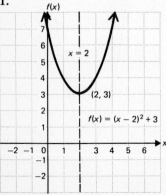

$x = 2$

(2, 3)

$f(x) = (x - 2)^2 + 3$

2.

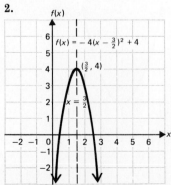

$f(x) = -4(x - \frac{3}{2})^2 + 4$

$(\frac{3}{2}, 4)$

$x = \frac{3}{2}$

3. $(1 - \sqrt{6}, 0), (1 + \sqrt{6}, 0)$

4. $(-1, 0), (3, 0)$ **5.** $(-4, 0)$ **6.** None

SOMETHING EXTRA—DEPRECIATION: THE SUM OF THE DIGITS METHOD, p. 404

1.

Year	Rate of depreciation	Annual depreciation	Total depreciation	Book value, V
4	$\frac{5}{36}$ or 13.9%	$541.67	$2816.67	$2383.33
5	$\frac{4}{36}$ or 11.1%	433.33	3250.00	1950.00
6	$\frac{3}{36}$ or 8.3%	325.00	3575.00	1625.00
7	$\frac{2}{36}$ or 5.6%	216.67	3791.67	1408.33
8	$\frac{1}{36}$ or 2.8%	108.33	3900.00	1300.00

2.

Year	Rate of depreciation	Annual depreciation	Total depreciation	Book value, V
0				$8700.00
1	$\frac{5}{15}$ or 33.3%	$2366.67	$2366.67	6333.33
2	$\frac{4}{15}$ or 26.7%	1893.33	4260.00	4440.00
3	$\frac{3}{15}$ or 20%	1420.00	5680.00	3020.00
4	$\frac{2}{15}$ or 13.3%	946.67	6626.67	2073.33
5	$\frac{1}{15}$ or 6.7%	473.33	7100.00	1600.00

EXERCISE SET 7.9, pp. 405–406

1.

3.

5.

7.

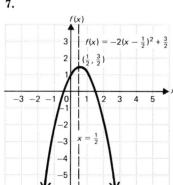

9. $(2 - \sqrt{3}, 0), (2 + \sqrt{3}, 0)$ **11.** $(-1, 0), (3, 0)$ **13.** None   **15.** $\left(-\frac{3}{2}, 0\right)$

17. $(3.1585884, 0), (-1.0103267, 0)$

MARGIN EXERCISES, SECTION 7.10, pp. 407–408

1. 225   **2.** 25 m by 25 m   **3.** $f(x) = x^2 - 2x + 1$ **4.** (a) $f(x) = 0.625x^2 - 50x + 1150$; (b) 510

EXERCISE SET 7.10, pp. 409–412

1. 19 ft by 19 ft; 361 ft^2 **3.** 121; 11 and 11   **5.** -4; 2 and -2 **7.** $-\dfrac{25}{4}$; $\dfrac{5}{2}$ and $-\dfrac{5}{2}$ **9.** $f(x) = 2x^2 + 3x - 1$

11. $f(x) = -3x^2 + 13x - 5$ **13.** (a) $f(x) = -4x^2 + 40x + 2$; (b) \$98   **15.** (a) $f(x) = 0.05x^2 - 10.5x + 650$; (b) 250

17. Minimum: -7.0234008

EXTENSION EXERCISES, pp. 413–414

1. $-8, -\dfrac{10}{3}, 0, \dfrac{7}{2}$ **3.** A: 15 km/h, B: 8 km/h   **5.** $\dfrac{22 \pm 4i\sqrt{31}}{7}$ **7.** $-\sqrt{3} \pm \sqrt{2}$ **9.** $\sqrt{3}, \dfrac{3 - \sqrt{3}}{2}$ **11.** $\dfrac{1}{2}$

13. $4 \pm 2\sqrt{2}$ **15.** $\pm\sqrt{2}$

17. The solutions of $ax^2 + bx + c = 0$ are $x = \dfrac{-b \pm \sqrt{b^2 - 4ac}}{2a}$.

(a) $\dfrac{-b + \sqrt{b^2 - 4ac}}{2a} + \dfrac{-b - \sqrt{b^2 - 4ac}}{2a} = \dfrac{-2b}{2a} = -\dfrac{b}{a}$;

(b) $\dfrac{-b + \sqrt{b^2 - 4ac}}{2a} \cdot \dfrac{-b - \sqrt{b^2 - 4ac}}{2a} = \dfrac{(-b)^2 - (\sqrt{b^2 - 4ac})^2}{4a^2} = \dfrac{b^2 - b^2 + 4ac}{4a^2} = \dfrac{4ac}{4a^2} = \dfrac{c}{a}$.

19. (a) $k = -\dfrac{3}{5}$; (b) $-\dfrac{1}{3}$ **21.** $x^2 - \sqrt{3}x + 8 = 0$ **23.** $h = -36, k = 15$ **25.** $|k| = 2$ **27.** $k = 6$ **29.** $\dfrac{432}{143}$

31. 9   **33.** y is multiplied by 8   **35.** $x = \dfrac{p + q}{2}$ **37.** $11\sqrt{2}$ ft

TEST OR REVIEW—CHAPTER 7, pp. 415–416

1. 9, 2   **2.** Length is 8.5 in.; width is 4 in.   **3.** $\dfrac{-1 \pm i\sqrt{3}}{2}$ **4.** 0.4, -4.4 **5.** 0, 2   **6.** 2 hours   **7.** Two nonreal

8. $x^2 - 4\sqrt{3}x + 9 = 0$ **9.** $r = \sqrt{\dfrac{3V}{\pi} - R^2}$ **10.** $\pm\sqrt{\dfrac{5 + \sqrt{5}}{2}}, \pm\sqrt{\dfrac{5 - \sqrt{5}}{2}}$ **11.** 6.664 cm^2

12.

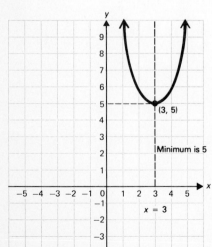

13. $2(x^2 + 2x + 1) - 8$, or $2(x + 1)^2 - 8$

14. $(3, 0)$ and $(-2, 0)$

15. -16; 4 and -4

16. $\dfrac{x^2}{5} - \dfrac{3x}{5} = f(x)$

CHAPTER 8

READINESS CHECK, p. 418

1. [6.2, ▪] 13   **2.** [6.3, ▪▪] $4\sqrt{3}$ **3.** [2.6, ▪]

4. [7.9, ▪]

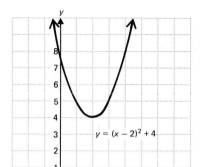

5. [3.2, ▪▪] (1, 2)   **6.** [7.6, ▪] $2, -2, 4, -4$

MARGIN EXERCISES, SECTION 8.1, pp. 418–420

1. $\sqrt{85} \approx 9.220$ **2.** 8   **3.** 14   **4.** 11   **5.** 10   **6.** $\sqrt{37} \approx 6.083$

EXERCISE SET 8.1, pp. 421–422

1. 5   **3.** 12   **5.** 10   **7.** $\sqrt{74} \approx 8.602$ **9.** $\sqrt{2} \approx 1.414$ **11.** $2\sqrt{3} \approx 3.464$ **13.** 5   **15.** $3\sqrt{2} \approx 4.242$

17. $4\sqrt{2} \approx 5.656$ **19.** Yes

21. Let (a, c) and (b, c) be two points on a horizontal line. Then $d = \sqrt{(a - b)^2 + (c - c)^2} = \sqrt{(a - b)^2} = |a - b|$. Let (c, a) and (c, b) be two points on a vertical line. Then $d = \sqrt{(c - c)^2 + (a - b)^2} = \sqrt{(a - b)^2} = |a - b|$.

MARGIN EXERCISES, SECTION 8.2, pp. 423–426

1.

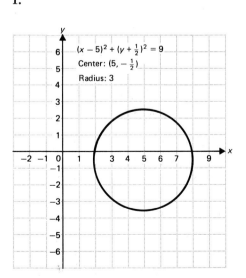

2. $(x + 3)^2 + (y - 1)^2 = 36$ **3.** $(0, 0), r = 8$

4. $(x + 4)^2 + (y - 5)^2 = 49$, Center: $(-4, 5)$, Radius: 7

5.

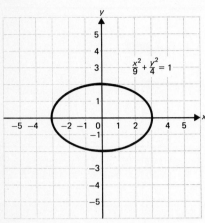

$\frac{x^2}{9} + \frac{y^2}{4} = 1$

6.

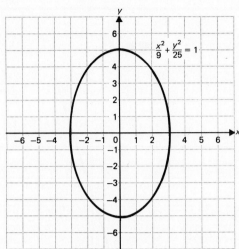

$\frac{x^2}{9} + \frac{y^2}{25} = 1$

EXERCISE SET 8.2, pp. 427–428

1.

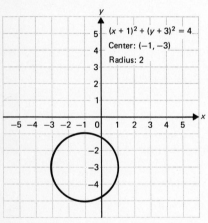

$(x + 1)^2 + (y + 3)^2 = 4$
Center: $(-1, -3)$
Radius: 2

3.

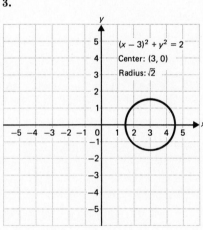

$(x - 3)^2 + y^2 = 2$
Center: $(3, 0)$
Radius: $\sqrt{2}$

5.

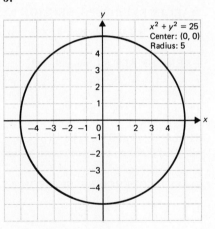

$x^2 + y^2 = 25$
Center: $(0, 0)$
Radius: 5

7. $x^2 + y^2 = 49$ **9.** $(x + 2)^2 + (y - 7)^2 = 5$ **11.** $(-4, 3), r = 2\sqrt{10}$ **13.** $(4, -1), r = 2$ **15.** $(2, 0), r = 2$

17.

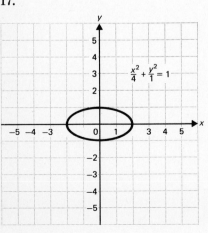

$\frac{x^2}{4} + \frac{y^2}{1} = 1$

19.

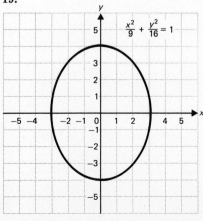

$\frac{x^2}{9} + \frac{y^2}{16} = 1$

MARGIN EXERCISES, SECTION 8.3, pp. 429–434

1.

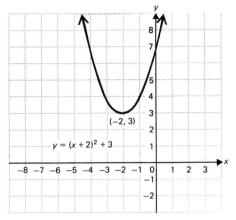

$y = (x + 2)^2 + 3$

$(-2, 3)$

2.

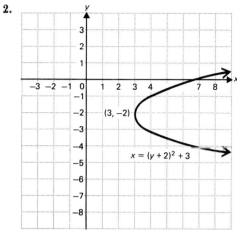

$(3, -2)$

$x = (y + 2)^2 + 3$

3.

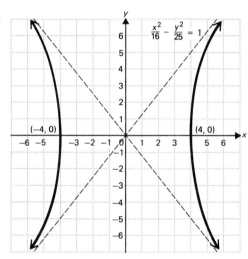

$\dfrac{x^2}{16} - \dfrac{y^2}{25} = 1$

$(-4, 0)$ $(4, 0)$

4.

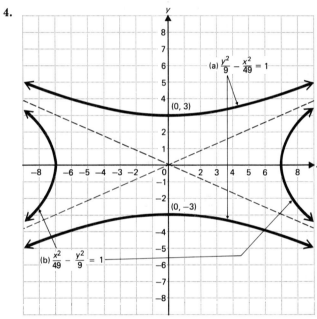

(a) $\dfrac{y^2}{9} - \dfrac{x^2}{49} = 1$

$(0, 3)$

$(0, -3)$

(b) $\dfrac{x^2}{49} - \dfrac{y^2}{9} = 1$

5.

$xy = -12$

6.

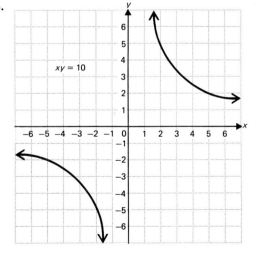

$xy = 10$

EXERCISE SET 8.3, pp. 435–436

1.

3.

5.

7.

9.

11.

13.

15.

17.

19.

MARGIN EXERCISES, SECTION 8.4, pp. 437–439

1. $(4, 3), (-3, -4)$　**2.** $(4, 7), (-1, 2)$　**3.** $\left(-\dfrac{5}{7}, \dfrac{22}{7}\right), (1, -2)$　**4.** 11 and 7　**5.** 12 m by 5 m

SOMETHING EXTRA—AN APPLICATION: HANDLING DIMENSION SYMBOLS (PART II), p. 440

1. 36 ft-lb　**2.** 6 km-hr　**3.** $4\,\dfrac{\text{in.-lb}}{\text{sec}}$　**4.** $24\,\dfrac{\text{man-hr}}{\text{da}}$

EXERCISE SET 8.4, pp. 441–442

1. $(-4, -3), (3, 4)$　**3.** $(0, 2), (3, 0)$　**5.** $(-2, 1)$　**7.** $\left(\dfrac{5 + \sqrt{70}}{3}, \dfrac{-1 + \sqrt{70}}{3}\right), \left(\dfrac{5 - \sqrt{70}}{3}, \dfrac{-1 - \sqrt{70}}{3}\right)$　**9.** 10 and 3

11. Length is 8 ft; width is 6 ft　**13.** $\left(x + \dfrac{5}{13}\right)^2 + \left(y - \dfrac{32}{13}\right)^2 = \dfrac{5365}{169}$

MARGIN EXERCISES, SECTION 8.5, pp. 443–440

1. $(2, 3), (2, -3), (-2, 3), (-2, -3)$　**2.** $(3, 2), (-3, -2), (2, 3), (-2, -3)$　**3.** Length is 2 in.; width is 1 in.

EXERCISE SET 8.5, pp. 445–446

1. $(-5, 0)$, $(4, 3)$, $(4, -3)$ **3.** $(3, 0)$, $(-3, 0)$ **5.** $(1, 2)$, $(-1, -2)$, $(2, 1)$, $(-2, -1)$ **7.** $(3, 2)$, $(-3, -2)$

9. 13 and 12, -13 and -12 **11.** Length is $\sqrt{3}$ m; width is 1 m

EXTENSION EXERCISES, pp. 447–448

1. (a) $\sqrt{49 + k^2}$; (b) $\sqrt{13}$; (c) $8\sqrt{m^2 + n^2}$; (d) $2\sqrt{3}c$; (e) $\dfrac{\sqrt{41}}{7}$; (f) $6\sqrt{2}$ **3.** (a) $\left(-\dfrac{15}{2}, 3\right)$; (b) $(0, b)$; (c) $(2, 4\sqrt{2})$; (d) $\left(-\dfrac{1}{6}, \dfrac{5}{88}\right)$

5. $(0, 4)$ **7.** (a) $\dfrac{x^2}{9} + \dfrac{y^2}{4} = 1$, C: $(0, 0)$, V: $(-3, 0)$, $(3, 0)$, $(0, 2)$, $(0, -2)$;

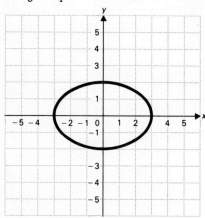

(b) $\dfrac{(x - 1)^2}{9} + \dfrac{(y + 2)^2}{4} = 1$, C: $(1, -2)$, V: $(-2, -2)$, $(4, -2)$, $(1, 0)$, $(1, -4)$;

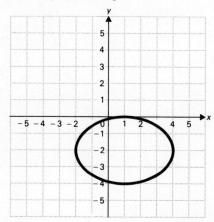

(c) $\dfrac{(x + 3)^2}{1} + \dfrac{(y - 4)^2}{16} = 1$, C: $(-3, 4)$, V: $(-2, 4)$, $(-4, 4)$, $(-3, 8)$, $(-3, 0)$;

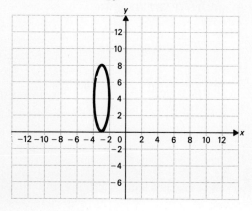

(d) $\dfrac{(x-1)^2}{25} + \dfrac{(y+1)^2}{4} = 1$, C: $(1, -1)$, V: $(6, -1)$, $(-4, -1)$, $(1, 1)$, $(1, -3)$.

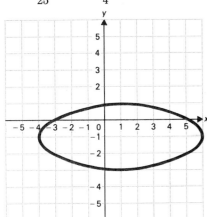

9. $\dfrac{x^2}{4} + \dfrac{y^2}{16} = 1$ **11.** (a) Circle; (b) Parabola; (c) Hyperbola; (d) Parabola; (e) Ellipse; (f) Circle

13. $(x+2)^2 + (y-1)^2 = 4$ **15.** $\dfrac{1}{2}, \dfrac{1}{4}$ **17.** 10 in. by 7 in. by 5 in.

TEST OR REVIEW—CHAPTER 8, pp. 449–450

1. $\sqrt{14} \approx 3.742$ **2.** $6\sqrt{5} \approx 13.416$ **3.** $(x-2)^2 + (y+7)^2 = 18$ **4.** Center: $(-4, 5)$; Radius: 7

5.

6.

7.

8.

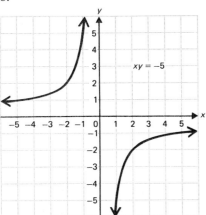

9. $(0, 3)$, $(4, 0)$ **10.** $(4, 0)$, $(-4, 0)$

CHAPTER 9

READINESS CHECK, p. 452

1. [6.8, ■] $\sqrt[5]{x^4}$ or $(\sqrt[5]{x})^4$ **2.** [1.6, ■] (a) 5; (b) 1   **3.** [1.6, ■] $\dfrac{1}{8^4}$ **4.** [6.9, ■] 49   **5.** [1.7, ■] x^{-2}

6. [1.7, ■] x^{-7} **7.** [1.7, ■] x^{-12} **8.** [6.1, ■] (a) 5.24×10^3; (b) 8.45×10^{-2}

MARGIN EXERCISES, SECTION 9.1, pp. 452–455

1. (a) $1, 3, 9, \dfrac{1}{3}, \dfrac{1}{9}, \dfrac{1}{27}$; (b)

2.

3.

4.

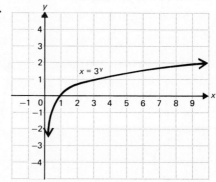

SOMETHING EXTRA—CALCULATOR CORNER: NUMBER PATTERNS, p. 456

1. 2, 2.25, 2.48832, 2.59374246, 2.691588029, 2.704813829, 2.715568521

2. 2, 0.5; 4, 0.25; 32, 0.03125; 1024, 0.00097656; 1.1259×10^{15}, 8.8818×10^{-16}; 1.2677×10^{30}, 7.8883×10^{-31}

EXERCISE SET 9.1, pp. 457–458

1.

3.

5.

7.

9.

11.

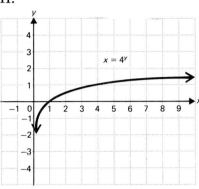

MARGIN EXERCISES, SECTION 9.2, pp. 459–461

1.

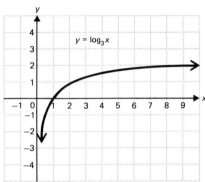

2. $0 = \log_6 1$ **3.** $-3 = \log_{10} 0.001$ **4.** $0.25 = \log_{16} 2$ **5.** $T = \log_m p$ **6.** $2^5 = 32$ **7.** $10^3 = 1000$ **8.** $a^7 = Q$

9. $t^x = M$ **10.** $10{,}000$ **11.** 3 **12.** 4 **13.** -4 **14.** 0

EXERCISE SET 9.2, pp. 463–464

1.

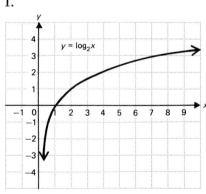

3. $3 = \log_{10} 1000$ **5.** $-3 = \log_5 \dfrac{1}{125}$ **7.** $\dfrac{1}{3} = \log_8 2$ **9.** $0.3010 = \log_{10} 2$

11. $3^t = 8$ **13.** $5^2 = 25$ **15.** $10^{-1} = 0.1$ **17.** $10^{0.845} = 7$ **19.** 9

21. 4 **23.** $\dfrac{1}{2}$ **25.** 2 **27.** 2 **29.** -1 **31.** 0 **33.** 4 **35.** -2 **37.** 0

9

MARGIN EXERCISES, SECTION 9.3, pp. 465–467

1. $\log_5 25 + \log_5 5$ **2.** $\log_b P + \log_b Q$ **3.** $\log_3 (7 \cdot 5)$ **4.** $\log_a (\text{CABIN})$ **5.** $5 \log_7 4$ **6.** $\dfrac{1}{2} \log_a 5$

7. $\log_b P - \log_b Q$ **8.** $\log_2 \dfrac{x}{25}$ **9.** $\dfrac{1}{2} (3 \log_a z - \log_a x - \log_a y)$ **10.** $\log_a \dfrac{x^5 z^{1/4}}{y}$

EXERCISE SET 9.3, pp. 469–470

1. $\log_2 32 + \log_2 8$ **3.** $\log_4 64 + \log_4 16$ **5.** $\log_c B + \log_c x$ **7.** $\log_a (6 \cdot 70)$ **9.** $\log_c (K \cdot y)$ **11.** $3 \log_a x$

13. $\log_a 67 - \log_a 5$ **15.** $\log_a \dfrac{15}{7}$ **17.** $2 \log_a x + 3 \log_a y + \log_a z$ **19.** $\log_b x + 2 \log_b y - 3 \log_b z$ **21.** $\log_a x^{2/3} y^{-1/2}$

23. $\log_a \dfrac{2x^4}{y^3}$ **25.** $\log_a (x + 2)$

MARGIN EXERCISES, SECTION 9.4, pp. 471–476

1. $2 \times 8 \approx 10^{0.3010} \times 10^{0.9031} \approx 10^{1.2041} \approx 16$ **2.** $\log (4 \times 2) = \log 4 + \log 2 \approx 0.6021 + 0.3010 \approx 0.9031$, antilog $0.9031 \approx 8$

3. $\log \dfrac{15}{3} = \log 15 - \log 3 \approx 1.1761 - 0.4771 \approx 0.6990$, antilog $0.6990 \approx 5$

4. $\log \sqrt[3]{8} = \dfrac{1}{3} \log 8 \approx \dfrac{1}{3}(0.9031) \approx 0.3010$, antilog $0.3010 \approx 2$ **5.** $\log 3^2 = 2 \log 3 \approx 2(0.4771) \approx 0.9542$, antilog $0.9542 \approx 9$

6. 0.4969   **7.** 0.9996   **8.** 0.6021   **9.** 5.74   **10.** 1   **11.** 3.62   **12.** $0.4609 + 2$ **13.** $0.4609 + (-4)$ **14.** 4.8312

15. 8.7846   **16.** $0.8932 + (-2)$ **17.** $0.0453 + (-4)$ **18.** 64,100   **19.** 64,100   **20.** 0.000425   **21.** 0.0159

22. $c = 4, m = 0.5529$ **23.** $c = -6, m = 0.9129$ **24.** 8482.3   **25.** 0.0000028

SOMETHING EXTRA—AN APPLICATION: FORGETTING, p. 476

1. 62   **2.** 46.79

EXERCISE SET 9.4, pp. 477–478

1. $\log (7 \times 2) = \log 7 + \log 2 \approx 0.8451 + 0.3010 \approx 1.1461$, antilog $1.1461 \approx 14$

3. $\log \dfrac{12}{3} = \log 12 - \log 3 \approx 1.0792 - 0.4771 \approx 0.6021$, antilog $0.6021 \approx 4$ **5.** 0.3909   **7.** 0.7251   **9.** 0.0294   **11.** 2.5403

13. 1.7202   **15.** 3.5877   **17.** $0.8062 + (-1)$ **19.** $0.6064 + (-5)$ **21.** 2330   **23.** 18   **25.** 0.613   **27.** 0.00346   **29.** 25.2

31. 0.00973   **33.** 0.000613

MARGIN EXERCISES, SECTION 9.5, pp. 479–480

1. $4.1589 - 10$ **2.** $24.1589 - 30$ **3.** $8.0530 - 10$ **4.** 0.05340   **5.** 13.51

EXERCISE SET 9.5, pp. 481–482

1. 189.6   **3.** 272.4   **5.** 8.770   **7.** 3.846   **9.** 3.637   **11.** 0.2560

MARGIN EXERCISES, SECTION 9.6, pp. 483–485

1. 3.6592   **2.** $8.3779 - 10$ **3.** 5956   **4.** 0.0002856

SOMETHING EXTRA—AN APPLICATION: WALKING SPEED, p. 486

1. 2.055 ft/sec   **2.** 2.963 ft/sec   **3.** 3.421 ft/sec   **4.** 4.262 ft/sec   **5.** 4.926 ft/sec   **6.** 5.377 ft/sec

EXERCISE SET 9.6, pp. 487–488

1. 1.6194   **3.** 0.4689   **5.** 2.8130   **7.** $9.1538 - 10$ **9.** $7.6291 - 10$ **11.** $9.2494 - 10$ **13.** 2.7786   **15.** 43.15

17. 4.444   **19.** 1.082   **21.** 0.7185   **23.** 0.002587   **25.** 0.01589   **27.** 4.834

MARGIN EXERCISES, SECTION 9.7, pp. 489–490

1. 3   **2.** 2   **3.** 7   **4.** 8.4

EXERCISE SET 9.7, pp. 491–492

1. 3.17   **3.** 3.32193   **5.** $\dfrac{5}{2}$ **7.** $-3, -1$ **9.** 1.4036   **11.** 10   **13.** $\dfrac{1}{3}$ **15.** 3   **17.** 7.27 yr   **19.** 6.7

EXTENSION EXERCISES, pp. 493–494

1.

3.

5.

7.

9.

11.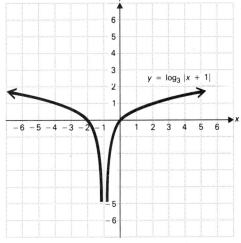

13. 25   **15.** $\dfrac{\sqrt{5}}{25}$ **17.** 6   **19.** 1296   **21.** $\dfrac{2}{3}$ **23.** x^2 **25.** 0   **27.** 1   **29.** -2 **31.** True   **33.** False

35. $\log_a (x^3 + y^3)$ **37.** $\dfrac{1}{2}\log_a (c - d) - \dfrac{1}{2}\log_a (c + d)$ **39.** 0.69   **41.** $\dfrac{10}{3}$ **43.** $\dfrac{12}{5}$ **45.** $\dfrac{2}{5}$ **47.** $\sqrt[3]{3}$ **49.** -1

51. $-3, -1$ **53.** 1, 100   **55.** $-625, 625$ **57.** $\dfrac{1}{2}$, 5000   **59.** $\dfrac{1}{a}$, a **61.** 1, $\dfrac{\log 5}{\log 3}$ **63.** $\dfrac{1}{2}$ **65.** True   **67.** -3

69. (a) \$1150.27, \$1323.13, \$4055.20; (b) 4.95 years; (c) $T = \dfrac{\ln 2}{k}$

TEST OR REVIEW—CHAPTER 9, pp. 495–496

1.

2.

3.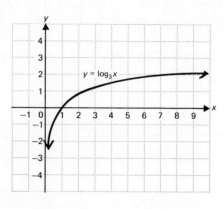

4. $-3 = \log_4 x$ **5.** $4^2 = 16$ **6.** $3 \log a + \frac{1}{2} \log b - 2 \log c$ **7.** $\log_a \frac{z^2 \sqrt[3]{x}}{y^3}$ **8.** 0.008358   **9.** $0.0899 + (-2)$ **10.** 445,000

11. 0.0000001825   **12.** $0.0913 + (-2)$ **13.** 0.0937   **14.** 9

CHAPTER 10

READINESS CHECK, p. 498

1. [1.1, ▣] 16   **2.** [1.1, ▣] -4 **3.** [2.1, ▣] 3   **4.** [2.1, ▣] -3

5. [2.6, ▬]   **6.** [2.6, ▬]   **7.** [2.2, ⬭] $-4, 2$ **8.** [2.2, ⬭] 0, 9

MARGIN EXERCISES, SECTION 10.1, pp. 498–501

1. $1 < 8$ **2.** $5 > -2$ **3.** $-3 > -8$ **4.** True   **5.** True   **6.** False   **7.** Yes   **8.** No   **9.** Yes

10. $x < 3$ **11.** $x \geqslant -2$

12.   **13.**

14. (a) $-2 < x + 4$ and $x + 4 < 7$, $-2 - 4 < x + 4 - 4$ and $x + 4 - 4 < 7 - 4$, $-6 < x$ and $x < 3$, $-6 < x < 3$;
(b) $-2 < x + 4 < 7$, $-2 - 4 < x + 4 - 4 < 7 - 4$, $-6 < x < 3$

15.   **16.**   **17.** $x < 1$ or $x > 7$

SOMETHING EXTRA—AN APPLICATION: HANDLING DIMENSION SYMBOLS (PART III) CHANGING UNITS, p. 502

1. 51.2 oz   **2.** 6200 m   **3.** 3080 ft/min   **4.** $0.26 per min

EXERCISE SET 10.1, pp. 503–504

1.   **3.**   **5.** $x \geqslant \dfrac{17}{8}$ **7.** $x \leqslant 3$

9.   **11.**

13. $-1 < x \leqslant 6$ **15.** $x \leqslant -9$ or $x > -5$ **17.** $x \leqslant -8.92$ or $x > -4.8003$

MARGIN EXERCISES, SECTION 10.2, pp. 505–508

1. $\{0, 1, 2, 3, 4, 5, 6, 7\}$ **2.** $\{1, 3, 5, 7, 9, \cdots\}$ **3.** True   **4.** False   **5.** $\{x \mid x \geqslant \sqrt{2}\}$ **6.** $\{2, -3\}$ **7.** $\{x \mid x > -5\}$

8. False   **9.** $\{a, b\}, \{a\}, \{b\}, \emptyset$ **10.** $\{0, 3\}$ **11.** $\{0, 1, 3, 4, 7, 9\}$

12.   **13.**

EXERCISE SET 10.2, pp. 509–510

1. $\{0, 2, 4, 6, 8\}$ **3.** $\{11, 13, 15, \cdots\}$ **5.** $\{x \mid x > \sqrt{3}\}$ **7.** $\{-5, 5\}$ **9.** $\{3, -2\}$ **11.** $\{x \mid x > -5\}$ **13.** True

15. False   **17.** False   **19.** True   **21.** True   **23.** True   **25.** $\{-4, 7\}, \{-4\}, \{7\}, \emptyset$ **27.** $\{6, 8\}$ **29.** $\{1, 4, 5, 6, 7, 8, 11\}$

31. $\emptyset$ **33.** $\{1, 2, 3, 4\}$ **35.** $\{4, 8, 11\}$ **37.**

39.   **41.** 2^n

MARGIN EXERCISES, SECTION 10.3, pp. 511–514

1. $\left\{y \mid y \leqslant \dfrac{3}{10}\right\}$ **2.** $\left\{y \mid y < -\dfrac{5}{12}\right\}$ **3.** $\{x \mid x \geqslant 12\}$ **4.** $\left\{y \mid y \leqslant -\dfrac{1}{5}\right\}$ **5.** $\left\{x \mid x < \dfrac{1}{2}\right\}$ **6.** $\left\{y \mid y \geqslant \dfrac{22}{13}\right\}$ **7.** $\{x \mid 3 \leqslant x < 4\}$

8. $\{x \mid -1 > x \geqslant -2\}$ **9.** 71 or higher   **10.** $n < 100$

EXERCISE SET 10.3, pp. 515–516

1. $\{x \mid x \leqslant 6\}$ **3.** $\left\{y \mid y < -\dfrac{1}{8}\right\}$ **5.** $\left\{x \mid x < -\dfrac{1}{56}\right\}$ **7.** $\left\{x \mid x \leqslant -\dfrac{2}{3}\right\}$ **9.** $\{x \mid 2 < x \leqslant 4\}$ **11.** 620.5 or less

13. $20,000   **15.** Gross sales greater than $15,000

MARGIN EXERCISES, SECTION 10.4, pp. 517–520

1. 29   **2.** 5   **3.** 20   **4.** 6, -6 **5.** $\dfrac{1}{2}, -\dfrac{1}{2}$

6.   **7.**   **8.**   **9.**
$-6 < x < 6$ $-\dfrac{1}{2} \leq x \leq \dfrac{1}{2}$ $x \leq -6$ or $x \geq 6$ $x < -\dfrac{1}{2}$ or $x > \dfrac{1}{2}$

10.   **11.**   **12.**
$x = -\dfrac{10}{3}$ or $x = 2$ $-4 < x < 1$ $x \leq -\dfrac{7}{3}$ or $x \geq 1$

EXERCISE SET 10.4, pp. 521–522

1. 34   **3.** 11   **5.** 33   **7.** 2   **9.** 3, -3,

11. $-3 < x < 3$ **13.** $x \leqslant -2$ or $x \geqslant 2$

15. $-9 < x < 15$ **17.** $-\dfrac{7}{2} \leqslant x \leqslant \dfrac{1}{2}$

10

19. $y < 1$ or $y > 6$

21. $x \leqslant -\dfrac{5}{4}$ or $x \geqslant \dfrac{23}{4}$ **23.** $-1 \leqslant x \leqslant 2$

MARGIN EXERCISES, SECTION 10.5, pp. 523–526

1. Yes   **2.**

3.

4.

5.

6.

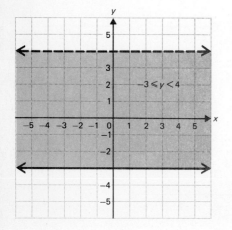

1. Yes   **3.** Yes

5.

7.

9.

11.

13.

15.

17.

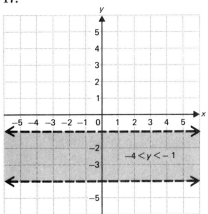

EXTENSION EXERCISES, pp. 529–530

1. $-\dfrac{1}{6} < x \leqslant \dfrac{37}{6}$ **3.** $x \leqslant \dfrac{1}{2}$ **5.** $10 < x \leqslant 18$ **7.** No solution   **9.** False   **11.** False   **13.** False   **15.** False

17. True   **19.** False   **21.** Z or $\{\ldots, -3, -2, -1, 0, 1, 2, 3, \ldots\}$ **23.** $\emptyset$ **25.** $\{x \mid -1 \leqslant x \leqslant 1\}$

27. A or $\{x \mid -4 < x \leqslant 2\}$ **29.** $\{x \mid x < -10 \text{ or } x \geqslant 9\}$ **31.** M or $\{0\}$ **33.** $\{m \mid m \leqslant 7\}$ **35.** $\{x \mid -4 < x \leqslant 1\}$ **37.** $\emptyset$

39. $\left\{x \mid \dfrac{2}{5} \leqslant x \leqslant \dfrac{4}{5}\right\}$ **41.** $\left\{x \mid -\dfrac{1}{8} < x < \dfrac{1}{2}\right\}$ **43.** True   **45.** False   **47.** False   **49.** $\left\{-\dfrac{7}{3}, \dfrac{7}{3}\right\}$ **51.** All reals

53. $\left\{x\middle|-\dfrac{13}{54} < x < -\dfrac{7}{54}\right\}$ **55.** All reals   **57.** $\left\{-\dfrac{1}{4},\ 1\right\}$

59.

61.
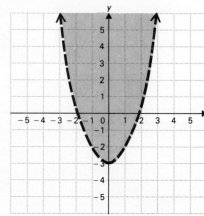

63. $\left\{x\middle|-2 < x < \dfrac{3}{2}\right\}$

65.

67.
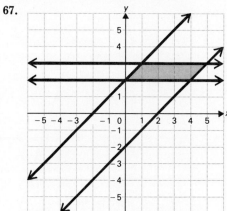

TEST OR REVIEW—CHAPTER 10, pp. 531–532

1.

2. $\{-2, 3\}$ **3.** $\{x\,|\,x > -5\}$ **4.** $\{x\,|\,x > -3\}$ **5.** a, b, c   **6.** $\{3, 5\}$

7. $\{1, 3, 5, 7, 9, 11, 13\}$ **8.**

9.   **10.** $-\dfrac{5}{3} < x < 3$

11. $x \leqslant -\dfrac{3}{2}$ or $x \geqslant \dfrac{9}{2}$

12.
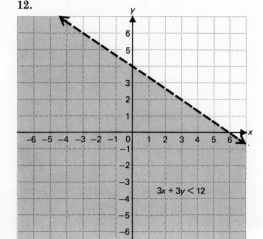

$3x + 3y < 12$

FINAL EXAMINATION, pp. 533–540

1. [5] 18   **2.** [21] $\dfrac{7}{6}$ **3.** [29] $-3y + 17$ **4.** [37] $-45x^6y^{-4}$ **5.** [55] $\dfrac{2}{3}$ **6.** [62] 64, 66, 68   **7.** [67] $r = \dfrac{A - P}{Pt}$

8. [93] $y = -x + 3$ **9.** [77, 85]

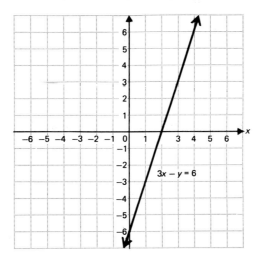

10. [137] -2 **11.** [111] $(1, -1)$ **12.** [125] $(2, -1, 1)$ **13.** [117] 25 and 19   **14.** [118] 56 L of beer A and 24 L of beer B

15. [165] $13x^3 - 7x^2 - 6x + 6$ **16.** [166] $3a^2 - 8ab - 15b^2$ **17.** [173] $6a^2 + 7a - 5$ **18.** [173] $9x^4 - 30x^2y + 25y^2$

19. [185] $(2x - 3)^2$ **20.** [195] $(3a - 2)(9a^2 + 6a + 4)$ **21.** [191] $(a + b + 5)(a + b - 5)$ **22.** [203] $8, -6$

23. [209] $x = \dfrac{c^2}{a - b}$ **24.** [214] 20   **25.** [259] $-\dfrac{6}{5}, 4$ **26.** [275] $R = \dfrac{E - IR}{I}$ **27.** [238] $\dfrac{4}{x + 2}$ **28.** [227] $\dfrac{x - 4}{3(x + 2)}$

29. [228] $\dfrac{(x + y)^2}{x^2 + y^2}$ **30.** [245] $x - a$ **31.** [255] $7x^3 + 9x^2 + 19x + 38$, R 72   **32.** [269] 350 mph   **33.** [267] $8\dfrac{2}{5}$ min

34. [292] 6.764×10^{-12} **35.** [304] $8x^2\sqrt{y}$ **36.** [309, 321] $\dfrac{\sqrt[3]{5xy}}{y}$ **37.** [323] $\dfrac{1 - 2\sqrt{x} + x}{1 - x}$ **38.** [333] $\sqrt[6]{x + 1}$

39. [336] 9   **40.** [342] $26 - 13i$ **41.** [353] $\pm i\sqrt{2}$ **42.** [361] $-2 \pm \sqrt{7}$ **43.** [376] $x^2 - 50 = 0$ **44.** [383] $5, -5, 2, -2$

45. [397]   **46.** [387] 20   **47.** [407] 1250 ft^2

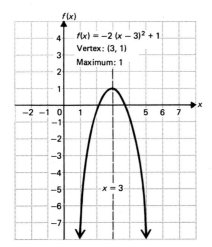

48. [443] $(\sqrt{5}, \sqrt{3}), (\sqrt{5}, -\sqrt{3}), (-\sqrt{5}, \sqrt{3}), (-\sqrt{5}, -\sqrt{3})$

49. [418, 439] 5 ft by 12 ft   **50.** [424] Center, $(2, -3)$; radius, 6

51. [426]

52. [459]

53. [467] $\log_a \dfrac{\sqrt[3]{x^2 \cdot z^5}}{\sqrt{y}}$ **54.** [460] $a^5 = c$ **55.** [489] 101   **56.** [473] 1.1492   **57.** [474] 0.006037   **58.** [489] 1.2920

59. [489] 1005   **60.** [507] $\{0, 2\}$ **61.** [518] $-2 \leqslant x \leqslant 3$

62. [518] $y < -5$ or $y > 2$

63. [524]

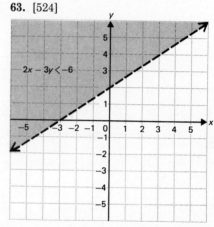

APPENDIX 2

MARGIN EXERCISES, pp. 543–547

1. (a) 27; (b) -23; (c) $5h - 3$; (d) $5a + 5h - 3$ **2.** (a) 9.41; (b) 54; (c) $4y^2 + 5$; (d) $h^2 + 5$; (e) $a^2 + 2ah + h^2 + 5$

3. (a) 13; (b) $3h^2 + 1$; (c) $3a^2 + 6ah + 3h^2 + 1$; (d) $6ah + 3h^2$ **4.** $g^{-1}(x) = \dfrac{x + 2}{3}$ **5.** $f^{-1}(x) = \sqrt{x + 1}$

6.

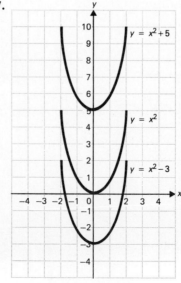

7.

EXERCISE SET, pp. 549–550

1. (a) 0; (b) 1; (c) 57; (d) $5t^2 + 4t$; (e) $5t^2 - 6t + 1$; (f) $10a + 5h + 4$ **3.** (a) 5; (b) -2; (c) -4; (d) $4|y| + 6y$;

(e) $2|a + h| + 3a + 3h$; (f) $\dfrac{2|a + h| + 3h - 2|a|}{h}$ **5.** (a) 3.14977; (b) 55.73147; (c) 3178.20675; (d) 1116.70323

7. (a) $\dfrac{2}{3}$; (b) $\dfrac{10}{9}$; (c) 0; (d) not possible   **9.** $f^{-1}(x) = \dfrac{x - 2}{2}$ **11.** $g^{-1}(x) = \dfrac{3x - 1}{2}$ **13.** $g^{-1}(x) = \sqrt{\dfrac{x - 3}{2}}$

15. $f^{-1}(x) = \log_{10}x$ **17.**

19.

21.

23.

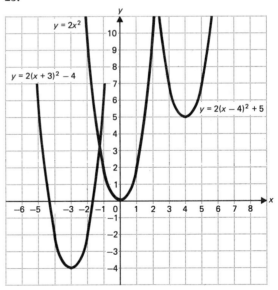

25. $\dfrac{-1}{x(x + h)}$ **27.** $\dfrac{1}{\sqrt{x + h} + \sqrt{x}}$